Moskovskoe matem. obschestvo.

# Transactions of the Moscow Mathematical Society for the year 1970

ТРУДЫ
МОСКОВСКОГО
МАТЕМАТИЧЕСКОГО
ОБЩЕСТВА
ТОМ 23 (1970)

**Volume 23**

This translation was prepared jointly by the
American Mathematical Society
and the
London Mathematical Society

Published by the American Mathematical Society
Providence, Rhode Island
1972

International Standard Book Number 0-8218-1623-3
Library of Congress Card Number 65-4713

*AMS (MOS) subject classifications* (1970).
Primary 28A65, 35Jxx, 35Kxx, 35Lxx, 54G05, 54H05



*Printed in the United States of America*

## TRANSACTIONS OF THE MOSCOW MATHEMATICAL SOCIETY

(Труды Московского Математического Общества Том 23, 1970)

## TABLE OF CONTENTS

Page

Anosov, D. V. and Katok, A. B. New examples in smooth ergodic theory. Ergodic diffeomorphisms. [Аносов, Д. В. и Каток, А. Б. Новые примеры в гладкой эргодической теории. Эргодические диффеоморфизмы, 3–36] ........ 1

Jakubov, S. Ja. Solvability of the Cauchy problem for abstract quasilinear hyperbolic equations of second order and their applications. [Якубов, С. Я. Разрешимость задачи коши для абстрактных квазилинейных гиперболических уравнений второго порядка и их приложения, 37–60] ........ 36

Pohožaev, S. I. On weakly nonlinear hyperbolic systems. [Похожаев, С. И. О слабо нелинейных гиперболических системах, 61–76] ........ 60

Demidov, A. S. Asymptotics of solutions of boundary value problems for a linear elliptic equation of second order with coefficients which have a "splash". [Демидов, А. С. Асимптотика решений краевых задач для линейного эллиптического уравнения 2-го порядка с коэффициентами, имеющими "всплеск", 77–112] ........ 75

Gluško, V. P. Estimates in $\mathfrak{L}_2$ and the solvability of general boundary value problems for degenerate elliptic second-order equations. [Глушко, В. П. Оценки в $\mathfrak{L}_2$ и разрешимость общих граничных задач для вырождающихся эллиптических уравнений второго порядка, 113–178] ........ 111

Èĭdel'man, S. D. and Ivasišen, S. D. Investigation of the Green matrix for a homogeneous parabolic boundary value problem. [Эйдульман, С. Д. и Ивасишен, С. Д. Исследование матрицы грина однородной параболической граничной задачи, 179–234] ........ 179

Efimov, B. A. Extremally disconnected compact spaces and absolutes. [Ефимов, Б. А. Экстремально несвязные бикомпакты и абсолюты, 235–276] ........ 243

Čoban, M. M. Many-valued mappings and Borel sets. II. [Чобан, М. М. Многозначные отображения и борелевские множества. II, 277–301] .... 286

Trudy Moskov. Mat. Obšč.
Tom 23 (1970)

Trans. Moscow Math. Soc.
Vol. 23 (1970)

# NEW EXAMPLES IN SMOOTH ERGODIC THEORY. ERGODIC DIFFEOMORPHISMS

*D. V. ANOSOV AND A. B. KATOK*

## Contents



## Introduction

1. In this paper we use a uniform construction method in order to obtain examples of measure-preserving ergodic $C^\infty$-diffeomorphisms defined on certain smooth manifolds which have various, sometimes unexpected metric properties. The manifolds to be considered will be compact, connected, smooth manifolds (with or without boundary) possessing a nontrivial smooth free group of circular rotations,[1)] or, as we shall call it, a periodic current. The invariant measure in our examples will have a positive smooth density. The metrical properties of the diffeomorphisms which we shall construct may have the following properties depending on the choice of the parameters in the construction:

a) a discrete spectrum with an arbitrary given (finite or infinite) number of basic (i.e. linearly independent over the ring **Z** of integers) frequencies;

b) a simple continuous singular spectrum in the absence of the mixing property;

c) properties a) and b) combined.

*AMS* 1970 *subject classifications*. Primary 28A65, 57D50; Secondary 54H20, 58F15.

[1)] Nontrivial in the sense that each point of the manifold is displaced by at least one element of the group.

These examples obviously show that [2] there is but a rather weak connection between the topological properties of a manifold and the ergodic properties of the diffeomorphisms defined on it.

2. It is well known (cf. [**6**]) that an arbitrary ergodic automorphism of a Lebesgue space with discrete spectrum generated by $h$ basic eigenfrequencies is metrically isomorphic to some shift transformation of the $h$-dimensional torus. The problem of the realization of such metric automorphisms by diffeomorphisms of smooth manifolds with smooth invariant measure is considerably more difficult. It is easy to prove that the manifold must necessarily be an $h$-dimensional torus if, in addition, the eigenfunctions are assumed to be smooth. The diffeomorphism is in this case smoothly equivalent to a shift operator. On the other hand, in the case of the two-dimensional torus A. N. Kolmogorov [**11**] has constructed a real-analytic diffeomorphism with an analytic invariant measure which is metrically isomorphic to the group shift and which possesses discontinuous eigenfunctions. (This implies that these diffeomorphisms can be not even topologically conjugate to a shift.) Until now no smooth realization of automorphisms with discrete spectrum having $h$ basic frequencies have been known on any manifolds other than the $h$-dimensional torus. There exists a conjecture (cf. [**10**]) to the effect that on a real-analytic $m$-dimensional manifold an ergodic, analytic diffeomorphism with analytic invariant measure may have a discrete spectrum with only $m$ basic frequencies. This problem is still open even though our results also show that the $C^\infty$ version of it is not true.

3. Even in the simple case of the two-dimensional circle

$$\mathbf{D}^2 = \{(x, y) \colon x^2 + y^2 \leqq 1\}$$

the problem is still open whether there exists an ergodic diffeomorphism conjugate to Lebesgue measure. However, for different special functions $\rho$ the topological properties of those mappings of the circle which are expressible in polar coordinates in the form

$$(r, \phi) \to (\rho(r, \phi), \phi + \leftarrow) \tag{0.1}$$

were analyzed a long time ago. An equivalent problem was first studied by Poincaré ([**19**], Chapter 19). In connection with the investigation of the neighborhood of a periodic orbit of a three-dimensional dynamic system, L. G. Šnirel′man [**20**] and A. S. Besicovitch [**2**] have constructed examples of topologically transitive[3] mappings of the form (0.1). Their examples were

---

[2] We refer to the case of two or more dimensions.

[3] Topological transitivity means that there exists an everywhere dense orbit (which must then be a continuum; cf. [**15**]). It is interesting that Šnirel′man and Besicovitch at first calculated as if all points of the circle, except the center and the points of its boundary, had everywhere dense orbits. The error was corrected in [**3**] and [**5**].

not smooth, but smoothness can be attained without essentially altering the construction (cf. [**21**]). However, it is easy to prove that a transformation of the form (0.1) of the circle cannot have an ergodic invariant measure equivalent to Lebesgue measure. A careful analysis of the examples given by Šnirel′man and Besicovitch has also been one of the starting points for the present paper.

4. The diffeomorphisms which we will exhibit form a nowhere dense set in the space of all diffeomorphisms of a given manifold which preserve a given measure, provided with the topology of $C^\infty$ or $C^n$ for some $n \geqq 1$. This is related to the fact that our diffeomorphisms belong to the closure of the set of periodic diffeomorphisms, which is nowhere dense in that space, and even to the set of those diffeomorphisms among them which are shifts with respect to the orbits of periodic flows. It is evident that even the set of all ergodic diffeomorphisms is not everywhere dense in the space described (in the two-dimensional case this follows from results by J. Moser [**13**], [**14**] under a sufficient smoothness assumption).

In the class of homeomorphisms the situation is different. In 1941 it was shown by Oxtoby and Ulam [**17**] that for any triangulated manifold[4] and for an arbitrary measure satisfying natural conditions, the ergodic homeomorphisms which preserve that measure form a set of the second category (an everywhere dense $G_\delta$ set) in the space of all (measure-preserving) homeomorphisms under the corresponding topology.[5] From the purely metrical point of view Halmos [**7**] (cf. [**6**]) showed that the situation is analogous: In the space of all automorphisms of a Lebesgue space provided with the weak topology, the ergodic automorphisms form a set of the second category. At the basis of both results there lies the same phenomenon: The periodic automorphisms (in the metric case) and the homeomorphisms which are periodic outside a set of sufficiently small measure (in the topological case) are everywhere dense in the corresponding basis. This is in good agreement with the fact that also in the smooth case the metric situation on the closure of the set of periodic diffeomorphisms resembles the topological case and the purely metrical case in many respects.

5. The methods of this paper are suitable for establishing also a number of additional results. Thus on manifolds with periodic flows A. B. Katok has constructed nonergodic diffeomorphisms which have an arbitrary finite or denumerably infinite number of ergodic components, where for each of them any one of the subcases a), b) or c) of subsection 1 is possible and, furthermore, the following modifications may occur:

---

[4] This is also true with respect to polyhedra of a somewhat more general type.

[5] An application of the uniform approximation technique allows one to derive sharper results from the work of Oxtoby and Ulam (cf. [**9**]).

a) As ergodic components one may choose open sets (with complementary sets of measure zero).

b) Each ergodic component intersects with any open set in a set of positive measure; hence, in particular, almost all orbits are everywhere dense.

As we have already stated, all these results are concerned with a special class of manifolds. Recently E. A. Sirodov [**21**] has proved that on many manifolds of dimension three or larger there exists a topologically transitive smooth flow (and consequently a diffeomorphism). Using some modifications of the methods of our paper, D. V. Anosov has proved that on an arbitrary manifold of dimension three or larger there exists an ergodic smooth flow and consequently a diffeomorphism (cf. [**1**]). (This was preceded by a more complicated construction of analogous flows with four or more dimensions due to A. B. Katok.) For the sake of completeness we also wish to mention that recently A. A. Blohin has constructed examples of smooth ergodic currents on all closed surfaces excepting spheres, projective planes and Klein bottle, on which no such flows exist. Similar examples were also investigated by J. Milnor.

6. In the first two sections we present here for convenience auxiliary material which is undoubtedly well known. §3 is the basic part of the paper. There we introduce an inductive process on an arbitrary manifold with periodic flow which enables us to construct, for a given smooth measure $\mu$, a sequence of periodic $\mu$-preserving diffeomorphisms which converges to an ergodic diffeomorphism under the $C^\infty$ topology. This construction possesses a significant nonuniqueness property. In §§4 to 6 it is shown that, under a suitable choice of the parameter in the construction, we may obtain a diffeomorphism with given metrical properties. Furthermore we can construct a diffeomorphism which is metrically isomorphic to a circular rotation. (For the construction of such a diffeomorphism, subsections 5, 6 and 7 of §3 are not needed except for the almost trivial verification of the inductive assumption at the end of subsection 5.) In §7 it is shown that in the closure of the set of diffeomorphisms belonging to a periodic flow, the ergodic diffeomorphisms form a set of the second category. The basic results of this paper have been announced in [**1**].

## §1. Background material about measures on smooth manifolds

1. A measure $\mu$ on an $n$-dimensional manifold $M^m$ of class $C^\infty$ is called a measure of class $C^\infty$ if on any coordinate neighborhood it induces an infinitely differentiable density relative to the local coordinates. For brevity such a measure will be called positive if this density does not vanish identically on any coordinate neighborhood. If $\mu_1$ and $\mu_2$ are two positive measures of class $C^\infty$ on $M^m$ then there exists a positive function $\rho(x)$ of class $C^\infty$ such that $\mu_1 = \rho(x)\mu_2$. Consequently all positive measures have the same class of null sets.

THEOREM 1.1. *Let $M^m$ be an m-dimensional open connected manifold of class $C^\infty$, and let $\mu_1$ and $\mu_2$ be positive measures of class $C^\infty$ on $M^m$ which are identical outside a compact set $N \subset M^m$, for which $\mu_1(M^m) = \mu_2(M^m)$. Then there exists a diffeomorphism $S$: $M^m \to M^m$ which is the identical mapping outside a compact set $N_1 \subset M^m$ such that $S^*\mu_1 = \mu_2$, using the notation*

$$(S^*\mu_1)(A) = \mu_1(S^{-1}(A)).$$

This theorem is formally different from Theorem 1 of Moser [**16**] inasmuch as we do not assume the manifold to be closed, but the theorem follows from the same Lemmas 1 and 2 of the paper [**16**].

By a "manifold" we shall henceforth always mean a connected, compact manifold of class $C^\infty$, closed or with boundary, unless something else is specified.

THEOREM 1.2. *Let $\mu_1$ and $\mu_2$ be two normalized positive measures of class $C^\infty$ on the manifold $M^m$. Then there exists a diffeomorphism $S$: $M^m \to M^m$ of class $C^\infty$ such that $S^*\mu_1 = \mu_2$.*

For closed manifolds this assertion reduces to the Theorem 1 from the paper [**16**] which we have already mentioned. For manifolds with boundary it follows from Theorem 1.1 if the measures $\mu_1$ and $\mu_2$ coincide on some neighborhood of the boundary. Therefore it suffices to prove the existence of a diffeomorphism $\hat{S}$ such that the measure $\hat{S}^*\mu_1$ coincides with $\mu_2$ on some neighborhood of the boundary.

In order to prove this we consider nonoverlapping tube-shaped neighborhoods $B_i$ of the connected components $A_i$ $(i = 1, \cdots, n)$ of the boundary. Then $B_i$ is diffeomorphic to the direct product $A_i \times [0, 10]$. Hence we can introduce coordinates $(y, t)$ on $B_i$, where $y \in A_i$, $0 \leqq t \leqq 10$. We consider some fixed positive measure $\lambda$ of class $C^\infty$ on $A_i$, and we let

$$d\mu_1(y, t) = \rho_1(y, t)\,dt\,d\lambda(y), \quad d\mu_2(y, t) = \rho_2(y, t)\,dt\,d\lambda(y).$$

Then we introduce the function $\alpha(y, t)$ by the relation

$$\int_0^t \rho_1(y, s)\,ds = \int_0^{\alpha(y,t)} \rho_2(y, s)\,ds.$$

The function $\alpha(y, t)$ is defined for any $y$ and for all sufficiently small positive $t$. Let $\delta \in (0, 3)$ be such that for $0 \leqq t \leqq \delta$ the function $\alpha(y, t)$ is defined and satisfies the condition $|\alpha| \leqq 1$. We construct functions $\phi$, $\psi \in C^\infty[0, 10]$ with the following properties:

The function $\phi$ is nonincreasing, $\phi(t) = 1$ for $t \in [0, 2\delta/3]$ and $\phi(t) = 0$ for $t \geqq \delta$, with $\phi'(t) > -4/\delta$ for all $t$.

The function $\psi$ is nondecreasing for $t \in [0, \delta]$, with $\psi(t) = 0$ for $t \in [0, \delta/3]$ and $\psi(t) = 4t/\delta$ for $t \in [2\delta/3, \delta]$, with $\psi'(t) > -1$ for $t \geqq \delta$ and $\psi(t) = 0$ for $t \geqq 8$.

It is easy to verify that the function

$$F(y, t) = \varphi(t)\,\alpha(y, t) + (1 - \varphi(t))\,t + \psi(t)$$

coincides with $\alpha(y,t)$ for $t \in [0, \delta/3]$ and with $t$ for $t \geqq 8$; furthermore, it satisfies $F_t'(x,t) > 0$. Therefore the mapping $\hat{S}_i\colon B_i \to B_i$ defined by the formula

$$\hat{S}_i(y,t) = (y, F(y,t)),$$

is the identity mapping for $t \geqq 8$; it may be extended to all points $M^m$ outside $B_i$ in a unique manner, and for $X \subset A_i \times [0, \delta/3]$ we have $\mu_1(X) = \mu_2(\hat{S}_i X)$. Hence the diffeomorphism $\hat{S} = \coprod_i \hat{S}_i$ has the property that $\hat{S}^*\mu_1 = \mu_2$ in some neighborhood of the boundary.

2. THEOREM 1.3. *Let $O^M$ be an open, connected manifold of class $C^\infty$, let $F_i$ and $G_i$ $(i = 1, \cdots, k)$ be two systems of open subsets from $O^m$ whose closures $\overline{F_i}$ and $\overline{G_i}$ are $C^\infty$-diffeomorphic to the m-dimensional sphere $\mathbf{D}^m$, with $\overline{F_i} \cap \overline{F_j} = \emptyset$ and $G_i \cap G_j = \emptyset$ for $i \neq j$. Then there exists a $C^\infty$-diffeomorphism $S^1$: $O^m \to O^m$ which is equal to the identity mapping outside some compact set $N_1 \subset O^m$ and which maps $F_i$ into $G_i$.*

This fact is well known and easily derived, for example, from [**18**].

In this paper we always denote by $\mathrm{Diff}^\infty(M^m, \mu)$ the space of those diffeomorphisms of class $C^\infty$ of the manifold $M^m$ which preserve a given positive finite measure $\mu$ of class $C^\infty$, provided with the natural topology.

LEMMA 1.1. *Under the conditions of Theorem 1.3 let $\mu$ be a given measure of class $C^\infty$ on the manifold $O^m$ such that $\mu(F_i) = \mu(G_i)$, $i = 1, \cdots, k$. Then for any $\epsilon > 0$ there exists a diffeomorphism $S \in \mathrm{Diff}^\infty(O^m, \mu)$ which coincides with the identity mapping outside a given compact set $N \subset O^m$ and which satisfies the inequality*[6] *$\mu(SF_i \triangle G_i) < \epsilon, i = 1, \cdots, k$.*

PROOF. We construct a diffeomorphism $S^1$ by means of Theorem 1.3. Let $(S^1)^*\mu = \rho(x)\mu = \mu_1$. By the assumptions of the lemma we have

$$\mu_1(G_i) = \int_{G_i} \rho(x)\,d\mu = \mu(G_i), \quad i = 1, \ldots, k.$$

We choose open sets $G_i^1$, $G_i^2$, $G_i^3$ whose closures are diffeomorphic to $\mathbf{D}^m$ and which furthermore satisfy

$$\overline{G_i^1} \subset G_i^2 \subset \overline{G_i^2} \subset G_i^3 \subset \overline{G_i^3} \subset G_i,$$

$$\mu(G_i^1) = \mu(G_i) - \delta, \quad \mu(G_i^2) < \mu(G_i) - \frac{2\delta}{3}, \quad \mu_1(G_i^3) > \mu_1(G_i) - \frac{\delta}{3},$$

where the number $\delta > 0$ is small enough so that

[6] As A. B. Krygin has shown, this lemma is in reality true with $SF_i = G_i$. But we shall not need this version.

$$\mu_1(G_i^1) > \mu(G_i) - \frac{\varepsilon}{2}. \tag{1.1}$$

We construct on $G_i$ a positive function $\rho_i(x)$ of class $C^\infty$ which satisfies

$$\rho_i(x) = 1 \text{ for } x \in G_i^2, \ \rho_i(x) = \rho(x) \text{ for } x \in G_i \setminus G_i^3,$$

$$\int_{G_i} \rho_i(x)\, d\mu = \mu(G_i).$$

(This is possible since $\mu(G_i^2) + \mu_1(G_i \setminus G_i^3) > \mu(G_i) - \delta/3$.)

Now we let $\mu^{(i)} = \rho_i(x)\mu$. By Theorem 1.1 there exists a diffeomorphism $S^{(i)}$: $G_i \to G_i$ which is the identity mapping along the boundary of the sphere $G_i$ and which satisfies the equation $S^{(i)*}\mu_1 = \mu^{(i)}$. Extending $S^{(i)}$ as the identity mapping on the entire manifold $O^m$, we obtain $\hat{S} = \prod_i S^{(i)}$. Now we consider the manifold $\hat{O}^m = O^m \setminus \bigcap_i G_i^1$. On $\hat{O}^m$ we have the two measures $\mu$ and $\hat{\mu} = \hat{S}^*\mu_1$. They coincide within the compact set $N_1 \cap (O^m \setminus \bigcup_i G_i^2)$ where $N_1$ stands for the same set as in Theorem 1.3. Furthermore, $\mu(\hat{O}^m) = \hat{\mu}(\hat{O}^m)$. Applying once more Theorem 1.1, we obtain a diffeomorphism $\widetilde{S}$: $\hat{O}^m \to \hat{O}^m$, equal to the identity mapping outside some compact set $N_2$, for which $\widetilde{S}^*\hat{\mu} = \mu$. We extend $\widetilde{S}$ as the identity mapping to all points of $O^m$, and we let $S = \widetilde{S}\hat{S}S^1$. Clearly $S^*\mu = \mu$. The diffeomorphism $S$ is the identity outside some compact set, since this is true for each of the diffeomorphisms $S^1$, $\hat{S}$ and $\widetilde{S}$.

Now we define $F_i^1 = (S^1)^{-1}G_i^1$. Obviously $SF_i^1 \subset G_i$, and therefore by (1.1) we have

$$\mu(F_i^1) = \mu_1(G_i^1) > \mu(G_i) - \frac{\varepsilon}{2},$$

i.e. $\mu(SF_i \triangle G_i) < \epsilon$. This proves the lemma.

## §2. Periodic flows

A flow $\{S_t\}$: $M^m \to M^m$ is called periodic if $S_\tau$ is the identity mapping for some $\tau > 0$. In the sequel we shall always assume that on the manifold $M^m$ there exists a nontrivial periodic flow of class $C^\infty$ (i.e. generated by a vector field of class $C^\infty$), and this flow will be denoted by $\{S^t\}$. For $x \in M^m$ we define

$$t(x) = \inf\{\tau : \tau > 0, \ S_\tau x = x\}.$$

Without loss of generality we may assume that $\max_x t(x) = 1$.

The following assertion follows immediately from a theorem by Bochner on the smooth operation of a compact group (cf. [**19**], §5.2, Theorem 1).

**Proposition** 2.1. *The closed set* $\Omega_1 = \{x \mid t(x) < 1\}$ *has measure zero and its complementary set is connected.*

Indeed, by Bochner's theorem, in the neighborhood of a fixed point, expressed in some coordinates, a group acts as a linear transformation. Therefore, the set of fixed points of our flow is a submanifold of codimension $\geqq 2$.

The points $x$ with $t(x) = 1/q$ $(q = 3, 4, \cdots)$ are fixed points under the cyclic group of mappings generated by $S_{1/q}$. Consequently they form locally a submanifold of codimension $\geqq 2$. Only the points with $t(x) = \frac{1}{2}$ have to be considered separately. They may form a submanifold $M^{m-1}$ of codimension one which is a local decomposition of $M^m$. But $S_{1/2}$ transforms the points on one side of $M^{-1}$ into points on the other side. Consequently these points are joined with each other by orbits of the flow $\{S_t\}$.

REMARK. Each periodic flow possesses a positive invariant measure of class $C^\infty$ on $M^m$. Let $S_t^*\mu = \rho_t(x)\mu$, and define

$$\hat{\mu} = \Big(\int_0^1 \rho_t(x)\, dt\Big)\mu_1.$$

Then the measure $\hat{\mu}$ is positive, it belongs to the class $C^\infty$ and is invariant with respect to $\{S_t\}$.

PROPOSITION 2.2. *Let $\mu$ be a positive measure of class $C^\infty$ on $M^m$. If there exists on $M^m$ a periodic flow $\{S_t\}$, then there exists also on $M^m$ a periodic flow with the same period which preserves the measure $\mu$.*

PROOF. We construct a positive measure $\hat{\mu}$ of class $C^\infty$ which is invariant with respect to $\{S_t\}$. We may assume that $\mu(M^m) = \hat{\mu}(M^m)$. By Theorem 1.2 there exists a diffeomorphism $S: M^m \to M^m$ satisfying $S^*\hat{\mu} = \mu$. Hence it follows that the measure $\mu$ is invariant with respect to the periodic flow $\{S^{-1}S_tS\}$.

We introduce some simple examples of periodic flows:

1) A one-parameter group of rotations of the unit sphere, or spherical surface, in $\mathbf{R}^{n+1}$ about a fixed $(n-1)$-dimensional linear subspace.

2) If on a Riemannian manifold there exists an oriented field of two-dimensional tangential planes, then this induces on a single tangential fibering a periodic flow of rotations about the orthogonal complement to that plane. For example, this is true on a tangential fibering of an oriented surface.

3) The existence of the periodic flow on a manifold $M^m$ allows us, in an obvious manner, to construct a periodic flow on the direct product $M^m \times Y$, where $Y$ is an arbitrary manifold of class $C^\infty$.

Now we consider a manifold $M^m$ with a periodic flow $\{S_t\}$. We need the following assertion on the existence of a special kind of mutually related fundamental domains for the mappings $S_{1/q}$ for the different values of $q$:

PROPOSITION 2.3. *There exists a system of sets $\Delta_q \subset M^m$ $(q = 1, 2, \cdots)$ with the following properties*:

2.3.1. *The set $\Delta_q$ is contained in the closure of the set of its interior points.*

2.3.2. *The boundary of $\Delta_q$ is a null set.*

2.3.3. *The set of interior points of $\Delta_q$ is open.*

2.3.4. $\bigcup_{k=0}^{q-1} S_{k/q}\Delta_q = M^m$.

2.3.5. $\bigcup_{k=1}^{q-1}(S_{k/q}\Delta_q \cap \Delta_q) \subset \Omega_1$ (*where* $\Omega_1$ *is the exceptional set from Proposition* 2.1).

2.3.6. *The intersection of* $\Delta_q$ *with any orbit is connected.*

2.3.7. $\Delta_q = \bigcup_{l=0}^{k-1} S_{l/kq}\Delta_{qk}$ *for any positive integer* $k$.

PROOF. Let $\Omega_2$ be the boundary of the manifold $M^m$. Then the set

$$\Omega = \Omega_1 \cup \Omega_2 \tag{2.1}$$

is obviously closed. The open manifold $M^m \setminus \Omega$ may be decomposed into equivalence classes each of which consists of the points belonging to the same orbit of the flow. On the factor-space $N^{m-1}$ so obtained we introduce in a natural way the structure of an open manifold of class $C^\infty$. The set $M^m \setminus \Omega$ is a fibering with the basis $N^{m-1}$ and the circumference $S^1$ as fiber. Let $\pi$ be the natural projection of $M^m \setminus \Omega$ onto $N^{m-1}$. Then the formula $\hat{\mu}(A) = \mu(\pi^{-1}(A))$ defines a positive measure of class $C^\infty$ on $N^{m-1}$. In $N^{m-1}$ there exists a closed subset $E$ of measure zero such that the closed submanifold $N^{m-1} \setminus E$ is connected and on it the fibering is a direct product. We construct a smooth decomposition $\Delta_0$ of this direct product. To the decomposition $\Delta_0$ we add the entire set $\Omega_1$ and one point belonging to $\pi^{-1}(E) \cup (\Omega_2 \setminus \Omega_1)$ from each orbit of the flow $\{S_t\}$, such that these points belong to the closure of $\Delta_0$. We denote the set so obtained by $\hat{\Delta}$, and we let $\Delta_q = \bigcup_{0 \leqq t < 1/q} S_t \hat{\Delta}$. The verification of properties 2.3.1–2.3.7 does not present any difficulty.

## §3. The basic construction

1. Suppose on the manifold $M^m$ there is given a (nontrivial) periodic flow which preserves a given positive normalized measure $\mu$ of class $C^\infty$. We shall augment the exposition by figures related to a simple case, namely the two-dimensional disk $\mathbf{D}^2$ with Lebesgue measure and with the flow $\{S_t\}$ which is expressible in the form $S_t(r,\phi) = (r, \phi + 2\pi t)$ in polar coordinates. As sets $\Delta_q$ we select the sectors $\Delta_q = \{(r,\phi):\ 0 \leqq \phi \leqq 2\pi/q\}$ which clearly satisfy the conditions 2.3.1–2.3.7.

2. Returning to the general case, we introduce the following notation. We let $S_{k/q}\Delta_q = \Delta_{k,q}$. The sets $\Delta_{k,q}$, $k = 0, \cdots, q-1$, form a decomposition mod 0 of the manifold $M^m$. We denote this decomposition by $\eta_q$. The automorphism $S_\alpha$, where $\alpha = p/q$ (with integer $p$), maps the decomposition $\eta_q$ into itself; if $p$ and $q$ are relatively prime then the factor-automorphism $S_\alpha/\eta_q$ permutes the $\Delta_{k,q}$ cyclicly. We observe that the interior $\operatorname{Int}\Delta_{k,q}$ is an open manifold for which

$$S_{\frac{k}{q}}(\operatorname{Int}\Delta_{0,q}) = \operatorname{Int}\Delta_{k,q}.$$

On $M^m$ we introduce an arbitrary Riemannian metric of class $C^\infty$ which is invariant under $\{S_t\}$. All distances which will be mentioned in the sequel are taken in this fixed metric. (In the case of the disk $\mathbf{D}^2$ we consider the

Euclidean metric.) By a standard method we introduce in the space $\mathrm{Diff}^\infty(M^m)$ of all diffeomorphisms $M^m$ of class $C^\infty$ a sequence of metrics $\rho_r(T_1, T_2)$ measuring the proximity of the $r$-flows of $T_1$ and $T_2$ as well as $T_1^{-1}$ and $T_2^{-1}$.

3. We construct inductively a sequence $T_n$ of diffeomorphisms of the manifold $M^m$ satisfying the following conditions which involve arbitrarily small numbers $\epsilon_n > 0$, positive integers $k_n$, $l_n$ and positive integers $a_n(i) \leqq q_n$, defined for $i = 0, \cdots, k_n - 1$, where $q_n$ is expressed inductively in terms of $k_s$ and $l_s$ with $s < n$.

*First Step.* Let $T_n = B_n^{-1} S_{\alpha_n} B_n$, where $B_n = A_n \cdots A_1$, and each $A_k$ is an element of $\mathrm{Diff}^\infty(M^m, \mu)$ which is the identity map within some neighborhood of the set $\Omega$ (cf. (2.1)). Here we define the numbers $\alpha_n = p_n/q_n$, where $p_n$ and $q_n$ are relatively prime positive integers. Finally, we let

$$\alpha_{n+1} = \alpha_n + \beta_n, \quad \text{where } \beta_n = \frac{1}{k_n l_n q_n^2}, \tag{3.1}$$

$$S_{\alpha_n} A_{n+1} = A_{n+1} S_{\alpha_n}.$$

Thus

$$p_{n+1} = k_n l_n q_n p_n + 1, \quad q_{n+1} = k_n l_n q_n^2,$$

$$T_{n+1} = B_n^{-1} S_{\alpha_n} A_{n+1}^{-1} S_{\beta_n} A_{n+1} B_n. \tag{3.2}$$

No conditions will be imposed directly on $k_n$ and $l_n$ but for the Second, Third, and Fourth Steps it will be necessary to choose these parameters sufficiently large (see subsections 4 and 5 below); furthermore, in §§5 and 6 additional stipulations will be made on the size of $k_n$ and $l_n$.

*Second Step.* Let $\xi_{n+1} = B_{n+1}^{-1} \eta_{q_{n+1}}$. There exists a set $E_{n+1} \subset M^m$ such that $\mu(E_{n+1}) > 1 - \epsilon_n$, where the diameter of the intersection between elements of the decomposition $\xi_{n+1}$ with the set $E_{n+1}$ is less than $1/2^{n+1}$ (which implies that $\xi_{n+1} \to \epsilon$ as $n$ tends to infinity provided, of course, that $\lim_{n\to\infty} \epsilon_n = 0$).

*Third Step.* $\rho_{[1/\epsilon_n]}(T_{n+1}, T_n) < \epsilon$ (which implies that the sequence $T_n$ converges under the topology of the space $\mathrm{Diff}^\infty(M^m, \mu)$ if $\sum_{k=1}^\infty \epsilon_k < \infty$).

*Fourth Step.* We let

$$R^{(n)} = \bigcup_{i=0}^{k_n - 1} \Delta_{a_n(i) k_n + i,\, k_n q_n}, \tag{3.3}$$

so that

$$\mu\,(A_{n+1}^{-1} R^{(n)} \triangle \Delta_{0, q_n}) < \varepsilon_n. \tag{3.4}$$

It should be noted that the arbitrariness in our construction, which will be used in §§4 to 6 in order to obtain diffeomorphisms $T = \lim T_n$ with different metric properties, arises from the possibility of choosing the parameters $a_n(i)$.

REMARK 3.1. We introduce the decomposition[7]

$$\eta_{n,\,n+1} = \{S_{\frac{k}{q_n}} R^{(n)},\ k = 0,\ 1,\ \ldots,\ q_n - 1\},$$

$$\xi_{n,\,n+1} = B_{n+1}^{-1}\,\eta_{n,\,n+1},\quad \xi_n' = B_{n+1}^{-1}\,\eta_{k_n q_n}.$$

It follows from the Second Step and (3.4) that $\xi_{n,n+1} \to \epsilon$ as $n$ tends to infinity. But since $\xi_{n,n+1} < \xi_n'$, this implies that $\xi_n' \to \epsilon$.

4. The number $\alpha_0 = p_0/q_0$ may be chosen arbitrarily. Suppose the numbers $\alpha_0, \cdots, \alpha_n$ and the diffeomorphisms $A_1, \cdots, A_n$ have already been defined in such a way that for each $k < n$ the conditions listed under Steps 1–4 are satisfied (with $n$ replaced by $k$).

Since the mapping $B_n^{-1}$ is uniformly continuous, there exists a number $\gamma_{n+1} > 0$ such that the diameter of the set $U$ is less than $1/2^{n+1}$ if the diameter of the set $B_n(U)$ is less than $\gamma_{n+1}$.

Let $k_n$ be a fixed, sufficiently large number (to be determined later) and consider the decomposition $\eta_{k_n, q_n}$. For any choice of $a_n(i)$ the set $R^{(n)}$ contains with any one point all the orbits of periodic diffeomorphisms $S_{\alpha_n}$ (or $S_{1/q_n}$, which amounts to the same) outside $\Omega_1$.

It is our aim to construct a diffeomorphism

$$A_{n+1} \in \mathrm{Diff}^\infty(M^m, \mu),$$

which is the identity mapping within some neighborhood of the set $\Omega$ and which possesses the properties (3.1), (3.4) and (3.5).

For a suitable positive integer $h_n$ and for every $s$, $0 \leqq s < h_n k_n$, there exists a set

$$R_s^{(n)} \subset \Delta_{a_n\left(\left[\frac{s}{h_n}\right]\right) k_n h_n + s,\ h_n k_n q_n},$$

(3.5) such that

$$\mu(R_s^{(n)}) > \frac{1-\varepsilon_n}{k_n h_n q_n} \quad \text{and} \quad \operatorname{diam} A_{n+1}^{-1} R_s^{(n)} < \gamma_{n+1}.$$

The existence of such a diffeomorphism will be proved in two steps. First we consider the special case where $R^{(n)} = \Delta_{0,q_n}$ (i.e. where $a_n(i) = 0$ for all $i = 0, \cdots, k_n - 1$). In this case it is possible to choose $h_n = 1$ in (3.5) as we have done. We select a number $k_n$ of mutually disjoint open sets $F_i \subset \operatorname{Int} \Delta_{0,q_n}$, $i = 0, \cdots, k_n - 1$, with the following properties:

---

[7] The notation $\xi_{n,n+1}$ and $\eta_{n,n+1}$ is explained by the fact that in subsection 6 of this section we will introduce decompositions $\xi_{n,m}$ and $\eta_{n,m}$ with $m > n$ of which the decompositions $\xi_{n,n+1}$ and $\eta_{n,n+1}$ are special cases.

$$\text{the numbers } \mu(F_i) \text{ are equal for all } i = 0, \ldots, k_n - 1;$$

$$\mu(F_i) > \frac{1 - \varepsilon_n/2}{k_n q_n}; \tag{3.6}$$

$$\overline{F}_i \text{ is diffeomorphic to } \mathbf{D}^m;$$

$$\operatorname{diam} F_i < \gamma_{n+1}.$$

On the other hand we select within each of the sets $\Delta_{r,k_nq_n}$, $r = 0, \cdots, k_n - 1$, an open set $G_i$ whose closure is diffeomorphic to $\mathbf{D}^m$ and which satisfies

$$\mu(G_i) = \mu(F_i).$$

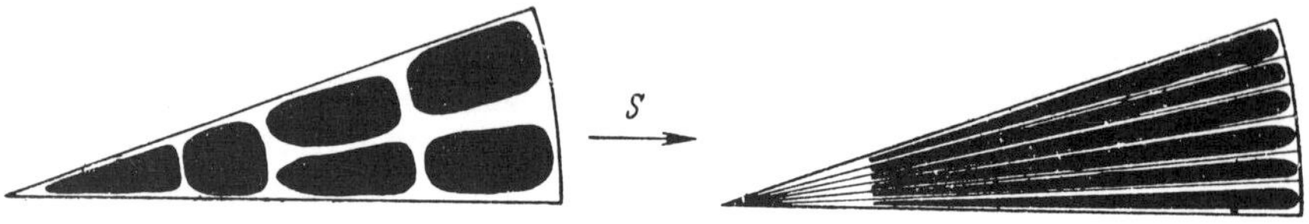


FIGURE 1

We apply Lemma 1.1 to the open set $O^m = \operatorname{Int} \Delta_{0,q_n}$ and the family of sets $F_i$, $G_i$, choosing $\epsilon = \epsilon_n/2k_nq_n$. We denote the set $SF_i \cap G_i$ by $R_i^{(n)}$. Since $S$ is the identity mapping outside the compact set $N \subset O^m$, we may extend it as the identity on the entire set $\Delta_{0,q_n}$ and subsequently we may define it on the entire set $M^m$ by the formula

$$Sx = S_{\frac{r}{q_n}} S S_{-\frac{r}{q_n}} x \text{ for } x \in \Delta_{r,q_n}, \quad r = 1, \ldots, q_n - 1. \tag{3.7}$$

Such an extension guarantees that $S$ and $S_{\alpha_n}$ commute. Letting $S = A_{n+1}$ we obtain a mapping which satisfies the conditions (3.1), (3.4) and (3.5).

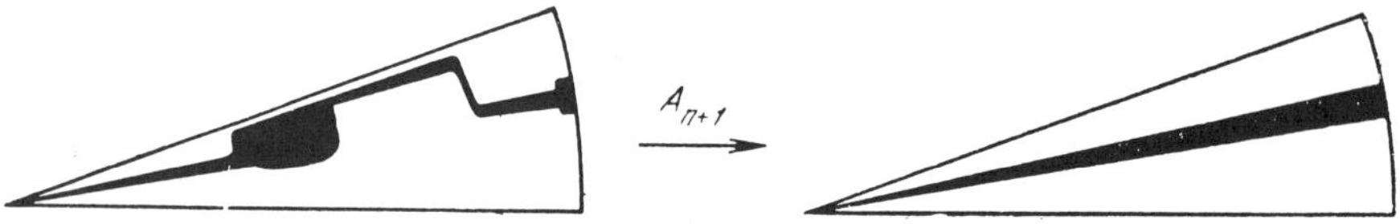


FIGURE 2

5. We now turn to the general case where not all $a_n(i)$ in (3.3) vanish. Let $b_r = q_n\mu(R^{(n)} \cap \Delta_{r,q_n})$, $r = 0, \cdots, q_n - 1$; then $b_r = k_{n,r}/k_n$, where

$$k_{n,r} = \#\{i : 0 \leqslant i < k_n,\ a_n(i) = r\}. \tag{3.8}$$

As in §2, we consider the factor space $N^{M-1}$ of the open manifold $M^M \setminus \Omega$ (see (2.1)). As a result we obtain an identification of the orbits of the flow $\{S_t\}$ and the measure $\hat{\mu}$ on them. We choose open sets $c_0, \cdots, c_{q_n-1} \subset N^{m-1}$ with mutually nonoverlapping closures. diffeomorphic to $\operatorname{Int}\mathbf{D}^m$ and such that $b_r \leqq \hat{\mu}(c_r) \geqq b_r - \epsilon_n/4k_nq_n$. We consider on $\bigcup_{r=0}^{q_n-1} c_r$ the function which is equal to $r/q_n$ if $x \in c^r$, $r = 0, \cdots, q_n - 1$. This function may be extended to the entire set $N^{m-1}$ in such a way that the resulting function $f$ is of class $C^\infty$ and vanishes outside a compact set $N \subset N^{m-1}$. We introduce on $M^m$ the function

$$f^*(x) = \begin{cases} f(\pi(x)), & \text{if } x \notin \Omega, \\ 0, & \text{if } x \in \Omega. \end{cases}$$

Clearly this function belongs to the class $C^\infty$. We define the mapping $\hat{S} \in \operatorname{Diff}^\infty(M^m, \mu)$ by $\hat{S}x = S_{f^*(x)}x$. Thus $\hat{S}$ is the identity mapping along $\Omega$. We note that

$$\left| \mu(\hat{S}\Delta_{0,q_n} \cap \Delta_{r,q_n}) - \frac{b_r}{q_n} \right| < \frac{\varepsilon_n}{4k_nq_n}. \tag{3.9}$$

Now let us consider the set $\Gamma_r = \Delta_{0,q_n} \cap \pi^{-1}c_r$. Obviously

$$\left| \mu(\Gamma_r) - \frac{b_r}{q_n} \right| < \frac{\varepsilon_n}{4k_nq_n^2},$$

so that for $b_r \neq 0$ (in which case $b_r \geqq 1/k_n$) one has

$$\mu(\Gamma_r) > \frac{1 - \varepsilon_n/4}{k_nq_n} k_{n,r}.$$

If the integer $h_n > 0$ is sufficiently large, then outside each of the sets $\Gamma_r$ we may choose $h_nk_{n,r}$ (see (3.8)) nonoverlapping open sets $F_j^r$ ($j = 0, \cdots, h_nk_{n,r} - 1$) in such a way that

the values $\mu(F_j^r)$ are equal for all $j$ and $r$;

$$\mu(F_j^r) > \frac{1 - \varepsilon_n/4}{h_nk_nq_n};$$

$\overline{F_j^r}$ is diffeomorphic to $\mathbf{D}^m$;

$$\operatorname{diam} F_j^r < \gamma_{n+1}.$$

We consider the decomposition $\eta_{h_nk_nq_n} = \{\Delta_{s,h_nk_nq_n}\}$. In each of the sets $\Delta_{s,h_nk_nq_n}$ with $s = 0, \cdots, h_nk_n - 1$ we choose an open set $G_s$ with $\mu(G_s) = \mu(F_j^r)$ whose closure is diffeomorphic to $\mathbf{D}^m$.

Let

$$0 \leqslant i_r(0) < i_r(1) < \ldots < i_r(k_{n,r} - 1) < k_n$$

be those indices $i$ for which $a_n(i) = r$. We associate with each set $F_j^r$ the set $G_s$ with the index

$$s = s(r, j) = h_n i_r\left(\left[\frac{i}{h_n}\right]\right) + j - h_n\left[\frac{j}{h_n}\right].$$

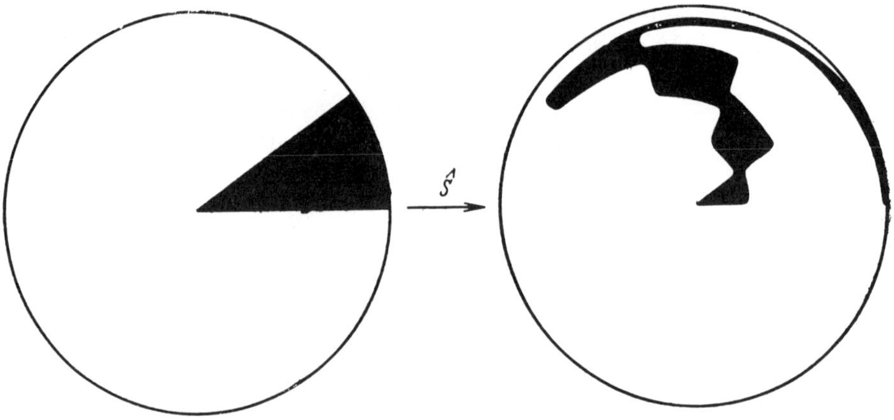


FIGURE 3

*Remark on the sketch.* For the disk $\mathbf{D}^2$ the factor space $N^{m-1}$ is identical with an interval. In this case the set $c_r$ is also an interval. This case is illustrated by Figure 3.

We apply Lemma 1.1 to the open manifold $\operatorname{Int}\Delta_{0,q_n}$ with the two systems of sets $F_j^r$ and $G_{s(r,j)}$ and with $\epsilon_n/4h_nk_nq_n$ instead of $\epsilon$.

The mapping $S$ so obtained can be extended as the identity on the entire set $\Delta_{0,q_n}$, and subsequently it can be extended to the entire set $M^m$ by formula (3.7).

We will show that we can make $A_{n+1} = S\hat{S}$ and that the conditions (3.1), (3.4) and (3.5) are satisfied.

Condition (3.1) holds for $A_{n+1}$ since it is satisfied with respect to the sets $\hat{S}$ and $S$.

Let us verify condition (3.4). Since $A_{n+1}$ preserves the measure $\mu$, it suffices to show that $\mu(A_{n+1}\Delta_{0,q_n}\triangle R^{(n)'}) < \epsilon_n$. It follows from our definitions that

$$R^{(n)} = \bigcup_{r=0}^{q_n-1} \bigcup_{j=0}^{h_nk_{n,r}-1} S_{\frac{r}{q_n}} \Delta_{s(r,j),h_nk_nq_n},$$

$$\mu\Big(\Delta_{0,q_n} \setminus \bigcup_{r=0}^{q_n-1} \bigcup_{j=0}^{h_nk_{n,r}-1} F_j^r\Big) < \frac{\varepsilon_n}{4q_n},$$

and also

$$\mu\left(\Delta_{s(r,j),h_nk_nq_n} \setminus G_{s(r,j)}\right) < \frac{\varepsilon_n}{4h_nk_nq_n}$$

Therefore

$$\mu\,(A_{n+1}\Delta_{0,q_n} \triangle R^{(n)}) \leqslant \frac{\varepsilon_n}{2q_n}$$

(3.10)
$$+\,\mu\,((A_{n+1}\bigcup_{r=0}^{q_n-1}\bigcup_{j=0}^{h_nk_{n,r}-1} F_j^r) \triangle (\bigcup_{r=0}^{q_n-1}\bigcup_{j=0}^{h_nk_{n,r}-1} S_{\frac{r}{q_n}}\Delta_{s(r,j),h_nk_nq_n})).$$

On the other hand, $F_j^r \subset \Gamma_r$, and on $\Gamma_r$ the mapping $\hat{S}$ equals $S_{r/q_n}$; thus

$$A_{n+1}F_j^r = S\hat{S}F_j^r = S_{\frac{r}{q_n}} SF_j^r,$$

Now the inequality $\mu(SF_j^r \triangle G_{s(r,j)}) < \epsilon_n/4h_nk_nq_n$ implies that $\mu(\cdots) < \epsilon_n/4q_n$ on the right-hand side of (3.10). Consequently the left-hand side is less than $\epsilon_n/q_n < \epsilon_n$.

Finally, condition (3.5) is satisfied if there exist numbers $r$ and $j$ for which $s = s(r,j)$, and as the set $R_s^{(n)}$ we may choose $SF_j^r \cap G_{s(r,j)}$.

Thus the construction of the diffeomorphism $A_{n+1}$ is completed, and it is evident that for any arbitrary multiple $l_n$ of $h_n$ formula (3.2) yields a diffeomorphism $T_{n+1}$ which satisfies the conditions of the First and the Fourth Step.

If we now choose as $l_n$ a sufficiently large multiple of $h_n$, then it follows from the condition of the Third Step that

$$\rho_{\left[\frac{1}{\varepsilon_n}\right]}(T_{n+1}, T_n) < \varepsilon_n$$

since in (3.2) $T_{n+1}$ converges to $T_n$ under the $C^\infty$ topology as $\beta_n$ tends to zero.

It remains to verify the conditions of the Second Step. Since $\operatorname{diam} A_{n+1}^{-1}R_s^{(n)} < \gamma_{n+1}$, we have $\operatorname{diam} B_{n+1}^{-1}R_s^{(n)} < 1/2^{n-1}$ (see the beginning of subsection 4). We take as $E_{n+1}$ the set

$$E_{n+1} = \bigcup_{k=0}^{q_n-1} S_{\frac{k}{q_n}} \bigcup_{s=0}^{h_nk_n-1} R_s^{(n)}.$$

It follows from (3.5) that $\mu(E_{n+1}) > 1 - \epsilon_n$. The intersection of an arbitrary element of the decomposition $\xi_{n+1}$ with $E_{n+1}$ is contained in some set of the form $S_{k/q_n}R_s^{(n)}$ (since an element of the decomposition $\eta_{q_{n+1}}$ is contained in some $\Delta_{m,h_nq_n}$), and since $S_{k/q_n}$ is an isometry, the diameter of this intersection does not exceed $1/2^{n+1}$.

6. The sequence $\xi_n = B_n^{-1}\eta_{q_n}$ of decomposition is, generally speaking, not monotonic if $R^{(n)} \neq \Delta_{0,q_n}$. We show how from this sequence a monotonic one may be obtained.

We associate with the element $\Delta_{k,q_n}$ of the decomposition $\eta_{q_n}$ the element $S_{k/q_n}R^{(n)}$ of the decomposition $\eta_{n,n+1}$ (see Remark 3.1). This correspondence defines a mapping

$$C_n : M^m / \eta_{q_n} \to M^m / \eta_{n,n+1}.$$

For a set $E$ consisting of elements of the decomposition $\eta_{q_n}$ we consider as image $E'$ under the mapping $C_n$ the projection of this set into $M^m/\eta_{q_n}$, and we denote by $C_nE$ the complete inverse image of the set $E'$ under the canonical projection $M^m \to M^m/\eta_{n,n+1}$. Since the elements of the decomposition $\eta_{n,m+1}$ consist of elements of the decomposition $\eta_{q_{m+1}}$, we may map the sets $C_{n-1}\cdots C_m\Delta_{k,q_m}$ for $m<n$. We denote by $\eta_{m,n}$ the decomposition $\{C_{n-1}\cdots C_m\Delta_{k,q_m}\}$. It is not difficult to see that $\eta_{m+1,n} > \eta_{m,n}$. Now we define $\xi_{m,n} = B_n^{-1}\eta_{m,n}$.

LEMMA 3.1. *If $\sum q_n\epsilon_n < \infty$,*[8] *then the sequence $\xi_{m,n}$ for fixed $m$ converges to a decomposition $\xi_{m,\infty}$ as $n$ tends to infinity. The sequence $\xi_{m,\infty}$ has the following properties*:

3.1.1. $\xi_{m+1,\infty} > \xi_{m,\infty}$, $m = 1, 2, \cdots$.

3.1.2. $T_m\xi_{m,\infty} = \xi_{m,\infty}$, *where the factor-automorphism $T_m/\xi_{m,\infty}$ is a cyclic permutation of the elements of the decomposition $\xi_{m,\infty}$.*

3.1.3. $\xi_{m,\infty} \to \epsilon$ *for* $m \to \infty$.

PROOF. We define a correspondence $P_n^m$ between the elements of the decompositions $\xi_m$ and $\xi_{m,n}$. Let $c \in \xi_m$ be the element $c = B_m^{-1}\Delta_{k,q_m}$. We define

$$P_n^m c = B_n^{-1}\ C_{n-1}\dots C_m\Delta_{k,q_m}.$$

For a set $E$ consisting of elements $c_i$ of the decomposition $\xi_m$ we let $P_n^mE = \bigcup P_n^m c_i$. It is easy to see that $P_n^m = P_n^l P_l^m$ for $n > l > m$. We will give an upper bound for

$$\sum_{c\in\xi_m} \mu(P_n^m c \triangle P_{n+1}^m c).$$

The element $d = C_{n-1}\cdots C_m\Delta_{k,q_m}$ consists of $q_n/q_m$ sets $\Delta_{l,q_n}$. Similarly, the element $d' = A_{n+1}^{-1}C_nC_{n-1}\cdots C_m\Delta_{k,q_n}$ consists of the sets $A_{n+1}^{-1}S_{l/q_n}R^{(n)}$ with the same $l$. From (3.1) and (3.4) we have

$$\mu(d \triangle d') \leqslant \frac{q_n}{q_m}\mu(\Delta_{0,q_n} \triangle A_{n+1}^{-1}R^{(n)}) \leqslant \frac{q_n}{q_m}\varepsilon_n.$$

Since the mapping $B_n^{-1}$ is measure preserving, the same inequality is valid for $\mu(P_n^m c \triangle P_{n+1}^m c)$ for any $c \in \xi_m$. Therefore

$$\sum_{c\in\xi_m} \mu(P_n^m c \triangle P_{n+1}^m c) < q_n\varepsilon_n. \tag{3.11}$$

From this we obtain for $n_2 > n_1 \geq m$ the inequality

[8] We choose the numbers $\epsilon_n$ after having determined the numbers $q_n$.

$$(3.12) \qquad \sum_{c\in\xi_m} \mu(P^m_{n_1}c \triangle P^m_{n_2}c) < \sum_{k=n_1}^{n_2-1} q_k \varepsilon_k.$$

which shows the convergence of the sequence $\xi_{m,n}$ as $n$ tends to infinity, and also the existence of the limit $\lim_{n\to\infty} P^m_n c$, which we denote by $P^m_\infty c$.

Property 3.1.1 is satisfied, i.e. $\xi_{m+1,n} > \xi_{m,n}$ for any $n$.

Property 3.1.2 is also satisfied, since the corresponding assertion holds true for the decomposition $\xi_{m,n}$ for arbitrary $n$.

Next we verify Property 3.1.3. From (3.12) we have

$$(3.13) \qquad \sum_{c\in\xi_m} \mu(c \triangle P^m_\infty c) < \sum_{l=m}^{\infty} q_l \varepsilon_l = \varepsilon'_m.$$

Let $A$ be an arbitrary measurable set. Since $\xi_m \to \epsilon$ for $m\to\infty$, there exist, in the Boolean $\sigma$-algebra $\mathfrak{B}(\xi_m)$, sets $A_m$ for which $\mu(A \triangle A_m)\to 0$. Let $A_m = \bigcup_{i\in I_m} c^{(i)}$, where $c^{(i)} \in \xi_m$, and let $A'_m = \bigcup_{i\in I_m} P^m_\infty c^{(i)}$. Clearly $A'_m \in \mathfrak{B}(\xi_{m,\infty})$. On the other hand, (3.13) implies

$$\mu(A \triangle A'_m) \leqslant \mu(A \triangle A_m) + \varepsilon'_m,$$

and consequently $\mu(A \triangle A'_m)$ converges to zero as $m$ tends to infinity. This proves the lemma.

REMARK 3.2. The mapping $P^\infty_n$: $M^m/\xi_n \to M^m/\xi_{n,\infty}$ is measure preserving and commutes with $T_n$.

7. Let $T = \lim_{n\to\infty} T_n$. By 3.1.2 and 3.1.3 the decompositions $\xi_{n,\infty}$ and the automorphisms $T_n$ define a cyclic approximation of the automorphism $T$ by periodic mappings (APM) with a certain speed of approximation in the sense of [8]. In order to estimate the speed of approximation we have to find a bound on the expression

$$\sum_{c\in\xi_{n,\infty}} \mu(Tc \triangle T_n c) \leqslant \sum_{k=n}^{\infty} \sum_{c\in\xi_{n,\infty}} \mu(T_{k+1}c \triangle T_k c) \leqslant \sum_{k=n}^{\infty} \sum_{c\in\xi_{k,\infty}} \mu(T_{k+1}c \triangle T_k c)$$

(here 3.1.1 is satisfied). In each term $c = P^k_\infty d$, where $d \in \xi_k$. We obtain

$$\sum_{d\in\xi_k} \mu(T_{k+1}P^k_\infty d \triangle T_k P^k_\infty d) \leqslant \sum_{d\in\xi_k} \mu(T_{k+1}P^k_\infty d \triangle T_{k+1} d)$$
$$+ \sum_{d\in\xi_k} \mu(T_k P^k_\infty d \triangle T_k d) + \sum_{d\in\xi_k} \mu(T_{k+1} d \triangle T_k d).$$

Since $T_{k+1}$ and $T_k$ are measure preserving, it follows from (3.13) that here the first two summations do not exceed $\epsilon'_k$.

Now we give an estimate for each individual term in the last summation. There $d$ has the form $d = B_k^{-1}\Delta_{l,q_k}$. Since (3.1), (3.2) and (3.4) hold, and since $B_k$ and $S_t$ are measure preserving, we have

$$\mu(T_{k+1}d \triangle T_k d) = \mu((B_k^{-1}T_{k+1}B_k\Delta_{l,q_k}) \triangle (B_k^{-1}T_kB_k\Delta_{l,q_k}))$$
$$= \mu((A_{k+1}^{-1}S_{\alpha_{k+1}}A_{k+1}\Delta_{l,q_k}) \triangle S_{\alpha_k}\Delta_{l,q_k}) = \mu((A_{k+1}^{-1}S_{\beta_k}A_{k+1}\Delta_{l,q_k}) \triangle \Delta_{l,q_k})$$
$$\leqslant \mu(S_{\beta_k}S_{\frac{l}{q_k}}R^{(k)} \triangle S_{\frac{l}{q_k}}R^{(k)}) + 2\varepsilon_k = \mu(S_{\beta_k}R^{(k)} \triangle R^{(k)}) + 2\varepsilon_k.$$

The set $R^{(n)}$ consists of $k_n$ sets of the form $\Delta_{s,k_nq_n}$. Clearly

$$\mu(\Delta_{s,k_kq_k} \triangle S_{\beta_k}\Delta_{s,k_kq_k}) = 2\beta_k,$$

i.e.

$$\mu(S_{\beta_k}R^{(k)} \triangle R^{(k)}) \leqslant 2k_k\beta_k = \frac{2}{l_kq_k^2}.$$

Finally,

$$\sum_{c\in\xi_{n,\infty}} \mu(Tc \triangle T_nc) \leqslant 2\sum_{k=n}^{\infty} \varepsilon_k' + 2\varepsilon_n' + \sum_{k=n}^{\infty}\frac{2}{l_kq_k} \leqslant 4\sum_{k=n}^{\infty}\varepsilon_k' + \sum_{k=n}^{\infty}\frac{2}{l_kq_k}. \tag{3.14}$$

From this inequality it is evident that if $l_n$ increases sufficiently rapidly and if $\epsilon_n$ decreases sufficiently rapidly, then the summation $\sum_{c\in\xi_{n,\infty}}\mu(Tc \triangle T_nc)$ may be made less than $f(q_n)$, where $f(n)$ is a sequence of positive numbers which can be made to grow arbitrarily fast. In other words, we can construct the automorphism $T$ in such a way that it possesses a cyclic APM with a speed $f(n)$ of approximation given in advance.

If, for example, $f(n) = o(1/n)$, then the diffeomorphism $T$ is ergodic (cf. [8]) but it is not mixing, and the shift operator $U_T$ in the space $L_2(M^m, \mu)$ has a simple singular spectrum.

## §4. A diffeomorphism which is metrically isomorphic to a circular rotation

1. It turns out that in the most natural case $R^{(n)} = \Delta_{0,q_n}$ we may obtain considerably more extensive information on the metric structure of the limiting diffeomorphism $T$ than what was given at the end of the preceding section.

THEOREM 4.1. *If in the mappings of §3 the set $R^{(n)}$ equals $\Delta_{0,q_n}$ for all $n$, then the limiting diffeomorphism $T$ is metrically isomorphic to the circular rotation by the angle $2\pi\alpha = \lim_{n\to\infty} 2\pi\alpha_n$.*

The proof is based on the following abstract lemma.

LEMMA 4.1. *Let $M_1$ and $M_2$ be Lebesgue spaces and let $\xi_n^{(i)}$ $(i = 1, 2)$ be monotonic sequences of finite measurable decompositions of the spaces $M_i$, $\xi_n^{(i)} \nearrow \epsilon$; furthermore, let $T_n^{(i)}$ be automorphisms of the spaces $M_i$ for which $T_n^{(i)}\xi_n^{(i)} = \xi_n^{(i)}$ and $\lim_{n\to\infty} T_n^{(i)} = T^{(i)}$, where the limit is taken in the weak topology of the space of automorphisms. Suppose there exist metric isomorphisms*

$$K_n : M_1/\xi_n^{(1)} \to M_2/\xi_n^{(2)},$$

*for which*

$$K_n^{-1}\, T_n^{(2)}/\xi_n^{(2)} K_n = T_n^{(1)}/\xi_n^{(1)}$$

*and*

(4.1) $$K_n \xi_{n-1}^{(1)} = \xi_{n-1}^{(2)}$$

*(in the last case,* $\xi_{n-1}^{(i)}$ *stands for the corresponding decomposition of the factor space* $M_i/\xi_n^{(i)}$*). Then the automorphisms* $T_1$ *and* $T_2$ *are metrically isomorphic.*

The proof of the lemma is shorter than its wording. First we observe that (4.1) implies the equation $K_n \xi_k^{(1)} = \xi_k^{(2)}$ for any $k < n$. Hence there follows also the existence of an automorphism $K$: $M_1 \to M_2$ for which $K/\xi_n^{(1)} = K_n$. [Indeed, let $x \in M_1$, $x \in c_n(x)$, $c_n(x) \in \xi_n^{(1)}$. Then $\{x\} = \bigcap_{n=1}^{\infty} c_n(x)$ for almost all $x \in M_1$. We let $Kx = \bigcap_{n=1}^{\infty} K_n(c_n(x))$.] It is easy to see that $K^{-1} T_n^{(2)} K/\xi_n^{(1)} = T_n^{(1)}$. Therefore

$$T^{(1)} = \lim T_n^{(1)} = \lim K^{-1} T_n^{(2)} K = K^{-1} \lim T_n^{(2)} K = K^{-1} T^{(2)} K.$$

Proof of Theorem 4.1. Under the conditions of Lemma 4.1, let $M_1 = M^m$ and $M_2 = \mathbf{S}^1$, and let $\xi_n^{(1)} = \xi_n$ and $\xi_n^{(2)}$ be decompositions of the circle $\mathbf{S}^{-1} = \{\zeta\colon\ \zeta \in \mathbf{C},\ |\zeta| = 1\}$ into the arcs

$$c_{n,k} = \left\{\zeta : \frac{2\pi k}{q_n} \leqslant \arg \zeta < \frac{2\pi\,(k+1)}{q_n}\right\},$$
$$k = 0, \ldots, q_n - 1;$$

furthermore let $T_n^{(1)} = T_n$, and let $T_n^{(2)}$ be the circular rotation by the angle $2\pi p_n/q_n$. By $K_n$ we denote the mapping which maps the element $B_n^{-1}\Delta_{k,q_n} \in \xi_n^{(1)}$ onto the element $c_{n,k} \in \xi_n^{(2)}$. The condition (4.1) turns out to be a consequence of the equation $R^{(n)} = \Delta_{0,q_n}$. Thus Theorem 4.1 follows from Lemma 4.1.

Remark 4.1. Theorem 4.1 may be reworded by stating that the shift operator $U_T$ in the space $L_2(M^m, \mu)$ has a discrete spectrum with one independent eigenfrequency, i.e. in the space $L_2(M^m, \mu)$ there exists a basis consisting of eigenfunctions

$$f_n(x) = (f_1(x))^n \text{ and } U_T f_n(x) = \exp(2\pi i n \alpha) f_n(x).$$

These eigenfunctions must necessarily be discontinuous. It can be shown that in the case of continuous eigenfunctions the manifold $M^m$ must be the circumference (but the functions $f_n(x)$ may, of course, be continuous at some points).

Remark 4.2. Even if the metrical structure of the diffeomorphisms constructed above is simple, their topological structure may turn out to be quite complicated. These diffeomorphisms may have closed invariant sets of a complicated type, singular invariant measures which are positive on any open set and relative to which the metrical properties of $T$ are completely different from those relative to $\mu$, etc.

## §5. A weakly mixing diffeomorphism

1. In this section we show that by a suitable choice of the set $R^{(n)}$ the limiting automorphism $T$ can in general be made not to have eigenfunctions. For this proof we need some facts from ergodic theory which are summarized in the following theorem.

THEOREM 5.1. *Let $T$ be an automorphism of a Lebesgue space $(M,\mu)$. Then the following assertions are equivalent.*

5.1.1. *The shift operator $U_T$ does not have eigenfunctions other than constants.*

5.1.2. *For an arbitrary pair of measurable sets $F$, $G\subset M$ there exists an increasing sequence $n_k$ of positive integers with density one such that*

$$\mu(T^{n_k}F\cap G)\to\mu(F)\cdot\mu(G)\quad \text{for}\quad k\to\infty.$$

5.1.3. *There exists a sequence of positive integers $n_k\to\infty$ such that for any pair of sets $F$, $G\subset M$*

$$\mu(T^{n_k}F\cap G)\to\mu(F)\cdot\mu(G)\quad \text{for}\quad k\to\infty.$$

5.1.4. *There exists a sequence of finite measurable decompositions $\xi_k\to\epsilon$ and a sequence of positive integers $n_k\to\infty$ such that*

$$\sum_{c_1,c_2\in\xi_k}|\mu(T^{n_k}c_1\cap c_2)-\mu(c_1)\mu(c_2)|\to 0$$

*as $k$ tends to infinity.*

The equivalence of the assertions 5.1.1 and 5.1.2 is the subject of the well-known "Weak Mixing Theorem", a proof of which may be found in [**6**], Russian pp. 58–60. The equivalence of 5.1.2 and 5.1.3 is a simple fact; a proof may be found, for example, in [**22**].

We now give a proof of the equivalence of the assertions 5.1.3 and 5.1.4 which is also very simple.

Suppose 5.1.3 is satisfied. We consider a fixed sequence of measurable decompositions $\xi_n\to\epsilon$. Let $c_1\cdots,c_{q_n}$ be the elements of the decomposition $\xi_n$. In view of 5.1.3, for each pair $(i,j)$, $1\leqq i,\ j\leqq q_n$ there exists a number $k_{i,j,n}$ such that

$$|\mu(T^{n_k}c_i\cap c_j)-\mu(c_i)\mu(c_j)|<\frac{1}{2^nq_n^2}$$

for $k\geqq k_{i,j,n}$. Let $k_n=\max_{i,j}k_{i,j,n}$. Then for $k\geqq k_n$ one has the inequality

$$\sum_{i,j=1}^{q_n}|\mu(T^{n_k}c_i\cap c_j)-\mu(c_i)\mu(c_j)|<\frac{1}{2^n}.$$

Conversely, suppose 5.1.4 is satisfied. We show that for any two measurable sets $F$, $G\subset M$ the sequence $\mu(T^{n_k}F\cap G)$ converges to $\mu(F)\mu(G)$. Let $\epsilon>0$ be fixed. We choose a number $K$ such that for every $k>K$ there exist sets $F_k$ and $G_k$ which are measurable relative to the decomposition $\xi_k$ and for which $\mu(F \triangle F_k) < \epsilon$ and $\mu(G \triangle G_k) < \epsilon$. Then we have

$$(5.1)\quad \begin{aligned} |\mu(T^{n_k}F \cap G) - \mu(F)\mu(G)| &\leqslant |\mu(T^{n_k}F_k \cap G_k) - \mu(F_k)\mu(G_k)| \\ &+ \mu(F \triangle F_k) + \mu(G \triangle G_k) + |\mu(F) - \mu(F_k)| \\ &+ |\mu(G) - \mu(G_k)| \leqslant |(T^{n_k}F_k \cap G_k) - \mu(F_k)\mu(G_k)| + 4\varepsilon. \end{aligned}$$

Suppose that sets $F_k$ and $G_k$ consist, respectively, of the elements $c_{i_1}, \cdots, c_{i_r}$ and $c_{j_1}, \cdots, c_{j_s}$ of the decomposition $\xi_k$. We use the abbreviation

$$|\mu(T^{n_k}c_i \cap c_j) - \mu(c_i)\mu(c_j)| = \omega_{ij}.$$

Adding the inequalities

$$\omega_{i_a j_b} \geqslant \mu(T^{n_k}c_{i_a} \cap c_{j_b}) - \mu(c_{i_a})\mu(c_{j_b}) \geqslant -\omega_{i_a j_b}$$

for $a = 1, \cdots, r$ and $b = 1, \cdots, s$ we obtain

$$\sum_{a=1}^{r}\sum_{b=1}^{s}\omega_{i_a j_b} \geqslant \mu(T^{n_k}F_k \cap G_k) - \mu(F_k)\mu(G_k) \geqslant -\sum_{a=1}^{r}\sum_{b=1}^{s}\omega_{i_a j_b}$$

and hence

$$|\mu(T^{n_k}F_k \cap G_k) - \mu(F_k)\mu(G_k)| \leqslant \sum_{i,j=1}^{q_n}\omega_{ij}.$$

It follows from 5.1.4 that here the right-hand side converges to zero as $k$ tends to infinity. Thus we obtain 5.1.3 from (5.1).

2. We show how the parameters in the construction of §3 have to be chosen in order for the limiting automorphism to have property 5.1.4. We shall apply the notation from §3. Inasmuch as in the First Step of the induction the numbers $p_{n+1}$ and $q_{n+1}$ are relatively prime, there exists for any $r$ with $0 \leqq r < q_{n+1}$ a number $k^r$, $0 \leqq k^r < q_{n+1}$, such that $k^r p_{n+1}/q_{n+1} \equiv r/q_{n+1} \pmod 1$. This means that

$$T_{n+1}^{k^r} = B_{n+1}^{-1}S_{k^r\alpha_{n+1}}B_{n+1} - B_{n+1}^{-1}S_{\frac{r}{q_{n+1}}}B_{n+1}.$$

Now we let $r_n = l_n q_n$. Then we have

$$T_{n+1}^{k^{r_n}} = B_{n+1}^{-1}S_{\frac{1}{k_n q_n}}B_{n+1}.$$

It follows from (3.3) that

$$(5.2)\quad S_{\frac{1}{k_n q_n}}R^{(n)} = \left(\bigcup_{i=0}^{k_n-2}\Delta_{a_n(i)k_n+i+1,\, k_n q_n}\right) \cup \Delta_{(a_n(k_n-1))+1)k_n,\, k_n q_n}.$$

We compute the measure of the intersection

$$(5.3)\quad S_{\frac{1}{k_n q_n}}R^{(n)} \cap S_{\frac{k}{q_n}}R^{(n)}.$$

For an arbitrary integer $a$ we denote by $\bar{a}$ the smallest nonnegative number which is congruent to $a$ modulo $q_n$. Clearly

$$S_{\frac{k}{q_n}} R^{(n)} = \bigcup_{i=0}^{k_n-1} \Delta_{\overline{(a_n(i)+k)k_n+i},\, k_n q_n}. \tag{5.4}$$

We denote by $Q_{n,k}$ the number of elements of the decomposition $\eta_{k_n q_n}$ in the intersection under consideration. Thus

$$\mu(S_{\frac{1}{k_n q_n}} R^{(n)} \cap S_{\frac{k}{q_n}} R^{(n)}) = \frac{Q_{n,k}}{k_n q_n}$$

and also

$$\mu(S_{\frac{1}{k_n q_n}} S_{\frac{l}{q_n}} R^{(n)} \cap S_{\frac{k}{q_n}} R^{(n)}) = \frac{Q_{n,\overline{k-l}}}{k_n q_n}.$$

3. We show that for sufficiently large $k_n$ there exists an $a_n(i)$ such that

$$\sum_{k=0}^{q_n-1} \left| \frac{Q_{n,k}}{k_n q_n} - \frac{1}{q_n^2} \right| < \frac{3}{q_n^2}. \tag{5.5}$$

Let $k_n = \lambda_n q_n + \mu_n$, where $0 \leqq \mu_n < q_n$. We define the number $a_n(i)$ for $i = 0, \cdots, k_n - 1$ as follows:

$$a_n(i) = \begin{cases} 0, & \text{if } i \text{ is even} \\ & \text{or if } \lambda_n q_n \leqq i \leqq k_n - 1 \text{ and } q_n \text{ is odd;} \\ r, & \text{where } r \text{ is defined by the condition} \\ & (2r-1)\lambda_n \leqslant i < (2r+1)\lambda_n,\ r = 0, \ldots, \left[\frac{q_n}{2}\right], \\ & \text{if } i \text{ is odd.} \end{cases} \tag{5.6}$$

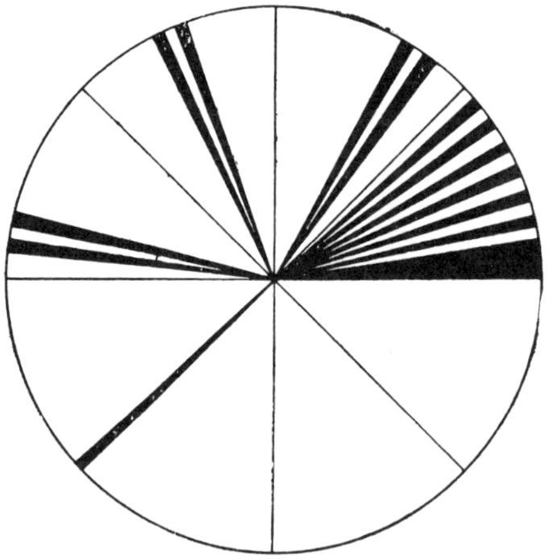

FIGURE 4

In this drawing we have taken $q_n = 8$, $k_n = 16$. The dark sectors belong to the set $R^{(n)}$.

For this system $a_n(i)$ the number $Q_{n,k}$ is $\lambda_n$ if $k$ is different from 0 and $q_n/2$. For example, for $0 < k < q_n/2$ the intersection (5.3) is obtained by only taking the sets $\Delta_{(a_n(i)+k)k_n+i,q_nk_n}$ with $a_n(i) = 0$ in (5.4) and $\Delta_{a_n(i)+i+1,k_nq_n}$ with $a_n(i) = k$ in (5.2), where this intersection contains exactly all $\lambda_n$ sets of the latter type. Conversely, for $q_n/2 < k \leqq q_n - 1$ the intersection (5.3) consists of all sets of the form $\Delta_{a_n(i)k_n+i+1,k_nq_n}$ with $a_n(i) = 0$ in (5.2) which overlap with sets $\Delta_{(a_n(i)+k)k_n+i,k_nq_n}$ with $a_n(i) = q_n - k$ in (5.4). This case also involves all $\lambda_n$ sets of the latter type. The reader may verify without difficulty that for the excluded values of $k$ the number $Q_{n,k}$ lies in the range between 0 and $2\lambda_n$.

Therefore we have

$$\sum_{k=0}^{q_n-1}\left|\frac{Q_{n,k}}{k_nq_n} - \frac{1}{q_n^2}\right| \leqslant q_n\left|\frac{\lambda_n}{q_n(\lambda_nq_n+\mu_n)} - \frac{1}{q_n^2}\right| + \frac{2}{q_n^2} \leqslant \frac{1}{q_n(\lambda_n+1)} + \frac{2}{q_n^2}.$$

Choosing $\lambda_n > q_n$, we obtain the inequality (5.5).

4. **Lemma** 5.1. *If $l_n$ increases sufficiently rapidly and if $\epsilon_n$ decreases sufficiently rapidly, then*

$$\sum_{c_1,\, c_2 \in \xi_{n,n+1}} |\mu(T^{k'_n}c_1 \cap c_2) - \mu(c_1)\mu(c_2)| \to 0.$$

**Proof.** The left side does not exceed

$$\sum_{c_1,c_2\in\xi_{n,n+1}} |\mu(T_{n+1}^{k'_n}c_1 \cap c_2) - \mu(c_1)\mu(c_2)| \tag{5.7}$$
$$+ \sum_{c_1,c_2\in\xi_{n,n+1}} |\mu(T^{k'_n}c_1 \cap c_2) - \mu(T_{n+1}^{k'_n}c_1 \cap c_2)|.$$

The relations (5.5), (3.1) and the fact $B_{n+1}$ is measure preserving imply that for $\lambda_n > q_n$ the first summation in (5.7) converges to zero as $n$ tends to infinity. On the other hand,

$$|\mu(T^{k'_n}c_1 \cap c_2) - \mu(T_{n+1}^{k'_n}c_1 \cap c_2)| \leqslant \mu((T^{k'_n}c_1 \triangle T_{n+1}^{k'_n}c_1) \cap c_2).$$

Therefore the second summation in (5.7) does not exceed

$$\sum_{c\in\xi_{n,\,n+1}} \mu(T^{k'_n}c \triangle T^{k'_n}c) \leqslant \sum_{c\in\xi_{n+1}} \mu(T^{k'_n}c \triangle T_{n+1}^{k'_n}c).$$

It remains to estimate the summation on the right-hand side of this last inequality. Since $T_{n+1}\xi_{n+1} = \xi_{n+1}$, it is not difficult to see that, for any integer $k$,

$$\sum_{c\in\xi_{n+1}} \mu(T^k c \triangle T_{n+1}^k c) \leqslant |k| \sum_{c\in\xi_{n+1}} \mu(Tc \triangle T_{n+1}c). \tag{5.8}$$

In order to estimate the last summation we use the mapping $P_\infty^{n+1}$ between the elements of the decompositions $\xi_{n+1}$ and $\xi_{n+1,\infty}$ as well as the inequalities

(3.13) and (3.14):

$$\sum_{c\in\xi_{n+1}} \mu(Tc \triangle T_{n+1}c) \leqslant \sum_{c\in\xi_{n+1}} \mu(Tc \triangle TP_\infty^{n+1}c)$$

$$+ \sum_{c\in\xi_{n+1}} \mu(TP_\infty^{n+1}c \triangle T_{n+1}P_\infty^{n+1}c) + \sum_{c\in\xi_{n+1}} \mu(T_{n+1}c \triangle T_{n+1}P_\infty^{n+1}c)$$

$$\leqslant 2\sum_{c\in\xi_{n+1}} \mu(c \triangle P_\infty^{n+1}c) + \sum_{c\in\xi_{n+1,\infty}} \mu(Tc \triangle T_{n+1}c)$$

$$\leqslant 2\varepsilon'_{n+1} + 4\sum_{k=n+1}^{\infty} \varepsilon'_k + 2\sum_{k=n+1}^{\infty} \frac{1}{q_k l_k}.$$

Thus, the second summation in (5.7) is not larger than

$$k^{r_n}\left(6\sum_{k=n+1}^{\infty} \varepsilon'_k + 2\sum_{k=n+1}^{\infty} \frac{1}{q_k l_k}\right).$$

Since $k^{r_n} < q_{n+1}$, this expression may be made arbitrarily small by a suitable choice of $\epsilon_{n+1}, \epsilon_{n+2}, \cdots$ and $l_{n+1}$ (which we select after having determined $q_{n+1}$). This completes the proof of Lemma 5.1.

5. It follows easily from (3.11) that $\xi_{n,n+1}$ converges to $\epsilon$ (using the fact that $\xi_n$ converges to $\epsilon$ and an argument analogous to that given in §3.6 for the convergence of $\xi_{n,\infty}$ to $\epsilon$). By Theorem 5.1, Lemma 5.1 and §3.7 we obtain the following theorem:

THEOREM 5.2. *If $a_n(i)$ is chosen according to (5.6), and if $l_n$ increases sufficiently rapidly and $\epsilon_n$ decreases sufficiently rapidly, then the shift operator $U_T$ on the space $L_2(M^m, \mu)$ has a continuous simple singular spectrum but it is not mixing.*

## §6. A diffeomorphism which is metrically isomorphic to a shift on the torus

1. In §4 we proved that on any manifold with a periodic flow there exist ergodic diffeomorphisms with discrete spectrum generated by a single eigenvalue. In the present section we generalize this result to the case of a discrete spectrum generated (over **Z**) by an arbitrary number $h$ of linearly independent eigenvalues.

As before, our procedure is based on a special choice of the sets $R^{(n)}$. However, this problem presents some additional difficulties in comparison with the problems solved in §§4 and 5. In §4 the isomorphism which existed between $T$ and a circular rotation was in a certain sense natural. The essence of the version of the limiting process used there lies in the fact that a sequence of decompositions $\eta_n$, converging to some decomposition $\eta$ whose elements are "transversal" to the orbits of the flow $\{S_t\}$ (in the case of $\mathbf{D}^2$ and the flow of rotations this is simply the decomposition into radii), is transformed into a sequence of decompositions $\xi_n$ which converges to $\epsilon$ such that the shifts $S_{\alpha_n}$ are transformed into automorphisms $T_n$ (precisely, $A_n \eta_k = \xi_k$ for $k \leq n$). Somehow we have joined together all ergodic components on the flow $\{S_t\}$ into one unit.

It is evident that this approach does not lead to any other automorphisms except those which are isomorphic to circular rotations. (In any case, for that purpose it would be necessary to start from flows with properties other than periodic ones.)

In §5 we obtained an automorphism which is not isomorphic to a circular rotation, but in that case we were concerned not with showing it to be metrically isomorphic to some automorphism of specified form, but only with a specific property of it. Now we shall obtain at each step of the construction an isomorphism between $T_n$ and some periodic shift on the torus such that the sequence of these shifts converges to an ergodic one. Here the isomorphisms at each step will necessarily be of a more "artificial" character than those in §4.

2. By $\mathbf{T}^h$ we denote the $h$-dimensional torus:

$$\mathbf{T}^h = \{\varphi = (\varphi_1, \dots, \varphi_h) : \varphi_i \in \mathbf{R}/\mathbf{Z}\}.$$

Furthermore, we denote the group shift on $\mathbf{T}^h$ by $T^\psi$: $\phi \to \phi + \psi$. We shall be concerned with a periodic flow $\{T^{\gamma t}\}$, where $\gamma = (\gamma_1, \cdots, \gamma_h)$ with relatively prime integers $\gamma_i$.

In $\mathbf{T}^h$ there is contained the $(h-1)$-dimensional torus

$$\mathbf{T}^{h-1} = \{\varphi = (\varphi_1, \dots, \varphi_{h-1}, 0)\}.$$

We consider the restriction of the shift $T^{\gamma/\gamma_h}$ to $\mathbf{T}^{h-1}$; this restriction turns out to be a shift on $\mathbf{T}^{h-1}$. Let $\Gamma_0$ be an open Dirichlet domain of the point $\bar{0} = (0, \cdots, 0)$. We enlarge $\Gamma_0$ by a part of the boundary in such a way as to obtain a fundamental domain, to be denoted by $\Gamma$. Obviously $\Gamma$ is also a fundamental domain for the flow $\{T^{\gamma t}\}$. For each natural number $k$ the set

$$\Gamma_k = \bigcup_{0 \leqslant t < \frac{1}{k}} T^{\gamma t}\, \Gamma$$

is a fundamental domain of the shift $T^{\gamma/k\gamma_h}$.

We show that under certain conditions a sequence $T^{\alpha^{(n)}}$ on the torus $\mathbf{T}^h$ converging to an ergodic shift $T^\alpha$ has a monotonic sequence of fundamental domains. The conditions which we shall now formulate are of course not necessary, but they are adapted to the inductive process by means of which we will construct on a manifold with a periodic flow a diffeomorphism which is metrically isomorphic to $T^\alpha = \lim T^{\alpha^{(n)}}$. The shift $T^{\alpha^{(n)}}$ will be included in the flow $\{T^{t\gamma^{(n)}}\}$.

LEMMA 6.1. *There exist sequences*

$$\alpha^{(n)} = (\alpha_1^{(n)}, \ldots, \alpha_h^{(n)}) \in \mathbf{T}^h, \ \gamma^{(n)} = (\gamma_1^{(n)}, \ldots, \gamma_h^{(n)}) \in \mathbf{Z}^h,$$

*with the following properties:*

$6.1_n$. *The greatest divisor of the* $\gamma_i^{(n)}$ *is* $(\gamma_1^{(n)}, \cdots, \gamma_h^{(n)}) = 1$.

$6.2_n$. *There exist relatively prime integers* $p_n$, $q_n$ *such that* $\alpha_i^{(n)} = p_n\gamma_i^{(n)}/q_n \pmod 1$.

$6.3_n$. *There exists an integer* $r_n$ *such that* $q_n = r_n\gamma_h^{(n)}$.

$6.4_n$. *There exists an integer* $s_{n-1}$ *such that* $\gamma_h^{(n)} = s_{n-1}\gamma_h^{(n-1)}$.

$6.5_n$. $\gamma_i^{(n)} \equiv \gamma_i^{(n-1)} \pmod{q_{n-1}}$, $i = 1, \ldots, h$.

$6.6_n$. *There exists an integer* $m_{n-1}$ *such that*

$$\frac{p_n}{q_n} = \frac{p_{n-1}}{q_{n-1}} + \frac{1}{m_{n-1}\, s_{n-1}\, q_{n-1}^2}$$

$6.7_n$. *Let* $\Gamma^{(n)} \subset \mathbf{T}^{h-1}$ *be a fundamental domain of the flow* $\{T^{t\gamma^{(n)}}\}$ *described above. Denote the diameter of* $\Gamma^{(n)}$ *by* $d_n$, *and by* $\sigma_n$ *the* $(h-1)$*-dimensional volume of the boundary of* $\Gamma^{(n)}$. *Then*

$$d_n < \frac{1}{2^{n-1}\gamma_h^{(n-1)}\sigma_{n-1}}.$$

$6.8_n$.

$$\left| \frac{1}{\gamma_h^{(n)}}\gamma^{(n)} - \frac{1}{\gamma_h^{(n-1)}}\gamma^{(n-1)} \right| < \frac{1}{2^{n-1}\sigma_{n-1}q_{n-1}}.$$

(*Distances, diameters and* $(h-1)$*-dimensional volumes are always taken relative to the Euclidean metric.*)

PROOF. Suppose numbers $\alpha^{(j)}$ and $\gamma^{(j)}$ satisfying $6.1_j$–$6.8_j$ have already been defined for $j \leqq n$. First we construct a number $\gamma^{(n+1)}$ satisfying the conditions $6.1_{n+1}$, $6.4_{n+1}$, $6.5_{n+1}$, $6.7_{n+1}$ and $6.8_{n+1}$. Here we shall use only the conditions $6.1_n$, $6.2_n$ and $6.3_n$.

We construct a matrix $A \in \mathbf{SL}(h, \mathbf{Z})$ whose last column coincides with the vector $\gamma^{(n)}$. This is possible in view of $6.1_n$ (see for example [**4**], Chapter 1, §2, Corollary 4). Here we can ascertain that the $(h-1)$-dimensional matrix $B$ obtained from the matrix $A$ by canceling the last column and the last row is nonsingular. Indeed, it is easy to see that in the case where $\det B = 0$, the vector $e_h = (0, \cdots, 0, 1)$ belongs to the lattice spanned by the first $h-1$ columns of the matrix $A$, and in this lattice we may choose a basis beginning with $e_h$. Let $e_h, a_1, \cdots, a_{k-1}$ be the vectors of this basis. Then we may replace $A$ by the matrix whose columns are

$$e_h + \gamma^{(n)}, \ a_1, \ \ldots, \ a_{h-2}, \ \gamma^{(n)}.$$

By applying the matrix $A$, the subspace $\mathbf{R}^{h-1} = \{(x_1, \cdots, x_{h-1}, 0)\}$ is mapped onto a subspace $A\mathbf{R}^{h-1}$ with equations of the form $x_h = \sum_{i=1}^{h-1} b_i x_i$. We observe that, denoting by $U_\delta$ a sphere of radius $\delta$ in the space $\mathbf{R}^{h-1}$, the cylinder $U_\delta \times \mathbf{R}$ intersects the subspace $A\mathbf{R}^{h-1}$ in an ellipsoid contained in the sphere of radius $b\delta$, where $b^2 = 1 + b_1^2 + \cdots + b_{h-1}^2$. The matrix $A$ transforms the unit cube $0 \leq x_i \leq 1$ of the space $\mathbf{R}^h$ into some parallelepiped in such a way that the edge $e_h$ of this cube is mapped onto the vector $\gamma^{(n)}$.

Let

$$\delta = \frac{1}{2^{n+1}\gamma_h^{(n)}\sigma_n b \,\|A^{-1}\|}.$$

We choose any vector $\vartheta = (\vartheta_1, \cdots, \vartheta_{h-1}, 0) \in \mathbf{Z}^h$ with $(\vartheta_1, \cdots, \vartheta_{h-1}) = 1$, using the property that in $\mathbf{T}^{h-1}$ every orbit of a periodic flow $\{T^{t\vartheta}\}$ intersects an arbitrary sphere of radius $\delta$. We let

$$\gamma^{(n+1)} = q_n A^{-1}\vartheta + (cq_n + 1)\gamma^{(n)} = A^{-1}(q_n\vartheta + (cq_n + 1)e_h) \tag{6.1}$$

and we show that the corresponding conditions are satisfied under a suitable choice of the integer $c$. Condition $6.1_{n+1}$ follows from (6.1) and the fact that the numbers $\vartheta_i$ are relatively prime; condition $6.4_{n+1}$ follows from (6.1) and $6.3_n$, and condition $6.5_{n+1}$ is an immediate consequence of (6.1). Let $c$ be so large that at the same time when the point $tq_n\vartheta$ is shifted by $\delta$, the point $(cq_n + 1)t$ is moved by not less than one. Then in $\mathbf{T}^h$ every orbit of the flow $\{T^{t(q_n\vartheta + (cq_n+1)e_h)}\}$ intersects with the torus $A\mathbf{T}^{h-1}$ in a $2b\delta$-net. It is easy to see that this ensures the validity of condition $6.7_{n+1}$. Finally, it is perfectly clear that in $6.8_{n+1}$ one has

$$\lim_{c\to\infty}\left|\frac{1}{\gamma_h^{(n+1)}}\gamma^{(n+1)} - \frac{1}{\gamma_h^{(n)}}\gamma^{(n)}\right| = 0.$$

The number $m_n$ which occurs in condition $6.6_{n+1}$ may be chosen arbitrarily. From $6.6_{n+1}$ we automatically obtain relatively prime numbers $p_{n+1}$, $q_{n+1}$, and then $6.2_{n+1}$ defines $\alpha^{(n+1)}$. Condition $6.3_{n+1}$ follows from $6.6_{n+1}$ and $6.4_{n+1}$. This completes the proof of the lemma.

3. The set $\Gamma_{q_n}^{(n)} = \bigcup_{0 \leq t < 1/q_n} T^{t\gamma^{(n)}}\Gamma^{(n)}$ is a fundamental domain for the shift $T^{\gamma^{(n)}q_n}$, and thus also for $T^{\alpha^{(n)}}$, since the numbers $p_n$ and $q_n$ are relatively prime.

We denote by $\zeta_n$ the decomposition of the torus $\mathbf{T}^h$ into the sets

$$\Gamma_{k,n} = T^{\frac{k}{q_n}\gamma^{(n)}}\Gamma_{q_n}^{(n)}, \quad k = 0, \ldots, q_n - 1,$$

and by $\zeta_n'$ the decomposition of $\mathbf{T}^{h-1}$ into the sets

$$\Gamma_{k,n}' = T^{\frac{k}{\gamma_h^{(n)}}\gamma^{(n)}}\Gamma^{(n)}, \quad k = 0, \ldots, \gamma_h^{(n)} - 1.$$

It follows from condition 6.3 that

$$\Gamma_{lr_n, n} = \bigcup_{0 \leq t < \frac{1}{q_n}} T^{t\gamma^{(n)}}\Gamma_{l,n}', \quad l = 0, \ldots, \gamma_h^{(n)} - 1.$$

Now we consider how the decompositions $\zeta_n$ and $\zeta_n'$ are related to the de- decompositions $\zeta_{n+1}$ and $\zeta_{n+1}'$. We begin with the decompositions $\zeta_n'$ and $\zeta_{n+1}'$.

It follows from conditions 6.3 and 6.5 that

$$T^{\frac{1}{\gamma_h^{(n)}}\gamma^{(n+1)}} = T^{\frac{1}{\gamma_h^{(n)}}\gamma^{(n)}}.$$

We shall consider this shift only on the torus $\mathbf{T}^{h-1}$. Denoting it by $V^{(n)}$, we obtain from conditions 6.4 and 6.5 the equation $(V^{(n+1)})^{s_n} = V^{(n)}$.

Let

$$K = \{k : 0 \leqslant k < \gamma_h^{(n+1)},\ \ (V^{(n+1)})^k \bar{0} \in \Gamma^{(n)}\}.$$

We define $\Gamma^{(n,n+1)} = \bigcup_{k\in K}\Gamma'_{k,n+1}$. Then $\Gamma^{(n,n+1)}$ is a fundamental domain for $V^{(n)}$. The decomposition into the sets $V^{(n)l}\Gamma^{(n,n+1)}$, $l = 0, \cdots, s_n - 1$, will be denoted by $\zeta'_{n,n+1}$. Clearly $\zeta'_{n,n+1} < \zeta'_{n+1}$. We estimate the $(h-1)$-dimensional Lebesgue measure $\mu_{h-1}(\Gamma^{(n)} \triangle \Gamma^{(n,n+1)})$. The set $\Gamma^{(n)}\setminus\Gamma^{(n,n+1)}$ is contained in a neighborhood of width $d_{n+1}$ of the boundary of $\Gamma^{(n)}$. Since $\Gamma^{(n)}$ is a convex set, one has

$$\mu_{h-1}(\Gamma^{(n)}\setminus\Gamma^{(n,n+1)}) \leqslant \sigma_n d_{n+1} < \frac{1}{2^n\gamma_h^{(n)}}$$

(in view of condition 6.7). But since $\mu_{h-1}(\Gamma^{(n)}) = \mu_{h-1}(\Gamma^{(n,n+1)})$, this implies the inequality

$$\mu_{h-1}(\Gamma^{(n)} \triangle \Gamma^{(n,n+1)}) < \frac{1}{2^{n-1}\gamma_h^{(n)}}. \tag{6.2}$$

Now we turn to the decompositions $\zeta_n$ and $\zeta_{n+1}$. We define

$$\Gamma_{q_n}^{(n,n+1)} = \bigcup_{0\leqslant t < \frac{1}{r_n\gamma_h^{(n+1)}}} T^{t\gamma^{(n+1)}}\Gamma^{(n,n+1)}.$$

The set $\Gamma_{q_n}^{(n,n+1)}$ consists of elements of the decomposition $\zeta_{n+1}$ (since $q_{n+1}$ is divisible by $r_n\gamma_h^{(n+1)}$) and is a fundamental domain of the shift $T^{\alpha^{(n)}}$. We note that the set $\Gamma_{q_n}^{(n,n+1)}$ depends only on the choice of the flow $T^{t\gamma^{(n+1)}}$ but not on the choice of $m_n$. The decomposition into the sets

$$T^{\frac{k}{q_n}\gamma^{(n)}}\Gamma_{q_n}^{(n,n+1)},\quad k = 0, \ldots, q_n - 1$$

will be denoted by $\zeta_{n,n+1}$. Evidently $\zeta_{n,n+1} < \zeta_{n+1}$. Now we estimate the $h$-dimensional Lebesgue measure $\mu_h(\Gamma_{q_n}^{(n,n+1)} \triangle \Gamma_{q_n}^{(n)})$.

We define

$$\hat{\Gamma}_{q_n}^{(n,n+1)} = \bigcup_{0\leqslant t\leqslant \frac{1}{q_n}} T^{t\gamma^{(n)}}\Gamma^{(n,n+1)}.$$

Since

$$\mu_h(\Gamma_{q_n}^{(n)} \triangle \hat{\Gamma}_{q_n}^{(n,n+1)}) = \frac{1}{r_n}\mu_{h-1}(\Gamma^{(n)} \triangle \Gamma^{(n,n+1)}),$$

it follows from (6.2) that

$$\mu_h(\Gamma_{q_n}^{(n)} \triangle \hat{\Gamma}_{q_n}^{(n,n+1)}) < \frac{1}{2^{n-1}q_n}. \tag{6.3}$$

The sets $T^{ts_n\gamma^{(n)}}\Gamma^{(n,n+1)}$ and $T^{t\gamma^{(n+1)}}\Gamma^{(n,n+1)}$ belong to the same torus $\phi_h = \text{const}$ by 6.4. Here the latter one of these sets is obtained from the former through a shift by the vector

$$t(\gamma_1^{(n+1)} - s_n\gamma_1^{(n)}, \ldots, \gamma_{h-1}^{(n+1)} - s_n\gamma_{h-1}^{(n)}).$$

In view of 6.8 the length of this vector does not exceed $t\gamma_h^{(n)}(2^n\sigma_n q_n)^{-1}$. For $t \leqq 1/q_n$ this length is not greater than $(2^n\sigma_n q_n r_n)^{-1}$. Thus for these $t$ the set $T^{ts_n\gamma^{(n)}}\Gamma^{(n,n+1)}\setminus T^{t\gamma^{(n+1)}}\Gamma^{(n,n+1)}$ is contained in the $(2^n\sigma_n q_n r_n)^{-1}$-neighborhood of the first of these sets, and the last one, as is evident from what was said above, lies the $(2^n\gamma_h^{(n)}\sigma_n)^{-1}$-neighborhood. Therefore we have

$$\mu_{h-1}(T^{ts_n\gamma^{(n)}}\Gamma^{(n,n+1)} \triangle T^{t\gamma^{(n+1)}}\Gamma^{(n,n+1)}) \leqslant \frac{4r_n}{2^n q_n}$$

and

$$\mu_h(\Gamma_{q_n}^{(n,n+1)} \triangle \hat{\Gamma}_{q_n}^{(n,n+1)}) \leqslant \frac{1}{2^{n-2} q_n}$$

(remembering that the $h$th coordinate is shifted by $r_n$). Finally, comparing this with (6.3), we find that

$$\mu_h(\Gamma_{q_n}^{(n,n+1)} \triangle \Gamma_{q_n}^{(n)}) < \frac{1}{2^{n-3} q_n}. \tag{6.4}$$

4. Now it is already easy to obtain from the sequence of decompositions $\zeta_n$ a monotonic sequence of decompositions $\zeta_{n,\infty}$ by a procedure which is analogous to that used in the proof of Lemma 3.1. For this purpose we establish a correspondence $Q_{n+1}^n$ between the elements of $\zeta_n$ and those of $\zeta_{n,n+1}$ by putting

$$Q_{n+1}^n \Gamma_{k,n} = T^{\frac{k}{q_n}\gamma^{(n)}} \Gamma_{q_n}^{(n,n+1)}. \tag{6.5}$$

In analogy to §3.6 we define for any set $E$ an image $Q_{n+1}^n E$ which consists of complete elements from the decomposition $\zeta_n$; furthermore, we introduce

$$Q_n^m = Q_n^{n-1} \ldots Q_{m+2}^{m+1} Q_{m+1}^m \text{ and } \zeta_{m,n} = Q_n^m \zeta_m \text{ for } n > m.$$

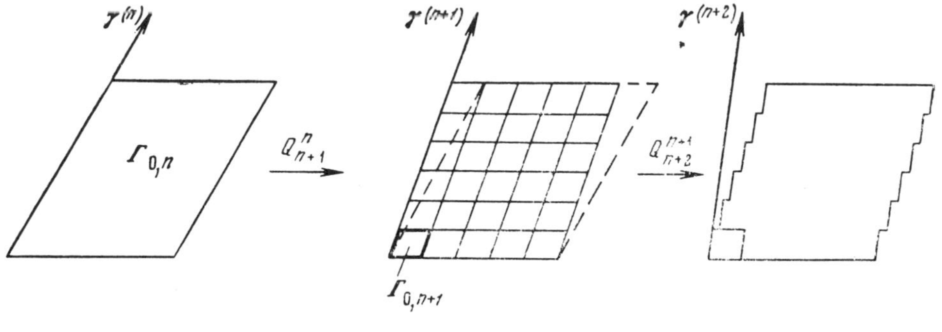


FIGURE 5

LEMMA 6.2. *For given m, the sequence of decompositions $\zeta_{m,n}$ converges to a decomposition $\zeta_{m,\infty}$. The sequence $\zeta_{m,\infty}$ has the following properties:*

6.2.1. $\zeta_{m+1,\infty} > \zeta_{m,\infty}$, $m = 1, 2, \cdots$.

6.2.2. $T^{\alpha^{(m)}}\zeta_{m,\infty}=\zeta_{m,\infty}$, *where the factor-automorphism* $T^{\alpha^{(m)}}/\zeta_{m,\infty}$ *is a cyclic permutation of the elements of the decomposition* $\zeta_{m,\infty}$.

6.2.3. $\zeta_{m,\infty}\rightarrow\epsilon$ *as* $m\rightarrow\infty$.

The proof of this lemma is completely analogous to the proof of Lemma 3.1, with the inequality (6.4) playing the role of (3.4).

We make some remarks which will be important in the sequel. The element $\lim_{n\rightarrow\infty}Q_n^m\Gamma_{k,m}$ of the decomposition $\zeta_{m,\infty}$ will be denoted by $Q_\infty^m\Gamma_{k,m}$. In the same manner we define the mapping

$$Q_\infty^m:\mathbf{T}^h/\zeta_m\rightarrow\mathbf{T}^h/\zeta_{m,\infty},$$

which is measure preserving and commutes with $T^{\alpha^{(m)}}$.

5. We shall construct simultaneously a sequence of shifts $T^{\alpha^{(n)}}$ of the torus $\mathbf{T}^h$ which satisfies Conditions 6.1–6.8 and a sequence of diffeomorphisms $T_n$ on the manifold $M^m$ with a periodic flow which satisfy the conditions of the First through the Fourth Step of §3. These sequences will be related to each other as follows: We introduce a mapping

$$K_n:\mathbf{T}^h/\zeta_n\rightarrow M^m/\xi_n$$

by the equations

$$K_n\Gamma_{k,n}=B_n^{-1}\Delta_{k,q_n},\quad k=0,1,\ldots,q_n-1. \tag{6.6}$$

Then the factor-automorphism satisfies

$$K_{n+1}/\zeta_{n,n+1}=P_{n+1}^nK_n(Q_{n+1}^n)^{-1}. \tag{6.7}$$

We show that in this case the limiting diffeomorphism $T$ on $M^m$ is metrically isomorphic to the limiting shift $T^\alpha$ on the torus $\mathbf{T}^h$. For this purpose we consider the mapping

$$\widetilde{K}_n:\mathbf{T}^h/\zeta_{n,\infty}\rightarrow M^m/\xi_{n,\infty},$$

where

$$\widetilde{K}_n=P_\infty^nK_n(Q_\infty^n)^{-1}.$$

It is easy to see that

1. $\widetilde{K}_nT^{\alpha^{(n)}}=T_n\widetilde{K}_n$.
2. The restriction of $\widetilde{K}_{n+1}$ to $\mathbf{T}^h/\zeta_{n,\infty}$ coincides with $\widetilde{K}_n$.

It follows from Lemma 4.1 that the automorphism $T^\alpha$ is metrically isomorphic to $T$.

It remains to construct the sequences $T_n$ and $T^{\alpha^{(n)}}$. We proceed by induction. Assuming the sequences to be defined for some $n$, we show how to define $T_{n+1}$ and $T^{\alpha^{(n+1)}}$ such that condition (6.7) is satisfied.

In order to define the shift $T^{\alpha^{(n+1)}}$ it suffices to construct a vector $\gamma^{(n+1)}$ and to exhibit a number $m_n$. Let the vector $\gamma^{(n+1)}$ be constructed by the relevant part of Lemma 6.1. Then, in particular, a number $s_n$ is defined by condition 6.4. In the First Step in the induction of §3 we let $k_n=s_n\gamma_h^{(n+1)}$.

We define the fundamental domain

$$\hat{\Gamma}^{(n+1)} = \bigcup_{0 \leqslant t < \frac{1}{k_n q_n}} T^{t\gamma^{(n+1)}} \Gamma^{(n+1)} \quad \text{of the shift } T^{\frac{1}{k_n q_n}\gamma^{(n+1)}}$$

and the decomposition $\hat{\zeta}_n$ into the sets

$$\hat{\Gamma}_{k,n} = T^{\frac{k}{k_n q_n}\gamma^{(n+1)}} \hat{\Gamma}^{(n+1)}, \quad k = 0, \ldots, k_n q_n - 1.$$

The set $\Gamma_{q_n}^{(n,n+1)}$, as we mentioned, does not depend on the choice of $m_n$; furthermore, it is easy to verify that this set consists of elements of the decomposition $\hat{\zeta}_n$, and thus $\hat{\zeta}_n > \hat{\zeta}_{n,n+1}$. We define a mapping $\hat{K}_n: \mathbf{T}^h/\hat{\zeta}_n \to M^m/\eta_{k_n q_n}$ by letting

$$\hat{K}_n \hat{\Gamma}_{k,n} = \Delta_{k,k_n q_n}, \quad k = 0, \ldots, k_n q_n - 1. \tag{6.8}$$

Now we introduce

$$R^{(n)} = \hat{K}_n \Gamma_{q_n}^{(n,n+1)}. \tag{6.9}$$

The set $R^{(n)}$ is a fundamental domain for $S_{1/q_n}$, such that we have

$$\hat{K}_n T^{\frac{1}{q_n}\gamma^{(n)}} /\hat{\zeta}_n = S_{\frac{1}{q_n}} \hat{K}_n \tag{6.10}$$

by (6.8). The quantities $k_n$ and $R^{(n)}$ thus defined determine the numbers $a_n(i)$ for $i = 0, \cdots, k_{n-1}$. Now we define a mapping $A_{n+1}$ and we choose a number $l_n$ such that the conditions of the Second, Third and Fourth Steps[9] are satisfied.

We introduce $m_n = \gamma_h^{(n+1)} l_n$ and we construct a decomposition $\zeta_{n+1}$. We have $\zeta_{n+1} > \hat{\zeta}_n$, where the fact that every element of the decomposition $\hat{\zeta}_n$ contains elements of the decomposition $\zeta_{n+1}$ may be accomplished in such a way that each following element is obtained from the preceding element by applying the mapping $T^{1/q_{n+1}r^{(n+1)}}$. Similarly the fact that every element of the decomposition $\eta_{k_n q_n}$ contains elements of the decomposition $\eta_{q_{n+1}}$ may be accomplished in such a way that each following element is obtained from the preceding element by applying the mapping $S_{1/q_{n+1}}$. Hence it follows from (6.8) that the diagram

$$\begin{array}{ccc} \mathbf{T}^h/\zeta_{n+1} & \xrightarrow{\overline{K}_n} & M^m/\eta_{q_{n+1}} \\ \downarrow & & \downarrow \\ \mathbf{T}^h/\hat{\zeta}_n & \xrightarrow{\hat{K}_n} & M^m/\eta_{k_n q_n} \end{array} \tag{6.11}$$

is commutative; here $\overline{K}_n \Gamma_{k,n+1} = \Delta_{k,q_{n+1}}$, and the vertical arrows stand for natural embeddings of the decompositions.

---

[9] Even though $l_n$ does not occur explicitly in these conditions, the possibility of satisfying them is related to the choice of $l_n$ in §3.5.

Now let $K_{n+1}$ be a mapping defined according to (6.6) with $n+1$ instead of $n$. We have to verify that $K_{n+1}$ satisfies (6.7), i.e.

(6.12) $$K_{n+1}Q^n_{n+1}\Gamma_{k,n} = P^n_{n+1}K_n\Gamma_{k,n}.$$

Here we rewrite the left-hand side, using the relations $K_{n+1} = B^{-1}_{n+1}\overline{K}_n$, (6.5), $\hat{\zeta}_n > \zeta_{n,n+1}$ and (6.11), in the form

$$B^{-1}_{n+1}\overline{K}_n T^{\frac{k}{q_n}\gamma^{(n)}}\Gamma^{(n,n+1)}_{q_n} = B^{-1}_{n+1}\widehat{K}_n T^{\frac{k}{q_n}\gamma^{(n)}}\Gamma^{(n,n+1)}_{q_n}.$$

To the right-hand side of (6.12) we apply (6.6), the definitions of $P^n_{n+1}$ and $C_n$ from §3.6, and (6.9), thus finding it equal to

$$P^n_{n+1}B^{-1}_n\Delta_{k,q_n} = B^{-1}_{n+1}C_n\Delta_{k,q_n} = B^{-1}_{n+1}S_{\frac{k}{q_n}}R^{(n)} = B^{-1}_{n+1}S_{\frac{k}{q_n}}\widehat{K}_n\Gamma^{(n,n+1)}_{q_n}.$$

Now the equation (6.12) follows from (6.10).

Thus the following theorem has been proved.

THEOREM 6.1. *Let $M^m$ be a manifold on which there exists a periodic flow of class $C^\infty$, let $h$ be an arbitrary positive integer and let $\mu$ be a positive measure of class $C^\infty$ on $M^m$. Then there exists a diffeomorphism $T \in \mathrm{Diff}^\infty(M^m, \mu)$ which is metrically isomorphic to some ergodic shift of the $h$-dimensional torus $\mathbf{T}^h$.*

REMARK 6.1. By means of minor modifications in our construction we may also obtain a diffeomorphism $T \in \mathrm{Diff}^\infty(M^m, \mu)$ which is metrically isomorphic to some ergodic shift of the infinite-dimensional torus $\mathbf{T}^\infty$ (i.e. a diffeomorphism with discrete spectrum generated over $\mathbf{Z}$ by a countable set of independent eigenvalues).

6. Finally, it is possible to construct also a diffeomorphism $T \in \mathrm{Diff}^\infty(M^m, \mu)$ which is metrically isomorphic to the direct product of some ergodic shift on $\mathbf{T}^h$ or $\mathbf{T}^\infty$ and some automorphism with continuous simple singular spectrum which is a realization of a diffeomorphism $\hat{T} \in \mathrm{Diff}^\infty(N^k, \nu)$ of an arbitrary manifold $N^k$ with a periodic flow by means of the construction from §5. In order to accomplish this one has to construct simultaneously a sequence $T^{\alpha^{(n)}}$ of shifts of the torus as described in Lemma 6.1, a sequence $T$ of periodic diffeomorphisms of manifold of $N^k$ as described in §5, and a sequence $\hat{T}_n$ of periodic diffeomorphisms of a manifold $M^m$ as described in §3. For this purpose one has to apply a certain lemma on the existence of a sequence $\epsilon_n \nearrow \epsilon$ of finite measurable decompositions which are invariant with respect to the automorphisms $T_n = T^1_n \times T^2_n$ of a Lebesgue space and which furthermore have the property that $T_n/\xi_n$ is cyclic if the sequences $T^i_n$ have invariant finite decompositions $\xi^i_n \nearrow \epsilon$, if these sequences themselves converge sufficiently rapidly under the weak topology and if the mappings $T^i_n/\xi^i_n$ are cyclic. Such a decomposition enables us to choose $a_n(i)$ for $T_{n+1}$ if $\gamma^{(n+1)}$ and $q_{n+1}$ have already been determined for $T^{\alpha^{(n+1)}}$ and if the corresponding quantities $\hat{a}_n(i)$ have been chosen for $\hat{T}_{n+1}$. Thereafter the number $l_n$ (for $T_{n+1}$) must be chosen sufficiently large in order to guarantee that the condition of the Third Step is satisfied simultaneously for $T_{n+1}$ and $\hat{T}_{n+1}$.

## §7. Ergodic diffeomorphisms contained in the closure of periodic diffeomorphisms

We consider all possible periodic flows of class $C^\infty$ on $M^m$ which preserve a given positive measure $\mu$ of class $C^\infty$, and we denote by $\mathfrak{P}(M^m, \mu)$ the set of all diffeomorphisms contained in such flows. The closure $\overline{\mathfrak{P}(M^m,\mu)}$ of this set in the space $\mathrm{Diff}^\infty(M^m,\mu)$ is a nowhere dense, perfect subset of this space. Any diffeomorphism $T$ obtained by the construction of §3 obviously belongs to the set $\overline{\mathfrak{P}(M^m,\mu)}$.

LEMMA 7.1. *The set of diffeomorphisms which as automorphisms of the Lebesgue space $(M^m,\mu)$ possess a cyclic APM with a given speed of approximation $f(n)$ is everywhere dense in $\mathfrak{P}(M^m,\mu)$.*

PROOF. It suffices to show that in any neighborhood of a diffeomorphism $S\in \mathfrak{P}(M^m, \mu)$ there exist diffeomorphisms which have a cyclic APM with a speed of approximation $f(n)$. Let $S = S_\alpha$, where $\{S_t\}$ is a periodic flow. We choose a fixed open neighborhood $\mathfrak{U}$ of the diffeomorphism $S$ in the space $\mathrm{Diff}^\infty(M^m,\mu)$ and we determine a rational number $\alpha_0$ such that $S_{\alpha_0}\in \mathfrak{U}$. Then we let $T_0 = S_{\alpha_0}$ in the construction of §3. The diffeomorphism $T$ may be constructed in such a manner that it is contained in any neighborhood of the automorphism $T_0$ given in advance (in order to accomplish this is only necessary to choose $\epsilon_n$ in the conditions of the Third Step sufficiently small). If, furthermore, $\epsilon_n$ and $l_n$ are suitably chosen such that the expression on the right-hand side of (3.14) becomes less than $f(q_n)$, then the diffeomorphism $T = \lim T_n$ possesses a cyclic AMP with the speed of approximation $f(n)$.

THEOREM 7.1. *The set of diffeomorphisms from $\overline{\mathfrak{P}(M^m,\mu)}$ which as automorphisms of the Lebesgue space $(M^m,\mu)$ possess a cyclic AMP with a given speed of approximation $f(n)$ is a set of the second category within $\overline{\mathfrak{P}(M^m,\mu)}$ (it contains an everywhere dense $G_\delta$-set).*

(We note that $\mathrm{Diff}^\infty(M^m,\mu)$ may be considered as a complete metric space with the metric

$$\rho(S, T) = \sum_{n=1}^{\infty} \frac{\rho_n(S, T)}{2^n(1+\rho_n(S, T))};$$

hence Baire's Theorem holds for any closed subset in the space.)

PROOF. We consider all possible sequences of diffeomorphisms $\{T_n\}$ which satisfy the inductive hypothesis of §3 with values of $\epsilon_n$ and $l_n$ for which the expression on the right-hand side of (3.14) does not exceed $f(q_n)$. Let $\mathfrak{U}(T_n)$ be a neighborhood of the diffeomorphism $T_n$ defined as follows:

$$\mathfrak{U}(T_n) = \{S : S\in \mathrm{Diff}^\infty(M^m, \mu),$$

$$\rho_{\left[\frac{1}{\varepsilon_n}\right]}(T_n, S) < 2\varepsilon_n, \quad \sum_{c\in\xi_{n,\infty}} \mu(Sc \triangle T_n c) < f(q_n)\}.$$

We denote by $\mathfrak{G}_n$ in the union of the neighborhoods $\mathfrak{U}(T_n)$ for all $T_n$ which occur in the sequence described above, and we denote by $\mathfrak{G}$ the set

$$\mathfrak{G} = \bigcap_n \bigcup_{s \geqslant n} \mathfrak{G}_s.$$

The set $\mathfrak{G}$ is a $G_\delta$ set since the set $\mathfrak{G}_n$ is open for every $n$. For any of the sequences $\{T_n\}$ under consideration, the limit $T$ belongs to $\mathfrak{G}$ (since the Third Step and (3.14) imply that $T \in \mathfrak{U}(T_n)$ for all $n$). Hence by Lemma 7.1 the set $\mathfrak{G}$ is everywhere dense in $\mathfrak{P}(M^m, \mu)$. We show that any diffeomorphism $S \in \mathfrak{G}$ possesses a cyclic APM with speed of approximation $f(n)$. Indeed, there exists a sequence $n_k \to \infty$ for which $T \in \mathfrak{G}_n$, i.e. $T \in \mathfrak{U}(T^{(k)}_{n_k})$, $k = 1, 2, \cdots$, where $T^{(k)}_{n_k}$ is the $n_k$th term of the sequence $\{T^{(k)}_n\}$.

All decompositions which correspond to the sequence $\{T^{(k)}_n\}$ will be labeled by the upper index $(k)$. By the conditions of the Second Step we have $\xi^{(k)}_{n_k} \to \epsilon$, and (3.13) implies that also $\xi^{(k)}_{n_k,\infty} \to \epsilon$. Hence it follows from the definition of $\mathfrak{U}(T^{(k)}_{n_k})$ that $S$ possesses a cyclic APM with speed of approximation $f(n)$.

By combining the theorem just proved with the results of [8] we obtain the following corollary:

COROLLARY. *The ergodic but nonmixing diffeomorphisms for which the shift operator $U_T$ in the space $L^2(M^m, \mu)$ has a simple singular spectrum lie in the space $\mathfrak{P}(M^m, \mu)$ and form a subset of the second category there.*

REMARK. The results of the present section carry over to the case of diffeomorphisms with finite smoothness.

The present article is not a joint paper in the usual sense; it was written by A. B. Katok. However, the construction described in §3 was preceded by a construction of a topologically transitive, (Lebesgue-) measure-preserving $C^\infty$-diffeomorphism of the two-dimensional circle due to D. V. Anosov. That construction involved the inductive conditions of the First and Third Step from §3, but instead of the Second and Fourth Step another argument was used. Mindful of this fact and also of the help rendered by D. V. Anosov for the writing of this paper through invaluable advise and comments, we thought it most fitting to publish the article under joint authorship. The ordering of the authors' names is alphabetical and does not have any additional significance.

Received 20 May 1969

## BIBLIOGRAPHY

[1] D. V. Anosov and A. B. Katok, *New examples of ergodic diffeomorphisms of smooth manifolds*, Uspehi Mat. Nauk **25** (1970), no. 4 (154), 173–174.

[2] A. S. Besicovitch, *A problem on topological transformations of the plane*, Fund. Math. **28** (1937), 61–65.

[3] ———, *A problem on topological transformations of the plane*. II, Proc. Cambridge Philos. Soc. **47** (1951), 38–45. MR **12**, 519.

[4] J. W. S. Cassels, *An introduction to the geometry of numbers*, Springer-Verlag, Berlin, 1959; Russian transl., "Mir", Moscow, 1965. MR **28** #1175; **31** #5841.

[5] W. H. Gottschalk, *Orbit-closure decompositions and almost periodic properties,* Bull. Amer. Math. Soc. **50** (1944), 915–919. MR **6**, 165.

[6] P. R. Halmos, *Lectures on ergodic theory,* Publ. Math. Soc. Japan, no. 3, Math. Soc. Japan, Tokyo, 1956; reprint, Chelsea, New York, 1960; Russian transl., IL, Moscow, 1959. MR **20** # 3958; **22** # 2677.

[7] ———, *Approximation theories for measure preserving transformations,* Trans. Amer. Math. Soc. **55** (1944), 1–18. MR **5**, 189.

[8] A. B. Katok and A. M. Stepin, *Approximations in ergodic theory,* Uspehi Mat. Nauk **22** (1967), no. 5 (137), 81–106 = Russian Math. Surveys **22** (1967), no. 5, 77–102. MR **36** # 2776.

[9] ———, *Metric properties of homeomorphisms that preserve measure,* Uspehi Mat. Nauk **25** (1970), no. 2 (152), 193–220 = Russian Math. Surveys **25** (1970), no. 2, 193–220. MR **41** # 5594.

[10] A. N. Kolmogorov, *Théorie générale des systèmes dynamiques et mécanique classique,* Proc. Internat. Congress Math. (Amsterdam, 1956), vol. 1, North-Holland, Amsterdam, 1957, pp. 315-333; Russian transl., Fizmatgiz, Moscow, 1961, pp. 187-208. MR **20** # 4066.

[11] ———, *On dynamical systems with an integral on the torus,* Dokl. Akad. Nauk SSSR **93** (1953), 763–766. (Russian) MR **16**, 36.

[12] D. Montgomery and L. Zippin, *Topological transformation groups,* Interscience Tracts in Pure and Appl. Math., no. 1, Interscience, New York, 1955. MR **17**, 383.

[13] J. Moser, *On invariant curves of area-preserving mappings of an annulus,* Sympos. (Colorado Springs, 1961), Nachr. Akad. Wiss. Göttingen Math.-Phys. Kl. II **1962**, 1–20; Russian transl., Matematika **6** (1962), no. 5, 51–67. MR **26** # 5255.

[14] ———, *A rapidly convergent iteration method and nonlinear differential equations.* I, II, Ann. Scuola Norm. Sup. Pisa (3) **20** (1966), 265–315, 499–535; Russian transl., Uspehi Mat. Nauk **23** (1968), no. 4 (142), 179–238. (Russian) MR **33** # 7667; **34** # 6280; **37** # 4382.

[15] ———, *On the volume elements on a manifold,* Trans. Amer. Math. Soc. **120** (1965), 286–294. MR **32** # 409.

[16] V. V. Nemyckiĭ and V. V. Stepanov, *Qualitative theory of differential equations,* 2nd ed., GITTL, Moscow, 1949; English transl., Princeton Math. Series, no. 22, Princeton Univ. Press, Princeton, N. J., 1960. MR **10**, 612; **22** # 12258.

[17] J. C. Oxtoby and S. Ulam, *Measure-preserving homeomorphisms and metrical transitivity,* Ann. of Math. (2) **42** (1941), 874–920. MR **3**, 211.

[18] R. S. Palais, *Local triviality of the restriction map for embeddings,* Comment. Math. Helv. **34** (1960), 305–312. MR **23** # A666.

[19] H. Poincaré, *Sur les courbes définies par les équations différentiels* III, IV, J. Math. Pures Appl. (4) **1** (1885), 167–244; ibid. **2** (1886), 151–217; Russian transl., GITTL, Moscow, 1947.

[20] L. G. Šnirel'man, *Example of a transformation of the plane,* Izv. Donsk. Politehn. Inst. Novočerkasske **14**, Naučn. Otdel. Fiz.-Mat. Čast' (1930), 64–74. (Russian)

[21] E. A. Sidorov, *Smooth topologically transitive dynamical systems,* Mat. Zametki **4** (1968), 751–759. (Russian) MR **39** # 939.

[22] M. D. Sklover, *Classical dynamical systems on the torus with continuous spectrum,* Izv. Vysš. Učebn. Zaved. Matematika **1967**, no. 10 (65), 113–124. (Russian) MR **37** # 1737.

Translated by:
B. Volkmann

Trudy Moskov. Mat. Obšč.
Tom 23 (1970)

Trans. Moscow Math. Soc.
Vol 23 (1970)

# SOLVABILITY OF THE CAUCHY PROBLEM FOR ABSTRACT QUASILINEAR HYPERBOLIC EQUATIONS OF SECOND ORDER AND THEIR APPLICATIONS

*S. Ja. JAKUBOV*

## Contents



## Introduction

In recent years, differential equations in abstract spaces with unbounded operator coefficients have been studied by many Soviet and foreign authors. The Cauchy problem and boundary value problems have been posed for these equations, and the boundedness and stability of the solutions have been studied, as well as periodicity of solutions and other properties.

Since many problems for partial differential equations, for infinite systems of ordinary differential equations and for integro-differential equations may be reduced to problems for differential equations with unbounded operator coefficients in an appropriate abstract space, the study of these abstract differential equations is one of the important problems of mathematics. As a result, abstract differential equations are studied by various methods.

One of the most illuminating methods turns out to be the method of semigroups of bounded linear operators in a Banach space which was developed in the well-known monograph by Hille and Phillips [**1**].

As important advantage of this method is that the problem for an abstract differential equation may be studied in the class of classical solutions and not in the class of generalized functions.

All papers published up to 1967 which dealt with correctly posed problems for linear differential equations in a Banach space have been taken into account in the book by Kreĭn [**2**]. In this paper we use the terminology from that book and also the results cited there.

Methods from the theory of semigroups are used in this paper to study the

---

*AMS* 1970 *subject classifications*. Primary 35L15, 35R20, 47H15.

solvability of the Cauchy problem for quasilinear hyperbolic equations

$$u''(t) + A(t)u'(t) + B(t)u(t) = f(t, u(t), u'(t)),$$
$$(1) \qquad u(0) = u_0,\ u'(0) = u_1,$$

in a Hilbert or Banach space. Conditions for both local and general solvability are determined. Abstract theorems are obtained and then applied to determine the solvability of boundary value problems for quasilinear partial differential equations of hyperbolic type.

## §1. The uniformly correct Cauchy problem

1. *The quasilinear equation of the first order.* Before studying the Cauchy problem for the quasilinear equation of the second order (1), we consider that of the first order:

$$(2) \qquad u'(t) = A(t)u(t) + f(t, u(t)),\ u(t_0) = u_0.$$

For this problem, the results are an extension of those obtained by Segal [**3**] and the author [**4**]. In the sequel, we shall often have cause to use a definition taken from the theory of unbounded operators [**4**]. Let $G$, $\widetilde{G}$ and $F$ be Banach spaces (real or complex) with topological inclusion $\widetilde{G} \subset G$. $B(G, F)$ will be used to denote the Banach space of bounded linear operators acting from $G$ into $F$.

DEFINITION. Let the operator $f(u)$ acting from $\widetilde{G}$ into $F$ have the Fréchet derivative $Df(u_0)$ at the point $u_0 \in \widetilde{G}$, where the operator $Df(u_0)$ (obviously belonging to $B(\widetilde{G}, F)$) can be extended so that its extension $\widetilde{D}f(u_0)$ belongs to $B(G, F)$. Then $\widetilde{D}f(u_0)$ is called the *$G$-extended Fréchet derivative* of the operator $f(u)$ at the point $u_0$.[1)]

The $G$-extended Fréchet derivative of $f(u)$ is usually the Fréchet derivative of the unbounded operator $f(u)$ acting from $G$ into $F$. But if the operator $f(u)$ acting from $G$ into $F$ is unbounded then its Fréchet derivative will in general also be an unbounded operator acting from $G$ into $F$. With the new definition, we select those operators whose derivatives are bounded from the class of operators (which includes those that are unbounded) acting from $G$ into $F$.

Let the operator $f(u)$, acting from a region $S$ of the space $\widetilde{G}$ into $F$, have a $G$-extended Fréchet derivative at every point of $S$. It is then clear that the operator $\widetilde{D}f(u)$ acts from $S$ into $B(G, F)$.

THEOREM 1. *Let the following assumptions hold.*

I. *The closed linear operator $A(t)$ has a domain of definition $\mathscr{D}(A(t)) = \mathscr{D}(A)$ independent of $t \in [0, T]$ and everywhere dense in $E$, and let it have a bounded inverse operator $A^{-1}(t)$ and be strongly continuously differentiable on $\mathscr{D}(A)$.*

---

[1)] We note that the $G$-extended Fréchet derivative is not in general uniquely defined. If, however, the set $\widetilde{G}$ is dense in the space $G$, uniqueness holds.

II. *The Cauchy problem for the equation* $u'(t) = A(t)u(t)$ *is uniformly correct* [2]; *that is*:

a) *for all* $s \in [0, T)$ *and* $u_0 \in \mathscr{D}(A)$ *a unique function* $u(t,s)$ *exists which is continuously differentiable and satisfies the equation on* $[s, T]$ *(it is called a solution of the equation) and also satisfies the condition* $u(s,s) = u_0$;

b) *the function* $u(t,s)$ *and its derivative* $u_t'(t,s)$ *are continuous for all values of both variables in* $T_\Delta$: $0 \leqq s \leqq t \leqq T$;

c) *the solution depends continuously on the initial conditions in the sense that if* $u_{0,n} \in \mathscr{D}(A)$ *converges uniformly to zero, then the corresponding solution* $u_n(t,s)$ *converges to zero and this follows uniformly for all t and s in* $T_\Delta$.

III. *The operator* $f(t,u)$ *acts from* $[0, T] \times E(A)$[2] *into E, and has a strongly continuous E-extended Fréchet derivative* $D_u f(t,u)$ *and a continuous derivative* $D_t f(t,u)$ *which satisfies a Lipschitz condition in u for every sphere in the space* $E(A)$.

IV. $u_0 \in \mathscr{D}(A)$.

*Then the Cauchy problem* (2) *has a unique solution* $u(t) \in C^1([0,t_0], E) \cap C([0,t_0], E(A))$ *which may be determined by the method of successive approximations.*

PROOF. In [2] it was shown that the solution of the problem (2) on the segment $[t_0, t_0 = h]$ reduces to the solution of the integral equation $u(t) = U(t,t_0)u_0 + \int_{t_0}^t U(t,\tau) f(\tau, u(\tau))\, d\tau$, where the solution $u(t)$ is contained in the Banach space $C_A^1[t_0, t_0 + h]$ of functions $u(t)$ that are prescribed on $[t_0, t_0 + h]$, continuously differentiable in $E$ and continuous in $E(A)$ under the norm $\| u(t) \| = \max[\| A(t)u(t) \| + \| u'(t) \|]$; the operator $U(t, \tau)$ is the evolution operator corresponding to $A(t)$. Then, using the formulas

$$
\begin{aligned}
A(t)u(t) &= W(t, t_0)A(t_0)u_0 + W(t, t_0)f(t_0, u(t_0)) - f(t, u(t)) \\
&\quad - \int_{t_0}^t W(t, \tau) A'(\tau) A^{-1}(\tau) f(\tau, u(\tau))\, d\tau + \int_{t_0}^t W(t, \tau) \frac{d}{d\tau} f(\tau, u(\tau))\, d\tau;
\end{aligned}
\tag{3}
$$

and

$$
\begin{aligned}
u'(t) &= W(t, t_0)A(t_0)u_0 + W(t, t_0)f(t_0, u(t_0)) \\
&\quad - \int_{t_0}^t W(t, \tau) A'(\tau) A^{-1}(\tau) f(\tau, u(\tau))\, d\tau \\
&\quad + \int_{t_*}^t W(t, \tau) \frac{d}{d\tau} f(\tau, u(\tau))\, d\tau; \quad W(t, \tau) = A(t)U(t, \tau)A^{-1}(\tau),
\end{aligned}
\tag{4}
$$

[2] $E(A)$ denotes the Banach space consisting of elements of $\mathscr{D}(A)$ under the norm $\| u \|_A = \| A(0)u \|$.

derived in [**2**], we prove that the principle of contraction mapping can be applied to the operator

$$z(t) = [Ku](t) = U(t, t_0)u_0 + \int_{t_0}^{t} U(t, \tau) f(\tau, u(\tau))\, d\tau,$$

acting in the space $C_A^1[t_0, t_0+h]$, where the positive number $h$ will be determined below.

Let us consider the set

$$\widetilde{S}_A^1(R_0, [t_0, t_0+h]) = \{u(\cdot) \in C_A^1[t_0, t_0+h];\ u(t_0) = u_0;\ \|u(\cdot)\| \leqslant R_0,$$
$$R_0 > 3(\|A(t_0)u_0\| + \|f(t_0, u_0)\|)\}.$$

It can be proved that there exists an $h(R_0) > 0$ in $S_A^1(R_0[t_0, t_0+h])$ such that the operator $K$ satisfies the conditions for the principle of contraction mapping.

If the operator $A(t) = A$ is independent of $t$, Segal [**3**] has considered the more general case, namely when $f(t,u)$ is defined on $\mathscr{D}(A^k)$ ($k$ being an integer greater than zero). Unfortunately Segal's result is only valid for $k=1$. The result was corrected in [**5**].

Let us consider the Cauchy problem

$$u'(t) = A(t)u(t) + B(t)u(t) + f(t, u(t)),\ u(t_0) = u_0. \tag{5}$$

**Theorem** 2. *Let conditions* I *to* IV *be satisfied, and also*

V. *The operators $B(t)$ and $A(0)B(t)A^{-1}(0)$ are strongly continuous.*

*Then the Cauchy problem* (5) *has a unique solution $u(t) \in C^1([0, t_0], E) \cap C([0, t_0], E(A))$, which may be determined by the method of successive approximations.*

The proof follows from a combination of Theorem 3.4 in Chapter II of [**2**] and Theorem 1 above.

We shall prove the solvability of problems (2) and (5) "in the small". In order that solvability "in the small" with a priori bounds should guarantee solvability "in the large", it is necessary to establish that for all initial conditions $u_0$ taken from a fixed sphere of the space $E(A)$, the length of the segment where solvability "in the small" is established is independent of $t_0$.

**Theorem** 3. *Let conditions* I *to* IV *be satisfied.*

*Then for any $r > 0$ there exists an $h(r) > 0$ such that for all initial conditions $(t_0, u_0)$ with $t_0 \in [0, T], u_0 \in \mathscr{D}(A)$ and $\|A(t_0)u_0\| \leqq r$ the problem* (2) *has a unique solution on $[t_0, t_0+h(r)] \cap [0, T]$.*

**Proof.** Since $\|A(t_0)u_0\| \leqq r$ implies

$$\|u_0\|_A = \|A(0)u_0\| \leqslant \|A(0)A^{-1}(t_0)\| \|A(t_0)u_0\| \leqslant \|A(0)A^{-1}(t_0)\| r,$$

from III we have $\|f(t_0, u_0)\| \leqq C(r)$. If we choose $R(r)$ from the relation

$$R(r) > \max_{0 \leqslant t_0 \leqslant t \leqslant T} (2\|W(t, t_0)\| r + 2\|W(t, t_0)\| C(r) + C(r))$$

and carry out the proof of Theorem 1 in the set $\widetilde{S}^1_A(R(r), [t_0, t_0+h])$, Theorem 3 is established.

**Remark.** An analogous theorem may be formulated for problem (5) using condition V.

In a Banach space with linear bounds for nonlinear operators an overall existence theorem may be proved. But in a Banach space with a differentiable norm (in Hilbert spaces or spaces $L_p$ with $p>1$, for example) a linear two-sided bound is replaced by a one-sided bound that plays a vital role in the sequel.

Let us suppose that $E$ is a Banach space with a differentiable norm. We denote the gradient of the square of the norm by $\Gamma u$.[3)]

**Theorem 4.** *Let conditions* I, III *and* IV, *and also the following conditions, hold.*

VI. *The operator $A(t)$ generates a strongly continuous contracting semigroup.*

VII. *For all functions $u(t)$ that are continuous in $E(A)$ and continuously differentiable in $E$ the following inequality holds*:

$$\operatorname{Re}\int_{t_0}^{t}(f(\tau, u(\tau)), \Gamma u(\tau))\,d\tau \leqslant C\left(1+\int_{t_0}^{t}\|u(\tau)\|^2\,d\tau\right).$$

VIII. *For all $u\in\mathscr{D}(A)$ such that $\|u\|\leqq R$ the following inequality holds*:

$$\|f'_t(t, u)\|\leqslant C(R)(1+\|A(0)u\|),\quad \|f'_u(t, u)\|_{B(E)}\leqslant C(R).$$

*Then problem* (2) *has a unique solution $u(t)$ contained in* $C^1([t_0, T], E) \cap C([t_0, T], E(A))$.

**Proof.** Let $u(t)$ be a solution of equation (2). Operating on both sides of equation (2) with the function $\Gamma u(t)$, we obtain

$$(u'(t), \Gamma u(t)) = (A(t)u(t), \Gamma u(t)) + (f(t, u(t)), \Gamma u(t)). \tag{6}$$

By definition we have

$$\operatorname{Re}(u'(t), \Gamma u(t)) = \frac{d}{dt}\|u(t)\|^2.$$

On the other hand, using VI and the results of [**6**] and [**7**], we obtain

$$\operatorname{Re}(A(t)u(t), \Gamma u(t)) \leqslant \omega\|u(t)\|^2.$$

Then from (6) we obtain the inequality

$$\frac{d}{dt}\|u(t)\|^2 \leqslant \omega\|u(t)\|^2 + \operatorname{Re}(f(t, u(t)), \Gamma u(t)).$$

Integrating from $t_0$ to $t$ and taking condition VII into account, we obtain the integral inequality

$$\|u(t)\| \leqslant C\left(1+\int_{t_0}^{t}\|u(\tau)\|^2 d\tau\right).$$

---

[3)] That is, for any $u$ and $h$ contained in $E$ we have $\|u+h\|^2 - \|u\|^2 = \operatorname{Re}(h, \Gamma u) + \omega(h)$, where $\lim_{|h|\to 0}|\omega(h)|/\|h\| = 0$.

From this the a priori bound $\|u(t)\| \leqq C$ follows for the solution of problem (2). Then from condition VIII we have

$$\left\|\frac{d}{dt} f(t, u(t))\right\| \leqslant C(1 + \|A(t)u(t)\| + \|u'(t)\|). \tag{7}$$

On the other hand, the solution of (2) satisfies the integral equation

$$u(t) = U(t, t_0)u_0 + \int_{t_0}^{t} U(t, \tau) f(\tau, u(\tau))\, d\tau.$$

Then from (3) and (4) we have the inequality

$$\|A(t)u(t)\| + \|u'(t)\| \leqslant 2\|A(t)U(t, t_0)u_0\| + 2\|W(t, t_0)f(t_0, u_0)\|$$
$$+ \|f(t, u(t))\| + 2\int_{t_0}^{t} \|W(t, \tau)\| \|A'(\tau)A^{-1}(\tau)\| \|f(\tau, u(\tau))\|\, d\tau$$
$$+ 2\int_{t_0}^{t} \|W(t, \tau)\| \left\|\frac{d}{d\tau} f(\tau, u(\tau))\right\| d\tau.$$

Then from (7) we obtain the integral inequality

$$\|A(t)u(t)\| + \|u'(t)\| \leqslant C\left[1 + \int_{t_0}^{t} (\|A(\tau)u(\tau)\| + \|u'(\tau)\|)\, d\tau\right],$$

and from this we obtain the a priori bound $\|A(t)u(t)\| \leqq C$ for the solution of problem (2). Hence, using Theorem 3, the local solution of (2), whose existence was proved in Theorem 1, may be extended on $[t_0, T]$.

REMARK. Bearing in mind further applications of Theorem 4, condition VII can be relaxed in the following way: condition VII need hold only for a solution $u(t)$ of (2).

REMARK. An analogous theorem may be formulated for problem (5), using condition V.

2. *The quasilinear hyperbolic equation of the second order.* We now apply these results to establish the solvability of the Cauchy problem

$$\begin{gathered} u''(t) + A(t)u'(t) + B(t)u(t) = f(t, u(t), u'(t)), \\ u(0) = u_0, \quad u'(0) = u_1. \end{gathered} \tag{1}$$

Stronger results that can be applied easily to partial differential equations are obtained in a Hilbert space $H$.

THEOREM 5. *Let the following assumptions hold.*

IX. *For every $t \in [0, T]$ the operator $B(t)$ is selfadjoint and positive definite in $H$, and for any $t \in [0, T]$ and any $u \in \mathscr{D}(B(t))$ the inequality $(B(t)u, u) \geqq \gamma(u, u)$ holds, where $\gamma$ is a constant greater than zero.*

X. *The operator $B^{1/2}(t)$ has a domain of definition $\mathscr{D}(B^{1/2}(t)) = \mathscr{D}(B^{1/2})$ which is independent of $t$; the operator $B^{1/2}(t)$ is strongly continuously differentiable on $\mathscr{D}(B^{1/2})$ and satisfies one of the two conditions:*

a) *the operator* $(B^{1/2}(t))'B^{-1/2}(t)$ *is strongly continuously differentiable on* $\mathcal{D}(B^{1/2})$,

b) *the operator* $B^{1/2}(t)\,(B^{1/2}(t))'B^{-1}(t)$ *exists, is bounded and is strongly continuous.*

XI. *The linear operator* $A(t)$ *is dissipative, i.e. for some real number* $\omega$ *and any* $u\in\mathcal{D}(A(t))$ *the inequality* $\mathrm{Re}(A(t)u,u)\geqq\omega(u,u)$ *holds; the operator* $A(t)$ *is subordinate to the operator* $B^{1/2}(0)$; *the operator* $A(t)$ *is strongly continuously differentiable on* $\mathcal{D}(B^{1/2})$.

XII. $u_0\in\mathcal{D}(B(0))$ *and* $u_1\in\mathcal{D}(B^{1/2})$).

XIII. *The operator* $f(t,u,w)$ *acts from* $[0,T]\times H(B^{1/2})\times H(B^{1/2})$ *into* $H$, *has a continuous derivative* $f_t'(t,u,w)$, *a strongly continuous Fréchet derivative* $f_u'(t,u,w)$ *and a strongly continuous H-extended Fréchet derivative* $f_w'(t,u,w)$ *which satisfies a Lipschitz condition in* $(u,w)$ *in every sphere of the space* $H(B^{1/2})\times H(B^{1/2})$.

*Then* (1) *has the unique solution*

$$u(t)\in C^2([0,t_0],H)\cap C^1([0,t_0],H(B^{1/2}))\cap C([0,t_0],H(B)),$$

*which may be determined by the method of successive approximations.*

PROOF. Using the substitution

$$\begin{cases} v_1(t)=\dfrac{1}{2}\,[iB^{1/2}(t)\,u(t)+u'(t)],\\[2mm] v_2(t)=\dfrac{1}{2}[-iB^{1/2}(t)\,u(t)+u'(t)]\end{cases}$$

(1) is reduced to the equivalent problem

$$v'(t)=\mathfrak{A}(t)\,v(t)+\mathfrak{B}(t)\,v(t)+F(t,v(t)),\quad v(0)=v_0, \tag{8}$$

where

$$\mathfrak{A}(t)=\begin{pmatrix} iB^{1/2}(t)-\dfrac{1}{2}A(t) & -\dfrac{1}{2}A(t)\\[2mm] -\dfrac{1}{2}A(t) & -iB^{1/2}(t)-\dfrac{1}{2}A(t)\end{pmatrix},$$

$$\mathfrak{B}(t)=\frac{1}{2}\begin{pmatrix} I & -I\\ -I & I\end{pmatrix}(B^{1/2}(t))'\,B^{-1/2}(t),$$

$$F(t,v)=\frac{1}{2}\begin{pmatrix} f(t,-iB^{-1/2}(t)(v_1-v_2),\,v_1+v_2)\\ f(t,-iB^{-1/2}(t)(v_1-v_2),\,v_1+v_2)\end{pmatrix},$$

$$v_0=\frac{1}{2}\begin{pmatrix} iB^{1/2}(0)\,u_0+u_1\\ -iB^{1/2}(0)\,u_0+u_1\end{pmatrix}.$$

It is easy to show that the operator $\mathfrak{A}(t)$ with domain of definition $\mathscr{D}(\mathfrak{A}(t)) = \mathscr{D}(B^{1/2}) \times \mathscr{D}(B^{1/2})$ is dissipative in the space $H^2 = H \times H$ and that $\mathrm{Re}(\mathfrak{A}(t)v, v) \leqq \omega_0(v,v)$ for all $v \in \mathscr{D}(\mathfrak{A}(t))$, where $\omega_0 = 0$ for $\omega \geqq 0$, $\omega_0 = -\omega$ for $\omega < 0$. On the other hand, the operator $\mathfrak{A}(t)$ has a bounded inverse

$$\mathfrak{A}^{-1}(t)$$
$$= \begin{pmatrix} -iB^{-1/2}(t) - {}^1\!/_2 B^{-1/2}(t)A(t)B^{-1/2}(t) & {}^1\!/_2 B^{-1/2}(t)A(t)B^{-1/2}(t) \\ {}^1\!/_2 B^{-1/2}(t)A(t)B^{-1/2}(t) & iB^{-1/2}(t) - {}^1\!/_2 B^{-1/2}(t)A(t)B^{-1/2}(t) \end{pmatrix}.$$

Thus the operator $\mathfrak{A}(t)$ generates a strongly continuous contracting subgroup for every $t$ contained in $[0, T]$. It is now easy to see that Kato's result [8] can be applied to the equation $v'(t) = \mathfrak{A}(t)v(t)$. From this and from condition X it follows that the linear part of (8) satisfies the conditions of Theorem 2. We shall show that the operator $F(t,v)$ satisfies condition III in the space $H^2$.

Since the operator $f(t,u,w)$ acts from $[0,T] \times H(B^{1/2}) \times H(B^{1/2})$ into $H$, so does the operator $f(t, -B^{-1/2}(t)(v_1 - v_2), v_1 + v_2)$.

Thus, using the equation $H^2(\mathfrak{A}) = H(B^{1/2}) \times H(B^{1/2})$, we can easily establish that $F(t,v)$ acts from $[0,T] \times H^2(\mathfrak{A})$ into $H^2$. From condition XIII it follows that $F(t,v)$ has an $H^2$-extended Fréchet derivative $F'_v(t,v)$ and

(9)
$$F'_v(t, v)$$
$$= \frac{1}{2} \begin{pmatrix} -if'_u(t, -iB^{-1/2}(t)(v_1 - v_2), v_1 + v_2)B^{-1/2}(t) + f'_w & if'_u B^{-1/2}(t) + f'_w \\ -if'_u B^{-1/2}(t) + f'_w & if'_u B^{-1/2}(t) + f'_w \end{pmatrix}.$$

Hence from XIII it follows that the operator $F'_v(t,v)$ acts from $[0,T] \times H^2(\mathfrak{A})$ into $B(H^2)$, is strongly continuous and satisfies the Lipschitz condition with respect to $v$ in every sphere of the space $H^2(\mathfrak{A})$. From condition XIII it also follows that a continuous derivative $F'_t(t,v)$ exists and satisfies the Lipschitz condition with respect to $v$ in every sphere in $H^2(\mathfrak{A})$. Thus the operator $F(t,v)$ satisfies condition III of Theorem 2. The theorem is proved.

Now we give an overall existence theorem for (1).

**Theorem 6.** *Let conditions* IX *to* XIII, *and also the following conditions, hold.*

XIV. *For any function $u(t)$ that is continuously differentiable in $H^2(B^{1/2})$ the following holds*:

$$\mathrm{Re} \int_0^t (f(\tau, u(\tau), u'(\tau)), u'(\tau))\, d\tau \leqslant C\left[1 + \int_0^t (\|B^{1/2}(0)u(\tau)\|^2 + \|u'(\tau)\|^2)\, d\tau\right].$$

XV. *For any $u$ and $w$ contained in $\mathscr{D}(B^{1/2})$ such that $\|B^{1/2}(0)u\| \leqq R$ and $\|w\| \leqq R$, the following inequalities hold*:

$$\|f'_t(t, u, w)\| \leqslant C(R)(1 + \|B^{1/2}(0)w\|),$$
$$\|f'_u(t, u, w)\|_{B(H(B^{1/2}),H)} \leqslant C(R), \tag{10}$$
$$(or\ \|f'_u(t, u, w)\|_{B(H)} \leqslant C(R)(1 + \|B^{1/2}(0)w\|)),$$
$$\|f'_w(t, u, w)\|_{B(H)} \leqslant C(R).$$

*Then problem* (1) *has the unique solution*

$$u(t) \in C^2([0, T], H) \cap C^1([0, T], H(B^{1/2})) \cap C([0, T], H(B)).$$

PROOF. We apply Theorem 4 to problem (8). Let $v(t)$ be a solution of (8). Then

$$\operatorname{Re}(F(t, v(t)), v(t))$$

$$= \frac{1}{2} \operatorname{Re}(f(t, -iB^{-1/2}(t)(v_1(t) - v_2(t)), v_1(t)+v_2(t)), v_1(t) + v_2(t)).$$

Since the function $u(t) = -iB^{-1/2}(t)(v_1(t) - v_2(t))$ is a solution of (1) and $u'(t) = v_1(t) + v_2(t)$, we obtain

$$\operatorname{Re}(F(t, v(t)), v(t)) = \frac{1}{2} \operatorname{Re}(f(t, u(t), u'(t)), u'(t)).$$

Hence from condition XIV it is easy to obtain the inequality

$$\operatorname{Re} \int_0^t (F(\tau, v(\tau)), v(\tau))\, d\tau \leqslant C\left(1 + \int_0^t \|v(\tau)\|^2 d\tau\right),$$

i.e. the operator $F(t, v)$ in the space $H^2$ satisfies condition VII.

We shall prove that $\|F_t'(t, v)\| \leqq C(R)(1 + \|\mathfrak{A}(0)v\|)$ and $\|F_v'(t, v)\|_{B(H)} \leqq C(R)$ for all $v \in H^2(\mathfrak{A})$ such that $\|v\| \leqq R$. From (9) it follows that this inequality holds if for all $v_1$ and $v_2$ contained in $\mathscr{D}(B^{1/2})$ such that $\|v_1\| \leqq R$ and $\|v_2\| \leqq R$ the following hold:

$$\|f_t'(t, -iB^{-1/2}(t)(v_1 - v_2), v_1 + v_2)\| \leqslant C(R)(1 + \|B^{1/2}(0)v_1\| + \|B^{1/2}(0)v_2\|),$$

$$\|f_u'(t, -iB^{-1/2}(t)(v_1 - v_2), v_1 + v_2)\|_{B(H(B^{1/2}),H)} \leqslant C(R),$$

$$\|f_w'(t, -iB^{-1/2}(t)(v_1 - v_2), v_1 + v_2)\|_{B(H)} \leqslant C(R).$$

These in turn follow from (10). We have therefore proved that the operator $F(t, v)$ in the space $H^2$ satisfies condition VIII of Theorem 4. The remaining conditions of Theorem 4 were verified in Theorem 5.

When $\mathscr{D}(B(t))$ is independent of $t$, we can prove stronger local and overall existence theorems for the solution of (1).

THEOREM 7. *Let conditions* IX, XI *and* XII, *and also the following conditions, hold.*

XVI. *The operator $B(t)$ has a domain of definition $\mathscr{D}(B)$ that is independent of $t$; the operator $B(t)$ is strongly continuously differentiable on $\mathscr{D}(B)$.*

XVII. *The operator $f(t, u, w)$ acts from $[0, T] \times H(B) \times H(B^{1/2})$ into $H$, and has a continuous derivative $f_t'(t, u, w)$, a strongly continuous $H(B^{1/2})$-extended Fréchet derivative $f_u'(t, u, w)$ and a strongly continuous $H$-extended Fréchet derivative $f_w'(t, u, w)$ which satisfies a Lipschitz condition with respect to $(u, w)$ in every sphere of the space $H(B) \times H(B^{1/2})$.*

*Then* (1) *has the unique solution* $u(t) \in C^2([0,t_0],H) \cap C^1([0,t_0],H(B^{1/2})) \cap C([0, t_0], H(B))$, *which can be found by the method of successive approximations.*

**Proof.** Since Daleckiĭ [9] has proved that under condition XVI the operator $B^{1/2}(t)$ is strongly continuously differentiable on $\mathcal{D}(B^{1/2})$, problem (1) may be reduced to the equivalent problem (8) as in Theorem 5. We apply Theorem 2 to problem (8). From condition XVI it also follows that the operator $B^{1/2}(t)\,(B^{1/2}(t))'B^{-1}(t)$ is bounded and strongly continuous. In fact, for any $u \in \mathcal{D}(B)$ we have

$$(B(t)\,u)' = (B^{1/2}(t))'\,B^{1/2}(t)\,u + B^{1/2}(t)\,(B^{1/2}(t))'\,u,$$

and we then obtain

$$B'(t)\,B^{-1}(t) = (B^{1/2}(t))'\,B^{-1/2}(t) + B^{1/2}(t)\,(B^{1/2}(t))'\,B^{-1}(t).$$

It may now be noted that the operators $\mathfrak{A}(t)$ and $\mathfrak{A}(t)\,\mathfrak{B}(t)\,\mathfrak{A}^{-1}(t)$ are strongly continuously differentiable; that is, the linear part of (8) satisfies the conditions of Theorem 2. The only condition still to be established is the one satisfied by the operator $F(t,v)$ in the space $H^2$. This may be done as it was in Theorem 5. It should moreover be noted that if $v \in H^2(\mathfrak{A})$, then $u = -iB^{-1/2}(t)\,(v_1 - v_2) \in H(B)$. In fact,

$$B(0)\,u = -iB(0)\,B^{-1}(t)\,B^{1/2}(t)\,B^{-1/2}(0)\,B^{1/2}(0)\,(v_1 - v_2) \in H.$$

Hence, as the operator $f(t,u,w)$ acts from $[0,T]\times H(B)\times H(B^{1/2})$ into $H$, it follows that the operator $f(t, -iB^{-1/2}(t)\,(v_1 - v_2), v_1 + v_2)$ acts from $[0,T]\times H(B^{1/2})\times H(B^{1/2})$ into $H$. It is now easy to note that the operator $F(t,v)$ acts from $[0,T]\times H^2(\mathfrak{A})$ into $H^2$. On the other hand, from condition XVII it follows that the operator $F(t,v)$ has an $H^2$-extended Fréchet derivative $F_v'(t,v)$ and that formula (9) is valid.

**Theorem** 8. *Let conditions* IX, XI, XII, XVI *and* XVII, *and also the following conditions, hold.*

XVIII. *For all functions $u(t)$ that are continuous in $H(B)$ and continuously differentiable in $H(B^{1/2})$, the following inequality holds:*

$$\operatorname{Re}\int_0^t (f(\tau, u(\tau), u'(\tau)), u'(\tau))\,d\tau \leqslant C\left[1 + \int_0^t (\|B^{1/2}(0)\,u(\tau)\|^2 + \|u'(\tau)\|^2)\,d\tau\right].$$

XIX. *For any $u \in \mathcal{D}(B)$ and $w \in \mathcal{D}(B^{1/2})$ such that $\|B^{1/2}(0)u\| \leq R$ and $\|w\| \leq R$ the following inequalities hold:*

$$\|f_t'(t, u, w)\| \leqslant C(R)\,(1 + \|B(0)\,u\| + \|B^{1/2}(0)\,w\|),$$

$$\|f_u'(t, u, w)\|_{B(H(B^{1/2}),H)} \leqslant C(R)^{4)}, \quad \|f_w'(t, u, w)\|_{B(H)} \leqslant C(R).^{4)}$$

---

[4] Or $\|f_u'(t,u,w)\|_{B(H)} \leq C(R)\,(1 + \|B(0)u\| + \|B^{1/2}(0)w\|)$.

*Then problem* (1) *has the unique solution*

$$u(t) \in C^2([0, T], H) \cap C^1([0, T], H(B^{1/2})) \cap C([0, T], H(B)).$$

The proof of this theorem is the same as that of Theorem 6. It remains only to note that the solution of equation (1) is continuous in the space $H(B)$.

We note that for $A(t) = 0$, $B(t) = B$ and $f(t, u(t), u'(t)) = f(t, u(t))$ Theorem 8 is a refinement of results by Browder [**10**], Browder and Strauss [**11**] and Segal [**3**], [**12**], as it was assumed in their papers that $f(t, u)$ acts from $[0, T] \times H(B^{1/2})$ into $H$, (but in Theorem 8 it is only from $[0, T] \times H(B)$ into $H$). An overall existence theorem for (1) with $A(t) = 0$ was established under other assumptions in the papers by Mamedov [**13**], Lions and Strauss [**14**], Medeiros [**15**] and Pogorelenko and Sobolevskiĭ [**16**]. In all of these papers it was assumed that $f(t, u)$ was defined on $H(B^{1/2})$ rather than on $H(B)$.

For $A(t) = 0$ the above result has been published by the author in [**17**] and [**18**]. Finally we note that analogous results can be obtained for a Banach space. For brevity we quote one theorem without proof for the existence of a solution of (1). In a Banach space, $B^{1/2}(t)$ denotes any square root of $B(t)$, i.e. $B(t) = B^{1/2}(t) \cdot B^{1/2}(t)$.

THEOREM 9. *Let the following conditions hold.*

XX. *The closed linear operator* $B^{1/2}(t)$ *has a domain of definition* $\mathscr{D}(B^{1/2}(t)) = \mathscr{D}(B^{1/2})$ *which is independent of* $t$, *and a bounded inverse* $B^{-1/2}(t)$.

XXI. *The Cauchy problem for the equations*

$$u'(t) = iB^{1/2}(t)u(t), \quad u'(t) = -iB^{1/2}(t)u(t)$$

*is uniformly correct.*

XXII. *The operator* $A(t)$ *is bounded and strongly continuously differentiable for* $t \in [0, T]$.

XXIII. *The operator* $B^{1/2}(t)$ *is strongly continuously differentiable on* $\mathscr{D}(B^{1/2})$ *and satisfies one of the two conditions*:

a) $B^{1/2}(t)$ *is twice strongly continuously differentiable on* $\mathscr{D}(B^{1/2})$;

b) *the operator* $B^{1/2}(t)(B^{1/2}(t))'B^{-1}(t)$ *is bounded and strongly continuous.*

XXIV. *The operator* $f(t, u, w)$ *acts from* $[0, T] \times E(B^{1/2}) \times E(B^{1/2})$ *into* $E$, *and has a continuous derivative* $f'_t(t, u, w)$, *a strongly continuous Fréchet derivative* $f'_u(t, u, w)$ *and a strongly continuous* $E$*-extended Fréchet derivative* $f'_w(t, u, w)$ *which satisfies the Lipschitz condition with respect to* $(u, w)$ *in every sphere of the space* $E(B^{1/2}) \times E(B^{1/2})$.

XXV. $u_0 \in \mathscr{D}(B(0))$ *and* $u_1 \in \mathscr{D}(B^{1/2})$.

*Then* (1) *has the unique solution*

$$u(t) \in C^2([0, t_0], E) \cap C^1([0, t_0], E(B^{1/2})) \cap C([0, t_0], E(B)).$$

An analogous result was proved by the author under more rigorous conditions in [**17**] and [**18**].

## §2. Boundary value problems for hyperbolic equations

1. *The quasilinear hyperbolic equation.* The abstract results proved in the first part of the paper enable us to establish existence and uniqueness theorems for boundary value problems for some classes of quasilinear partial differential equations. All of the results are obtained subject to the condition $n < 4m$. First we prove solvability "in the small" with respect to $t$. We consider two cases: $n < 2m$ and $2m \leqq n < 4m$.

For $n < 2m$ we consider the equation

$$D_t^2 u(t, x) + \sum_{|\alpha| \leqslant m} A_\alpha(t, x) D^\alpha D_t u(t, x) + \sum_{|\alpha| \leqslant 2m} B_\alpha(t, x) D^\alpha u(t, x) = F(t, x, u, \ldots, D^m u, D_t u), \tag{11}$$

where

$$x = (x_1, \ldots, x_n) \in \Omega; \ \alpha = (\alpha_1, \ldots, \alpha_n); \ |\alpha| = \alpha_1 + \ldots + \alpha_n;$$

$$D^\alpha = D_1^{\alpha_1} \ldots D_n^{\alpha_n}; \ D_k = i \frac{\partial}{\partial x_k} \ (k = \overline{1, n}); \ D_t = \frac{\partial}{\partial t},$$

with the boundary conditions

$$E_j(x, D) u|_\Gamma = \sum_{|\alpha| \leqslant k_j} e_{\alpha j}(x) D^\alpha u(t, x)|_\Gamma = 0, \ k_j \leqslant 2m - 1, \ j = \overline{1, m} \tag{12}$$

and initial conditions

$$u(0, x) = u_0(x), \quad u_t'(0, x) = u_1(x). \tag{13}$$

We assume that the linear part of (11) satisfies the following conditions:

XXVI. $B_\alpha(t, x) \in C^{|\alpha|}(\overline{\Omega})$ and $B(t, x, D) = \sum_{|\alpha| \leqq 2m} B_\alpha(t, x) D^\alpha$ are formally selfadjoint positive definite elliptic operators in $\overline{\Omega}$, i.e. $B(t, x, D)$ coincides with the formally adjoint operator

$$B'(t, x, D) u = \sum_{|\alpha| \leqslant 2m} D^\alpha (\overline{B_\alpha(t, x)} u(x)) \text{ and } \sum_{|\alpha| = 2m} B_\alpha(t, x) \xi^\alpha > 0$$

for every real $\xi \neq 0$, $x \in \Omega$ and $t \in [0, T]$, where $\xi^\alpha = \xi_1^{\alpha_1} \cdots \xi_n^{\alpha_n}$.

XXVII. The system of boundary operators $\{E_j\}_1^m$ is normal [**19**], i.e. the boundary $\Gamma$ is not a characteristic surface of the operators $\{E_j\}$ and the orders of the operators $E_j$ are different; $0 \leqq k_1 < k_2 < \cdots < k_m \leqq 2m - 1$; the system is connected with the operator $B(t, x, D)$ by the Šapiro-Lopatinskiĭ condition (see [**20**] and [**21**]).

XXVIII. The operator $B(t)$ generated by the system $(B(t, x, D), \{E_j\}_1^m)$ is selfadjoint in $L_2(\Omega)$ with domain of definition $\mathscr{D}(B(t)) = W_2^{2m}(\Omega, \{E_j\}_1^m)$[5)] and the Višik-Gårding inequality holds for $B(t)$, i.e. $(B(t)u, u) \geqq C_1 \|u\|_{W_2^m}^2 - C_2 \|u\|_{L_2}^2$, $C_1 > 0$, for all $u \in W_2^{2m}(\Omega, \{E_j\}_1^m)$.

---

[5)] We use $W_2^l(\Omega, \{E_j\}_1^p)$ for $l > k_j$, $j = 1, \cdots, p$, to denote the subspace of $W_2^l(\Omega)$ in which all the $l$-times continuously differentiable functions in $\overline{\Omega}$ that satisfy the boundary conditions $E_j u|_\Gamma = 0$ for $j = 1, \cdots, p$ form a compact set.

XXIX. $A_\alpha(t,x)$ is contained in $C(\overline{\Omega})$, and the operator $A(t)$ generated by the system $(A(t,x,D) = \sum_{|\alpha| \leq m} A_\alpha(t,x) D^\alpha, \{E_j\}_1^r)$, where $k_r \leq m-1$ and $k_{r+1} \geqq m$, satisfies a dissipative condition in $L_2(\Omega)$, i.e. $\operatorname{Re}(A(t)u,u) \leqq \omega \|u\|_{L_2}^2$ for all $u \in W_2^m(\Omega, \{E_j\}_1^r)$.

We note that propositions XXVI to XXIX hold if $\{E_j\}_1^m$ is a Dirichlet system, i.e. $E_j = \partial^{j-1}/\partial n^{j-1}$; the operator

$$B(t, x, D) = \sum_{|\alpha|,|\beta| \leqslant m} D^\alpha B_{\alpha\beta}(t, x) D^\beta, \quad B_{\alpha\beta} = \overline{B_{\beta\alpha}},$$

is formally selfadjoint and the operator $A(t,x,D)$ is positive elliptic in $\overline{\Omega}$, i.e. $\operatorname{Re}\sum_{|\alpha|=m} A_\alpha(t,x)\xi^\alpha > 0$, with $\xi \neq 0$ and $m$ even.

In [**22**] it was proved that some problems arising in the theory of vibrating discs satisfy condition XXVIII. We note that in these problems the boundary operators are different from a Dirichlet system.

The existence of selfadjoint boundary conditions $\{E_j\}$ differing from a Dirichlet system was established in [**23**].

THEOREM 10. *Let conditions* XXVI *to* XXIX, *and also the following conditions, hold.*

XXX. $A_\alpha(t,x) \in C^{1,0}([0,T] \times \overline{\Omega})$; $B_\alpha(t,x) \in C^{1,|\alpha|}([0,T] \times \overline{\Omega})$; $u_0(x)$ *and* $u_1(x)$ $\in W_2^{2m}(\Omega, \{E_j\}_1^m)$.

XXXI. $n < 2m$; *the function* $F(t,x,z)$ *is continuous together with its derivatives with respect to* $t$ *and* $z_k$ *in the region* $\{t \in [0,T], x \in \overline{\Omega}, z_k \in Z\}$, *where* $Z$ *is the complex plane, and these derivatives satisfy a Lipschitz condition with respect to* $z_k$ *in every bounded set* $|z_k| \leqq R$ *with the Lipschitz constant depending only on* $R$.[6)]

*Then the problem defined by* (11), (12) *and* (13) *has a unique solution*

$$u(t, x) \in C([0, t_0], W_2^{2m}(\Omega, \{E_j\}_1^m)) \cap C^1([0, t_0], W_2^m(\Omega, \{E_j\}_1^r)) \cap C^2([0, t_0], L_2(\Omega)),$$

*which can be determined by the method of successive approximations.*

PROOF. From condition XXVIII it follows that for some $\omega > 0$ the operator $B(t) + \omega I$ is selfadjoint and positive definite in $L_2(\Omega)$ (for brevity we designate this by $B(t)$ as well). Moreover, from the inequality of Višik and Gårding it follows that $(B(t)u,u) \geqq C\|u\|^2_{W_2^m(\Omega)}$ for all $u \in \mathscr{D}(B(t)) = W_2^{2m}(\Omega, \{E_j\}_1^m)$.

---

6) The function $F(z)$ is said to be differentiable with respect to $z$ if it is differentiable with respect to $u$ and $w$, where $z = u + iw$. We thereby set

$$F'(z) = \begin{pmatrix} \dfrac{\partial F_1}{\partial u} & \dfrac{\partial F_1}{\partial w} \\ \dfrac{\partial F_2}{\partial u} & \dfrac{\partial F_2}{\partial w} \end{pmatrix}, \quad \text{where} \quad F(z) = F_1(u, w) + iF_2(u, w).$$

This means that we have the topological inclusion $H(B^{1/2}) \subset W_2^m(\Omega, \{E_j\}_1^r)$, where $k_r \leqq m-1$, $k_{r+1} \geqq m$. In fact, $\mathscr{D}(A(t)) = W_2^m(\Omega, \{E_j\}_1^r) \supset \mathscr{D}(B^{1/2})$. On the other hand, since $A(t)$ is a differential operator of order not higher than $m$, we have $\|A(t)u\|_{L_2} \leqq C\|u\|_{W_2^m} \leqq C\|B^{1/2}(0)u\|_{L_2}$ for all $u \in \mathscr{D}(B^{1/2}(0))$. Thus the operator $A(t)$ is subordinate to the operator $B^{1/2}(0)$. On the other hand, since the coefficients of the differential expressions $A(t)$ and $B(t)$ are continuously differentiable functions of $t$, $A(t)$ and $B(t)$ will be strongly continuously differentiable.

Thus the linear part of the abstract problem

$$u''(t) + A(t)u'(t) + B(t)u(t) = f(t, u(t), u'(t)), \ u(0) = u_0, \ u'(0) = u_1,$$

to which the problem defined in (11), (12) and (13) has been reduced, satisfies the conditions of Theorem 7 in the space $L_2(\Omega)$. Now we check that the nonlinear operator $f(t,u,v)$ defined by

$$f(t, u, v) = F(t, x, u(x), \ldots, D^m u(x), v(x)) + \omega u(x),$$

satisfies condition XVII of Theorem 7.

In order to verify this condition we note that $H(B) = W_2^{2m}(\Omega, \{E_j\}_1^m)$ and $H(B^{1/2}) \subset W_2^m(\Omega)$. Then the operator $f(t,u,v)$ will act from $[0,T] \times H(B) \times H(B^{1/2})$ into $H = L_2(\Omega)$ if it acts from $[0,T] \times W_2^{2m}(\Omega) \times W_2^m(\Omega)$ into $L_2(\Omega)$. This fact follows from Sobolev's imbedding theorem: $W_2^{2m}(\Omega) \subset C^m(\bar{\Omega})$ (as $n < 2m$).

It is easy to show that the operator $f(t,u,v)$ acting from $[0,T] \times W_2^{2m}(\Omega) \times W_2^m(\Omega)$ into $L_2(\Omega)$ has the derivative

$$f_t'(t, u, v) = F_t'(t, x, u(x), \ldots, D^m u(x), v(x))$$

and Fréchet derivatives

$$(14) \qquad \begin{aligned} f_u'(t, u, v) &= \sum_{|\alpha| \leqslant m} F_{u_\alpha}'(t, x, u, \ldots, D^m u, v)\, D^\alpha + \omega, \\ f_v'(t, u, v) &= F_v'(t, x, u(x), \ldots, D^m u(x), v(x)). \end{aligned}$$

From (14) it follows that the bounded linear operator $f_u'(t,u,v)$ acting from $W_2^{2m}(\Omega)$ into $L_2(\Omega)$ may be extended up to a bounded operator acting from $W_2^m(\Omega)$ into $L_2(\Omega)$, i.e. the operator $f(t,u,v)$ has a $W_2^m(\Omega)$-extended Fréchet derivative with respect to $u$.

On the other hand, since the function $H(t,x) = F_v'(t,x,u,\cdots,D^m u,v)$ is continuous with respect to $x \in \bar{\Omega}$ for any $u(x) \in W_2^{2m}(\Omega) \subset C^m(\bar{\Omega})$ and $v(x) \in W_2^m(\Omega) \subset C(\bar{\Omega})$, the operator of multiplication on the function $H(t,x)$ is bounded in $L_2(\Omega)$, i.e. the operator $f(t,u,v)$ has an $L_2$-extended Fréchet derivative with respect to $v$. Similarly the smoothness of the derivatives is established. The theorem is proved.

For $2m \leqq n < 4m$ we have to examine the quasilinear equation

$$D_t^2 u(t, x) + \sum_{|\alpha| < m} A_\alpha (t, x) D^\alpha D_t u(t, x) + \sum_{|\alpha| \leqslant 2m} B_\alpha (t, x) D^\alpha u(t, x)$$

$$\tag{15} + \sum_{|\alpha| \leqslant m} C_\alpha (t, x) D^\alpha u(t, x) = F(t, x, u, \ldots, D^\beta u) D_t u(t, x)$$

$$+ \Phi(t, x, u, \ldots, D^\beta u).$$

THEOREM 11. *Assume that conditions* XXVI *to* XXX *are satisfied, and also that*

XXXII. $2m \leqq n < 4m$; $|\beta| = \beta_1 + \beta_2 + \cdots + \beta_n < (4m - n)/2$; $C_\alpha(t, x) \in C^{1,0}([0, T] \times \Omega)$; *the functions* $F(t, x, z)$ *and* $\Phi(t, x, z)$ *are continuous together with their derivatives with respect to* $t$ *and* $z_k$ *in the region* $\{t \in [0, T], x \in \overline{\Omega}, z_k \in Z\}$; *furthermore, these derivatives satisfy a Lipschitz condition with respect to* $z_k$ *in every bounded set* $|z_k| \leqq R$, $0 \leqq r_k \leqq R$ *with a Lipschitz constant depending only on* $R$.

*Then the problem defined by* (15), (12) *and* (13) *has the unique solution*

$$u(t, x) \in C([0, t_0], W_2^{2m}(\Omega, \{E_j\}_1^m)) \cap C^1([0, t_0]; W_2^m(\Omega, \{E_j\}_1^r)) \cap C^2([0, t_0], L_2(\Omega)),$$

*which can be determined by the method of successive approximations.*

PROOF. We consider the nonlinear operator

$$F(t, u, v) = - \sum_{|\alpha| \leqslant m} C_\alpha (t, x) D^\alpha u(x) + F(t, x, u(x), \ldots, D^\beta u(x)) v(x)$$

$$+ \Phi(t, x, u(x), \ldots, D^\beta u(x)).$$

Since $|\beta| < 2m - n/2$, we have $W_2^{2m}(\Omega) \subset C^{|\beta|}(\Omega)$. Then by XXXII the operator $F(t, u, v)$ acts from $[0, T] \times W_2^{2m}(\Omega) \times W_2^m(\Omega)$ into $L_2(\Omega)$ and is bounded (in fact it acts from $[0, T] \times W_2^{2m}(\Omega) \times L_2(\Omega)$ into $L_2(\Omega)$ and is bounded). It is obvious that $f(t, u, v)$ as an operator acting from $[0, T] \times W_2^{2m}(\Omega) \times W_2^m(\Omega)$ into $L_2(\Omega)$ has a derivative

$$f_t'(t, u, v) = - \sum_{|\alpha| \leqslant m} \frac{\partial C_\alpha (t, x)}{\partial t} D^\alpha u(x) + F_t'(t, x, u, \ldots, D^\beta u) v$$

$$+ \Phi_t'(t, x, u(x), \ldots, D^\beta u(x))$$

and Fréchet derivatives

$$f_u'(t, u, v) = - \sum_{|\alpha| \leqslant m} C_\alpha (t, x) D^\alpha + v(x) \sum_{|\alpha| \leqslant |\beta|} F_{u_\alpha}'(t, x, u, \ldots, D^\beta u) D^\alpha$$

$$+ \sum_{|\alpha| \leqslant |\beta|} \Phi_{u_\alpha}'(t, x, u, \ldots, D^\beta u) D^\alpha,$$

$$f_v'(t, u, v) = F(t, x, u, \ldots, D^\beta u).$$

Since $|\beta| < (4m - n)/2$, from Sobolev's imbedding theorem it follows that $W_2^m(\Omega) \subset W_{n/m}^{|\beta|}(\Omega)$. Then for any $v, h \in W_2^m(\Omega)$ the following inequality holds:

$$\sum_{|\alpha|\leqslant|\beta|}\int_\Omega |v(x)\,D^\alpha h(x)|^2\,dx \leqslant \sum_{|\alpha|\leqslant|\beta|}\Big(\int_\Omega |v(x)|^{2\cdot\frac{n}{n-2m}}\Big)^{\frac{n-2m}{n}}\Big(\int_\Omega |D^\alpha h(x)|^{2\cdot\frac{n}{2m}}\Big)^{\frac{2m}{n}}$$

$$\leqslant \sum_{|\alpha|\leqslant|\beta|}\|v\|^2_{L_{\frac{2n}{n-2m}}}\|D^\alpha h\|^2_{L_{\frac{n}{m}}} \leqslant C\|v\|^2_{W_2^m}\|h\|^2_{W^{|\beta|}_{\frac{n}{m}}} \leqslant C\|v\|^2_{W_2^m}\|h\|^2_{W_2^m}.$$

It then follows that the bounded linear operator $f_u'(t,u,v)$ acting from $W_2^{2m}(\Omega)$ into $L_2(\Omega)$ may be extended up to a bounded operator acting from $W_2^m(\Omega)$ into $L_2(\Omega)$. Hence the operator $f(t,u,v)$ has a $W_2^m(\Omega)$-extended Fréchet derivative $f_u'(t,u,v)$.

On the other hand, since $W_2^{2m}(\Omega)\subset C^{|\beta|}(\overline{\Omega})$, the function $H(t,x) = F(t,x,u(x),\cdots,Du(x))$ is continuous with respect to $x$ for any $u\in W_2^{2m}(\Omega)$. Hence $f(t,u,v)$ has also a $L_2(\Omega)$-extended Fréchet derivative $f_v'(t,u,v)$. The smoothness of the derivatives follows from Sobolev's imbedding theorem and condition XXXII. The theorem is proved.

Now we establish a theorem on overall solvability of boundary value problems for quasilinear partial differential equations of hyperbolic type.

We shall investigate two cases: 1) $n<2m$ and 2) $2m\leqq n<4m$.

In the first case we consider the quasilinear equation

$$\begin{aligned} &D_t^2u(t,x)+\sum_{|\alpha|\leqslant m}A_\alpha(t,x)D^\alpha D_tu(t,x)+\sum_{|\alpha|\leqslant 2m}B_\alpha(t,x)D^\alpha u(t,x)\\ &\quad+\sum_{|\alpha|\leqslant m}C_\alpha(t,x)D^\alpha u(t,x)=F(t,x,|u|^2,\dots,|D^\beta u|^2)D_tu(t,x)\\ &\quad+\Phi'(|u|^2)u(t,x)+f(t,x). \end{aligned}\tag{16}$$

**Theorem** 12. *Let conditions* XXVI *to* XXX, *and also the following conditions, hold.*

XXXIII. $|\beta|<(2m-n)/2$; $C_\alpha(t,x)\in C^{1,0}([0,T]\times\overline{\Omega})$; *the function* $F(t,x,r)$ *is continuous together with its derivatives with respect to* $t$ *and* $r_k$ *in the region* $\{t\in[0,T];\ x\in\overline{\Omega};\ r_k\in[0,\omega)\}$, *and in addition these derivatives satisfy a Lipschitz condition with respect to* $r_k$ *in every bounded set* $0\leqq r_k\leqq R$ *with a Lipschitz constant depending only on* $R$; $\Phi(r)$ *is a real-valued function twice continuously differentiable on* $[0,\infty)$ *and* $\Phi''(r)$ *satisfies a Lipschitz condition in* $r$ *in every bounded set* $0\leqq r\leqq R$ *with the Lipschitz constant depending on* $R$.

XXXIV. $\operatorname{Re}F(t,s,r)\leqq C$; $\Phi(r)\leqq C(1+r)$.

*Then the problem defined by* (16), (12) *and* (13) *has the unique solution*

$$u(t,x)\in C([0,T],W_2^{2m}(\Omega,\{E_j\}_1^m))\cap C^1([0,T],W_2^m(\Omega,\{E_j\}_1^r))\cap C^2([0,T],L_2(\Omega)).$$

**Proof.** Local solvability of (16), (12) and (13) follows from Theorem 10. We shall prove that the nonlinear operator

$$f(t, u, v) = -\sum_{|\alpha|\leqslant m} C_\alpha(t, x) D^\alpha u(x) + F(t, x, |u|^2, \ldots, |D^\beta u|^2) v(x) \tag{17}$$
$$+ \Phi'(|u(x)|^2) u(x) + \omega u(x) + f(t, x),$$

in the space $L_2(\Omega)$ satisfies conditions XVIII and XIX of Theorem 8.

Let the function $u(t)$ with values from $W_2^{2m}(\Omega, \{E_j\}_1^m)$ be continuous in the metric of $W_2^{2m}(\Omega)$ and continuously differentiable in the metric of $H(B^{1/2}) \subset W_2^m(\Omega)$. The set of such functions will be denoted by $S$.

Taking into account that $|\beta| < (2m-n)/2 < 2m - n/2$ and Sobolev's imbedding theorem, we find that the functions $D^\alpha u(t, x)$ with $|\alpha| \leqq |\beta|$ are continuous in the region $\bar{Q}$ if the corresponding function $u(t)$ is contained in $S$.

First we prove that the following one-sided bound holds for a function $u(t)$ from the set $S$:

$$\operatorname{Re}\int_0^t (f(\tau, u(\tau), u'(\tau)), u'(\tau))\, d\tau \leqslant C\left[1 + \int_0^t (\|B_0^{1/2} u(\tau)\|^2 + \|u'(\tau)\|^2)\, d\tau\right]. \tag{18}$$

From (17) we obtain the equation

$$\operatorname{Re}\int_0^t (f(\tau, u(\tau), u'(\tau)), u'(\tau))\, d\tau = -\operatorname{Re}\int_0^t\int_\Omega \sum_{|\alpha|\leqslant m} C_\alpha(\tau, x) D^\alpha u(\tau, x)$$
$$\times \overline{\frac{\partial u(\tau, x)}{\partial \tau}}\, dx d\tau + \operatorname{Re}\int_0^t\int_\Omega F(\tau, x, |u(\tau, x)|^2, \ldots, |D^\beta u(\tau, x)|^2) \left|\frac{\partial u(\tau, x)}{\partial \tau}\right|^2 dx d\tau$$
$$+ \operatorname{Re}\int_0^t\int_\Omega \frac{d\Phi(|u(\tau, x)|^2)}{dr} u(\tau, x) \overline{\frac{\partial u(\tau, x)}{\partial \tau}}\, dx d\tau \tag{19}$$
$$+ \operatorname{Re}\omega \int_0^t\int_\Omega u(\tau, x) \overline{\frac{\partial u(\tau, x)}{\partial \tau}}\, dx d\tau + \operatorname{Re}\int_0^t\int_\Omega f(\tau, x) \overline{\frac{\partial u(\tau, x)}{\partial \tau}}\, dx d\tau.$$

It is obvious that

$$\operatorname{Re}\int_0^t\int_\Omega \sum_{|\alpha|\leqslant m} C_\alpha(\tau, x) D^\alpha u \cdot \frac{\partial u}{\partial \tau}\, dx d\tau \leqslant C\int_0^t \sum_{|\alpha|\leqslant m} \|D^\alpha u(\tau)\|^2\, d\tau \tag{20}$$
$$+ \int_0^t \|u'(\tau)\|^2\, d\tau \leqslant C\int_0^t (\|B_0^{1/2} u(\tau)\|^2 + \|u'(\tau)\|^2)\, d\tau,$$

and

$$\operatorname{Re}\int_0^t\int_\Omega f(\tau, x) \overline{\frac{\partial u(\tau, x)}{\partial \tau}}\, dx d\tau \leqslant C\left(1 + \int_0^t \|u'(\tau)\|^2\, d\tau\right). \tag{21}$$

Then from XXXIV we have

(22) $$\operatorname{Re}\int_0^t\int_\Omega F(\tau, x, |u|^2, \ldots, |D^\beta u|^2)\left|\frac{\partial u(\tau, x)}{\partial\tau}\right|^2 dxd\tau \leqslant C\int_0^t \|u'(\tau)\|^2 d\tau.$$

In addition

$$\operatorname{Re}\int_0^t\int_\Omega \frac{d\Phi(|u(\tau, x)|^2)}{dr}\, u(\tau, x)\, \overline{\frac{\partial u(\tau, x)}{\partial\tau}}\, dxd\tau$$

$$= \frac{1}{2}\int_0^t\int_\Omega \frac{d\Phi(|u(\tau, x)|^2)}{dr}\,\frac{\partial|u(\tau, x)|^2}{\partial\tau}\, dxd\tau$$

(23) $$= \frac{1}{2}\int_0^t\int_\Omega \frac{\partial\Phi(|u(\tau, x)|^2)}{\partial\tau}\, dxd\tau = \frac{1}{2}\int_\Omega [\Phi(|u(t, x)|^2) - \Phi(|u(0, x)|^2)]\, dx$$

$$\leqslant C\int_\Omega (1 + |u(t, x)|^2)\, dx \leqslant C(1 + \|u(t)\|^2)$$

$$\leqslant C\left(1 + \|u_0\|^2 + \int_0^t \|u'(\tau)\|^2 d\tau\right) \leqslant C\left(1 + \int_0^t \|u'(\tau)\|^2 d\tau\right).$$

We note that inequality (23) has been established only for functions $u(t) \in S$ for which $u(0) = u_0$. This will be sufficient for our purposes. Substituting inequalities (20) to (23) in (19), we obtain the required bound (18), i.e. condition XVIII is satisfied.

We shall now prove that the operator $f(t, u, v)$ defined by (17) also satisfies condition XIX.

In Theorem 10 it was proved that the operator $f(t, u, v)$ defined by (17) acts from $[0, T] \times W_2^{2m}(\Omega) \times W_2^m(\Omega)$ into $L_2(\Omega)$, has the derivative

$$f_t'(t, u, v) = -\sum_{|\alpha|\leqslant m} \frac{\partial C_\alpha(t, x)}{\partial t} D^\alpha u(x) + F_t'(t, x, |u|^2, \ldots$$

$$\ldots, |D^\beta u|^2)\, v(x) + f_t'(t, x),$$

the $W_2^m(\Omega)$-extended Fréchet derivative

$$f_u'(t, u, v) = -\sum_{|\alpha|\leqslant m} C_\alpha(t, x) D^\alpha + 2v(x)\sum_{|\alpha|\leqslant|\beta|} F_{r_\alpha}'(t, x, |u|^2, \ldots$$

$$\ldots, |D^\beta u|^2)\, D^\alpha u D^\alpha + 2\Phi''(|u|^2)\, u^2 + \Phi'(|u|^2) + \omega,$$

and the $L_2(\Omega)$-extended Fréchet derivative

$$f_v'(t, u, v) = F(t, x, |u|^2, \ldots, |D^\beta u|^2).$$

Hence for any $u \in W_2^{2m}(\Omega)$ and $v \in W_2^m(\Omega)$ such that $\|u\|_{W_2^m} \leqq R$ and $\|v\|_{L_2} \leqq R$, the inequality

$$\|f_t'(t, u, v)\| \leqslant C(R),\ \|f_u'(t, u, v)\|_{B(W_2^m, L_2)} \leqslant C(R),\ \|f_v'(t, u, v)\|_{B(L_2)} \leqslant C(R),$$

holds if and only if

$$\max_{\bar{Q}} |F_t'(t, x, \dots, |D^\beta u|^2)| \leqslant C(R); \quad \max_{\bar{Q}} |F_{r_\alpha}'(t, x, \dots, |D^\beta u|^2)| \leqslant C(R);$$

$$\max_{\bar{\Omega}} |\Phi''(|u|^2)| \leqslant C(R).$$

And these inequalities follow from condition XXXIII using the inclusion $W_2^m(\Omega) \subset C^{|\beta|}(\bar{\Omega})$ and the inequality

$$\sum_{|\alpha| \leqslant |\beta|} \int_\Omega | vF_{r_\alpha}'(t, x, |u|^2, \dots, |D^\beta u|^2) D^\alpha u D^\alpha h|^2 \, dx$$

$$\leqslant C(R) \sum_{|\alpha| \leqslant |\beta|} \|D^\alpha h\|_C^2 \leqslant C(R) \|h\|_{W_2^m}^2 .$$

Thus Theorem 8 may be applied to the abstract Cauchy problem

$$u''(t) + A(t) u'(t) + B(t) u(t) = f(t, u(t), u'(t)), \; u(0) = u_0, \; u'(0) = u_1,$$

which is derived from the problem defined in (16), (12) and (13). The theorem is proved.

We now study the case $2m \leqq n < 4m$. We consider the quasilinear equation of hyperbolic type

$$\text{(24)} \quad \begin{aligned} &D_t^2 u(t, x) + \sum_{|\alpha| \leqslant m} A_\alpha(t, x) D^\alpha D_t u(t, x) + \sum_{|\alpha| \leqslant 2m} B_\alpha(t, x) D^\alpha u(t, x) \\ &+ \sum_{|\alpha| \leqslant m} C_\alpha(t, x) D^\alpha u(t, x) = \Phi'(|u(t, x)|^2) u(t, x) + f(t, x) \end{aligned}$$

with boundary conditions (12) and initial conditions (13).

THEOREM 13. *Assume that conditions* XXVI *to* XXX *are satisfied, and also that*:

XXXV. $C_\alpha(t,x) \in C^{1,0}([0, T] \times \bar{\Omega})$; *the real-valued function* $\Phi(r)$ *is twice continuously differentiable on* $[0, \infty)$ *and in addition* $\Phi''(r)$ *satisfies a Lipschitz condition in every set* $0 \leqq r \leqq R$; *the following bounds hold*:

$$\Phi(r) \leqslant C(1 + r),$$
$$|\Phi'(r)| \leqslant C(1 + r^q),$$
$$|\Phi''(r)| \leqslant C(1 + r^{q-1}),$$

*where* $q = m(n - 2m)^{-1}$ *for* $n > 2m$ *and* $q$ *is arbitrary for* $n = 2m$.

*Then the problem defined by* (24), (12) *and* (13) *has the unique solution*

$$u(t, x) \in C([0, T], W_2^{2m}(\Omega, \{E_j\}_1^m)) \cap C^1([0, T], W_2^m(\Omega, \{E_j\}_1^r))$$
$$\cap C^2([0, T], L_2(\Omega)).$$

PROOF. Local solvability of the problem defined by (24), (12) and (13) follows from Theorem 11. It remains to prove that the nonlinear operator

$$\Phi(u) = \Phi'(|u(x)|^2)\, u(x) \tag{25}$$

in the space $L_2(\Omega)$ satisfies conditions XVIII and XIX of Theorem 8. We select a function $u(t)$ in the set $S$. Then from (23) we have

$$\operatorname{Re}\int_0^t (\Phi(u(\tau)),\ u'(\tau))\, d\tau = \operatorname{Re}\int_0^t\int_\Omega \frac{d\Phi(|u(\tau,x)|^2)}{dr}\, u(\tau,x)\, \frac{\overline{\partial u(\tau,x)}}{\partial\tau}\, dx\, d\tau$$

$$\leqslant C\left(1 + \int_0^t \|u'(\tau)\|^2\, d\tau\right),$$

i.e. $\Phi(u)$ satisfies the one-sided bound in condition XVIII. In order to verify condition XIX it is sufficient to prove that the $W_2^m(\Omega)$-extended Fréchet derivative

$$\Phi'_u(u) = 2\Phi''(|u(x)|^2)\, u^2(x) + \Phi'(|u(x)|^2)$$

satisfies the bound

$$\|\Phi'_u(u)\|_{B(W_2^m, L_2)} \leqslant C(R), \tag{26}$$

when $u \in W_2^{2m}(\Omega)$ and $\|u\|_{W_2^m} \leqq R$. Let $h \in W_2^m(\Omega)$. Then

$$\|\Phi'_u(u)h\|^2_{L_2} \leqslant C\left(\int_\Omega |\Phi''(|u|^2)|^2\, |u(x)|^4\, |h(x)|^2\, dx\right.$$

$$\left. + \int_\Omega |\Phi'(|u|^2)|^2\, |h(x)|^2\, dx\right) \leqslant C\int_\Omega |u(x)|^{4q}\, |h(x)|^2\, dx$$

$$\leqslant C\left(\int_\Omega |u(x)|^{4q\cdot\frac{n}{2m}}\, dx\right)^{\frac{2m}{n}} \left(\int_\Omega |h(x)|^{2\cdot\frac{n}{n-2m}}\, dx\right)^{\frac{n-2m}{n}}$$

$$\leqslant C\|u\|_{L_{\frac{2n}{n-2m}}}^{\frac{4m}{n-2m}}\, \|h\|^2_{L_{\frac{2n}{n-2m}}} \leqslant C\|u\|_{W_2^m}^{\frac{4m}{n-2m}}\, \|h\|^2_{W_2^m}.$$

Thus, in particular, we obtain the bound (26). Hence the operator defined by (25) satisfies condition XIX. Consequently Theorem 8 may be applied to the problem defined by (24), (12) and (13), and from this Theorem 13 follows.

In papers [**10**]–[**12**] particular cases of equations (16) and (24) were considered, namely $A(t) = 0$, $C(t) = 0$, and $F = 0$.

REMARK. We note that if the functions $A_\alpha(t,x)$, $B_\alpha(t,x)$, $C_\alpha(t,x)$, $e_\alpha(x)$, $F(t,x,r)$ and the initial functions $u_0(x)$ and $u_1(x)$ are real-valued, then the solutions of the problems (16), (12) and (13) and (24), (12) and (13) will also be real-valued functions $u(t,x)$. This is a consequence of the fact that $u(t,x)$ and $\overline{u(t,x)}$ are solutions of the same equation. As the solution is unique, these solutions must be identical.

**Remark.** All of these results may be carried over to the corresponding systems of equations without any essential changes.

2. *The quasilinear hyperbolic system of the second order.* In this section we shall show that the semi-abstract conditions XXVI to XXX hold for classical hyperbolic equations of the second order. The results are also given for a system of hyperbolic equations of the second order. We present only one theorem, as similar theorems may be formulated and proved in an analogous way.

We consider the quasilinear hyperbolic system

$$\frac{\partial^2 u(t,x)}{\partial t^2} + \sum_{i=1}^{n} A_i(t,x)\frac{\partial^2 u(t,x)}{\partial x_i \partial t} - \sum_{i,j=1}^{n} \frac{\partial}{\partial x_i}\left(B_{ij}(t,x)\frac{\partial u(t,x)}{\partial x_j}\right) \tag{27}$$
$$+ C(t,x)\frac{\partial u(t,x)}{\partial t} + \sum_{i=1}^{n} d_i(t,x)\frac{\partial u(t,x)}{\partial x_i} = \Phi'(|u|^2)u(t,x) + f(t,x)$$

with the conditions

$$u|_\Gamma = 0,\ u(0,x) = u_0(x),\ u_t(0,x) = u_1(x). \tag{28}$$

Here $u(t,x)$ is an $N$-dimensional vector function equal to $(u_1(t,x),\cdots,u_N(t,x))$.

**Theorem 14.** *Let the following conditions be satisfied.*

XXXVI. *$A_i(t,x)$ is an $N$-dimensional Hermitian matrix; $B_{ij}(t,x) = B_{ji}^*(t,x)$, $A_i(t,x) \in C^1(\bar{Q})$, $B_{ij}(t,x) \in C^2(\bar{Q})$ and $\partial A_i(t,x)/\partial x_i \leqq 0$; for every real $\xi \in l_n$ and complex $\zeta \in l_N$ the following inequality holds:*

$$\left(\sum_{i,j=1}^{n} B_{ij}(t,x)\xi_i\xi_j\zeta,\zeta\right) \geqslant b\sum_{i=1}^{n}\xi_i^2(\zeta,\zeta),\quad b = \text{const} > 0.$$

XXXVII. *The elements of the $N$-dimensional matrices $C(t,x)$ and $d_i(t,x)$, and also of the vector function $f(t,x)$, together with their first order derivatives in $t$, are continuous in $\bar{Q}$.*

XXXVIII. *For $n \leqq 3$ the real function $\Phi(r)$ is twice continuously differentiable on $[0,\infty)$, and $\Phi''(r)$ satisfies a local Lipschitz condition; the following bounds hold:*

$$\Phi(r) \leqslant C(1+r),$$

$$|\Phi^{(l)}(r)| \leqslant C(1+r^{q-l+1}),\ l = 1,\ 2,$$

*where $q=1$ for $n=3$, $q$ is arbitrary for $n=2$ and $\Phi(r)$ has no restrictions on its growth for $n=1$.*

XXXIX. *$u_0 \in W_{2,0}^2(\Omega)$ and $u_1(x) \in W_{2,0}^1(\Omega)$.*

*Then the problem defined by* (27) *and* (28) *has the unique solution*

$$u(t,x) \in C([0,T],W_{2,0}^2(\Omega)) \cap C^1([0,T],W_{2,0}^1(\Omega)) \cap C^2([0,T],L_2(\Omega)).$$

**Proof.** Let $B(t)$ be the operator generated by the system $\{B(t,x,D);\ u|_\Gamma = 0\}$, where

$$B(t, x, D)u = -\sum_{i,j=1}^{n} \frac{\partial}{\partial x_i}\left(B_{ij}(t, x)\frac{\partial u(x)}{\partial x_j}\right),$$

then it is selfadjoint in $L_2^N(\Omega)$ (see [24] and [25]) and $\mathcal{D}(B(t)) = W_{2,0}^2(\Omega)$. On the other hand, the Višik-Gårding inequality holds (see [26] and [27]):

$$(B(t, x, D)u, u) \geqslant C_1 \| u \|_{W_2^1}^2 - C_2 \| u \|_{L_2}^2, \quad C_1 > 0,$$

for any $u \in W_{2,0}^2(\Omega)$. Thus the operator $B(t)$ satisfies condition XXVIII.

We shall show that the operator $A(t)$ defined by the differential expression

$$A(t, x, D)u = \sum_{i=1}^{n} A_i(t, x)\frac{\partial u(x)}{\partial x_i}$$

and having domain of definition $\mathcal{D}(A) = W_{2,0}^1(\Omega)$, is dissipative in the space $L_2^N(\Omega)$. Let $u(x)$ be contained in $\mathcal{D}(A) = W_{2,0}^1(\Omega)$, and let

$$A_i(t, x) = \| a_{ks}^i(t, x) \|_{k,s=1}^N,$$

where

$$a_{ks}^i(t, x) = \overline{a_{sk}^i(t, x)}.$$

Then

$$(A(t)u, u) = \int_\Omega \sum_{i=1}^{n} \sum_{k,s=1}^{N} a_{ks}^i(t, x)\frac{\partial u_k(x)}{\partial x_i}\overline{u_s(x)}\,dx$$

$$= \int_\Gamma \sum_{i=1}^{n} \sum_{k,s=1}^{N} a_{ks}^i(t, x)u_k(x)\overline{u_s(x)}\cos(n, x_i)\,dx$$

$$-\int_\Omega \sum_{i=1}^{n} \sum_{k,s=1}^{N} u_k(x)\frac{\partial}{\partial x_i}(a_{k,s}^i(t, x)\overline{u_s(x)}\,dx$$

$$= -\int_\Omega \sum_{i=1}^{n} \sum_{k,s=1}^{N} u_k(x)\overline{a_{sk}^i(t, x)}\,\overline{\frac{\partial u_s(x)}{\partial x_i}}\,dx$$

$$-\int_\Omega \sum_{i=1}^{n} \sum_{k,s=1}^{N} u_k(x)\frac{\partial}{\partial x_i}(a_{ks}^i(t, x))\overline{u_s(x)}\,dx$$

$$= -(u, A(t)u) - \int_\Omega \sum_{i=1}^{n} \sum_{k,s=1}^{N} \frac{\partial}{\partial x_i}(a_{ks}^i(t, x))u_k(x)\overline{u_s(x)}\,dx.$$

Thus for any $u \in \mathcal{D}(A) = W_{2,0}^1(\Omega)$ we have

$$2\mathrm{Re}\,(A(t)u, u) = -\int_\Omega \sum_{i=1}^{n} \sum_{k,s=1}^{N} \frac{\partial}{\partial x_i}(a_{ks}^i(t, x))u_k(x)\overline{u_s(x)}\,dx \geqslant 0,$$

as $\partial A_i(t,x)/\partial x_i \leqq 0$, i.e. $A(t)$ satisfies condition XXIX.

The remainder of the proof is analogous to that of Theorem 13.

REMARK. It is also possible to investigate the case $n \leqq 4m$, but for this it is necessary to limit the growth of nonlinearity even for local existence theorems.

In conclusion, I should like to express my gratitude to Professor S. G. Kreĭn for his valuable advice and for discussion of the results.

Received 1 OCT 1968

## BIBLIOGRAPHY

[1] E. Hille and R. S. Phillips, *Functional analysis and semi-groups*, rev. ed., Amer. Math. Soc. Colloq. Publ., vol. 31, Amer. Math. Soc., Providence, R.I., 1957; Russian transl., IL, Moscow, 1962. MR **19**, 664.

[2] S. G. Kreĭn, *Linear differential equations in a Banach space*, "Nauka", Moscow, 1967; English transl., Transl. Math. Monographs, vol. 29, Amer. Math. Soc., Providence, R.I., 1971. MR **40** # 508.

[3] I. E. Segal, *Non-linear semi-groups*, Ann. of Math. (2) **78** (1963), 339–364. MR **27** # 2879.

[4] S. Ja. Jakubov, *Quasi-linear differential equations in abstract spaces*, Akad. Nauk Azerbaĭdžan. SSR Dokl. **22** (1966), no. 8, 8–12. (Russian) MR **34** # 6267.

[5] D. A. Ahundov and S. Ja. Jakubov, *The uniformly correct Cauchy problem for abstract first order quasilinear differential equations in a Banach space*, Akad. Nauk Azerbaĭdžan. SSR Dokl. **25** (1969), no. 2, 3–7. (Russian) MR **42** # 897.

[6] G. Lumer and R. S. Phillips, *Dissipative operators in a Banach space*, Pacific J. Math. **11** (1961), 679–698. MR **24** # A2248.

[7] V. G. Maz′ja and P. E. Sobolevskiĭ, *On generating operators of semi-groups*, Uspehi Mat. Nauk **17** (1962), no. 6 (108), 151–154. (Russian) MR **27** # 1838.

[8] T. Kato, *Integration of the equation of evolution in a Banach space*, J. Math. Soc. Japan **5** (1953), 208–234. MR **15**, 437.

[9] Ju. L. Daleckiĭ, *On the problem of fractional powers of selfadjoint operators*, Trudy Sem. Funkcional. Anal. **1958**, no. 6, 44–48. (Russian)

[10] F. E. Browder, *On non-linear wave equations*, Math. Z. **80** (1962), 249–264. MR **26** # 5283.

[11] F. E. Browder and W. A. Strauss, *Scattering for non-linear wave equations*, Pacific J. J. Math. **13** (1963), 23–43. MR **27** # 6047.

[12] I. E. Segal, *Non-linear relativisitic partial differential equations*, Proc. Internat. Congress Math. (Moscow, 1966), "Mir", Moscow, 1968, pp. 681–690. MR **38** # 5444.

[13] Ja. D. Mamedov, *Certain properties of solutions of nonlinear hyperbolic equations in Hilbert space*, Dokl. Akad. Nauk SSSR **158** (1964), 45–48 = Soviet Math. Dokl. **5** (1964), 1176–1179. MR **29** # 3759.

[14] J.-L. Lions and W. A. Strauss, *Some non-linear evolution equations*, Bull. Soc. Math. France **93** (1965), 43–96. MR **33** # 7663.

[15] L. Adauto Medeiros, *Temporally inhomogeneous nonlinear wave equations in Hilbert spaces*, An. Acad. Brasil. Ci. **37** (1965), 193–199. MR **33** # 1759.

[16] V. A. Pogorelenko and P. E. Sobolevskiĭ, *Non-local existence theorem of solutions of non-linear hyperbolic equations in a Hilbert space*, Ukrain. Mat. Ž. **19** (1967), no. 1, 113-115. . (Russian) MR **34** # 7928.

[17] S. Ja. Jakubov, *The Cauchy problem for second-order differential equations in Banach space*, Dokl. Akad. Nauk SSSR **168** (1966), 759–762 = Soviet Math. Dokl. **7** (1966), 735–739. MR **33** # 6178.

[18] ———, *Nonlinear abstract hyperbolic equations*, Dokl. Akad. Nauk SSSR **176** (1967), 46–49 = Soviet Math. Dokl. **8** (1967), 1055-1059. MR **36** # 4178.

[19] M. Schechter, *General boundary value problems for elliptic partial differential equations*, Comm. Pure Appl. Math. **12** (1959), 457–486. MR **23** # A2626.

[20] Z. Ja. Šapiro, *On general boundary problems for equations of elliptic type*, Izv. Akad. Nauk SSSR Ser. Mat. **17** (1953), 539–562. (Russian) MR **16**, 42.

[21] Ja. B. Lopatinskiĭ, *A method of reducing boundary problems for a system of differential equations of elliptic type to regular integral equations*, Ukrain. Mat. Ž. **5** (1953), no. 2, 123–151; English transl., Amer. Math. Soc. Transl. (2) **89** (1970), 149–183. MR **17**, **494**.

[22] I. N. Valeškevič, *Some mixed problems for the biharmonic wave equation*, Vesci Akad. Navuk BSSR Ser. Fiz.-Tèhn. Navuk **1963**, no. 3, 14–24. (Russian) MR **28** # 337.

[23] N. Aronszajn and A. Milgram, *Differential operators on Riemannian manifolds*, Rend. Circ. Mat. Palermo. (2) **2** (1953), 266–325. MR **16**, 252.

[24] O. A. Ladyženskaja, *On the closure of the elliptic operator*, Dokl. Akad. Nauk SSSR **79** (1951), 723–725. (Russian) MR **14**, 280.

[25] O. V. Guseva, *On boundary problems for strongly elliptic systems*, Dokl. Akad. Nauk SSSR **102** (1955), 1069–1072. (Russian) MR **17**, 161.

[26] M. I. Višik, *On strongly elliptic systems of differential equations*, Mat. Sb. **29** (**71**) (1951), 615–676. (Russian) MR **14**, 174.

[27] L. Gårding, *Dirichlet's problem for linear elliptic partial differential equations*, Math. Scand. **1** (1953), 55–72. MR **16**, 366.

Translated by:
H. J. Norton

Trudy Moskov. Mat. Obšč.
Tom 23 (1970)

Trans. Moscow Math. Soc.
Vol. 23 (1970)

# ON WEAKLY NONLINEAR HYPERBOLIC SYSTEMS

*S. I. POHOŽAEV*

## Contents



## Introduction

We shall consider weakly nonlinear hyperbolic (in the sense of Friedrichs) systems of first-order differential equations. We shall prove existence and uniqueness theorems for solutions of some problems for such systems.

In §1 we shall consider the nonlinear equation

$$A(u) = w, \tag{*}$$

where $A$ is a (nonlinear) operator from its domain $\mathfrak{D}(A)$, which is a subset of a Banach space $X$, into an uniformly convex Banach space $Y$. The spaces $X$ and $Y$ are either real or complex. With regard to the operator $A$, we shall assume that its domain $\mathfrak{D}(A)$ is a linear manifold and that the operator $A(u)$ has on $\mathfrak{D}(A)$ the differential $A'(u)$ in the sense which will be indicated in §1. For equation (*) with such an operator we shall prove the following theorem.

THEOREM. *Let the range of the operator $A$ considered on $\mathfrak{D}(A)$ be closed in $Y$. Let the range of the linear operator $A'(u)$ with the domain $\mathfrak{D}(A)$ be everywhere dense in $Y$ for any $u$ in $\mathfrak{D}(A)$.*

*Then for any $w$ in $Y$ equation (*) has a solution $u$ in $\mathfrak{D}(A)$.*

In §2, on the basis of this theorem we shall prove a theorem on the existence and uniqueness of a solution of a mixed problem for a weakly nonlinear hyperbolic symmetrizable system. The mixed problem will be considered in the space $L_2$.

In §3 we shall consider a Cauchy problem with periodic boundary conditions in the space $W_2^l$ for a weakly nonlinear hyperbolic symmetrizable system. We shall prove a theorem on the existence and uniqueness of a solution of this problem. This theorem will contain the condition of the existence of an a priori estimate. In conclusion, we shall present an example in which this condition is satisfied.

---

*AMS (MOS) subject classifications* (1970). 35L50.

## §1. Basic theorem

Let $X$ and $Y$ be Banach spaces (real or complex). Let $Y$ be a uniformly convex space. This means that the function $\delta(\epsilon)$ which is defined for all $\epsilon$ with $0 < \epsilon \leq 2$ by the formula

$$\delta(\varepsilon) = \inf\left\{1 - \frac{1}{2}\|u + v\| \,\middle|\, \|u - v\| \geqslant \varepsilon,\ \|u\| \leqslant 1,\ \|v\| \leqslant 1\right\}$$

is strictly positive for all such $\epsilon$.

We consider a nonlinear and, generally speaking, unbounded operator $A$ with the domain $\mathfrak{D}(A) \subseteq X$ which takes on its values in $Y$. With regard to the operator $A$, we shall assume the following. Let the domain $\mathfrak{D}(A)$ be a linear manifold. Let the operator $A(u)$ be Gâteaux differentiable at every point $u \in \mathfrak{D}(A)$ in the following sense:

For any $u$ and $v$ in $\mathfrak{D}(A)$ the (strong) limit

$$\lim_{t \to 0} \frac{A(u + tv) - A(u)}{t} = A'(u)v \quad (t \in \mathbf{R}^1)$$

exists, and $A'(u)v$ with the domain $\mathfrak{D}(A'(u)) \equiv \mathfrak{D}(A)$ is linear in $v$, so that to every point $u$ in $\mathfrak{D}(A)$ there corresponds a linear operator $A'(u)$ with domain $\mathfrak{D}(A)$ which does not depend on $u$.

We consider the equation

$$A(u) = w. \tag{1}$$

THEOREM 1. *Let $A(u)$ be the operator defined above. Let the range of $A$ considered on $\mathfrak{D}(A)$ be closed in $Y$. Let the range of the linear operator $A'(u)$ with the domain $\mathfrak{D}(A)$ be everywhere dense in $Y$ for any $u$ in $\mathfrak{D}(A)$.*

*Then for any $w$ in $Y$ equation (1) has a solution $u_*$ in $\mathfrak{D}(A)$.*

PROOF. We denote by $S$ the range of $A$ considered on $\mathfrak{D}(A)$. Since the set $S$, according to the condition, is closed in $Y$ and obviously nonempty, for any $w \in Y$, there exists, according to a theorem of Edelstein [1], a sequence $\{w_n\}$ such that $w_n \to w$ strongly in $Y$ as $n \to \infty$, and for every $w_n$ $(n = 1, 2, \cdots)$ there exists a point $y_n \in S$ such that

$$\|y_n - w_n\| = \inf\{\|y - w_n\| \mid y \in S\}.$$

We set $y_n = A(u_n)$, where $u_n \in \mathfrak{D}(A)$, $n = 1, 2, \cdots$.

Now let us prove that $A(u_n) - w_n = 0$ for any $n = 1, 2, \cdots$.

Assume the contrary. Then we have $A(u_{n_0}) - w_{n_0} \neq 0$ for some $n_0$. Since $u_{n_0}$ is a minimum point of the functional $\|A(u) - w_{n_0}\|$, we obtain $\|A(u_{n_0}) - w_{n_0}\| \leq \|A(u) - w_{n_0}\|$ for any $u$ in $\mathfrak{D}(A)$.

We introduce the following notation: $u_0 = u_{n_0}$, $w_0 = w_{n_0}$, and $z_0 = A(u_{n_0}) - w_{n_0}$. Then we have

$$z_0 = A(u_0) - w_0 \neq 0 \tag{2}$$

and

(3) $$\|A(u_0)-w_0\| \leqslant \|A(u)-w_0\| \quad \text{for any} \quad u \in \mathfrak{D}(A).$$

By virtue of the condition of the Gâteaux differentiability (in the sense indicated above), for $u_0$ and for any fixed $x$ in $\mathfrak{D}(A)$ we have

(4) $$A(u_0+tx) = A(u_0) + tA'(u_0)x + \omega(u_0, x; t),$$

where

$$\frac{\|\omega(u_0, x; t)\|}{t} \to 0 \quad \text{as} \quad t \to 0 \quad (t \in \mathbf{R}^1).$$

Since the range of the operator $A'(u_0)$ is dense in $Y$, for $\delta_0 = \|z_0\|/2 > 0$ there exists an element $z_1$ in the range of $A'(u_0)$ such that

(5) $$\|z_1 - z_0\| < \delta_0.$$

Let $x_0$ be a fixed element of $\mathfrak{D}(A)$ such that

(6) $$A''(u_0)x_0 = z_1.$$

Now let us consider $u = u_0 + tx_0$, for fixed $z_0$ and $x_0$, where $t \in \mathbf{R}^1$.

Then for any $t$ we have $u \in \mathfrak{D}(A)$, and from inequality (3), taking (4) into account, we obtain

$$\|A(u_0)-w_0\| \leqslant \|A(u)-w_0\| = \|A(u_0)-w_0+tA'(u_0)x_0+\omega(u_0, x_0; t)\|.$$

Hence we have, making use of the notation $z_0 = A(u_0) - w_0$ and of inequality (5),

$$\|z_0\| \leqslant \|z_0 + tz_0 + t(z_1 - z_0) + \omega(u_0, x_0; t)\|$$
$$\leqslant |1+t|\,\|z_0\| + \frac{|t|}{2}\|z_0\| + \|\omega(u_0, x_0; t)\|.$$

By the assumption, $\|z_0\| > 0$. Therefore this implies

(7) $$1 \leqslant |1+t| + \frac{|t|}{2} + \frac{\|\omega(u_0, x_0; t)\|}{\|z_0\|}.$$

Since $z_0$ does not depend on $t$ and $\|\omega(u_0, x_0; t)\|/t \to 0$ as $t \to 0$, inequality (7) is contradictory for sufficiently small negative $t$.

This proves that $w_n = A(u_n)$ for any $n = 1, 2, \cdots$, where $u_n \in \mathfrak{D}(A)$, i.e. $w_n$ belongs to the range $S$ of $A$ considered on $\mathfrak{D}(A)$.

Since $w_n \to w$ strongly in $Y$ as $n \to \infty$, and since by the assumption of the theorem $S$ is closed in $Y$, we hence obtain that $w$ belongs to $S$, i.e. that there exists a $u_* \in \mathfrak{D}(A)$ such that $w = A(u_*)$.

The theorem has been proved.

## §2. Mixed problem in the space $L_2$

In this section we consider mixed problems for hyperbolic systems of first-order differential equations. On the basis of Theorem 1 we prove Theorem 2 on the (unique) solvability of these problems. In this connection we substantially use the linear theory of systems of first-order differential equations which was developed in [**2**]–[**9**].

Let $G$ be a bounded domain in the space $\mathbf{R}^n$ with a sufficiently smooth boundary $\Gamma$, and let $\Omega = [0, T]\times G$ and $\Omega' = [0, T]\times\Gamma$, where $T < \infty$.

We denote by $L_2(G)$, $L_2(\Omega)$ and $L_2(\Omega')$ the Hilbert spaces which consist of real vector-valued functions whose moduli are square integrable on $G$, $\Omega$ and $\Omega'$, respectively.

We shall now recall some results of the linear theory.

Let $\mathfrak{L}_0$ be the linear operator defined by the relation

(8) $$\mathfrak{L}_0 u \equiv \frac{\partial u}{\partial t} - \sum_{j=1}^{n} A_j(t, x)\frac{\partial u}{\partial x_j} + B_0 u,$$

where $A_j(t,x)$ are real square matrices of order $m$ and $B_0$ is a bounded linear operator from $L_2(\Omega)$ into $L_2(\Omega)$. We shall consider the operator $\mathfrak{L}_0$ subject to the following condition.

**Condition** $(A_1)$ (see [**9**]). Let elements of the matrices $A_j(t,x)$ $(j = 1, \cdots, n)$ belong to class $C^3(\overline{\Omega})$. Let there exist a matrix $s(t,x,\xi) \in C^{1,3,\infty}_{t,x,\xi}(\overline{\Omega}\times\Sigma)$, where $\xi = (\xi_1, \cdots, \xi_n)$, $\Sigma$ is the unit ball $|\xi| = 1$, and the matrix $s(t,x,\xi)$ satisfies the following conditions.

1) The matrix $s(t,x,\xi)\cdot\sum_{j=1}^n A_j(t,x)\xi_j$ is Hermitian.

2) The matrix $s(t,x,\xi)$ is Hermitian positive definite for $(t,x)\in\overline{\Omega}$ and $\xi\neq 0$.

3) The matrix $s(t,x,\xi)$ is positively zero-order homogeneous with respect to $\xi$.

We denote by $M(t,x)$ a smooth real $m\times m$ matrix given on $\Omega' = [0,T]\times\Gamma$ and consider the problem

(9) $$\mathfrak{L}_0 u = f(t, x),$$

(10) $$M(t, x)\, u = g(t, x) \quad \text{for} \quad (t, x)\in\Omega',$$

(11) $$u(0, x) = h(x),$$

where $f(t,x)$, $g(t,x)$, and $h(x)$ are vector-valued functions which take on their values in $\mathbf{R}^m$.

We set $A_\nu(t,x) = \sum_{i=1}^n A_j(t,x)\nu_j$, where $(\nu_1, \cdots, \nu_n) = \bar{\nu}$ is the unit vector of the interior normal to $\Omega' = [0,T]\times\Gamma$ at a point $(t,x)\in\Omega'$.

We consider mixed problem (9)–(11) subject to Condition $(A_1)$ and to the following conditions.

**Condition** $(A_2)$. Let $s(t,x,\xi) = I$ for $(t,x)\in\Omega'$, so that the matrices $A_1(t,x), \cdots, A_n(t,x)$, subject to Condition $(A_1)$, are symmetric on $\Omega'$.

Let the matrix $A_\nu(t,x)$ be nonsingular for $(t,x)\in\Omega'$.

**Condition** $(A_3)$. Let the matrix $A_\nu(t,x)$ for $(t,x)\in\Omega'$ admit the representation $A_\nu(t,x) = B_-(t,x) + B_+(t,x)$, where the matrices $B_-(t,x)$ and $B_+(t,x)$ satisfy the following conditions

1) $B_+$ and $B_-$ are real smooth $(m\times m)$ matrices.

2) The matrix $(B_+ - B_-)/2$ has a symmetric positive definite part.

3) The kernels $N_+$ and $N_-$ of the matrices $B_+$ and $B_-$, respectively, satisfy the relation

$$N_+ \oplus N_- = \mathbf{R}^m.$$

Let the limit matrix $M$ be such that $M = -B_-$.

As is well known, Conditions ($A_2$) and ($A_3$) imply the condition of maximal positiveness for the matrix $A_\nu(t,x)$. This condition was used in [**9**].

To problem (9)–(11) we assign the operator $\mathfrak{A}_0$ according to the formula

$$\mathfrak{A}_0 u = \{\mathfrak{L}_0 u,\ Mu,\ u(0,x)\}.$$

This is an operator from the domain $\mathfrak{D}(\mathfrak{A}_0) = C^1(\overline{\Omega})$ into the Hilbert space $H = L_2(\Omega) \oplus L_2'(\Omega') \oplus L_2(G)$ (the square of the norm in $H$ is equal to the sum of the squares of the corresponding norms), where $L_2'(\Omega')$ is the closed subspace of $L_2(\Omega')$ which consists of vectors $g(t,x)$ belonging to the range of the matrix $M(t,x)$. The operator $\mathfrak{A}_0$ admits the closure $\overline{\mathfrak{A}}_0$ with the domain $\mathfrak{D}(\overline{\mathfrak{A}}_0)$ which is considered in the space

$$L_2(\Omega) \oplus L_2(\Omega') \oplus L_2(G_T), \text{ where } G_T = T \times G$$

(see [**9**]). A solution of the equation $\overline{\mathfrak{A}}_0 u = V$ with $V = \{f,g,h\} \in H$ is said to be a strong solution.

It follows from [**9**] that if the above conditions are satisfied, then problem (9)–(11) has a unique strong solution for any $\{f,g,h\}$ in $L_2(\Omega) \oplus L_2'(\Omega') \oplus L_2(G)$, i.e. the solution in $\mathfrak{D}(\overline{\mathfrak{A}}_0)$.

For a vector-valued function $u(t,x) \in C^1(\overline{\Omega})$ the following inequality holds:

$$\|u\|^2_{L_2(G_\tau)} + \int_0^t [\|u\|^2_{L_2(G_\tau)} + \|u\|^2_{L_2(\Gamma_\tau)}]\,d\tau$$

(12)
$$\leqslant c(t)\left\{\int_0^t [\|\mathfrak{L}_0 u\|^2_{L_2(G_\tau)} + \|Mu\|^2_{L_2(\Gamma_\tau)}]\,d\tau + \|u\|^2_{L_2(G_0)}\right\},$$

where $c(t)$ does not depend on $u \in C^1(\overline{\Omega})$ and is continuous in $t$ for all $t \in [0,T]$, $G_\tau = \tau \times G$, and $\Gamma_\tau = \tau \times \Gamma$ (see [**9**]).

Now we shall consider the following problem in the domain $\Omega$ defined above:

(13)
$$\frac{\partial u}{\partial t} - \sum_{j=1}^n A_j(t,x)\frac{\partial u}{\partial x_j} + b(t,x,u) = f(t,x),$$

(14)
$$M(t,x)u = g(t,x) \quad \text{for } (t,x) \in \Omega',$$

(15)
$$u(0,x) = h(x).$$

Here $b(t,x,u)$ is a vector-valued function which takes on its values in $\mathbf{R}^m$ and which is assumed to satisfy the following condition:

**Condition** (B). Let the function $b(t,x,u)$ be continuous in $(t,x,u) \in [0,T] \times \overline{G} \times \mathbf{R}^m$.

Let the components $b_i(t,x,u)$ $(i = 1, \cdots, m)$ of the vector-valued function $b(t,x,u)$ have continuous partial derivatives $\partial b_i/\partial u_k$ $(k = 1, \cdots, m;\ i = 1, \cdots, m)$ which are assumed to be bounded:

$$\left|\frac{\partial b_i(t,x,u)}{\partial u_k}\right| \leqslant K_1, \quad (i, k = 1, \ldots, m),$$

for all $(t,x,u) \in [0,T]\times G\times \mathbf{R}^m$, where $K_1$ is a constant.

We introduce the operator $\mathfrak{A}(u) = \{\mathfrak{L}_1 u + B(u), Mu, u(0,x)\}$, where $\mathfrak{L}_1 u \equiv \partial u/\partial t - \sum_{j=1}^n A_j(t,x)\,\partial u/\partial x_j$ and $B(u)$ is the operator from $L_2(\Omega)$ into $L_2(\Omega)$ defined by the function $b(t,x,u)$, $B(u,(t,x)) \equiv b(t,x,u(t,x))$. We assign the linear operator $\mathfrak{A}_1(u) = \{\mathfrak{L}_1 u, Mu, u(0,x)\}$ to the operator $\mathfrak{A}(u)$ and extend the latter on the domain $\mathfrak{D}(\overline{\mathfrak{A}}_1)$ of the closure $\overline{\mathfrak{A}}_1$. Since $B(u)$, by virtue of Condition (B), is a bounded continuous operator from $L_2(\Omega)$ into $L_2(\Omega)$, this (strong) extension of $\mathfrak{A}(u)$ is entirely defined. Everywhere in the sequel we shall understand by $\mathfrak{A}(u)$ this extension.

LEMMA 1. *Let Conditions* ($A_1$), ($A_2$), ($A_3$) *and* (B) *be satisfied. Then for any vectors* $u_1(t,x)$ *and* $u_2(t,x)$ *in* $\mathfrak{D}(\overline{\mathfrak{A}}_1)$ *the following inequality holds:*

$$\begin{aligned}(16)\quad &\|u_2-u_1\|^2_{L_2(G_T)} + \int_0^T [\|u_2-u_1\|^2_{L_2(G_t)} + \|u_2-u_1\|^2_{L_2(\Gamma_t)}]\,dt \\ &\leqslant c_T\left\{\int_0^T [\|f_2-f_1\|^2_{L_2(G_t)} + \|g_2-g_1\|^2_{L_2(\Gamma_t)}]\,dt + \|h_2-h_1\|^2_{L_2(G_0)}\right\},\end{aligned}$$

*where* $c_T$ *does not depend on* $u_1, u_2 \in \mathfrak{D}(\overline{\mathfrak{A}}_1)$,

$$\{f_l, g_l\, h_l\} = \mathfrak{A}(u_l) \quad (l = 1, 2)$$

*and*

$$G_t = t\times G, \quad \Gamma_t = t\times\Gamma.$$

To prove this, it is sufficient to use inequality (12) and Condition (B).

THEOREM 2. *Let Conditions* ($A_1$), ($A_2$), ($A_3$) *and* (B) *be satisfied. Then for any element* $\{f,g,h\}$ *in* $L_2(\Omega)\oplus L_2'(\Omega')\oplus L_2(G)$ *problem* (13)–(15) *has a unique strong solution, i.e. the solution in* $\mathfrak{D}(\overline{\mathfrak{A}}_1)$.

PROOF. The facts that the linear operator $\mathfrak{A}_1 u$ is closed and the nonlinear operator $B(u)$ from $L_2(\Omega)$ into $L_2(\Omega)$ is continuous imply that the operator $\mathfrak{A}(u)$ considered on $\mathfrak{D}(\overline{\mathfrak{A}}_1)$ is closed. Then it follows from Lemma 1 that the operator $\mathfrak{A}(u)$ with the domain $\mathfrak{D}(\overline{\mathfrak{A}}_1)$ has a range that is closed in $H$. Indeed, let $u_n \in \mathfrak{D}(\overline{\mathfrak{A}}_1)$, and let

$\mathfrak{L}_1 u_n + B(u_n) \equiv f_n \to f$ strongly in $L_2(\Omega)$, $Mu_n \equiv g_n \to g$ strongly in $L_2(\Omega')$

and

$$u_n(0,x) \equiv h_n \to h \text{ strongly in } L_2(G) \text{ as } n\to\infty.$$

Then by virtue of inequality (16) the sequence $\{u_n(t,x)\}$ is a Cauchy sequence in the space $L_2(\Omega)\oplus L_2(\Omega')\oplus L_2(G_T)$, and its limit $\{u(t,x), \bar u(t,x'), u(T,x)\}$ is a strong solution $u(t,x)$ of the equation $\mathfrak{A}(u) = \{f,g,h\}$, because $\mathfrak{A}(u)$ is closed. Therefore $\{f,g,h\}$ belongs to the range of $\mathfrak{A}(u)$ considered on $\mathfrak{D}(\overline{\mathfrak{A}}_1)$.

Condition (B) implies the Gâteaux differentiability of the nonlinear operator $B(u)$ at any point $u \in L_2(\Omega)$, and therefore the differentiability of $\mathfrak{A}(u)$ for any $u \in \mathfrak{D}(\overline{\mathfrak{A}}_1)$, in the sense indicated in §1. The differential $\mathfrak{A}'(u)$ of the operator $\overline{\mathfrak{A}}(u)$ has the form $\mathfrak{A}'(u)v = \{\mathfrak{L}_1 v + B'(u)\, v, Mv, v(0,x)\}$, where the differential $B'(u)$ of the operator $B(u)$ corresponds to the matrix $(\partial b_i/\partial u_k)$.

For a fixed $u \in L_2(\Omega)$ the operator $B'(u)v$ is linear in $v$, and is a bounded operator from $L_2(\Omega)$ into $L_2(\Omega)$. Then it follows from [**9**] that if the indicated conditions are satisfied, for any fixed $u \in \mathfrak{D}(\overline{\mathfrak{A}}_1)$ the linear problem $\mathfrak{A}'(u)v = \{\bar{f}, \bar{g}, \bar{h}\}$ has a solution $v \in \mathfrak{D}(\overline{\mathfrak{A}}_1)$ for any element $\{\bar{f}, \bar{g}, \bar{h}\} \in L_2(\Omega) \oplus L_2'(\Omega') \oplus L_2(G)$.

Then it follows from Theorem 1 that for any element $\{f, g, h\}$ of $L_2(\Omega) \oplus L_2'(\Omega') \oplus L_2(G)$ problem (13)–(15) has a solution which belongs to $\mathfrak{D}(\overline{\mathfrak{A}}_1)$.

The uniqueness of the solution of problem (13)–(15) follows from (16).

Theorem 2 has been proved.

Remark. Theorem 2 contains a theorem on the solvability of mixed problems for hyperbolic (in the sense of Friedrichs) symmetric weakly nonlinear systems (13), since in this case the matrices $A_j(t,x)$ $(j = 1, \cdots, n)$ are symmetric, and it is sufficient to choose $s(t,x,\xi) \equiv I$ for $S(t,x,\xi)$.

## §3. Cauchy problem with periodic boundary conditions in the space $W_2^l$

Let $\Omega = [0, T] \times G$, $G = J_{2\pi}^n$, where $J_{2\pi} = [0, 2\pi] \subset \mathbf{R}^1$ and $T < \infty$. We consider the following Cauchy problem in $\Omega$:

$$\frac{\partial u}{\partial t} - \sum_{j=1}^{n} A_j(t,x) \frac{\partial u}{\partial x_j} + b(t,x,u) = f(t,x), \tag{17}$$

$$u(0,x) = h(x), \tag{18}$$

where $u = u(t,x)$ is a vector-valued function which takes on its values in $\mathbf{R}^m$. Here $A_j(t,x)$ are symmetric matrices of order $m$ whose elements are functions of class $C^{l+1}(\bar{\Omega})$, periodic with period $2\pi$ in all of the space variables $x_1, \cdots, x_n$, and $b(t,x,u)$ is a vector-valued function which takes on its values in $\mathbf{R}^m$ and whose projections $b_i(t,x,u)$ $(i = 1, \cdots, m)$ are functions of class $C^{l+1}(\bar{\Omega} \times \mathbf{R}^m)$, periodic with period $2\pi$ in all $x_1, \cdots, x_n$.

The Cauchy problem (17)–(18) is considered in the Sobolev space $W_2^l(\Omega)$ [**10**] of real vector-valued functions of $(t,x) \in \Omega$ which are periodic with period $2\pi$ in all $x_1, \cdots, x_n$.

Let $W_2^l(G)$ ($l$ is an integer) be the Sobolev space [**10**] of vector-valued functions which are defined on $G$, take on their values in $\mathbf{R}^m$, and are periodic with period $2\pi$ in all $x_1, \cdots, x_n$.

With regard to an integer $l$, we assume that $2l > n + 1$.

*The domain.* Let us consider the following linear Cauchy problem:

$$\frac{\partial u}{\partial t} - \sum_{j=1}^{n} A_j(t, x) \frac{\partial u}{\partial x_j} = f(t, x), \tag{19}$$

$$u(0, x) = h(x), \tag{20}$$

where the matrices $A_j(t,x)$ $(j = 1, \cdots, n)$ were defined above.

The linear operator which corresponds to the Cauchy problem (19)–(20) (under the condition of the periodicity in $x$) acts from the space of sufficiently smooth functions periodic in $x$ defined on the cylinder $\bar{\Omega}$ into the space $W_2^l(\Omega) \oplus W_2^l(G)$. This operator admits a closure with its domain of definition considered in the space $W_2^l(\Omega) \oplus W_2^l(G_T)$, $G_T = T \times G$. Let $\mathfrak{D}$ be the domain of this closure.

Let $B(u)$ be the operator determined by the vector-valued function $b(t,x,u)$, $B(u(t,x)) \equiv b(t,x,u(t,x))$. The operator $B(u)$ is defined on $W_2^l(\Omega)$ with $2l > n+1$, and it is a continuous operator from $W_2^l(\Omega)$ into $W_2^l(\Omega)$. Therefore the operator $\{\partial u/\partial t - \sum_{j=1}^{n} A_j \partial u/\partial x_j + B(u),\ u(0,x)\}$ admits a closure with the domain of definition $\mathfrak{D}$. Henceforth as the operator

$$\left\{\frac{\partial u}{\partial t} - \sum_{j=1}^{n} A_j \frac{\partial u}{\partial x_j} + B(u),\ u(0, x)\right\}$$

we understand this closure.

We now assume that there exists an a priori estimate of solutions of the Cauchy problem (17)–(18), i.e. that for any solution $u$ in $\mathfrak{D}$ of this problem the following inequality holds:

$$\int_{G_T} \sum_{|\alpha|<l} |D_x^{\alpha} u|^2 \, dx + \int_0^T \int_{G_t} \sum_{|\alpha|\leqslant l} |D^{\alpha} u|^2 \, dx dt \leqslant C(r), \tag{21}$$

if $\|f\|_{W_2^l(\Omega)}$ and $\|h\|_{W_2^l(\Omega)} \leqq r$, where $C(r)$ is finite for finite $r \geqq 0$.

Here, $D^{\alpha} = D_0^{\alpha_0} D_1^{\alpha_1} \cdots D_n^{\alpha_n}$, where $D_0 = \partial/\partial t, D_k = \partial/\partial x_k$ $(k = 1, \cdots, n)$, and $D_x^{\alpha} = D_1^{\alpha_1} \cdots D_n^{\alpha_n}$.

THEOREM 3. *Let the conditions indicated in* §3 *be satisfied. Then for any vector-valued functions* $f(t,x)$ *in* $W_2^l(\Omega)$ *and* $h(x)$ *in* $W_2^l(G)$ *there exists a solution* $u(t,x)$ *in* $\mathfrak{D}$ *of the Cauchy problem* (17)–(18). *This solution is unique for fixed* $f(t,x)$ *and* $h(x)$.

A proof of the existence of a solution of the Cauchy problem (17)–(18) consists of several parts. In each of these parts we shall prove that the corresponding assumption of Theorem 1 holds.

1. *Differentiability*. In order to prove the differentiability (in the sense indicated in §1) of the operator which corresponds to the Cauchy problem (17)–(18), it is sufficient to verify that the operator $B(u)$ from $W_2^l(\Omega)$ into $W_2^l(\Omega)$ is differentiable.

The (Fréchet) differentiability of $B(u)$ follows from Sobolev's theorem [10] on embedding the space $W_2^l(\Omega)$ into the corresponding spaces with $2l > n+1$, and from the formula for higher-order derivatives of the superposition of two functions (see [11]). The matrix of the elements $\partial b_i/\partial u_k$ $(i,k = 1,\cdots,m)$ corresponds to the differential $B'(u)$ of $B(u)$. The operator $B'(u)v$, for fixed $u$ in $W_2^l(\Omega)$, is a bounded operator linear in $v$ from $W_2^l(\Omega)$ into $W_2^l(\Omega)$.

2. *Solvability of the linear problem.* Let us consider the following linear Cauchy problem with an arbitrary fixed $u$ in $\mathfrak{D}$:

$$\frac{\partial v}{\partial t} - \sum_{j=1}^{n} A_j \frac{\partial v}{\partial x_j} + B'(u)v = f, \tag{22}$$

$$v(0, x) = h(x). \tag{23}$$

If $v \in \mathfrak{D}$ is a solution of this problem (under the condition of the periodicity in $x$), then the following estimate holds:

$$\|v\|^2_{W_2^l(G_T)} + \|v\|^2_{W_2^l(\Omega)} \leqslant c\,[\,\|f\|^2_{W_2^l(\Omega)} + \|h\|^2_{W_2^l(G)}],$$

where $c$ is a positive number which depends on $\|u\|_{W_2^l(\Omega)}$ and does not depend on $v \in \mathfrak{D}$, $f \in W_2^l(\Omega)$, and $h \in W_2^l(G)$ $(2l > n+1)$. A similar estimate holds also for "negative" norms of the corresponding solutions of the Cauchy problem formally adjoint to the Cauchy problem (22)–(23) on the interval $[0, T]$ (under the condition of the periodicity in $x$). It follows from these estimates and from the conditions indicated in §3 regarding the matrices $A_j(t,x)$ $(j = 1,\cdots,n)$ and the vector-valued functions $b(t,x,u)$ that for any vectors $f$ in $W_2^l(\Omega)$ and $h$ in $W_2^l(G)$ the Cauchy problem (22)–(23) (under the condition of the periodicity in $x$ and for $2l > n+1$) has a solution $v$ in $\mathfrak{D}$, and, moreover, that this solution is unique for fixed $f$ and $h$ (see [12], [13]).

3. *Closedness of the domain.* Let $u_1$ and $u_2$ be two solutions in $\mathfrak{D}$ of the Cauchy problem (17)–(18) which correspond to $f_1$, $h_1$ and $f_2, h_2$. We set $w = u_2 - u_1$ and consider the following function $J(\lambda)$:

$$J(\lambda) = \int_0^t \int_G \sum_{|\alpha| \leqslant l} D^\alpha (B(u_1 + \lambda w) - B(u_1))\, D^\alpha w\, dx\, d\tau$$

for any fixed $t$ in $(0, T]$. Here $B(u) = B(u(t,x)) \equiv b(t,x,u(t,x))$. Let $\Omega_t = [0,t] \times G$, $0 < t \leqq T$.

Since the operator $B(u)$ from $W_2^l(\Omega_t)$ into $W_2^l(\Omega_t)$ is (Fréchet) differentiable for $2l > n+1$, the function $J(\lambda)$ is differentiable and its derivative $J'(\lambda)$ is

$$J'(\lambda) = \int_0^t \int_G \sum_{|\alpha| \leqslant l} D^\alpha \{B'(u_1 + \lambda w)\, w\}\, D^\alpha w\, dx\, d\tau. \tag{24}$$

The differential $B'(\bar{u})w$ is a bounded linear operator from $W_2^l(\Omega_t)$ into

$W_2^l(\Omega_t)$ such that

$$\|B'(\bar{u})\cdot w\|_{W_2^l(\Omega_t)} \leqslant C_0(R)\cdot\|w\|_{W_2^l(\Omega_t)}$$

for $\|\bar{u}\|_{W_2^l(\Omega)} \leqq R$, where $C_0(R)$ is finite if $R \geqq 0$ is finite.

For $\lambda \in [0,1]$ and $\|u_1\|_{W_2^l(\Omega)}, \|u_2\|_{W_2^l(\Omega)} \leqq R$ we have

$$\|u_1+\lambda w\|_{W_2^l(\Omega_t)} \leqslant \|u_1+\lambda w\|_{W_2^l(\Omega)} \leqslant R.$$

It then follows from (24) that

$$|J'(\lambda)| \leqslant C_0(R)\|w\|^2_{W_2^l(\Omega_t)}$$

for $\|u_1\|_{W_2^l(\Omega)}, \|u_2\|_{W_2^l(\Omega)} \leqq R$ and $\lambda \in [0,1]$.

Taking into account that $J(0)=0$, we obtain

$$(25)\quad |J(1)| = \Big|\int_0^t\int_G \sum_{|\alpha|\leqslant l} D^\alpha(B(u_1+w)-B(u_1))D^\alpha w\,dx\,d\tau\Big| \leqslant C_0(R)\|w\|^2_{W_2^l(\Omega_t)}$$

for

$$\|u_1\|_{W_2^l(\Omega)},\ \|u_2\|_{W_2^l(\Omega)} \leqslant R.$$

The vector-valued function $w$ belongs to $\mathfrak{D}$ and is a solution of the following (periodic in $x$) Cauchy problem:

$$(26)\qquad \frac{\partial w}{\partial t} - \sum_{j=1}^n A_j\frac{\partial w}{\partial x_j} + B(u_1+w) - B(u_1) = f_2 - f_1,$$

$$(27)\qquad w(0,x) = h_2(x) - h_1(x).$$

Hence and from inequality (25) it follows that

$$(28)\qquad \begin{aligned} &\int_{G_T}\sum_{|\alpha|\leqslant l}|D_x^\alpha w|^2\,dx + \int_0^T\int_G\sum_{|\alpha|\leqslant l}|D^\alpha w|^2\,dx\,dt \\ &\leqslant C_1(R)\Big[\int_{G_0}\sum_{|\alpha|\leqslant l}|D_x^\alpha w|^2\,dx + \int_0^T\int_G\sum_{|\alpha|\leqslant l}|D^\alpha(f_2-f_1)|^2\,dx\,dt\Big], \end{aligned}$$

if $\|u_1\|_{W_2^l(\Omega)}, \|u_2\|_{W_2^l(\Omega)} \leqq R$, where $C_1(R) < \infty$ for $0 \leqq R < \infty$.

Making use of (26) and the initial condition (27), we obtain

$$\begin{aligned} \int_{G_0}\sum_{|\alpha|\leqslant l}|D_x^\alpha w|^2\,dx \leqslant C_2(R)\Big[&\int_{G_0}\sum_{|\alpha|\leqslant l}|D_x^\alpha(h_2-h_1)|^2\,dx \\ &+\int_0^T\int_G\sum_{|\alpha|\leqslant l}|D^\alpha(f_2-f_1)|^2\,dx\,dt\Big] \end{aligned}$$

for $\|u_1\|_{W_2^l(\Omega)}, \|u_2\|_{W_2^l(G)} \leqq R$, where $C_2(R) < \infty$ for $0 \leqq R < \infty$.

It then follows from (28) that

$$\int\limits_{G_T}\sum_{|\alpha|\leqslant l}|D_x^\alpha w|^2\,dx+\int\limits_0^T\int\limits_G\sum_{|\alpha|\leqslant l}|D^\alpha w|^2\,dxdt \tag{29}$$
$$\leqslant C_3(R)\Big[\int\limits_{G_0}\sum_{|\alpha|\leqslant l}|D_x^\alpha(h_2-h_1)|^2\,dx+\int\limits_0^T\int\limits_G\sum_{|\alpha|\leqslant l}|D^\alpha(f_2-f_1)|^2\,dxdt\Big]$$

for $\|u_1\|_{W_2^l(\Omega)}$, $\|u_2\|_{W_2^l(\Omega)}\leqq R$, where $C_3(R)$ is a positive number, finite for finite $R\geqq 0$. Now let there exist a sequence of solutions $u_n\in\mathfrak{D}$ $(n=1,2,\cdots)$ of the Cauchy problem (17)–(18) which correspond to $f_n$ and $h_n$ $(n=1,2,\cdots)$. Let $f_n\to f$ strongly in $W_2^l(\Omega)$ and $h_n\to h$ strongly in $W_2^l(G)$ as $n\to\infty$. It then follows from (21) that there exists a constant $R>0$ such that

$$\int\limits_{G_T}\sum_{|\alpha|\leqslant l}|D_x^\alpha u_n|^2\,dx+\int\limits_0^T\int\limits_G\sum_{|\alpha|\leqslant l}|D^\alpha u_n|^2\,dxdt\leqslant R,\quad (n=1,2,\ldots).$$

It then follows from (29) that

$$\int\limits_{G_T}\sum_{|\alpha|\leqslant l}|D_x^\alpha(u_n-u_m)|^2\,dx+\int\limits_0^T\int\limits_G\sum_{|\alpha|\leqslant l}|D^\alpha(u_n-u_m)|^2\,dxdt$$
$$\leqslant C_3(R)\Big[\int\limits_{G_0}\sum_{|\alpha|\leqslant l}|D_x^\alpha(h_n-h_m)|^2\,dx+\int\limits_0^T\int\limits_G\sum_{|\alpha|\leqslant l}|D^\alpha(f_n-f_m)|^2\,dxdt\Big].$$

Thus the sequence $\{u_n(x,t),u_n(x,T)\}$ converges strongly in the space $W_2^l(\Omega)\oplus W_2^l(G_T)$, the square of whose norm is equal to the sum of the squares of the corresponding norms. The operator $B(u)$ from $W_2^l(\Omega)$ into $W_2^l(\Omega)$ $(2l>n+1)$ is continuous; therefore $B(u_n)\to B(u)$ strongly in $W_2^l(\Omega)$ as $n\to\infty$, where $u$ in $W_2^l(\Omega)$ is the limit of the sequence $\{u_n\}$.

We then have

$$\frac{\partial u_n}{\partial t}-\sum_{j=1}^n A_j\frac{\partial u_n}{\partial x_j}=F_n, \tag{30}$$

$$u_n(0,x)=h_n, \tag{31}$$

where $F_n=(-B(u_n)+f_n)\to(-B(u)+f)$ strongly in $W_2^l(\Omega)$, $h_n\to h$ strongly in $W_2^l(\Omega)$ as $n\to\infty$, and $u_n\in\mathfrak{D}$. Since the linear operator which corresponds to the linear Cauchy problem (30)–(31) is closed, there exists a solution $u$ in $\mathfrak{D}$ of the following Cauchy problem:

$$\frac{\partial u}{\partial t}-\sum_{j=1}^n A_j\frac{\partial u}{\partial x_j}+B(u)=f,$$
$$u(0,x)=h(x).$$

We remind the reader that a solution of the Cauchy problem (17)–(18) is understood to be a vector-valued function $u$ in the domain of the operator which is the closure of the linear operator corresponding to the linear Cauchy problem (19)–(20) (under the condition of the periodicity in $x$).

Thus we have proved that the domain of the closed nonlinear operator which corresponds to the Cauchy problem (17)–(18) is closed.

To complete the proof of Theorem 3 we now use Theorem 1. The uniqueness of the solution $u$ in $\mathfrak{L}$ of the Cauchy problem (17)–(18) (under the condition of the periodicity in $x$) follows from (29). Theorem 3 has been proved.

In conclusion we present an example of the Cauchy problem (17)–(18) such that there exists an a priori estimate (21) for its solutions.

EXAMPLE. We consider the Cauchy problem (17)–(18) subject to the condition of the periodicity in $x$ for $n = 2$ and $l = 2$.

We assume that all of the conditions which were indicated in §3 are satisfied, with the exception of the condition of the existence of the a priori estimate (21).

Assume that the vector-valued function $b(t,x,u)$ does not depend on the variables $t$ and $x$, i.e. $b = b(u)$, and that it satisfies the following inequalities:

$$ub(u) \geqslant -c_1|u|^2 - c_0, \tag{32}$$

$$v\cdot b'_u(u)v \geqslant -c_2|v|^2, \tag{33}$$

$$|w\cdot(b''_u(u)v_1)\cdot v| \leqslant c_3|v|\cdot|v_1|\cdot|w| \tag{34}$$

for any vectors $u, v, v_1$ and $w$ in $\mathbf{R}^n$, where $c_0, c_1, c_2$ and $c_3$ are positive constants. We note that (32) follows from (33).

*Derivation of the a priori estimate.* Let $u(t,x)$ be a sufficiently smooth periodic in $x$ function defined on $\overline{\Omega}$. We set $f = L(u)$, where

$$L(u) \equiv \frac{\partial u}{\partial t} - \sum_{j=1}^{2} A_j(t,x)\frac{\partial u}{\partial x_j} + b(u),$$

and $h(x) = u(0,x)$.

Using the condition of the periodicity in $x$ and the symmetry of the matrices $A_j(t,x)$ $(j = 1,2)$, which belong to class $C^3(\overline{\Omega})$, we obtain, integrating by parts,

$$\begin{aligned}
&\frac{1}{2}\int_{G_t}|D^\alpha u|^2\,dx - \frac{1}{2}\int_{G_0}|D^\alpha u|^2\,dx \\
&+\frac{1}{2}\int_0^t\int_G\sum_{j=1}^{2} D^\alpha u\cdot\frac{\partial A_j}{\partial x_j}\cdot D^\alpha u\,dx\,d\tau \\
&-\int_0^t\int_G\sum_{j=1}^{2}\sum_{\beta\leqslant\alpha,|\beta|\geqslant 1}\frac{\alpha!}{\beta!\,(\alpha-\beta)!}D^\alpha u\cdot D^\beta A_j\cdot D^{\alpha-\beta}\frac{\partial u}{\partial x_j}\,dx\,d\tau \\
&+\int_0^t\int_G D^\alpha u\cdot D^\alpha b(u)\,dx\,d\tau = \int_0^t\int_G D^\alpha u\cdot D^\alpha f\,dx\,d\tau,
\end{aligned}\tag{35}$$

where the inequality $\beta \leqq \alpha$ means $\beta_i \leqq \alpha_i$ for $i = 0, 1, 2$ and

$$(\alpha - \beta) = (\alpha_0 - \beta_0, \ \alpha_1 - \beta_1, \ \alpha_2 - \beta_2), \ |\alpha| \leqslant 2.$$

Applying this relation in the case

$$D^\alpha = D_x^\alpha \equiv D_1^{\alpha_1} D_2^{\alpha_2} \left( D_k = \frac{\partial}{\partial x_k} \quad \text{for} \quad k = 1, 2 \right)$$

and using inequalities (32) and (33), for $t \in [0, T]$ we obtain

$$\int_{G_t} \sum_{|\alpha| \leqslant 1} |D_x^\alpha u|^2 \, dx \leqslant k_1 \Big[ \int_{G_0} \sum_{|\alpha| \leqslant 1} |D_x^\alpha h|^2 \, dx + \int_0^t \int_{G_\tau} \sum_{|\alpha| \leqslant 1} |D_x^\alpha f|^2 \, dx \, d\tau + (2\pi)^2 \cdot T \cdot c_0, \tag{36}$$

where $k_1$ is a positive number which depends only on the matrices $A_j(t, x)$ $(j = 1, 2)$ of class $C^3(\overline{\Omega})$ with $\overline{\Omega} = [0, T] \times J_{2\pi}^2$, where $J_{2\pi} = [0, T]$, on the constants $c_1$ and $c_2$ from (32) and (33), and on $T$.

From (35) with $D^\alpha = D_x^\alpha$ and $|\alpha| = 2$, taking into account inequalities (32)-(34), we obtain

$$\int_{G_t} \sum_{|\alpha|=2} |D_x^\alpha u|^2 \, dx \leqslant \int_{G_0} \sum_{|\alpha|=2} |D_x^\alpha h|^2 \, dx + \int_0^t \int_{G_\tau} \sum_{|\alpha|=2} |D_x^\alpha f|^2 \, dx \, d\tau + k_2 \int_0^t \int_{G_\tau} \sum_{|\alpha| \leqslant 2} |D_x^\alpha u|^2 \, dx \, d\tau + k_3 \int_0^t \int_{G_\tau} \sum_{|\nu|=1} |D_x^\nu u|^4 \, dx \, d\tau, \tag{37}$$

where $k_2$ and $k_3$ are positive numbers which does not depend on $u$ in $C^2(\overline{\Omega})$ and $t$ in $[0, T]$.

To estimate the integral $\int_{G_t} \sum_{|\nu|=1} |D_x^\nu u|^4 dx$ with $n = 2$ we use the Gagliardo-Nirenberg inequality [**14**], applying it to the functions $v = D_x u$. Then we obtain

$$\int_{G_t} \sum_{|\nu|=1} |D_x^\nu u|^4 \, dx \leqslant k_4 \int_{G_t} \sum_{|\alpha| \leqslant 2} |D_x^\alpha u|^2 \, dx \cdot \int_{G_t} \sum_{|\beta| \leqslant 1} |D_x^\beta u|^2 \, dx, \tag{38}$$

where the positive number $k_4$ does not depend on $u(t, \cdot) \in W_2^2(G_t)$ and $t \in [0, T]$.

By virtue of (36) the integral $\int_{G_t} \sum_{|\beta| \leq 1} |D_x^\beta u|^2 dx$ is bounded for any $t$ in $[0, T]$.

Integrating (38), we obtain

$$\int_0^t \int_{G_\tau} \sum_{|\nu|=1} |D_x^\nu u|^4 \, dx \, d\tau \leqslant k_5 \Big[ \int_{G_0} \sum_{|\alpha| \leqslant 1} |D_x^\alpha h|^2 \, dx + \int_0^t \int_{G_\tau} \sum_{|\alpha| \leqslant 1} |D_x^\alpha f|^2 \, dx \, d\tau + (2\pi)^2 \cdot T \cdot c_0 \Big] \int_0^t \int_{G_\tau} \sum_{|\alpha| \leqslant 2} |D_x^\alpha u|^2 \, dx \, d\tau, \tag{39}$$

where $k_5 = k_4 k_1$.

From (36), (37) and (39) it follows that

$$(40)\quad \int_{G_t} \sum_{|\alpha|\leqslant 2} |D_x^\alpha u|^2\,dx + \int_0^t \int_{G_\tau} \sum_{|\alpha|\leqslant 2} |D_x^\alpha u|^2\,dx\,d\tau \leqslant \Phi\left(\|f\|_{W_2^2(\Omega)} + \|h\|_{W_2^2(G)}\right)$$

for $0 < t \leqq T$, where $\phi(\rho)$ $(\rho \geqq 0)$ is a real-valued positive function, finite for finite values of its argument $\rho \geqq 0$.

In order to estimate the integral $\int_0^T \int_{G_t} \sum_{|\alpha|\leqq 2} |D^\alpha u|^2 dxdt$ in the case $D^\alpha = D_0^{\alpha_0} D_1^{\alpha_1} D_2^{\alpha_2}$ with $D_0 = \partial/\partial t$, we should use the equation $L(u) = f$. Differentiating this equation successively with respect to $t$ and using Sobolev's embedding theorems [**10**], we estimate the integral $\int_{G_t} \sum_{|\alpha|\leqq 2} |D^\alpha u|^2 dx$ by the integral $\int_{G_t} \sum_{|\alpha|\leqq 2} |D_x^\alpha u|^2 dx$ and the norm $\|f\|_{W_2^2(\Omega)}$. We then obtain from inequality (40)

$$(41)\quad \int_{G_T} \sum_{|\alpha|\leqslant 2} |D_x^\alpha u|^2\,dx + \int_0^T \int_{G_t} \sum_{|\alpha|\leqslant 2} |D^\alpha u|^2\,dx\,dt \leqslant \Phi_1\left(\|f\|_{W_2^2(\Omega)} + \|h\|_{W_2^2(G)}\right),$$

where $\Phi_1(\rho)$ $(\rho \geqq 0)$ is a real-valued positive function, finite for finite values of its argument $\rho \geqq 0$.

Now let $u$ in $\mathfrak{D}$ be the solution of the Cauchy problem (17)–(18) which corresponds to $f$ in $W_2^2(\Omega)$ and $h$ in $W_2^2(G)$. Then, making use of the estimate just obtained and passing to the limit with a sequence of sufficiently smooth, periodic in $x$, vector-valued functions $u_n(t,x)$ such that

$$\|u_n - u\|_{W_2^2(G_T)} + \|u_n - u\|_{W_2^2(\Omega)} + \|L(u_n) - f\|_{W_2^2(\Omega)} + \|u_n - h\|_{W_2^2(G_0)} \to 0$$

as $n \to \infty$, we obtain that

$$\int_{G_T} \sum_{|\alpha|\leqslant 2} |D_x^\alpha u|^2\,dx + \int_0^T \int_{G_t} \sum_{|\alpha|\leqslant 2} |D^\alpha u|^2\,dx\,dt \leqslant C(r),$$

if $\|f\|_{W_2^2(\Omega)}, \|h\|_{W_2^2(G)} \leqq r$, where $C(r)$ is finite for finite $r \geqq 0$.

Thus the condition of the existence of a priori estimate (21) holds in this example.

REMARK. In the general case the a priori estimate (21) follows from an a priori estimate of $\max_{\bar{\Omega}} |u|$, where $u$ is the solution of the Cauchy problem (17) – (18) (subject to the condition of the periodicity in $x$). To prove this fact, one should use Moser's lemma [**11**] on estimating higher-order derivatives of superpositions of smooth functions and Sobolev's embedding theorems.

Received 2 DEC 1968

## BIBLIOGRAPHY

[1] M. Edelstein, *On nearest points of sets in uniformly convex Banach spaces*, J. London Math. Soc. **43** (1968), 375–377. MR **17**, 1213.

[2] A. A. Dezin, *Mixed problems for certain symmetric hyperbolic systems*, Dokl. Akad. Nauk SSSR **107** (1956), 13–16. (Russian) MR **17**, 1213.

[3] K. O. Friedrichs, *Symmetric positive linear differential equations*, Comm. Pure Appl. Math. **11** (1958), 333–418. MR **20** # 7147.

[4] A. A. Dezin, *Boundary problems for some first-order symmetric systems*, Mat. Sb. **49** (**91**) (1959), 459–484. (Russian)

[5] P. D. Lax and R. S. Phillips, *Local boundary conditions for dissipative symmetric linear differential operators*, Comm. Pure Appl. Math. **13** (1960), 427–455. MR **22** # 9718.

[6] L. Sarason, *On weak and strong solutions of boundary value problems*, Comm. Pure Appl. Math. **15** (1962), 237–288. MR **27** # 460.

[7] K. O. Friedrichs and P. D. Lax, *Boundary value problems for first order operators*, Comm. Pure Appl. Math. **18** (1965), 355–388. MR **30** # 5186.

[8] R. S. Phillips and L. Sarason, *Singular symmetric positive first order differential operators*, J. Math. Mech. **15** (1966), 235–271. MR **32** # 4357.

[9] M. S. Agranovič, *Positive boundary-value problems for some systems of first order*, Trudy Moskov. Mat. Obšč. **16** (1967), 3–24 = Trans. Moscow Math. Soc. **1967**, 1–26. MR **37** # 1791.

[10] S. L. Sobolev, *Applications of functional analysis in mathematical physics*, Sibirsk. Otd. Akad. Nauk SSSR, Novosibirsk, 1962; English transl., of 1st ed., Transl. Math. Monographs, vol. 7, Amer. Math. Soc., Providence, R. I., 1963. MR **29** # 2624.

[11] J. Moser, *A rapidly convergent iteration method and non-linear partial differential equations*. I, Ann. Scuola Norm. Sup. Pisa (3) **20** (1966), 265–315. MR **33** # 7667.

[12] P. D. Lax, *On Cauchy's problem for hyperbolic equations and the differentiability of solutions of elliptic equations*, Comm. Pure Appl. Math. **8** (1955), 615-633. MR **17**, 1212.

[13] M. Nagumo, *Lectures on the contemporary theory of partial differential equations*, Kyōritsu Shuppan, Tokyo, 1957 (Japanese); Russian transl., "Mir", Moscow, 1967. MR **38** # 6205.

[14] L. Nirenberg, *On elliptic partial differential equations*, Ann. Scuola Norm. Sup. Pisa (3) **13** (1959), 115–162. MR **22** # 823.

Translated by:
K Makowski

Trudy Moskov. Mat. Obšč.
Tom 23 (1970)

Trans. Moscow Math. Soc.
Vol. 23 (1970)

# ASYMPTOTICS OF SOLUTIONS OF BOUNDARY VALUE PROBLEMS FOR A LINEAR ELLIPTIC EQUATION OF SECOND ORDER WITH COEFFICIENTS WHICH HAVE A "SPLASH"

*A. S. DEMIDOV*

## Contents



## Introduction[1)]

In the present article we investigate boundary value problems in a bounded domain $G \subset R^n$ for an elliptic equation of second order

$$Lu \equiv \sum_{i,j=1}^{n} a_{ij} \frac{\partial^2 u}{\partial x_i \, \partial x_j} + \sum_{k=1}^{n} b_k \frac{\partial u}{\partial x_k} + cu = f, \tag{1}$$

the smallest cofficients and right side of which have a "splash" along some $(n-1)$-dimensional surface $\gamma$ lying within $G$. (The surface $\gamma$ is sufficiently smooth, possibly with a boundary.) Let us call an $\epsilon$-neighborhood of $\gamma$ a "splash base", which we will designate by $T_\epsilon$. We now represent "splash height" in the following way ($x \in T_\epsilon$):

$$\begin{aligned} b_k(x) &= \frac{\beta_0^k(x)}{\varepsilon^{\sigma_1}} + \beta_1^k(x), \\ c(x) &= \frac{\gamma_0(x)}{\varepsilon^{2\sigma_2}} + \gamma_1(x), \\ f(x) &= \frac{\eta_0(x)}{\varepsilon^{2\sigma_3}} + \eta_1(x), \end{aligned} \tag{1'}$$

where the $\sigma_i$ are rational numbers,[2)] $\sigma = \max(\sigma_1, \sigma_2, \sigma_3) \leqq 1$, and $\beta_1^k$, $\gamma_1$ and $\eta_1$ are continuations (for example, smooth and without "splashes") of the coefficients of the operator $L | G \setminus T_\epsilon$. In this article the asymptotic behavior is given for a solution of problem (1)–(1′) as $\epsilon \to 0$ (relative to powers of $\epsilon^{1/Q}$, where $Q$ is a multiple of the denominators of $\sigma_i$, $i = 1, 2, 3$).

---

*AMS (MOS) subject classifications* (1970). 35J25.

1) Here we are speaking about the principal features of the problem; an exact formulation is given in §1.

2) For the case of irrational $\sigma_i$ see Remark 2.7.

The case $\sigma_1 = \lambda\sigma_2 = \mu\sigma_3 = 1$, where $\lambda$, $\mu = 1, 2$, was investigated by M. I. Višik and L. A. Ljusternik in [**1**]. By means of methods developed in [**1**] and [**2**] we will analyze here the case $\sigma < 1$ and, in a more general form than in [**1**], the case $\sigma = 1$.

By working through the following example we can get a representation of the influence of $\sigma$ on the initial term of the asymptotics.

**Introductory example.** Consider the problem

$$\frac{d^2u_\varepsilon}{dv^2} - c_\varepsilon(v)\, u_\varepsilon = 0, \quad -1 < v < 1, \tag{2}$$
$$u_\varepsilon(-1) = 1, \quad u_\varepsilon(1) = 2,$$

where $c_\epsilon(\nu) = (a/\epsilon^\sigma)^2$ for $|\nu| < \epsilon$ and is zero for $|\nu| > \epsilon$ (Figure 1). A graph of the solution of (2) is given in Figure 2 (for the case that $a$ is real).

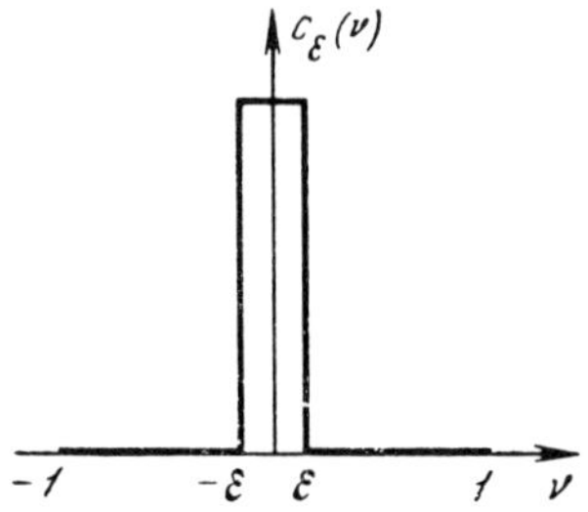


FIGURE 1

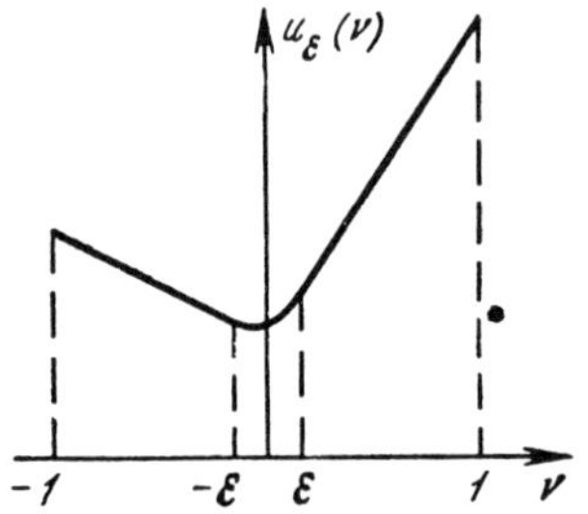


FIGURE 2

Now let $\epsilon$ go to zero. By an easy calculation we can show that $u_0 = \lim_{\epsilon\to 0} u_\epsilon$ is a solution of the following problem (on the interval $[-1, 1]$):

a) for $0 < \sigma < \frac{1}{2}$

$$\frac{d^2u_0}{dv^2} = 0, \quad v \neq 0,$$
$$u_0(+0) - u_0(-0) = 0,$$
$$u_0'(+0) - u_0'(-0) = 0,$$
$$u_0(-1) = 1, \; u_0(1) = 2,$$

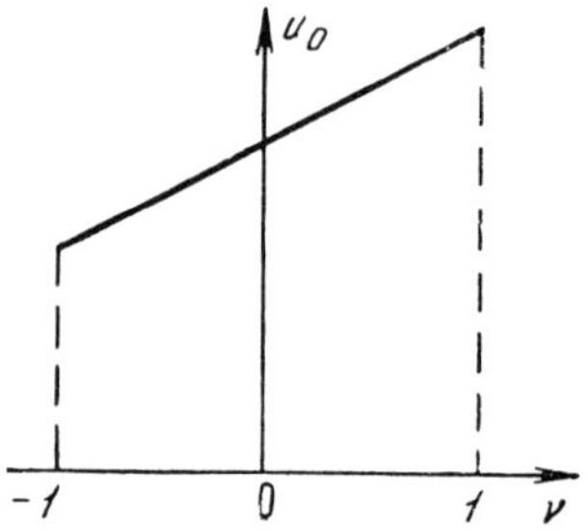


FIGURE 3a

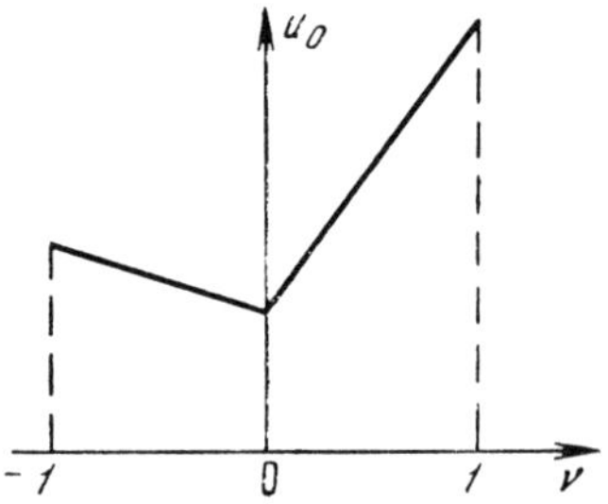


FIGURE 3b

i.e. the "splash" $c_\epsilon(\nu)$ has no influence on the limiting solutions of (2) (see Figure 3a);

b) for $\sigma = \frac{1}{2}$

$$\frac{d^2u_0}{d\nu^2} = 0, \quad \nu \neq 0,$$

$$u_0(+0) - u_0(-0) = 0,$$

$$u_0'(+0) - u_0'(-0) - 2a^2u_0(0) = 0,$$

$$u_0(-1) = 1, \quad u_0(1) = 2,$$

i.e. $u_0(\nu)$ (see Figure 3b for the case $a^2 > 0$) is a continuous function which is a solution of the following problem for an equation with a generalized coefficient:

$$\frac{d^2u_0}{d\nu^2} - 2a^2\,\delta u_0 = 0,$$

$$u_0(-1) = 1, \quad u_0(1) = 2,$$

where $\delta$ is Dirac's $\delta$-function.[3)]

This is a natural result because $c_\epsilon(\nu) \to 2a^2\delta$ (in the topology $D'$) for $\sigma = \frac{1}{2}$ and the fundamental solution of the operator $d^2/d\nu^2$ is a continuous function at zero. In the case of a discontinuous fundamental solution the analogous result does not hold; for example, the only solution of $du/d\nu + \delta u = 0$ is identically zero, and the limiting solution of $du_\epsilon/d\nu + \delta_\epsilon u_\epsilon = 0$ ($\delta_\epsilon \to \delta$ in $D'$) is $C\exp(-\theta)$, where $\theta = \theta(\nu)$ is the Heaviside function (characteristic function of the halfaxis $\nu > 0$).

We also observe that $c_\epsilon(\nu) \to 0$ in $D'$ for $\sigma < \frac{1}{2}$, so that the result in the previous case is also natural. However, a singularity which is weaker than the $\delta$-function is not always "erased" in the limit. In fact, we consider the following problem (with an operator having a discontinuous fundamental solution):

$$\frac{\partial^2 u_\varepsilon}{\partial x_1^2} + \frac{\partial^2 u_\varepsilon}{\partial x_2^2} + \frac{\partial^2 u_\varepsilon}{\partial x_3^2} - c_\varepsilon(\rho)\, u_\varepsilon = 0,$$

$$u_\varepsilon \big|_{\rho=1} = 1,$$

where

$$\rho = \sqrt{x_1^2 + x_2^2 + x_3^2}, \quad c_\varepsilon(\rho) = \left(\frac{a}{\varepsilon}\right)^2 \text{ for } |\rho| < \varepsilon \text{ and } c_\varepsilon(\rho) = 0 \text{ for } |\rho| > \varepsilon.$$

---

[3)] The equation $d^2u_0/d\nu^2 - 2a^2\delta u_0 = 0$ is understood in the generalized sense, i.e. we are seeking a function (continuous at zero) satisfying the condition

$$\int_{-1}^{1} u_0 \frac{d^2\varphi}{d\nu^2}\, d\nu - 2a^2 u_0(0)\, \varphi(0) = 0$$

for any $\phi \in C_0^\infty(-1, 1)$.

We show that for $u_0 = \lim_{\epsilon\to 0} u_\epsilon = 1/\rho$ for $a = i(k\pi + \pi/2)$ $(k = 0, \pm 1, \pm 2, \cdots)$ (at the same time as $c_\epsilon(\rho) \to 0$ in $D'$). This example corresponds to problems with "splashes" along an $(n-3)$-dimensional surface $\gamma$.

We now investigate the case in which the singularity becomes "stronger" than the $\delta$-function in the limit, i.e. when $\sigma > \frac{1}{2}$.

c) $\frac{1}{2} < \sigma < 1$; in this case $u_0$ is a solution of the problem

$$\frac{d^2 u_0}{d\nu^2} = 0, \quad \nu \neq 0,$$
$$u_0(+0) = u_0(-0) = 0,$$
$$u_0(-1) = 1, \quad u_0(+1) = 2,$$

i.e. $c_\epsilon(\nu)$ splits the limiting problem into two boundary value problems without "splashes" and "drives" $u_0$ to zero at the point $\nu = 0$ (see Figure 3c).

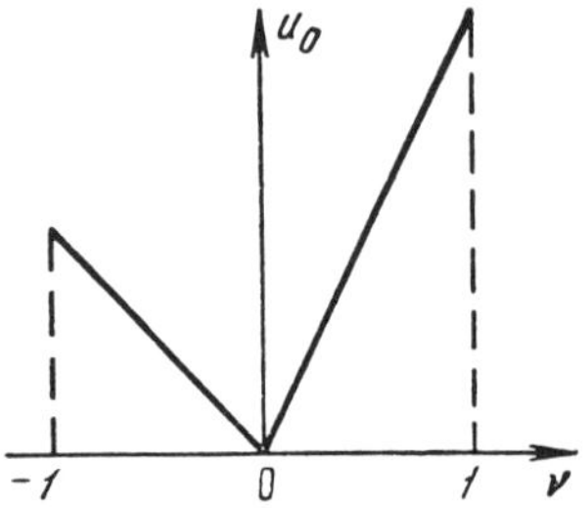


FIGURE 3c

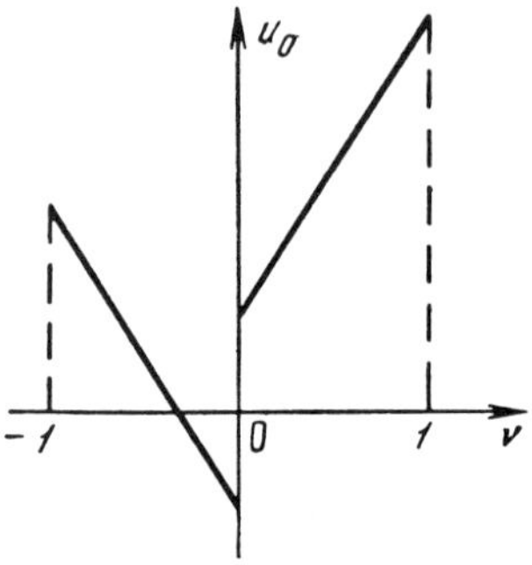


FIGURE 3d′

d) $\sigma = 1$; here there are two subcases:

$\alpha^\circ$. $a \neq ik\pi/2$; then $u_0$ is a solution of the problem of the previous case;

$\beta^\circ$. $a = ik\pi/2$; then $u_0$ is a solution of the following problem:

$$\frac{du_0}{d\nu^2} = 0, \quad \nu \neq 0;$$
$$u_0(+0) - (-1)^k u_0(-0) = 0;$$
$$u_0'(+0) - (-1)^k u_0'(-0) = 0;$$
$$u_0(-1) = 1, \quad u_0(1) = 2;$$

i.e. when $k$ is even the singularity vanishes in the limit (Figure 3a), and when $k$ is odd the limiting solution is even discontinuous (see Figure 3d′).

e) $\sigma > 1$, $a^2 > 0$; in this case $u_0$ is a solution of the problem in c) (see Figure 3c); $\lim_{\epsilon\to 0} u_\epsilon$ does not exist for $a^2 < 0$ (here, as above, we have pointwise convergence in mind).

We will now make two further observations. First, from the graphs of the solutions of the limiting problems it is apparent that for $\sigma = \frac{1}{2}$, $u_0(\nu)$ approaches a solution of the limiting problem in the best way. This is clearly an important property of an approximate solution of problems with "splash" type coefficients. However, in this article we do not take up this question. The second observation relates to the simulation of boundary value problems with certain conjugation conditions on surfaces lying inside the domain (for example, on surfaces of discontinuity of the coefficients). The corresponding result, which follows from Corollary 2.2, shows that these problems can be treated as the limit of boundary value problems with coefficients having a "splash". Simulation by the small parameter method is sufficiently rapid; see, for example, [**1**], [**3**], [**4**].

In §1 we give an exact formulation of the problem, and we also formulate some results required below on the solvability of certain problems. In §2 it is shown that the description resulting from the introductory example is essentially valid in the general case.

This article was written under the guidance of M. I. Višik, to whom the author is sincerely grateful for his frequent consultations and constant attention.

The author also wishes to thank Ju. A. Dubinskiĭ for a series of observations which helped to improve the article.

## §1. Formulation of the problem. Theorems on solvability

Let $G$ be a bounded domain with regular[4)] boundary $\Gamma$, and let $\gamma$ be a sufficiently smooth $(n-1)$-dimensional surface lying inside $G$. Without loss of generality (see Remark 1.1) we will assume that $\gamma$ is homeomorphic to a sphere and consequently splits $G$ into two subdomains $G^-$ and $G^+$ (Figure 4).

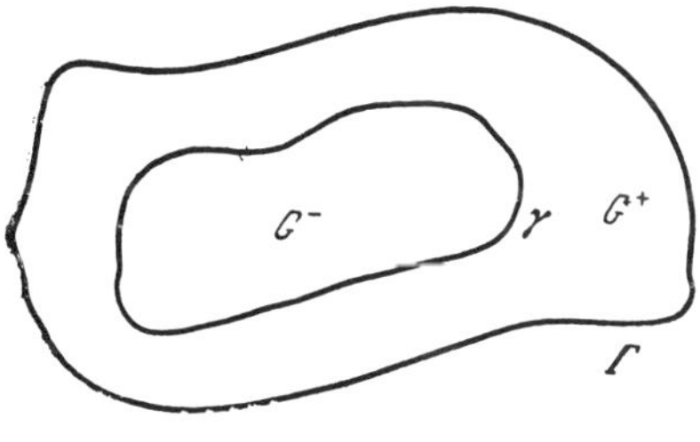


Figure 4

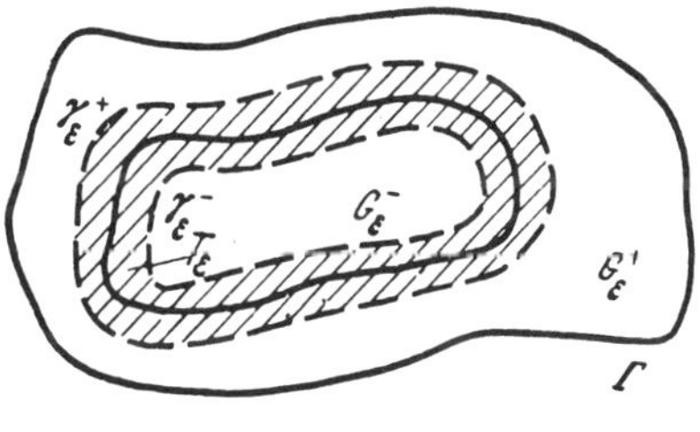


FIGURE 5

Into a neighborhood of $\gamma$ we introduce the local coordinates $(\nu;\tau)=(\nu;\tau_2,\cdots,\tau_n)$ such that $\gamma=\{(\nu;\tau):\nu=0\}$; let $\nu>0$ correspond to the points of $G^+$ and $\nu<0$ to the points of $G^-$.

Set $T_\epsilon=\{(\nu;\tau):|\nu|<\epsilon\}$, $\gamma_\epsilon=\{|\nu|=\epsilon\}$ and $G_\epsilon=G_\epsilon^-\cup G_\epsilon^+$ (Figure 5).

Suppose that in $G\backslash\gamma$ we have given the elliptic operator

4) This means that each point $a$ of the boundary $\Gamma$ has a tangent plane and a sphere which lies in $G$ and the boundary of which contains this point $a$ (see [**5**], and also [**6**])).

$$L(G\setminus\gamma)=\sum a_{ij}\frac{\partial^2}{\partial x_i\partial x_j}+\sum b_k\frac{\partial}{\partial x_k}+c, \tag{1.1}$$

the coefficients of which are sufficiently smooth in $G\setminus\gamma$ up to $\gamma$ (a certain minimal smoothness is necessary for the solvability of the problem; for the construction of asymptotics we may require a higher degree of smoothness in $T_\epsilon\setminus\gamma$). Being elliptic means that, in $\bar{G}^{\pm}$ $\sum a_{ij}\xi_i\xi_j>0$ when $\xi_1^2+\cdots+\xi_n^2>0$.

Let there be another elliptic operator in $T_\epsilon\setminus\gamma$ given (for $\epsilon>0$) in terms of local coordinates $(\nu;\tau)=(\nu;\tau_2,\cdots,\tau_n)$:

$$L_\varepsilon(T_\varepsilon\setminus\gamma)=\sum_{|\alpha|\leqslant 2}\frac{1}{\delta^{(2-|\alpha|)l}}\,a_\alpha^\varepsilon(\nu;\tau)\,\partial^\alpha, \tag{1.2}$$

where

$$\alpha=(\alpha_1,\ldots,\alpha_n);\quad |\alpha|=\alpha_1+\ldots+\alpha_n;\quad \partial^\alpha=\frac{\partial^{|\alpha|}}{\partial\nu^{\alpha_1}\partial\tau_2^{\alpha_2}\ldots\partial\tau_n^{\alpha_n}};$$

$$\delta=\varepsilon^{\frac{1}{2k}};\quad k={}^1/_2,\,1,\,2,\,3,\,\ldots,\,l=1,\,2,\,3,\,\ldots,$$

and the coefficients $a_\alpha^\epsilon(\nu;\tau)$ allow an asymptotic expansion

$$a_\alpha^\varepsilon(\nu;\tau)=\overset{0}{a}_\alpha(\nu;\tau)+\delta\overset{1}{a}_\alpha(\nu;\tau)+\ldots+\delta^M\overset{M}{a}_\alpha(\nu;\tau)+\delta^{M+1}\overset{M+1}{a}_{\alpha;\varepsilon}(\nu;\tau), \tag{1.3}$$

and the $\overset{j}{a}_\alpha(\nu;\tau)$ are sufficiently smooth in $T_\epsilon\setminus\gamma$ up to $\gamma$ (this is necessary for solvability and for asymptotics).

REMARK 1.1. If the surface $\gamma$ has a boundary, we extend $\gamma$ smoothly up to the boundaryless surface $\gamma\cup\gamma'$; in this case in $T'_\omega=\{(\nu;\tau):|\nu|<\epsilon,\ \tau\in\gamma'\}$ the coefficients $a_\alpha^\epsilon(\nu;\tau)$ will be equal to the corresponding coefficients of the operator $L(G\setminus\gamma)$ (in particular $\overset{j}{a}_\alpha(\nu;\tau)=0$ in $T'_\epsilon$ for $j\neq(2-|\alpha|)l$).

REMARK 1.2. The case of a finite number of nonintersecting surfaces homeomorphic to spheres is completely analogous to the case of one such surface; it is however more cumbersome to put in writing.

We now introduce the operator

$$L_\varepsilon=\begin{cases}L(G\setminus\gamma)\text{ in }G_\varepsilon,\\ L_\varepsilon(T_\varepsilon\setminus\gamma)\text{ in }T_\varepsilon\setminus\gamma.\end{cases} \tag{1.4}$$

REMARK 1.3. The coefficients of the operator $L_\epsilon$ necessarily permit a discontinuity on $\gamma_\epsilon$ (see Figure 5); we investigate the discontinuity of coefficients on $\gamma$ for completeness (for example, for the case of a one-sided neighborhood of $\gamma$; see Example 2.1).

EXAMPLE 1.1.

$$L_\varepsilon=\begin{cases}\displaystyle\sum\frac{\partial^2}{\partial x_i^2}\text{ in }G_\varepsilon,\\ \displaystyle\frac{\partial^2}{\partial\nu^2}+\frac{\partial^2}{\partial\tau_2^2}+\ldots+\frac{\partial^2}{\partial\tau_n^2}+\frac{c}{\varepsilon^{3/2}}\text{ in }T_\varepsilon;\end{cases}$$

here $k=2$, $l=3$ (see 1.2).

Let there be given in $G\backslash\gamma$ a sufficiently smooth function $f(x)$ and in $T_\epsilon\backslash\gamma$ the function $h_\epsilon(\nu;\tau)$, which allows an asymptotic expansion in $\delta=\epsilon^{1/2k}$ $(k=\frac{1}{2},1,2,3,\cdots)$:

$$(1.5)\qquad h_\varepsilon(\nu;\tau)=h_0(\nu;\tau)+\delta h_1(\nu;\tau)+\ldots+\delta^M h_M(\nu;\tau)+\delta^{M+1}h_{M+1}^\varepsilon(\nu;\tau),$$

where $h_j(\nu;\tau)$ are sufficiently smooth functions up to $\gamma$. Now set

$$(1.6)\qquad f_\varepsilon=\begin{cases} f(x) & \text{in } G_\varepsilon,\\ \dfrac{1}{\delta^{2l}}h_\varepsilon(\nu;\tau) & \text{in } T_\varepsilon\backslash\gamma.\end{cases}$$

Finally we come to the problem which is the main concern of this article. Consider the following elliptic problem in $G$:

$$(1.7)\qquad \begin{aligned} &L_\varepsilon u_\varepsilon\equiv\sum a_\varepsilon^{ij}\frac{\partial^2 u_\varepsilon}{\partial x_i\partial x_j}+\sum b_\varepsilon^k\frac{\partial u_\varepsilon}{\partial x_k}+c_\varepsilon u_\varepsilon=f_\varepsilon \text{ in } G_0,\\ &\left[u_\varepsilon|_{\gamma_i}\equiv u_\varepsilon|_{\gamma_i+0}-u_\varepsilon|_{\gamma_i-0}=0,\right.\\ &\left[\frac{\partial u_\varepsilon}{\partial\nu}\Big|_{\gamma_i}\equiv\frac{\partial u_\varepsilon}{\partial\nu}\Big|_{\gamma_i+0}-\frac{\partial u_\varepsilon}{\partial\nu}\Big|_{\gamma_i-0}=0,\right.\\ &Bu_\varepsilon|_\Gamma\equiv\left(t\frac{\partial u_\varepsilon}{\partial\nu}+su_\varepsilon\right)\Big|_\Gamma=\varphi,\end{aligned}$$

where the operator $L_\epsilon$ is given by (1.4), the function $f_\epsilon$ is defined by (1.6); $G_0=G\backslash(\gamma\cup\gamma_\epsilon^+\cup\gamma_\epsilon^-)$; the surfaces $\gamma$, $\gamma_\epsilon^+$, and $\gamma_\epsilon^-$ will be designated respectively by $\gamma_i$ $(i=0,1,2)$ (the coefficients of operator $L_\epsilon$ break on these surfaces); $\gamma_i\pm0$ designate respectively the outer and inner sides of surface $\gamma_i$. By $\nu$ (for each surface $\gamma_0,\gamma_1,\gamma_2,\Gamma$) we mean a direction forming an acute angle with the direction of the outer normal to the corresponding surface. We will investigate Dirichlet's operator $(t\equiv0,\ s\equiv1)$ and Neumann's operator $(t\equiv0, s\geqq0,$ but $s\not\equiv0)$ as boundary operators. We assume the following smoothness for the parameters of the problem: $s,\phi\in C^0(\Gamma)$; $a_\epsilon^{ij},b_\epsilon^k,c_\epsilon,\ f_\epsilon\in C^0(\overline{G_0^\alpha})$, where $G_0^\alpha$ is a connectivity component of $G_0=G\backslash(\gamma_0\cup\gamma_\epsilon)$; $a_\epsilon^{ij}\in C^1(G_0)$. That $L_\epsilon$ is elliptic means that in the closure of each connectivity component of $G_0$ we have $\sum a_\epsilon^{ij}\xi_i\xi_j>0$ for $\xi_1^2+\cdots+\xi_n^2>0$.

DEFINITION 1.1. A *solution of problem* (1.7) is a function $u_\epsilon(x)$ defined in $G_0$ which has the necessary smoothness (i.e. has all derivatives required by the conditions of the problem) and which satisfies all the relations of problem (1.7).

In [8] (see also [5]) we find the following theorem on an a priori estimate (uniform in $\epsilon$) for (1.7).

THEOREM 1.1. *Let the coefficients of $L_\epsilon$ satisfy the condition $c_\epsilon\leqq0$ in $G_0$, and in $T_\epsilon\backslash\gamma$ let either $b_\epsilon^1(x)\geqq\min(\alpha c_\epsilon(x),-\beta)$ or $b_\epsilon^1(x)\leqq\max(-\alpha c_\epsilon(x),\beta)$, where $\alpha$ and $\beta$ are fixed constants. Then*

(1.8) $$\|u_\varepsilon\| \leqslant C(\|L_\varepsilon u_\varepsilon\| + \|Bu_\varepsilon|_\Gamma\|'),$$

*where* $\|\cdot\|$ *and* $\|\cdot\|'$ *are respectively norms of maximum modulus in* $G$ *and on* $\Gamma$, *and the constant* $C$ *depends only on* $\alpha$ *and* $\beta$ *(and the size of the domain in the direction* $x_1$*).*

In §2 we encounter a problem with more general conjugation conditions on the surface $\gamma_i$ than in (1.7). Thus we consider the following elliptic problem (with a boundary operator which is either a Dirichlet operator or a Neumann operator):

(1.9)
$$\begin{gathered}
Lu \equiv \sum a_{ij}\frac{\partial^2 u}{\partial x_i \partial x_j} + \sum b_k \frac{\partial u}{\partial x_k} + cu = f,\\
P_i u|_{\gamma_i} \equiv u|_{\gamma_i+0} - p_i u|_{\gamma_i-0} = \psi_i,\\
Q_i u|_{\gamma_i} \equiv \frac{\partial u}{\partial \nu}\Big|_{\gamma_i+0} - \left(q_i \frac{\partial u}{\partial \nu} + r_i u\right)\Big|_{\gamma_i-0} = \chi_i,\\
Bu|_\Gamma \equiv \left(t\frac{\partial u}{\partial \nu} + su\right)\Big|_\Gamma = \varphi,
\end{gathered}$$

where $p_i \neq 0, q_i \neq 0$, and $p_i, \psi_i \in C^1(\gamma_i)$; $q_i$, $r_i$, $\chi_i \in C^0(\gamma_i)$, and the remaining parameters of the problem satisfy the same conditions as in (1.7).

First we consider (1.9) in the following function space. A solution is sought in the space $H_2(G;\hat\gamma)$ which is the closure in the norm

$$\|u\|_2 = \Big(\sum_j \sum_{|\alpha|\leqslant 2} \int_{G_0^j} |D^\alpha u|^2 dx\Big)^{1/2}$$

of the set of functions defined in $\overline{G_0}$ and infinitely differentiable in each $\overline{G_0^j}$ (recall that $\overline{G_0^j}$ is the $j$th connectivity component of $G_0 = G \setminus (\bigcup_{i=0}^m \gamma_i)$). By the space of right sides we mean the space $\mathscr{H}_{(0)}(G;\hat\gamma) = \{H_0(G);\ H_{3/2}(\hat\gamma);\ H_{1/2}(\hat\gamma);\ H_{1/2+\kappa}(\Gamma)\}$, where $H_s(\Omega)$ is a Sobolev-Slobodeckiĭ space on the manifold $\Omega$, $\hat\gamma = \bigcup_{i=0}^m \gamma_i$, $\kappa = 1$ for Dirichlet's problem and $\kappa = 0$ for Neumann's problem.

Theorem 1.2 on the null index of operator $\mathfrak{A} = \{L;\ P_1|_{\gamma_1}, \cdots, P_m|_{\gamma_m};\ Q_1|_{\gamma_1}, \cdots, Q_m|_{\gamma_m};\ B|_\Gamma\}$ in the spaces $H_2(G;\hat\gamma)$, $\mathscr{H}_{(0)}(G;\hat\gamma)$ follows from [**9**] (see [**8**]).

THEOREM 1.2. *The operator* $\mathfrak{A}: H_2 \to \mathscr{H}_{(0)}$ *is a continuous* $\Phi$*-operator (in the terminology of Gohberg and Kreĭn* [**10**]*) having null index, i.e.* 1) $\mathfrak{A}$ *is continuous,* 2) $\operatorname{Im}\mathfrak{A} = \overline{\operatorname{Im}\mathfrak{A}}$, 3) $\alpha = \dim\operatorname{Ker}\mathfrak{A} < \infty$, 4) $\beta = \dim\operatorname{Coker}\mathfrak{A} < \infty$, *and* 5) $\alpha = \beta$.

From this theorem we get Theorem 1.3 on the unique solvability of (1.9) in the classical sense. A proof is in [**8**]; see also [**11**] and [**12**]. Note that from theorems on the smoothness of generalized solutions of elliptic equations [**7**] it follows that if the data of problem (1.9) are sufficiently smooth, then its solution is also smooth up to the boundary.

DEFINITION 1.2. A *classical solution of* (1.9) is a function $u(x)$ which is defined in $G_0$, has the necessary smoothness (i.e. has all derivatives required by the conditions of the problem), and satisfies all the relations of problem (1.9)

THEOREM 1.3. *Let* $c \leqq 0$, $p_i q_i > 0$, $q_i r_i \geqq 0$, $p_i = \text{const}$. *Then* (1.9) (*and in particular* (1.7)) *has a unique solution in the classical sense.*

## §2. Asymptotic expansions of a solution of a problem with coefficients having a "splash"

This section is the main part of the article. Here we consider (1.7) for different rational $\sigma = 1/2k \leq 1$. For each such $\sigma$ the asymptotic behavior of a solution is given in powers of $\delta = \epsilon^{1/2k}$. It turns out that the initial term of the asymptotics depends on $\sigma$ just like the limiting solution of the introductory example (see the Introduction). The basic reason for this is that in a neighborhood of $\gamma$ the operator $L_\epsilon$ is defined basically by increasing "splash" coefficients and the direction in $\nu$ (since the "splash" occurs in this direction and consequently the solution depends primarily on $\nu$ and the direction in $\tau$ influences only the lowest terms of the asymptotics). Thus the "principal part" of the operator "stratifies itself" into ordinary differential operators relative to a transverse cross-section to $\gamma$; moreover, these operators contain information on "splashes".

A proper separation of the "principal part" of the operator $L_\epsilon(T_\epsilon \setminus \gamma)$ is carried out with the help of the substitution $t = \nu/\delta^l$ (see [1]). Specifically, expanding the coefficients

$$a_\alpha^\varepsilon(\nu;\ \tau) = a_\alpha^\varepsilon(\delta^l t;\ \tau) = \overset{0}{a}_\alpha(\delta^l t;\ \tau) + \delta \overset{1}{a}_\alpha(\delta^l t;\ \tau) + \dots \tag{2.1}$$

of $L_\epsilon(T_\epsilon \setminus \gamma)$ (see (1.2), (1.3)) into a Taylor series at the points $(\pm 0; \tau)$ and observing that $\partial^\beta/\partial\nu^\beta = (1/\delta^{l\beta})\, \partial^\beta/\partial t^\beta$, we get the following "decomposition" of $L_\epsilon(T_\epsilon \setminus \gamma)$ (see [1]):

$$L_\varepsilon(T_\varepsilon \setminus \gamma) = \frac{1}{\delta^{2l}}(M_0 + \delta M_1 + \dots + \delta^M M_M + \delta^{M+1} M_{M+1,\varepsilon}), \tag{2.2}$$

where

$$M_0 = \sum_{|\alpha| \leqslant 2} \overset{0}{a}_\alpha(\pm 0;\ \tau)\, \partial^\alpha, \quad \alpha = (\beta;\ 0), \tag{2.3}$$

i.e., setting $a^\pm_\beta(\tau) = \overset{0}{a}_{(\beta;0)}(\pm 0; \tau)$, we get

$$M_0 = \sum_{\beta \leqslant 2} a^\pm_\beta(\tau) \frac{\partial^\beta}{\partial t^\beta}, \quad 0 < |t| < \delta^{2k-l}. \tag{2.3'}$$

In this way $M_0$ is an ordinary differential operator in $t$ (with coefficients which are constant for each $\tau$ and each $t$).

We get an asymptotic expansion of the solution $u_\epsilon$ of (1.7) in the following way (see [1]).

In $G\setminus\gamma$ there will be functions $u_i(x)$ such that

$$u_\varepsilon|_{G_\varepsilon} \equiv w_\varepsilon|_{G_\varepsilon}, \tag{2.4}$$

where

$$w_\varepsilon|_{G\setminus\gamma} \equiv u_0 + \delta u_1 + \ldots + \delta^N u_N + O(\delta^{N+1}), \tag{2.5}$$

and

$$\begin{gathered} L(G\setminus\gamma)\, w_\varepsilon = f, \\ Bw_\varepsilon|_\Gamma = \varphi. \end{gathered} \tag{2.6}$$

In $T_\varepsilon\setminus\gamma$ we can construct functions $v_j(t;\tau)$ such that

$$u_\varepsilon|_{T_\varepsilon\setminus\gamma} \equiv v_\varepsilon, \tag{2.7}$$

where

$$v_\varepsilon(t;\tau) \equiv v_0 + \delta v_1 + \ldots + \delta^M v_M + O(\delta^{N+1}). \tag{2.8}$$

Hence it follows in particular (see (1.4)–(1.7) and (2.2)) that $v_\varepsilon$ satisfies the following equation:

$$\begin{aligned} L_\varepsilon(T_\varepsilon\setminus\gamma)\, v_\varepsilon &\equiv \frac{1}{\delta^{2l}}(M_0 + \delta M_1 + \ldots)(v_0 + \delta v_1 + \ldots) \\ &= \frac{1}{\delta^{2l}}(h_0^\pm(\tau) + \delta h_1^\pm(\tau) + \ldots), \end{aligned} \tag{2.9}$$

where $(h_0^\pm(\tau) + \delta h_1^\pm(\tau) + \cdots)$ is a Taylor series expansion of the function $h_\varepsilon(\delta^l t;\tau)$ at the point $(\pm 0;\tau)$ (see (1.5)). Equation (2.9) means that the functions $v_j(t;\tau)$ satisfy the equations $(0 < |t| < \delta^{2k-l})$

$$M_0 v_0 = h_0^\pm, \tag{2.10;0}$$

$$\cdots\cdots$$

$$M_0 v_j = h_j^\pm - M_1 v_{j-1} - \ldots - M_j v_0. \tag{2.10;j}$$

From (2.5) and (2.6) it follows that the functions $u_i(x)$ should satisfy the following equations and boundary conditions:

$$\begin{gathered} L(G\setminus\gamma)\, u_0 = f, \\ Bu_0|_\Gamma = \varphi, \end{gathered} \tag{2.11;0}$$

$$\cdots\cdots\cdots$$

$$\begin{gathered} L(G\setminus\gamma)\, u_i = 0, \quad i \geqslant 1, \\ Bu_i|_\Gamma = 0. \end{gathered} \tag{2.11;i}$$

Later we will give an algorithm for finding functions $u_i$ and $v_j$ which satisfy requirements (2.4)–(2.8). This algorithm is based, naturally, on consequences of conditions (2.4)–(2.8), i.e. on (2.10) and (2.11), and also on coalescence conditions for the solution $u_\varepsilon$ of (1.7) on the surfaces $\gamma_0(\gamma), \gamma_1(\gamma_\varepsilon'), \gamma_2(\gamma_\varepsilon'')$, i.e. on the conditions

$$\left.\frac{\partial^\beta u_\varepsilon}{\partial \nu^\beta}\right|_{\gamma_i+0} = \left.\frac{\partial^\beta u_\varepsilon}{\partial \nu^\beta}\right|_{\gamma_i-0};\ \beta = 0,\, 1;\ i = 0,\, 1,2. \tag{2.12}$$

We now analyze condition (2.12). First it follows from (2.12) and from (2.7)–(2.8) that on the surface $\gamma_0 = \{(t;\tau): t=0\}$ the functions $v_j(t;\tau)$ coalesce continuously with derivatives in $t$, i.e.

$$\frac{\partial^\beta v_j}{\partial t^\beta}\bigg|_{t=+0} = \frac{\partial^\beta v_j}{\partial t^\beta}\bigg|_{t=-0} = \frac{\partial^\beta v_j}{\partial t^\beta}(0), \quad \beta = 0, 1. \tag{2.13}$$

Secondly, from (2.4)–(2.8) and from (2.12) we have ($\beta = 0, 1$)

$$\frac{\partial^\beta w_\varepsilon(\nu;\tau)}{\partial \nu^\beta}\bigg|_{\nu=\pm\varepsilon} = \frac{1}{\delta^{l\beta}}\,\frac{\partial^\beta v_\varepsilon(t;\tau)}{\partial t^\beta}\bigg|_{t=\pm\delta^{2k-l}}. \tag{2.14}$$

To shorten our notation, instead of (2.14) we will write

$$w_\varepsilon^{(\beta)}(\pm\varepsilon) = \frac{1}{\delta^{l\beta}}\,\frac{\partial^\beta v_\varepsilon}{\partial t^\beta}(\pm\delta^{2k-l}), \quad \beta = 0, 1. \tag{2.14'}$$

Condition (2.14) (conjugation condition) establishes a relationship between

$$w_\varepsilon(x) = u_0 + \delta u_1 + \dots + \delta^N u_N + O(\delta^{N+1}) \tag{2.5}$$

and

$$v_\varepsilon(t;\tau) = v_0 + \delta v_1 + \dots + \delta^M v_M + O(\delta^{N+1}) \tag{2.8}$$

on the surfaces $\gamma_\epsilon^\pm$, depending on $\epsilon$. But $v_i$ and $v_j$ are sufficiently smooth in $T_\epsilon\backslash\gamma$ up to $\gamma$, since $v_j$ and $u_i$ satisfy elliptic equations (2.10; $j$) and (2.11; $i$) with sufficiently smooth coefficients and right sides. Thus, applying Taylor's theorem to the left side and also (for $\sigma = 1/2k < 1$) to the right side of (2.14′), we get as consequences of (2.5) and (2.8) asymptotic expansions of $w_\epsilon^{(\beta)}(\pm\epsilon)$ and $(1/\delta^{l\beta})(\partial^\beta v_\epsilon/\partial t^\beta)(\pm\delta^{2k-l})$ in powers of $\delta$. Specifically, designating by $[\kappa]$ the integer part of the number $\kappa$, we have

$$\begin{aligned} w_\varepsilon^{(\beta)}(\pm\varepsilon) &= u_0^{(\beta)}(\pm\varepsilon) + \delta u_1^{(\beta)}(\pm\varepsilon) + \dots = u_0^{(\beta)}(\pm 0) + \delta u_1^{(\beta)}(\pm 0) \\ &\quad + \dots + \delta^{2k-1}u_{2k-1}^{(\beta)}(\pm 0) + \delta^{2k}\left(u_{2k}^{(\beta)}(\pm 0) \pm u_0^{(\beta+1)}(\pm 0)\right) + \dots \\ &= u_0^{(\beta)}(\pm 0) + \dots + \delta^p \sum_{s=0}^{\left[\frac{p}{2k}\right]} \frac{(\pm 1)^s}{s!} u_{p-(2k)s}^{(s+\beta)}(\pm 0) + \dots, \end{aligned} \tag{2.15}$$

$$\begin{aligned} \frac{1}{\delta^{l\beta}}\frac{\partial^\beta v_\varepsilon}{\partial t^\beta}(\pm\delta^{2k-l}) &= \frac{1}{\delta^{l\beta}}\frac{\partial^\beta v_0}{\partial t^\beta}(\pm 0) + \dots \\ &\quad + \delta^{r-l\beta}\sum_{s=0}^{\left[\frac{r}{2k-l}\right]} \frac{(\pm 1)^s}{s!}\,\frac{\partial^{s+\beta} v_{r-(2k-l)s}}{\partial t^{s+\beta}}(\pm 0) + \dots \quad \text{for } \frac{l}{2k} < 1, \end{aligned} \tag{2.16}$$

$$\begin{aligned} \frac{1}{\delta^{l\cdot\beta}}\frac{\partial^\beta v_\varepsilon}{\partial t^\beta}(\pm\delta^{2k-l}) &= \frac{1}{\delta^{2k\cdot\beta}}\left(\frac{\partial^\beta v_0}{\partial t^\beta}(\pm 1) + \dots\right. \\ &\quad \left.\dots + \delta^r \frac{\partial^\beta v_r}{\partial t^\beta}(\pm 1) + \dots\right), \quad \frac{l}{2k} = 1. \end{aligned} \tag{2.17}$$

Hence by (2.14′) and (2.13) we have the following conjugation conditions for $u_i$ and $v_j$ (on a surface $\gamma$ not depending on $\epsilon$).

In the case $l/2k < 1$:

$$(2.18;0)\text{–}(2.18;l-1) \qquad \frac{\partial v_0}{\partial t}(0)=0,\ \dots,\ \sum_{s=0}^{\left[\frac{l-1}{2k-l}\right]} \frac{(\pm 1)^s}{s!}\frac{\partial^{s+1}v^\pm}{\partial t^{s+1}}=0,$$

$$(2.19;i) \qquad u_i^\pm(0)=v_i(0)+\sum_{s=1}^{\left[\frac{i}{2k-l}\right]}\frac{(\pm 1)^s}{s!}\frac{\partial^s v^\pm}{dt^s}-\sum_{s=1}^{\left[\frac{l}{2k}\right]}\frac{(\pm 1)^s}{s!}u^{\pm(s)},$$

$$(2.20;i) \qquad u_i^{\pm\prime}(0)+\sum_{s=1}^{\left[\frac{i}{2k}\right]}\frac{(\pm 1)^s}{s!}u^{\pm(s+1)}=\frac{\partial v_{i+l}}{\partial t}(0)+\sum_{s=1}^{\left[\frac{i+l}{2k-l}\right]}\frac{(\pm 1)^s}{s!}\frac{\partial^{s+1}v^\pm}{\partial t^{s+1}}.$$

REMARK 2.1. In (2.18)–(2.20) and in the future we will shorten the notation by writing $\sum_{s=s_0}^{|K-L|}F^\pm(0)$ (or even $\sum_{s=s_0}^{|K-L|}F^\pm$) instead of $\sum_{s=s_0}^{|K-L|}F_{K-L,s}(\pm 0)$.

In the case $l/2k=1$ we find from (2.14′), (2.15), and (2.17) that

$$(2.21;0)\text{–}(2.21;2k-1) \qquad \frac{\partial v_0}{\partial t}(\pm 1)=0,\ \dots,\ \frac{\partial v_{2k-1}}{\partial t}(\pm 1)=0,$$

$$(2.22;i) \qquad u_i^\pm(0)=v_i(\pm 1)-\sum_{s=1}^{\left[\frac{i}{2k}\right]}\frac{(\pm 1)^s}{s!}u^{\pm(s)},$$

$$(2.23;i) \qquad \frac{\partial v_{i+2k}}{\partial t}(\pm 1)=\sum_{s=0}^{\left[\frac{i}{2k}\right]}\frac{(\pm 1)^s}{s!}u^{\pm(s+1)}.$$

REMARK 2.2. We will refer to conjugation conditions (2.18; 0)–(2.18; $l-1$) ((2.21; 0)–(2.21; $2k-1$)) as initial conditions, and we will refer to conditions (2.19;$i$), (2.20; $i$) ((2.22; $i$), (2.23; $i$)) as basic conjugation conditions for $l/2k<1$ (for $l/2k=1$).

We have proved the following theorem.

THEOREM 2.1. *If the function $u_i(x)$ and $v_j(t;\tau)$ satisfy conditions* (2.4)–(2.8), *then $u_i$ and $v_j$ will also satisfy equations* (2.11; $i$), (2.10; $j$) *and conjugation conditions* (2.18)–(2.20) ((2.21)–(2.23)) *for $l/2k<1$ (for $l/2k=1$), and $v_j$ will also satisfy coalescence condition* (2.13) *on $\gamma$.*

ASSUMPTION 2.1. Henceforth (if not specified to the contrary) we will assume that:

a) The conditions of Theorem 1.1 are satisfied for the operator $L_\epsilon$ (see (1.14)), and consequently there is an a priori estimate (uniform in $\epsilon$) for (1.7):

(1.8) $$\| u_\varepsilon \| \leqslant C ( \| L_\varepsilon u_\varepsilon \| + \| B u_\varepsilon |_\Gamma \|' );$$

b) For the operator

$$L(G\setminus\gamma) = \sum a_{ij} \frac{\partial^2}{\partial x_i \partial x_j} + \sum b_k \frac{\partial}{\partial x_k} + c$$

(see (1.1)) the condition $c \leqq 0$ holds. Consequently by Theorem 1.3 the problem

(1.9′) $$\begin{gathered} L(G\setminus\gamma)\, u = f, \\ u^+|_\gamma - pu^-|_\gamma = \psi, \\ u^{+\prime}|_\gamma - (qu^{-\prime} + ru^-)|_\gamma = \chi, \\ Bu|_\Gamma = \varphi \end{gathered}$$

has a unique solution for $pq > 0$, $qr \geqq 0$, $p = \text{const}$.

Under these assumptions we get the basic theorem of this article, namely, the converse of Theorem 2.1. That is, conditions (2.10; $j$), (2.11; $i$), (2.13), (2.18)–(2.23) uniquely define certain functions $u_i(x)$ and $v_j(t;\tau)$, and it turns out that $u_i$ and $v_j$ satisfy (2.4) – (2.8) (asymptotics of the solution $u_\varepsilon$ of (1.7) are also constructed).

With respect to the construction of asymptotics of a solution of (1.7) we consider separately cases involving different values of $\sigma = l/2k$, namely $0 < \sigma < \frac{1}{2}$, $\sigma = \frac{1}{2}$, $\frac{1}{2} < \sigma < 1$, $\sigma = 1$ (see the introductory example).

REMARK 2.3. For $\sigma < 1$ condition (2.13) concerning coalescence of $v_j$ on $\gamma$ plays an essential role.

Consider the first case, i.e.

1°. $0 < \sigma < \frac{1}{2}$. We write the initial conjugation conditions in the following way:

(2.18; 0)–(2.18; $l-1$) $$\frac{\partial v_0}{\partial t}(0) = 0, \ \ldots, \ \frac{\partial v_{l-1}}{\partial t}(0) = 0.$$

We now do the first step of an iteration process. The basic conjugation condition gives

(2.19; 0) $$u_0^+(0) = u_0^-(0) = v_0(0),$$

(2.20; 0) $$u_0^{+\prime}(0) = u_0^{-\prime}(0) = \frac{\partial v_l}{\partial t}(0).$$

From this and from (2.11); 0) we get for $u_0(x)$ the following problem with conjugation on $\gamma$:

(2.24; 0) $$\begin{gathered} L(G\setminus\gamma)\, u_0 = f, \\ u_0^+|_\gamma - u_0^-|_\gamma = 0, \\ u_0^{+\prime}|_\gamma - u_0^{-\prime}|_\gamma = 0, \\ Bu_0|_\Gamma = \varphi. \end{gathered}$$

Having solved it (see Assumption 2.1), we get from (2.10; 0), (2.19; 0) and (2.18; 0) Cauchy's problem for $v_0(t;\tau)$:

$$M_0 v_0 = h_0^{\pm}, \quad 0 < |t| < \delta^{2k-l},$$
(2.25; 0)
$$v_0(0) = u_0^{+}(0), \quad \frac{\partial v_0}{\partial t}(0) = 0.$$

The first step of the iteration is thus completed.

We now do the $i$th step. From the basic conditions (2.19; $i$), (2.20; $i$) and from (2.11; $i$) we get for $u_i(x)$ the following problem with conjugation on $\gamma$:

(2.24; $i$)
$$\begin{aligned} L(G\setminus\gamma)\, u_i &= 0, \\ u_i^+|_\gamma - u_i^-|_\gamma &= \psi_i, \\ u_i^{+'}|_\gamma - u_i^{-'}|_\gamma &= \chi_i, \\ Bu_i|_\Gamma &= 0, \end{aligned}$$

here $\psi_i$ and $\chi_i$ are functions of $\tau$ known up to the $i$th step, namely:

(2.26; $i$)
$$\psi_i = \sum_{s=1}^{\left[\frac{i}{2k-l}\right]} \left(\overline{s|}\_ \frac{\partial^s v}{\partial t^s}\right) - \sum_{s=1}^{\left[\frac{i}{2k}\right]} (\overline{s|}\_ u^{(s)}),$$

(2.27; $i$)
$$\chi_i = \sum_{s=1}^{\left[\frac{i+l}{2k-l}\right]} \left(\overline{s|}\_ \frac{\partial^{s+1} v}{\partial t^{s+1}}\right) - \sum_{s=1}^{\left[\frac{i}{2k}\right]} (\overline{s|}\_ u^{(s+1)}),$$

where the operator $\overline{s|}\_$ (we call it an "$s$-jump" operator) behaves according to the formula

(2.28)
$$\overline{s|}\_ F = \frac{1}{s!}\left(F^+(0) - (-1)^s F^-(0)\right).$$

The notation $\overline{s|}\_$ is introduced for brevity; for example, if we write out the expression $\sum_{s=1}^{[i/2k]}(\overline{s|}\_ u^{(s)})$ in full, using the notation of 2.1, we get

$$\sum_{s=1}^{\left[\frac{i}{2k}\right]} (\overline{s|}\_ u^{(s)}) = \sum_{s=1}^{\left[\frac{i}{2k}\right]} \frac{1}{s!}\left(u^{+(s)}(0) - (-1)^s u^{-(s)}(0)\right).$$

The initial conjugation conditions yield the second Cauchy datum for $v_i(t;\tau)$ only for $i < l$; if $i \geqq l$, then $(\partial v_i/\partial t)(0)$ is defined by (2.20; $i-l$); namely,

(2.18; $i$)
$$\frac{\partial v_i}{\partial t}(0) = \sum_{s=0}^{\left[\frac{i-l}{2k}\right]} \frac{1}{s!}\, u^{+(s+1)} - \sum_{s=1}^{\left[\frac{i}{2k-l}\right]} \frac{1}{s!} \frac{\partial^{s+1} v^+}{\partial t^{s+1}}.$$

Having solved problem (2.24; $i$) (see Assumption 2.1), we get from (2.10; $i$), (2.19; $i$) and (2.18; $i$) the following Cauchy problem for $v_i(t;\tau)$ ($|t| < \delta^{2k-l}$):

(2.25; $i$)
$$M_0 v_i = h_i^{\pm} - M_1 v_{i-1} - \dots - M_i v_0,$$
$$v_i(0) = \xi_i,$$
$$\frac{\partial v_i}{\partial t}(0) = \eta_i,$$

here $\xi_i$ and $\eta_i$ are already known functions of $\tau$; namely,

(2.29; $i$)
$$\xi_i = \sum_{s=0}^{\left[\frac{i}{2k}\right]} \frac{1}{s!} u^{+(s)} - \sum_{s=1}^{\left[\frac{i}{2k-l}\right]} \frac{1}{s!} \frac{\partial^s v^+}{\partial t^s},$$

(2.30; $i$)
$$\eta_i = \sum_{s=0}^{\left[\frac{i-l}{2k}\right]} \frac{1}{s!} u^{+(s+1)} - \sum_{s=1}^{\left[\frac{i}{2k-l}\right]} \frac{1}{s!} \frac{\partial^{s+1} v^+}{\partial t^{s+1}}.$$

Thus the $i$th stage of the iteration process is completed.

2°. $\sigma = \frac{1}{2}$. In this case we will assume in addition that the leading coefficient of the operator $M_0$ (see (2.3)) is continuous, i.e.

(2.31)
$$a_2^+(\tau) = a_2^-(\tau) = a_2(\tau)$$

and consequently in this case (see (2.3′))

(2.3″)
$$M_0 = a_2(\tau)\frac{\partial^2}{\partial t^2} + a_1^{\pm}(\tau)\frac{\partial}{\partial t} + a_0^{\pm}(\tau), \quad 0 < |t| < \delta^{2k-l}.$$

The initial conjugation conditions for $\sigma = \frac{1}{2}$ are as follows:

(2.18; 0)–(2.18; $k-1$)
$$\frac{\partial v_0}{\partial t}(0) = 0, \dots, \frac{\partial v_{k-1}}{\partial t}(0) = 0.$$

We now do the first step of the iteration. The basic conjugation conditions give

(2.19; 0)
$$u_0^+(0) = v_0(0),$$
$$u_0^-(0) = v_0(0),$$

(2.20; 0)
$$u_0^{+\prime}(0) = \frac{\partial v_k}{\partial t}(0) + \frac{\partial^2 v_0^+}{\partial t^2}(0),$$
$$u_0^{-\prime}(0) = \frac{\partial v_k}{\partial t}(0) - \frac{\partial^2 v_0^-}{\partial t^2}(0).$$

From the last condition it follows that

(2.32; 0)
$$u_0^{+\prime}(0) - u_0^{-\prime}(0) = \frac{\partial^2 v_0^+}{\partial t^2}(0) + \frac{\partial^2 v_0^-}{\partial t^2}(0).$$

But by (2.10; 0) and (2.3″) we have

$$a_2(\tau)\frac{\partial^2 v_0^+}{\partial t^2}(0)+a_1^+(\tau)\frac{\partial v_0}{\partial t}(0)+a_0^+(\tau)v_0(0)=h_0^+,$$

(2.33; 0)

$$a_2(\tau)\frac{\partial^2 v_0^-}{\partial t^2}(0)+a_1^-(\tau)\frac{\partial v_0}{\partial t}(0)+a_0^-(\tau)v_0(0)=h_0^-;$$

thus

$$\frac{\partial^2 v_0}{\partial t^2}(0)+\frac{\partial^2 v_0^-}{\partial t^2}(0)=\frac{1}{a_2}\cdot\{(h_0^++h_0^-)$$

(2.34; 0)
$$-(a_1^++a_1^-)\frac{\partial v_0}{\partial t}(0)-(a_0^++a_0^-)v_0(0)\}.$$

We know the value of $(\partial v_0/\partial t)(0)$ by virtue of the initial conjugation conditions (2.18; 0); namely, $(\partial v_0/\partial t)(0)=0$; conjugation condition (2.19; 0) gives information about $v_0(0)$: $v_0(0)=u_0^-(0)$. Thus from (2.11; 0), (2.19; 0), (2.32; 0), and (2.34; 0) we get for $u_0(x)$ the following problem with conjugation on $\gamma$:

(2.35; 0)
$$\begin{aligned} L(G\setminus\gamma)u_0&=f,\\ u_0^+|_\gamma-u_0^-|_\gamma&=0,\\ u_0^{+\prime}|_\gamma-(u_0^{-\prime}+ru_0^-)|_\gamma&=\chi_0,\\ Bu_0|_\Gamma&=\varphi,\end{aligned}$$

where

(2.36)
$$r=-\frac{a_0^++a_0^-}{a_2},$$

(2.37; 0)
$$\chi_0=\frac{h_0^++h_0^-}{a_2}.$$

Since $r\geqq 0$ (see Assumption 2.1), problem (2.35; 0) has a unique solution. Having solved this problem, we get from (2.10; 0), (2.19; 0), and (2.18; 0) the following Cauchy problem for $v_0(t;\tau)$ $(0<|t|<\delta^k)$:

(2.38; 0)
$$\begin{aligned} M_0v_0&=h_0^\pm,\\ v_0(0)&=u_0^+(0),\\ \frac{\partial v_0}{\partial t}(0)&=0.\end{aligned}$$

The initial step of the iteration is finished.

Now we do the $i$th step. The basic conjugation conditions for $\sigma=\frac{1}{2}$ are as follows:

(2.19; $i$)
$$u_i^\pm(0)=v_i(0)+\alpha_i^\pm,$$

(2.20; $i$)
$$u_i^{\pm\prime}(0)=\frac{\partial v_{i+k}}{\partial t}(0)\pm\frac{\partial^2 v_i^\pm}{\partial t^2}(0)+\beta_i^\pm,$$

where $\alpha_i^{\pm}$ and $\beta_i^{\pm}$ are known functions of $\tau$ up to the $i$th step; namely,

(2.39; $i$)
$$\alpha_i^{\pm} = \sum_{s=1}^{\left[\frac{i}{k}\right]} \frac{(\pm 1)^s}{s!} \frac{\partial^s v^{\pm}}{\partial t^s} - \sum_{s=1}^{\left[\frac{i}{2k}\right]} \frac{(\pm 1)^s}{s!} u^{\pm(s)},$$

(2.40; $i$)
$$\beta_i^{\pm} = \sum_{s=2}^{\left[\frac{i+k}{k}\right]} \frac{(\pm 1)^s}{s!} \frac{\partial^{s+1} v^{\pm}}{\partial t^{s+1}} - \sum_{s=1}^{\left[\frac{i}{2k}\right]} \frac{(\pm 1)^s}{s!} u^{\pm(s+1)}.$$

From (2.20; $i$) it follows that

(2.32; $i$)
$$u_i^{+\prime}(0) - u_i^{-\prime}(0) = \frac{\partial^2 v_i^{+}}{\partial t^2}(0) + \frac{\partial^2 v_i^{-}}{\partial t^2}(0) + (\beta_i^{+} - \beta_i^{-}).$$

But by (2.10; $i$) and (2.3″) we have

(2.33; $i$)
$$a_2 \frac{\partial^2 v_i^{\pm}}{\partial t^2}(0) + a_1^{\pm} \frac{\partial v_i}{\partial t}(0) + a_0^{\pm} v_i(0) = \widehat{h}_i^{\pm},$$

where

(2.41; $i$)
$$\widehat{h}_i^{\pm} = h_i^{\pm} - (M_1 v_{i-1} + \dots + M_i v_0)|_{t=\pm 0}.$$

Thus

(2.34; $i$)
$$\frac{\partial^2 v_i^{+}}{\partial t^2}(0) + \frac{\partial^2 v_i^{-}}{\partial t^2}(0) = \frac{1}{a_2}\Big\{ (\widehat{h}_i^{+} + \widehat{h}_i^{-}) - (a_1^{+} + a_1^{-}) \frac{\partial v_i}{\partial t}(0) - (a_0^{+} - a_0^{-}) v_i(0) \Big\}.$$

The value of $(\partial v_i/\partial t)(0)$ for $i < k$ is known from the initial conjugation conditions, and for $i \geqq k$ we find $(\partial v_i/\partial t)(0)$ from (2.20; $i-k$); namely

(2.18; $i$)
$$\frac{\partial v_i}{\partial t}(0) = \zeta_i,$$

where

(2.42; $i$)
$$\zeta_i = \sum_{s=0}^{\left[\frac{i-k}{2k}\right]} \frac{1}{s!} u^{+(s+1)} - \sum_{s=1}^{\left[\frac{i}{k}\right]} \frac{1}{s!} \frac{\partial^{s+1} v^{+}}{\partial t^{s+1}}$$

is a known function of $\tau$ up to the $i$th step. Conjugation condition (2.19; $i$) gives information on $v_i(0)$: $v_i(0) = u_i^{-}(0) - \alpha_i^{-}$. Thus from (2.11; $i$), (2.19; $i$), (2.32; $i$), and (2.34; $i$) we get for $u_i(x)$ the following problem with conjugation on $\gamma$:

(2.35; $i$)
$$\begin{aligned} &L(G\diagdown\gamma)u_i = 0,\\ &u_i^{+}|_{\gamma} - u_i^{-}|_{\gamma} = \psi_i,\\ &u_i^{+\prime}|_{\gamma} - (u_i^{-\prime} + r u_i^{-})|_{\gamma} = \chi_i,\\ &Bu_i|_{\Gamma} = 0, \end{aligned}$$

where

$$\psi_i = a_i^+ - a_i^-, \tag{2.43; i}$$

$$\chi_i = \frac{1}{a_2}\{(\hat{h}_i^+ + \hat{h}_i^-) + a_i^-(a_0^+ + a_0^-) - \zeta_i(a_i^+ + a_i^-)\}, \tag{2.37; i}$$

$$r = -\frac{1}{a_2}(a_0^+ + a_0^-). \tag{2.36}$$

Since $r \geqq 0$ (see Assumption 2.1), problem (2.35; $i$) has a unique solution. Having solved this problem, we get from (2.10; $i$), (2.19; $i$) and (2.18; $i$) the following Cauchy problem for $v_i(t;\tau)$ $(0 < |t| < \delta^k)$:

$$\begin{aligned} &M_0 v_i = h_i^\pm - M_1 v_{i-1} - \ldots - M_i v_0, \\ &v_i(0) = u_i^+(0) - a_i^+, \\ &\frac{\partial v_i}{\partial t}(0) = \zeta_i. \end{aligned} \tag{2.38; i}$$

The $i$th step of the iteration process for the case $\sigma = \frac{1}{2}$ is now finished.

3°. $\frac{1}{2} < \sigma < 1$. In this case, as in the preceding one, we will assume in addition that

$$a_2^+(\tau) = a_2^-(\tau) = a_2(\tau), \tag{2.31}$$

i.e. (see (2.3′))

$$M_0 = a_2(\tau)\frac{\partial^2}{\partial t^2} + a_1^\pm(\tau)\frac{\partial}{\partial t} + a_0^\pm(\tau), \quad 0 < |t| < \delta^{2k-l}. \tag{2.3″}$$

We require further that

$$a_0^+(\tau) + a_0^-(\tau) \neq 0. \tag{2.44}$$

Analogously to the previous case we shorten our notation by letting

$$\hat{h}_i^\pm = h_i^\pm - (M_1 v_{i-1} + \ldots + M_i v_0)|_{t=\pm 0}. \tag{2.41; i}$$

We also let

$$\hat{h}_i = \frac{1}{2}(\hat{h}_i^+ + \hat{h}_i^-), \tag{2.45; i}$$

$$a_\beta = \frac{1}{2}(a_\beta^+(\tau) + a_\beta^-(\tau)), \quad \beta = 0, 1, 2. \tag{2.46}$$

We now proceed to the construction of terms of the asymptotics. Set $l = k + \alpha$, where $0 < \alpha < k$ (note that $k \geqq 2$). The initial conjugation conditions give

$$\frac{\partial v_0}{\partial t}(0) = 0, \tag{2.18; 0}$$

$$\cdots\cdots$$

$$\frac{\partial v_{k-\alpha-1}}{\partial t}(0) = 0; \tag{2.18; k - α - 1}$$

$$\frac{\partial v_{k-\alpha}}{\partial t}(0) = \mp \frac{\partial^2 v_0^{\pm}}{\partial t^2}(0), \tag{2.18'; $k-\alpha$}$$

. . . . . . . . . . . . . . . . . . . . . . . . . . . .

$$\frac{\partial v_{(k+\alpha)-1}}{\partial t}(0) = \mp \frac{\partial^2 v_{2\alpha-1}^{\pm}}{\partial t^2}(0) - \sum_{s=2}^{\left[\frac{l-1}{2k-l}\right]} \frac{(\pm 1)^s}{s!} \frac{\partial^{s+1} v^{\pm}}{\partial t^{s+1}}. \tag{2.18'; $k+\alpha-1$}$$

Consider the second group of initial conditions. From (2.18'; $k-\alpha$) it follows that

$$\frac{\partial^2 v_0^{+}}{\partial t^2}(0) + \frac{\partial^2 v_0^{-}}{\partial t^2}(0) = 0, \tag{2.47; 0}$$

thus for (2.10; 0), (2.3″) and (2.46) we get

$$a_1 \frac{\partial v_0}{\partial t}(0) + a_0 v_0(0) = \hat{h}_0, \tag{2.48; 0}$$

and since $a_0 \neq 0$ (see (2.44)), we have

$$v_0(0) = \frac{1}{a_0}\left\{\hat{h}_0 - a_1 \frac{\partial v_0}{\partial t}(0)\right\}. \tag{2.49; 0}$$

But $(\partial v_0/\partial t)(0) = 0$ by (2.18; 0). Thus from (2.10; 0), (2.49; 0) and (2.18; 0) we get the following Cauchy problem for $v_0(t;\tau)$ $(0 < |t| < \delta^{2k-l})$:

$$\begin{aligned} M_0 v_0 &= h_0^{\pm}, \\ v_0(0) &= \frac{1}{a_0}\cdot \hat{h}_0, \\ \frac{\partial v_0}{\partial t}(0) &= 0. \end{aligned} \tag{2.50; 0}$$

Having defined $v_0(t;\tau)$ by this formula, we find the second Cauchy datum for $v_{k-\alpha}(t;\tau)$; namely,

$$\frac{\partial v_{k-\alpha}}{\partial t}(0) = -\frac{\partial^2 v_0^{+}}{\partial t^2}(0). \tag{2.18; $k-\alpha$}$$

We now go to the last condition in the second group of initial conditions, i.e. we consider (2.18'; $k+\alpha-1$). We have

$$\frac{\partial^2 v_{2\alpha-1}^{+}}{\partial t^2}(0) + \frac{\partial^2 v_{2\alpha-1}^{-}}{\partial t^2}(0) = \theta_{2\alpha-1}, \tag{2.47; $2\alpha-1$}$$

where $\theta_{2\alpha-1}$ is an already known function of $\tau$:

$$\theta_{2\alpha-1} = -\sum_{s=2}^{\left[\frac{l-1}{2k-l}\right]} \left(\overline{s|}_{-} \frac{\partial^{s+1} v}{\partial t^{s+1}}\right) \tag{2.51; $2\alpha-1$}$$

($\overline{s|}_{-}$ is defined by (2.28)). It is not now difficult to find the first Cauchy datum for the function $v_{2\alpha-1}(t;\tau)$; note that we found the second datum earlier (while considering initial conjugation condition (2.18′; $2a-1$). We now have

(2.18; $2\alpha-1$) $$\frac{\partial v_{2\alpha-1}}{\partial t}(0)=\Psi_{2\alpha-1},$$

where

(2.52; $2\alpha-1$) $$\Psi_{2\alpha-1}=-\sum_{s=1}^{\left[\frac{2\alpha-1}{k-\alpha}\right]}\frac{1}{s!}\frac{\partial^{s+1}v^{+}}{\partial t^{s+1}}.$$

The value of $v_{2\alpha-1}(0)$ is found in the following way. From (2.10; $2\alpha-1$), (2.3″), (2.45; $2\alpha-1$), (2.46), (2.47; $2\alpha-1$), and (2.18; $2\alpha-1$) we get

(2.48; $2\alpha-1$) $$a_1\Psi_{2\alpha-1}+a_0v_{2\alpha-1}(0)=\widehat{h}_{2\alpha-1}-\frac{a_2}{2}\theta_{2\alpha-1},$$

and since $a_0\neq 0$ (see (2.44)), we have

(2.49; $2\alpha-1$) $$v_{2\alpha-1}(0)=\Phi_{2\alpha-1},$$

where

(2.53; $2\alpha-1$) $$\Phi_{2\alpha-1}=\frac{1}{a_0}\left\{\widehat{h}_{2\alpha-1}-\frac{a_2}{2}\theta_{2\alpha-1}-a_1\Psi_{2\alpha-1}\right\}.$$

Thus (see (2.10; $2\alpha-1$), (2.49; $2\alpha-1$), (2.18; $2\alpha-1$)) we get for $v_{2\alpha-1}(t;\tau)$ the following Cauchy problem ($0<|t|<\delta^{2k-l}$):

(2.50; $2\alpha-1$) $$\begin{gathered}M_0v_{2\alpha-1}=h^{\pm}_{2\alpha-1}-M_1v_{2\alpha-2}-\dots-M_{2\alpha-1}v_0,\\ v_{2\alpha-1}(0)=\Phi_{2\alpha-1},\\ \frac{\partial v_{2\alpha-1}}{\partial t}(0)=\Psi_{2\alpha-1}.\end{gathered}$$

Having solved this problem, we define

(2.52; $l-1$) $$\Psi_{k+\alpha-1}=-\sum_{s=1}^{\left[\frac{l-1}{2k-l}\right]}\frac{1}{s!}\frac{\partial^{s+1}v^{+}}{\partial t^{s+1}},$$

and by (2.18′; $k+\alpha-1$) we get

(2.18; $k+\alpha-1$) $$\frac{\partial v_{k+\alpha-1}}{\partial t}(0)=\Psi_{k+\alpha-1}.$$

So from the initial conjugation conditions we have found the functions

(2.54; 0)–(2.54; $2\alpha-1$) $$v_0(t;\tau),\ \dots,\ v_{2\alpha-1}(t;\tau)\quad(|t|<\delta^{2k-l}),$$

and we have also established that

$$\frac{\partial v_0}{\partial t}(0)=\Psi_0,\ \dots,\ \frac{\partial v_{2\alpha-1}}{\partial t}(0)=\Psi_{2\alpha-1},\ \dots,\ \frac{\partial v_{l-1}}{\partial t}(0)=\Psi_{l-1},$$

(2.18; 0)–(2.18; $l-1$)

where $\Psi_j$ are defined by formulas (2.52; $j$).

We now do the $i$th step of the iteration ($i \geqq 0$). The basic conjugation conditions give

$$u_i^{\pm}(0) = \varphi_i^{\pm}, \tag{2.19; $i$}$$

$$\frac{\partial v_{i+k+\alpha}}{\partial t}(0) \pm \frac{\partial^2 v_{i+2\alpha}^{\pm}}{\partial t^2}(0) = u_i^{\pm\prime}(0) + \gamma_i^{\pm}, \tag{2.20; $i$}$$

where $\phi_i^{\pm}$, $\gamma_i^{\pm}$ are quantities known up to the $i$th step; namely:

$$\varphi_i^{\pm} = \sum_{s=0}^{\left[\frac{i}{k-\alpha}\right]} \frac{(\pm 1)^s}{s!} \frac{\partial^s v^{\pm}}{\partial t^s} - \sum_{s=1}^{\left[\frac{i}{2k}\right]} \frac{(\pm 1)^s}{s!} u^{\pm(s)}, \tag{2.55; $i$}$$

$$\gamma_i^{\pm} = \sum_{s=1}^{\left[\frac{i}{2k}\right]} \frac{(\pm 1)^s}{s!} u^{\pm(s+1)} - \sum_{s=2}^{\left[\frac{i+l}{k-\alpha}\right]} \frac{(\pm 1)^s}{s!} \frac{\partial^{s+1} v^{\pm}}{\partial t^{s+1}}. \tag{2.56; $i$}$$

Thus from (2.19; $i$) and (2.11; $i$) we get for $u_i(x)$ the following two boundary value problems in $G^-$ and $G^+$:

$$\begin{gathered} L(G\setminus\gamma)u_i = f \quad (\text{for } i > 0,\ f \equiv 0), \\ u_i|_{\gamma\pm 0} = \varphi_i^{\pm}, \\ Bu_i|_\Gamma = \varphi \quad (\varphi \equiv 0 \text{ for } i > 0). \end{gathered} \tag{2.57; $i$}$$

From Assumption 2.1 it follows easily that (2.57; $i$) has a unique solution. Having solved it, we find that

$$\frac{\partial^2 v_{i+2\alpha}^{+}}{\partial t^2}(0) + \frac{\partial^2 v_{i+2\alpha}^{-}}{\partial t^2}(0) = \theta_{i+2\alpha}, \tag{2.47; $i+2\alpha$}$$

where $\theta_{i+2\alpha}$ is an already known function of $\tau$; namely,

$$\theta_{i+2\alpha} = (u_i^{+\prime}(0) - u_i^{-\prime}(0)) + (\gamma_i^{+} - \gamma_i^{-}). \tag{2.51; $i+2\alpha$}$$

Now it is not difficult to find the first Cauchy datum for the function $v_{i+\alpha}(t;\tau)$. Note that we found the second datum earlier (see (2.20; $i+\alpha-k$)). We have

$$\frac{\partial v_{i+2\alpha}}{\partial t}(0) = \Psi_{i+2\alpha}, \tag{2.18; $i+2\alpha$}$$

where

$$\Psi_{i+2\alpha} = \sum_{s=0}^{\left[\frac{i+\alpha-k}{2k}\right]} \frac{1}{s!} u^{+(s+1)} - \sum_{s=1}^{\left[\frac{i+2\alpha}{k-\alpha}\right]} \frac{1}{s!} \frac{\partial^{s+1} v^{+}}{\partial t^{s+1}}.$$

The value of $v_{i+2\alpha}(0)$ is found in the following way. From (2.10; $i+2\alpha$), (2.3″), (2.45; $i+2\alpha$), (2.46), (2.47; $i+2\alpha$) and (2.18; $i+2\alpha$) we get

$$a_1\Psi_{i+2\alpha} + a_0 v_{i+2\alpha}(0) = \widehat{h}_{i+2\alpha} - \frac{a_2}{2}\theta_{i+2\alpha}, \tag{2.48; $i+2\alpha$}$$

and since $a_0 \neq 0$ (see (2.44)), we have

$$v_{i+2\alpha}(0) = \Phi_{i+2\alpha},$$

where

$$\Phi_{i+2\alpha} = \frac{1}{a_0}\left\{\hat{h}_{i+2\alpha} - \frac{a_2}{2}\,\theta_{i+2\alpha} - a_1\Psi_{i+2\alpha}\right\}. \qquad (2.53; i+2\alpha)$$

Thus (see 2.10; $i+2\alpha$), (2.49; $i+2\alpha$) and (2.18; $i+2\alpha$)) we get for $v_{i+2\alpha}(t;\tau)$ the following Cauchy problem ($0<|t|<\delta^{2k-l}$):

$$\begin{aligned} M_0 v_{i+2\alpha} &= h^{\pm}_{i+2\alpha} - M_1 v_{i+2\alpha-1} - \dots - M_{i+2\alpha} v_0,\\ v_{i+2\alpha}(0) &= \Phi_{i+2\alpha},\\ \frac{\partial v_{i+2\alpha}}{\partial t}(0) &= \Psi_{i+2\alpha}. \end{aligned} \qquad (2.50; i+2\alpha)$$

Having solved this problem, we define

$$\Psi_{i+l} = \sum_{s=0}^{\left[\frac{i}{2k}\right]} \frac{1}{s!}\, u^{+(s+1)} - \sum_{s=1}^{\left[\frac{i+l}{2k-l}\right]} \frac{1}{s!}\, \frac{\partial^{s+1} v^+}{\partial t^{s+1}}, \qquad (2.52; i+l)$$

and by virtue of (2.20; $i$) we get

$$\frac{\partial v_{i+l}}{\partial t}(0) = \Psi_{i+l}. \qquad (2.18; i+l)$$

So after the $i$th step we have the following functions:

(2.58; 0)–(2.58; $i$) $\qquad u_0(x), \dots, u_i(x)$,

(2.45; 0)–(2.54; $i+2\alpha$) $\qquad v_0(t;\tau), \dots, v_i(t;\tau), \dots, v_{i+2\alpha}(t;\tau)$,

and also

$$\frac{\partial v_0}{\partial t}(0) = \Psi_0, \dots, \frac{\partial v_{i+2\alpha}}{\partial t}(0) = \Psi_{i+2\alpha}, \dots, \frac{\partial v_{i+l}}{\partial t}(0) = \Psi_{i+l},$$

(2.18; 0)–(2.18; $i+l$)

where $\Psi_j$ is defined by (2.52; $j$).

4°. $\sigma = 1$.

REMARK 2.4. The case $l = 2k = 1$ ($\sigma = l/2k$) was investigated (if $T$ is a one-sided neighborhood of $\gamma$) by M. I. Višik and L. A. Ljusternik in [1].

Recall that for $\sigma = 1$ (2.21; 0)–(2.21; $2k-1$) are the initial conjugation conditions on $\gamma$ for $u_i$ and $v_j$ and that (2.22; $i$), (2.23; $i$) are the basic conditions.

Now consider the homogeneous Neumann boundary value problem on the interval $[-1, 1]$

$$\begin{aligned} M_0 v &= 0,\\ \frac{\partial v}{\partial t}(-1) &= 0,\\ \frac{\partial v}{\partial t}(+1) &= 0 \end{aligned} \qquad (2.59)$$

for an ordinary differential operator

(2.3′)
$$M_0 = \sum_{\beta\leqslant 2} a_\beta^\pm(\tau)\frac{\partial^\beta}{\partial t^\beta}, \quad 0<|t|<1.$$

*Case* A°. Problem (2.59) is not on a spectrum (i.e. for any $\tau$ (2.59) has only a trivial solution). In this case the iteration process proceeds as follows. From (2.10; 0) and (2.21; 0) we get the following boundary value problem for the function $v_0(t;\tau)$:

(2.60; 0)
$$M_0 v_0 = h_0^\pm$$
$$\frac{\partial v_0}{\partial t}(\pm 1) = 0.$$

Since the corresponding homogeneous problem, i.e. (2.59), has only a null solution, problem (2.60; 0) has a unique solution. Having solved this problem, we get from (2.11; 0) and (2.22; 0) the following boundary value problems for $u_0(x)$ in $G^-$ and in $G^+$:

(2.61; 0)
$$L(G\setminus\gamma)u_0 = f,$$
$$u_0|_{\gamma\pm 0} = v_0(\pm 1),$$
$$Bu_0|_\Gamma = \varphi.$$

It follows from Assumption 2.1 that these two boundary value problems have unique solutions. The initial step of the iteration is now finished.

Now do the $i$th step. For $i<2k$, $(\partial v_i/\partial t)(\pm 1) = 0$ by (2.21; $i$); for $i \geqq 2k$, we find the Neumann datum from (2.23; $i-2k$):

(2.21; $i$)
$$\frac{\partial v_i}{\partial t}(\pm 1) = \chi_i^\pm,$$

where $\chi_i^\pm$ is a known function of $\tau$ up to the $i$th step; namely,

(2.62; $i$)
$$\chi_i^\pm = \sum_{s=0}^{\left[\frac{i-2k}{2k}\right]} \frac{(\pm 1)^s}{s!} u^{\pm(s+1)}.$$

From (2.21; $i$) and (2.10; $i$) we get for $v_i(t;\tau)$ the following boundary value problem ($|t|<1$):

(2.60; $i$)
$$M_0 v_i = h_i^\pm - M_1 v_{i-1} - \dots - M_i v_0,$$
$$\frac{\partial v_i}{\partial t}(\pm 1) = \chi_i^\pm.$$

By hypothesis, problem (2.59) does not lie on a spectrum; thus (2.60; $i$) has a unique solution. Having solved it, we get from (2.11; $i$) and (2.22; $i$) the following boundary value problems for $u_i(x)$ in $G^-$ and in $G^+$:

(2.61; $i$)
$$L(G\setminus\gamma)u_i = 0,$$
$$u_i|_{\gamma\pm 0} = v_i(\pm 1) - \sum_{s=1}^{\left[\frac{i}{2k}\right]} \frac{(\pm 1)^s}{s!} u^{\pm(s)},$$
$$Bu_i|_\Gamma = 0.$$

These problems have unique solutions by virtue of Assumption 2.1. The $i$th step of the iteration for case A° is now finished.

CASE B°. Problem (2.59) lies on a spectrum.

REMARK. 2.5. Our treatment of this case is entirely analogous to that of [1] when $l = 1$ (i.e. $k = \frac{1}{2}$ and $\delta = \epsilon$).

However, the "splash base" $T_\epsilon$ is not necessarily a one-sided neighborhood of $\gamma$, as it was in [1]. The geometry of $T_\epsilon$ for $\sigma = 1$ has an essential effect on the asymptotics of a solution of (1.7), since the asymptotics depend on the presence or absence of a spectrum for the problem connected with $T_\epsilon$ (see Example 2.1 below).

We now do the initial step of the iteration. From (2.10; 0) and (2.21; 0) it follows that $v_0$ is a solution of the following problem:

$$\text{(2.63; 0)} \qquad \begin{gathered} M_0 v_0 = h_0^{\pm}, \\ \frac{\partial v_0}{\partial t}(\pm 1) = 0. \end{gathered}$$

The solvability condition of this problem consists in the orthogonality of the right side to the kernel of conjugations of the problem, i.e.

$$\text{(2.64)} \qquad (h_0^{\pm}, y) \equiv \int_{-1}^{1} h_0^{\pm} \cdot y \, dt = 0,$$

where $y = y(t; \tau)$ is the base solution of the homogeneous problem:

$$\text{(2.59*)} \qquad \begin{gathered} M_0^* y \equiv a_2^{\pm} \frac{\partial^2 y}{\partial t^2} - a_1^{\pm} \frac{\partial y}{\partial t} + a_0^{\pm} y = 0, \quad 0 < |t| < 1, \\ y(+0) - \frac{a_2^-}{a_2^+} y(-0) = 0, \\ \frac{\partial y}{\partial t}(+0) - \left[ \frac{a_2^-}{a_2^+} \frac{\partial y}{\partial t}(-0) + \left( a_1^+ \frac{a_2^-}{a_2^+} - a_1^- \right) y(-0) \right] = 0, \\ \left( a_2^{\pm} \frac{\partial y}{\partial t} - a_1^{\pm} y \right)_{t=\pm 1} = 0. \end{gathered}$$

Problem (2.59*) is conjugate to (2.59) because

$$\text{(2.65)} \qquad \begin{aligned} (M_0 v, y) - (v, M_0^* y) &= v \left( a_1^{\pm} y - a_2^{\pm} \frac{\partial y}{\partial t} \right) \Big|_{t=-1}^{t=1} \\ &+ a_2^{\pm} y \frac{\partial v}{\partial t} \Big|_{t=-1}^{t=1} + v \left( a_1^{\pm} y - a_2^{\pm} \frac{\partial y}{\partial t} \right) \Big|_{t=+0}^{t=-0} + a_2^{\pm} y \frac{\partial v}{\partial t} \Big|_{t=+0}^{t=-0} \end{aligned}$$

REMARK 2.6. If the coefficients $a_2^{\pm}$ and $a_1^{\pm}$ are continuous, then the condition of special conjugation of a solution on $\gamma$ in (2.59*) is transformed into a coalescence condition of class $C^1$, and the problem itself becomes an ordinary boundary value problem on the interval $[-1, 1]$:

(2.59*)
$$M_0^* y \equiv a_2 \frac{\partial^2 y}{\partial t^2} - a_1 \frac{\partial y}{\partial t} + a_0^{\pm} y = 0$$
$$\left( a_2 \frac{\partial y}{\partial t} - a_1 y \right)\Big|_{t=\pm 1} = 0.$$

We will assume that condition (2.64) is satisfied, i.e. that (2.63; 0) is solvable. Its solution has the form

(2.66; 0)
$$v_0 = \widetilde{v}_0 + c_0 \omega,$$

$\tilde{v}_0 = \tilde{v}_0(t; \tau)$ is a particular solution of (2.63; 0), $\omega = \omega(t; \tau)$ is a characteristic vector of (2.59), and $c_0 = c_0(\tau)$ is an as yet unknown constant (with respect to $t$). The basic conjugation condition gives us

(2.22; 0)
$$u_0^{\pm}(0) = v_0(\pm 1),$$

so that

(2.67; 0)
$$c_0 = \frac{u_0^-(0) - \widetilde{v}_0(-1)}{\omega(-1)}.$$

Furthermore, from (2.66; 0) and (2.22; 0) we get the first conjugation condition for $u_0$ on $\gamma$: namely,

(2.68; 0)
$$u_0^+(0) - \frac{\omega(+1)}{\omega(-1)} u_0^-(0) = \left( - \frac{\omega(+1)}{\omega(-1)} \widetilde{v}_0(-1) + \widetilde{v}_0(+1) \right).$$

To get the second condition, consider the problem for which $v_i(t; \tau)$ is a solution. From (2.10; 1) and (2.23; 0) it follows that

(2.63; 1)
$$M_0 v_1 = h_1^{\pm} - M_1 v_0,$$
$$\frac{\partial v_1}{\partial t}(\pm 1) = u_0^{\pm\prime}(0).$$

The solvability condition (2.63; 1) consists in the fact that (see (2.65))

(2.69; 0)
$$(h_1^{\pm} - M_1 v_0, y) = a_2^{\pm} y . \frac{\partial v_1}{\partial t}\Big|_{t=-1}^{t=1},$$

where $y$ is the base solution of (2.59*). By virtue of (2.23; 0), (2.66; 0) and (2.67; 0), condition (2.69; 0) gives the following second conjugation $u_0(x)$ in $\gamma$:

(2.70; 0)
$$q^+ u_0^{+\prime}|_\gamma - (q^- u_0^{-\prime} + r u_0^-)|_\gamma = \chi_0,$$

where

(2.71)
$$q^{\pm} = a_2^{\pm} y(\pm 1),$$

(2.72)
$$r = - \frac{(M_1 \omega, y)}{\omega(-1)},$$

(2.73; 0)
$$\chi_0 = -(M_1 \widetilde{v}_0, y) + \frac{\widetilde{v}_0(-1)}{\omega(-1)} (M_1 \omega, y) + (h_1^{\pm}, y).$$

Thus from (2.11; 0), (2.68; 0), and (2.70; 0) we get for $u_0(x)$ the following boundary value problem with conjugation on $\gamma$:

$$
\begin{aligned}
&L(G\setminus\gamma)\,u_0 = f,\\
&u_0^+|_\gamma - p u_0^-|_\gamma = \psi_0,\\
&q^+u_0^{+\prime}|_\gamma - (q^-u_0^{-\prime} + r u_0^-)|_\gamma = \chi_0,\\
&Bu_0|_\Gamma = \varphi,
\end{aligned}
\tag{2.74; 0}
$$

where

$$
p = \frac{\omega(+1)}{\omega(-1)},
\tag{2.75}
$$

$$
\psi_0 = -\frac{\omega(+1)}{\omega(-1)}\,\tilde{v}_0(-1) + \tilde{v}_0(+1).
\tag{2.76; 0}
$$

ASSUMPTION 2.2. Let the homogeneous problem

$$
\begin{aligned}
&L(G\setminus\gamma)\,u = 0,\\
&u^+|_\gamma - p u^-|_\gamma = 0,\\
&q^+u^{+\prime}|_\gamma - (q^-u^{-\prime} + r u^-)|_\gamma = 0,\\
&Bu|_\Gamma = 0
\end{aligned}
\tag{2.74°}
$$

have only the null solution (for example, if the hypotheses of Theorem 1.3 are satisfied: $pq^+q^- > 0$, $q^+q^-r \geqq 0$, $p = \text{const}$). Then by Theorem 1.2 the corresponding nonhomogeneous problem (in particular, problem (2.74; 0) has a unique solution. At the same time we know $u_0(x)$, and consequently we know $c_0$ (see (2.67; 0)) and $v_0(t;\tau)$ (see (2.66; 0)). The initial step of the iteration is now finished.

Now we do the $i$th step. From (2.10; $i$) and (2.23; $i-1$) it follows that $v_i$ is a solution of the following problem:

$$
\begin{aligned}
&M_0 v_i = h_i^{\pm} - M_1 v_{i-1} - \dots - M_i v_0,\\
&\frac{\partial v_i}{\partial t}(\pm 1) = \sum_{s=0}^{i-1} \frac{(\pm 1)^s}{s!} u^{\pm(s+1)}.
\end{aligned}
\tag{2.63; i}
$$

This problem is solvable, as was shown in the preceding step. A general solution of (2.63; $i$) has the form

$$
v_i = \tilde{v}_i + c_i\omega,
\tag{2.66; i}
$$

where $\tilde{v}_i = \tilde{v}_i(t;\tau)$ is a particular solution of this problem, $\omega = \omega(t;\tau)$ is a characteristic vector of (2.59) (see above), and $c_i = c_i(\tau)$ is an as yet unknown constant (with respect to $t$). The basic conjugation condition gives

$$
\sum_{s=0}^{i} \frac{(\pm 1)^s}{s!} u^{\pm(s)} = \tilde{v}_i(\pm 1) + c_i\omega(\pm 1).
\tag{2.22; i}
$$

whence

$$c_i = \frac{1}{\omega(\pm 1)} \left\{ \sum_{s=0}^{i} \frac{(\pm 1)^s}{s!} u^{\pm(s)} - \tilde{v}_i(\pm 1) \right\} \tag{2.67; i}$$

and consequently the first conjugation condition for $u_i$ on $\gamma$ is as follows:

$$u_i^+ |_\gamma - p u_i^- |_\gamma = \psi_i, \tag{2.68; i}$$

where

$$\psi_i = \left\{ p \left( \sum_{s=1}^{i} \frac{(-1)^s}{s!} u^{-(s)} - \tilde{v}_i(-1) \right) - \left( \sum_{s=1}^{i} \frac{1}{s!} u^{+(s)} - \tilde{v}_i(+1) \right) \right\}. \tag{2.67; i}$$

To get the second conjugation condition for $u_i$ on $\gamma$ we consider the problem for which $v_{i+1}(t; \tau)$ should be a solution. From (2.10; $i+1$) and (2.23; $i$) it follows that

$$\begin{aligned} &M_0 v_{i+1} = -M_1 v_i + (h_{i+1}^{\pm} - M_2 v_{i-2} - \ldots - M_i v_0), \\ &\frac{\partial v_{i+1}}{\partial t}(\pm 1) = u_i^{\pm\prime}(0) + \sum_{s=1}^{i} \frac{(\pm 1)^s}{s!} u^{\pm(s+1)}. \end{aligned} \tag{2.63; i+1}$$

The solvability condition for (2.63; $i+1$) consists in the fact that (see (2.65))

$$(-M_1 v_i, y) + (h_{i+1}^{\pm} - M_2 v_{i-1} - \ldots - M_{i+1} v_0, y) = a_2^{\pm} y \frac{\partial v_{i+1}}{\partial t} \Big|_{t=-1}^{t=1}, \tag{2.69; i}$$

where $y$ is a solution of (2.59*). By (2.23; $i$), (2.66; $i$), and (2.67; $i$) condition (2.69; $i$) yields the following second conjugation of the function $u_i$ on $\gamma$:

$$q^+ u_i^{+\prime} |_\gamma - (q^- u_i^{-\prime} + r u_i^-) |_\gamma = \chi_i, \tag{2.70; i}$$

where

$$\begin{aligned} \chi_i = &-(M_1 \tilde{v}_i, y) + \frac{1}{\omega(-1)} \left( \tilde{v}_i(-1) - \sum_{s-1}^{i} \frac{(-1)^s}{s!} u^{-(s)} \right) (M_1 \omega, y) \\ &+ (h_{i+1}^{\pm} - M_2 v_{i-1} - \ldots - M_0 v_{i+1}, y) - a_2^+ y(+1) \sum_{s=1}^{i} \frac{1}{s!} u^{+(s+1)} \\ &+ a_2^- y(-1) \sum_{s=1}^{i} \frac{(-1)^s}{s!} u^{-(s+1)}. \end{aligned} \tag{2.73; i}$$

Thus from (2.11; $i$), (2.68; $i$), and (2.70; $i$) we get for $u_i(x)$ the following boundary value problem with conjugation on $\gamma$:

$$\begin{aligned} &L(G \setminus \gamma) u_i = 0, \qquad q^+ u_i^{+\prime} |_\gamma - (q^- u_i^{-\prime} + r u_i^-) |_\gamma = \chi_i, \\ &u_i^+ |_\gamma - p u_i^- |_\gamma = \psi_i, \qquad B u_i |_\Gamma = 0. \end{aligned} \tag{2.74; i}$$

Since, according to Assumption 2.2, the corresponding homogeneous problem (2.74°) has only the null solution, problem (2.74; $i$) has a unique solution (see Theorem 1.2). At the same time we know $u_i(x)$, and consequently we know $c_i$ (see (2.67; $i$)) and $v_i(t;\tau)$ (see (2.66; $i$)). Thus the $i$th step of the iteration for Case B° is finished.

We constructed earlier a formal expansion of the solution $u_\varepsilon$ of (1.7) in powers of $\delta$. We now show that this is an asymptotic expansion. We introduce the function

$$u_\varepsilon^N = \begin{cases} u_0 + \dots + \delta^N u_N \text{ в } G_\varepsilon, \\ v_0 + \dots + \delta^N v_N + \dots + \delta^{N+2l} v_{N+2l} \text{ in } T_\varepsilon. \end{cases} \tag{2.77}$$

We define the behavior of this function on $\gamma_\varepsilon^\pm$. We have

$$u_\varepsilon^N(\pm\varepsilon\pm 0) = u_0(\pm\varepsilon) + \dots + \delta^N u_N(\pm\varepsilon)$$
$$= \sum_{i=0}^{N} \delta^i \sum_{s=0}^{\left[\frac{i}{2k}\right]} \frac{(\pm 1)^s}{s!} u^{\pm(s)} + O(\delta^{N+1}),$$

$$u_\varepsilon^N(\pm\varepsilon\mp 0) = v_0(\pm\delta^{2k-l}) + \dots + \delta^{N+2l} v_{N+2l}(\pm\delta^{2k-l})$$
$$= \sum_{i=0}^{N} \delta^i \sum_{s=0}^{\left[\frac{i}{2k-l}\right]} \frac{(\pm 1)^s}{s!} \frac{\partial^s v^\pm}{\partial t^s} + O(\delta^{N+1}) \text{ for } \sigma<1,$$

$$u_\varepsilon^N(\pm\varepsilon\mp 0) = \sum_{i=0}^{N} \delta^i v_i(\pm 1) + O(\delta^{N+1}) \text{ for } \sigma = 1.$$

Hence in view of the fact that

$$\sum_{s=0}^{\left[\frac{i}{2k}\right]} \frac{(\pm 1)^s}{s!} u^{\pm(s)} = \sum_{s=0}^{\left[\frac{i}{2k-l}\right]} \frac{(\pm 1)^s}{s!} \frac{\partial^s v^\pm}{\partial t^s} \text{ for } \sigma<1 \tag{2.19; i}$$

and

$$\sum_{s=0}^{\left[\frac{i}{2k}\right]} \frac{(\pm 1)^s}{s!} u^{\pm(s)} = v_i(\pm 1) \text{ for } \sigma = 1 \tag{2.22; i}$$

we get

$$\urcorner\!\!\lfloor\, u_\varepsilon^N(\pm\varepsilon) = O(\delta^{N+1}). \tag{2.78}$$

Moreover,

$$(u_\varepsilon^N)'(\pm\varepsilon\pm 0) = u_0'(\pm\varepsilon) + \dots + \delta^N u_N'(\pm\varepsilon) =$$

$$= \sum_{i=0}^{N} \delta^i \sum_{s=0}^{\left[\frac{i}{2k}\right]} \frac{(\pm 1)^s}{s!} u^{\pm(s+1)} + O(\delta^{N+1}).$$

For $(u_\epsilon^N)'(\pm\epsilon\mp 0)$ we have for $\sigma < 1$ (taking (2.18; 0)–(2.18; $l-1$) into account) that

$$(u_\varepsilon^N)'(\pm\varepsilon\mp 0) = \sum_{i=0}^{N} \delta^i \sum_{s=0}^{\left[\frac{i+l}{2k-l}\right]} \frac{(\pm 1)^s}{s!} \frac{\partial^{s+1} v^{\pm}}{\partial t^{s+1}} + O(\delta^{N+1}),$$

and for $\sigma = 1$ (using (2.21; 0)–(2.21; $2k-1$)) we get

$$(u_\varepsilon^N)'(\pm\varepsilon\mp 0) = \sum_{i=0}^{N} \delta^i \frac{\partial v_{i+2k}}{\partial t}(\pm 1) + O(\delta^{N+1}).$$

Hence (see (2.20; $i$) for $\sigma < 1$ and (2.23; $i$) for $\sigma = 1$) we have

(2.79) $$\rceil\!\!\lfloor\, (u_\varepsilon^N)'(\pm\varepsilon) = O(\delta^{N+1}).$$

We now modify the function $u_\epsilon^N$ so that it will be possible to apply a priori estimate (1.8) to the modified $\tilde{u}_\epsilon^N = u_\epsilon^N + \delta^{N+1}\alpha_{N,\epsilon}$ (see (2.78), (2.79)), i.e. $\tilde{u}_\epsilon^N$ satisfies on $\gamma_\epsilon^\pm$ a coalescence condition of class $C^1$:

$$\left.\frac{\partial^\beta \tilde{u}_\varepsilon^N}{\partial \nu^\beta}\right|_{\gamma_\varepsilon^\pm+0} = \left.\frac{\partial^\beta \tilde{u}_\varepsilon^N}{\partial \nu^\beta}\right|_{\gamma_\varepsilon^\pm-0}, \quad \beta = 0, 1;$$

moreover $\alpha_{N,\epsilon}$ equals zero in $T_\epsilon$ and is different from zero near $T_\epsilon$. Obviously $L_\epsilon|_{G_\epsilon}\alpha_{N,\epsilon} = O(1)$.

Let $u_\epsilon$ be a solution of the original problem (1.7). Apply to $(u_\epsilon - \tilde{u}_\epsilon^N)$ the a priori estimate (1.8):

(1.8′)
$$\begin{aligned}\|u_\varepsilon - \tilde{u}_\varepsilon^N\| &\leqslant C\{\|L_\varepsilon(u_\varepsilon - \tilde{u}_\varepsilon^N)\|_{G_0} + \|B(u_\varepsilon - \tilde{u}_\varepsilon^N)|_\Gamma\|'\}\\ &= C\{\|L_\varepsilon(u_\varepsilon - u_\varepsilon^N)\|_{T_\varepsilon} + \|L_\varepsilon(u_\varepsilon - \tilde{u}_\varepsilon^N)\|_{G_\varepsilon} + \|B(u_\varepsilon - u_\varepsilon^N)|_\Gamma\|'\}.\end{aligned}$$

We evaluate $\|L_\epsilon(u_\epsilon - u_\epsilon^N)\|_{T_\epsilon}$. Since $u_\epsilon$ is a solution,

$$L_\varepsilon|_{T_\varepsilon} u_\varepsilon = \frac{1}{\delta^{2l}}(h_0^\pm + \ldots + \delta^{N+2l} h_{N+2l}^\pm) + \delta^{N+1} h_{N+2l;\,\varepsilon}^\pm.$$

Further, using a "decomposition" of the operator $L_\epsilon(T_\epsilon \setminus \gamma)$ (see (2.2)), we have

$$\begin{aligned}L_\varepsilon(T_\varepsilon \setminus \gamma)\, u_\varepsilon^N = \frac{1}{\delta^{2l}}(M_0 + \ldots + \delta^{N+2l} M_{N+2l}\\ + \delta^{N+2l+1} M_{N+2l+1;\varepsilon})(v_0 + \ldots + \delta^{N+2l} v_{N+2l}) =\end{aligned}$$

$$= \frac{1}{\delta^{2l}} [(M_0 v_0) + \dots + \delta^{N+2l} (M_0 v_{N+2l} + \dots + M_{N+2l} v_0)]$$

$$+ \delta^{N+1} [(M_1 v_{N+2l} + \dots + M_{N+2l+1;\varepsilon} v_0) + \dots + \delta^{N+2l} (M_{N+2l+1;\varepsilon} v_{N+2l})].$$

Hence in view of (2.10; $j$) we have

$$L_\varepsilon (T_\varepsilon \setminus \gamma)(u_\varepsilon - u_\varepsilon^N) = O(\delta^{N+1})$$

and consequently

$$\| L_\varepsilon (u_\varepsilon - u_\varepsilon^N) \|_{T_\varepsilon} = O(\delta^{N+1}).$$

Analogously (with regard to $L_\varepsilon |_{G_\varepsilon} \alpha_{N;\varepsilon} = O(1)$) we get

$$\| L_\varepsilon (u_\varepsilon - \tilde{u}_\varepsilon^N) \|_{G_\varepsilon} = O(\delta^{N+1}),$$

$$\| B (u_\varepsilon - u_\varepsilon^N) |_\Gamma \|' = O(\delta^{N+1}).$$

Thus by virtue of estimate (1.8′) we have

$$\| u_\varepsilon - \tilde{u}_\varepsilon^N \| = O(\delta^{N+1})$$

and finally

(2.80) $$u_\varepsilon = u_\varepsilon^N + \delta^{N+1} z_{N;\varepsilon}, \text{ where } \| z_{N;\varepsilon} \| = O(1).$$

We have thus proved the following basic theorem.

THEOREM 2.2. *Let the conditions of Assumption 2.1 be fulfilled for operators $L_\varepsilon$ and $L(G \setminus \gamma)$. Then the solution $u_\varepsilon$ of (1.7) has the form*

(2.80) $$u_\varepsilon = u_\varepsilon^N + \delta^{N+1} z_{N;\varepsilon}, \quad \| z_{N;\varepsilon} \| = O(1),$$

(2.77) $$u_\varepsilon^N = \begin{cases} u_0 + \dots + \delta^N u_N & \text{in } G_\varepsilon, \\ v_0 + \dots + \delta^N v_N + \dots + \delta^{N+2l} v_{N+2l} & \text{in } T_\varepsilon, \end{cases}$$

*where the functions $u_i$ and $v_j$ are defined in the following manner (depending on $\sigma = l/2k$).*

*1°. $0 < \sigma < \frac{1}{2}$. $u_i = u_i(x)$ are solutions of the following problems with conjugation on $\gamma$:*

(2.24; $i$)
$$\begin{aligned} &L(G \setminus \gamma) u_i = f \quad (f \equiv 0 \ \textit{for} \ i \geqslant 1), \\ &u_i |_{\gamma+0} - u_i |_{\gamma-0} = \psi_i, \\ &\left.\frac{\partial u_i}{\partial \nu}\right|_{\gamma+0} - \left.\frac{\partial u_i}{\partial \nu}\right|_{\gamma-0} = \chi_i, \\ &B u_i |_\Gamma = \varphi \quad (\varphi \equiv 0 \ \textit{for} \ i \geqslant 1), \end{aligned}$$

*where $\psi_i$ and $\chi_i$ are given by formulas (2.26; $i$) and (2.27; $i$). The functions $v_j = v_j(\nu/\delta^l; \tau)$ are defined in $T_\varepsilon$ as solutions of the following Cauchy problems for an ordinary differential operator $M_0$ (see (2.3′)):*

$$M_0 v_j = h_j^{\pm} - M_1 v_{j-1} - \ldots - M_j v_0,$$

(2.25; $j$)
$$v_j(0) = \xi_j,$$
$$\frac{\partial v_j}{\partial t}(0) = \eta_j,$$

*where* $\xi_j$ *and* $\eta_j$ *are given by* (2.29; $j$) *and* (2.30; $j$).

2°. $\sigma = \frac{1}{2}$. *We assume in addition that the leading coefficient of* $M_0$ *is continuous, i.e.*

(2.31)
$$a_2^{+}(\tau) = a_2^{-}(\tau) = a_2(\tau).$$

*In this case the functions* $u_i = u_i(x)$ *are solutions of problems with conjugation on* $\gamma$:

$$L(G \setminus \gamma) u_i = f \quad (f \equiv 0 \ \text{for} \ i \geqslant 1)$$
$$u_i|_{\gamma+0} - u_i|_{\gamma-0} = \psi_i,$$

(2.35; $i$)
$$\left.\frac{\partial u_i}{\partial \nu}\right|_{\gamma+0} - \left.\left(\frac{\partial u_i}{\partial \nu} + r u_i\right)\right|_{\gamma-0} = \chi_i,$$
$$Bu_i|_\Gamma = \varphi \ (\varphi \equiv 0 \ \text{for} \ i \geqslant 1),$$

*where* $\psi_i$, $\chi_i$ *and* $r$ *are given by* (2.43; $i$), (2.37; $i$) *and* (2.36). *The functions* $v_j = v_j(\nu/\sqrt{\epsilon}; \tau)$ *are defined in* $T_\epsilon$ *as solutions of the following Cauchy problems*:

$$M_0 v_j = h_j^{\pm} - M_1 v_{j-1} - \ldots - M_j v_0,$$

(2.38; $j$)
$$v_j(0) = u_j^{+}(0) - \alpha_j^{+},$$
$$\frac{\partial v_j}{\partial t}(0) = \zeta_j,$$

*where* $\alpha_j^{+}$, $\zeta_j$ *are given by* (2.39; $j$) *and* (2.42; $j$).

3°. $\frac{1}{2} < \sigma < 1$. *We assume in addition that the coefficients of the operator*

(2.3′)
$$M_0 = \sum_{\beta \leqslant 2} a_\beta^{\pm}(\tau) \frac{\partial^\beta}{\partial t^\beta}, \quad 0 < |t| < \delta^{2k-l}$$

*satisfy the conditions*

(2.31)
$$a_2^{+}(\tau) = a_2^{-}(\tau) = a_2(\tau),$$

(2.44)
$$a_0^{+}(\tau) + a_0^{-}(\tau) \neq 0.$$

*In this case the functions* $u_i = u_i(x)$ *are solutions of the following boundary value problems in* $G^-$ *and in* $G^+$:

$$L(G \setminus \gamma) u_i = f \ (f \equiv 0 \ \text{for} \ i \geqslant 1),$$

(2.57; $i$)
$$u_i|_{\gamma \pm 0} = \varphi_i^{\pm},$$
$$Bu_i|_\Gamma = \varphi \ (\varphi \equiv 0 \ \text{for} \ i \geqslant 1),$$

*where* $\phi_i^{\pm}$ *are given by* (2.55; $i$). *The functions* $v_j = v_j(\nu/\delta^l; \tau)$ *are defined in*

$T_\epsilon$ *as solutions of the following Cauchy problems:*

$$
(2.50; j) \qquad \begin{aligned} &M_0 v_j = h_j^{\pm} - M_1 v_{j-1} - \ldots - M_j v_0, \\ &v_j(0) = \Phi_j, \\ &\frac{\partial v_j}{\partial t}(0) = \Psi_j, \end{aligned}
$$

*where* $\Phi_j$ *and* $\Psi_j$ *are given by* (2.53; $j$) *and* (2.52; $j$).

4°. $\sigma = 1$. *Case* A°, *i.e. the problem*

$$
(2.59) \qquad \begin{aligned} &M_0 v = 0, \quad -1 < t < 1, \\ &\frac{\partial v}{\partial t}(\pm 1) = 0 \end{aligned}
$$

*has only the null solution. In this case the functions* $u_i = u_i(x)$ *are solutions of the following boundary value problems in* $G^-$ *and in* $G^+$:

$$
(2.61; i) \qquad \begin{aligned} &L(G \setminus \gamma) u_i = f \ (f \equiv 0 \ \textit{for} \ i \geqslant 1), \\ &u_i|_{\gamma \pm 0} = v_i(\pm 1) - \sum_{s=1}^{\left[\frac{i}{2k}\right]} \frac{(\pm 1)^s}{s!} u^{\pm(s)}, \\ &Bu_i|_\Gamma = \varphi \ (\varphi \equiv 0 \ \textit{for} \ i \geqslant 1). \end{aligned}
$$

*The functions* $v_j = v_j(\nu/\epsilon; \tau)$ *are defined as solutions of Neumann problems for the operator* $M_0$:

$$
(2.60; j) \qquad \begin{aligned} &M_0 v_j = h_j^{\pm} - M_1 v_{j-1} - \ldots - M_j v_0, \\ &\frac{\partial v_j}{\partial t}(\pm 1) = \chi_j^{\pm}, \end{aligned}
$$

*where* $\chi_j^{\pm}$ *are given by* (2.62; $j$).

*Case* B°, *i.e. problem* (2.59) *lies on a spectrum.*[5] *In addition we assume that Assumption* 2.2 *holds. Let the condition*

$$
(2.64) \qquad \int_{-1}^{1} h_0^{\pm} y \, dt = 0,
$$

*where* $y = y(t; \tau)$ *is a solution of* (2.59*), *be satisfied.*[6] *In this case* $u_i = u_i(x)$ *are solutions of the following boundary value problems with conjugation on* $\gamma$:

---

[5] Case B° is investigated for $l = 1$ (i.e. $k = \frac{1}{2}$, $\delta = \epsilon$).

[6] If (2.64) is not satisfied, then the solution $u_\epsilon$ of (1.7) is sought (see [1]) in the form $u_\epsilon = \bar{u}_\epsilon / \epsilon$, where $\bar{u}_\epsilon = u_0 + \epsilon u_1 + \cdots$ in $G_\epsilon$ and $\bar{u}_\epsilon = v_0 + \epsilon v_1 + \cdots$ in $T_\epsilon$. Thus the problem reduces to the case under consideration.

(2.74; $i$)
$$\begin{gathered} L(G\setminus\gamma)\, u_i = f \ (f \equiv 0 \ \text{for} \ i \geqslant 1), \\ u_i|_{\gamma+0} - pu_i|_{\gamma-0} = \psi_i, \\ q^+ \frac{\partial u_i}{\partial \nu}\Big|_{\gamma+0} - \left(q^- \frac{\partial u_i}{\partial \nu} + ru_i\right)\Big|_{\gamma-0} = \chi_i, \\ Bu_i|_\Gamma = \varphi \ (\varphi \equiv 0 \ \text{for} \ i \geqslant 1), \end{gathered}$$

*where* $p$, $q^\pm$, $r$, $\psi_i, \chi_i$ *are given by* (2.75), (2.71), (2.72), (2.76; $i$), (2.73; $i$). *The functions* $v_j = v_j(\nu/\epsilon; \tau)$ *are defined as solutions of Neumann problems for the operator* $M_0$:

(2.63; $j$)
$$\begin{gathered} M_0 v_j = h_j^\pm - M_1 v_{j-1} - \dots - M_j v_0, \\ \frac{\partial v_j}{\partial t}(\pm 1) = \sum_{s=0}^{i-1} \frac{(\pm 1)^s}{s!} u^{\pm(s+1)}. \end{gathered}$$

Two corollaries (the second is basic) follow from Theorem 2.2.

COROLLARY 2.1. *Let* $0 < \sigma_1 < \sigma_2 < \frac{1}{2}$ *or* $\frac{1}{2} < \sigma_1 < \sigma_2 < 1$, *where* $\sigma_1$ *and* $\sigma_2$ *are rational numbers. Let* $Q$ *be the least common multiple of the denominators of* $\sigma_1$ *and* $\sigma_2$. *Then for any* $\sigma = l/Q$ *such that* $\sigma_1 \leqq \sigma \leqq \sigma_2$, *asymptotic expansions of a solution* $u_\epsilon$ *of* (1.7) *in powers of* $\epsilon^{1/Q}$ *coincide in some of the first terms.*

COROLLARY 2.2. *Let the hypotheses of Theorem 2.2 be satisfied. Then the limiting solution* $u_0 = \lim_{\epsilon\to 0} u_\epsilon$ *of* (1.7) *is defined in the following way (depending on* $\sigma$*).*

1°. $0 < \sigma < \frac{1}{2}$.

(2.24; 0)
$$\begin{gathered} L(G\setminus\gamma)\, u_0 = f, \\ u_0|_{\gamma+0} - u_0|_{\gamma-0} = 0, \\ \frac{\partial u_0}{\partial \nu}\Big|_{\gamma+0} - \frac{\partial u_0}{\partial \nu}\Big|_{\gamma-0} = 0, \\ Bu_0|_\Gamma = \varphi. \end{gathered}$$

2°. $\sigma = \frac{1}{2}$.

(2.35; 0)
$$\begin{gathered} L(G\setminus\gamma)\, u_0 = f, \\ u_0|_{\gamma+0} - u_0|_{\gamma-0} = 0, \\ \frac{\partial u_0}{\partial \nu}\Big|_{\gamma+0} - \left(\frac{\partial u_0}{\partial \nu} + ru_0\right)\Big|_{\gamma-0} = \chi_0, \\ Bu_0|_\Gamma = \varphi, \end{gathered}$$

*where*

(2.36)
$$r = -\frac{1}{a_2(0;\tau)}(a_0(+0;\tau) + a_0(-0;\tau)),$$

(2.37; 0)
$$\chi_0 = \frac{1}{a_2(0;\tau)}(h_0(+0;\tau) + h_0(-0;\tau)).$$

3°. $\frac{1}{2} < \sigma < 1$.

(2.57; 0)
$$\begin{aligned} L(G\setminus\gamma)\, u_0 &= f, \\ u_0|_\gamma &= \varphi_0, \\ Bu_0|_\Gamma &= \varphi, \end{aligned}$$

*where*

(2.55; 0)
$$\varphi_0 = \frac{h_0(+0;\tau)+h_0(-0;\tau)}{a_0(+0;\tau)+a_0(-0;\tau)}.$$

4°. $\sigma = 1$. *Case* A°.

(2.61; 0)
$$\begin{aligned} L(G\setminus\gamma)\, u_0 &= f, \\ u_0|_{\gamma\pm 0} &= v_0(\pm 1;\tau), \\ Bu_0|_\Gamma &= \varphi \end{aligned}$$

*where $v_0(t;\tau)$ is a solution of the boundary value problem*

(2.60; 0)
$$M_0 v_0 \equiv \sum_{\beta\leqslant 2} a_\beta^\pm(\tau)\frac{\partial^\beta v_0}{\partial t^\beta} = h_0^\pm, \quad |t|<1,$$
$$\frac{\partial v_0}{\partial t}(\pm 1) = 0.$$

*Case* B°.

(2.74; 0)
$$\begin{aligned} L(G\setminus\gamma)\, u_0 &= f, \\ u_0|_{\gamma+0} - pu_0|_{\gamma-0} &= 0, \\ q^+\frac{\partial u_0}{\partial \nu}\Big|_{\gamma+0} - \left(q^-\frac{\partial u_0}{\partial \nu} + ru_0\right)\Big|_{\gamma-0} &= 0, \\ Bu_0|_\Gamma &= \varphi, \end{aligned}$$

*where*

(2.75)
$$p = \frac{\omega(+1)}{\omega(-1)},$$

(2.71)
$$q^\pm = a_2^\pm y(\pm 1),$$

(2.72)
$$r = -\frac{1}{\omega(-1)}\int_{-1}^{1} M_1\omega y\,dt,$$

*$\omega$ is a characteristic vector of the problem*

(2.59)
$$M_0\omega = 0,$$
$$\frac{\partial\omega}{\partial t}(\pm 1) = 1,$$

*$y$ is a solution of the conjugate problem (2.59*), and $M$ is the second operator coefficient in "decomposition" (2.2) of the operator $L_\epsilon(T_\epsilon\setminus\gamma)$ in powers of $\epsilon$.*

REMARK 2.7. From Corollary 2.1 it follows, in particular, that Corollary 2.2 is valid for any real $\sigma$, i.e. when the "splashes" of coefficients of the limiting problem are represented in the form (1.1′) for any real $\sigma$.

EXAMPLE 2.1. Consider the problem

$$\begin{cases} L_\varepsilon u_\varepsilon = f, \\ u_\varepsilon|_\Gamma = \varphi, \end{cases}$$

where

$$L_\varepsilon = \begin{cases} \Delta = \sum_{i=1}^{n} \frac{\partial^2}{\partial x_i^2} \text{ in } G_\varepsilon, \\ \frac{\partial^2}{\partial \nu^2} + \sum_{j=2}^{n} \frac{\partial^2}{\partial \tau_j^2} + \left(\frac{a(\nu;\tau)}{\varepsilon}\right)^2 \text{ in } T_\varepsilon. \end{cases}$$

We have (see (2.2), (2.3′))

$$M_0 = \frac{\partial^2}{\partial t^2} + (a^\pm(\tau))^2, \quad M_1 = 0.$$

Consider the case in which the problem

$$\text{(2.59)} \qquad M_0\omega = 0, \quad -1 < t < 1, \qquad \frac{\partial \omega}{\partial t}(\pm 1) = 0$$

lies on a spectrum. It is clear that in this example problem (2.59) and its conjugate (2.59*) coincide, so that the characteristic vectors of these problems are proportional. Consider two cases.

1°. $a(+0;\tau) = a(-0;\tau) = a$ (see the introductory example). In this case problem (2.59) is on a spectrum only for

$\alpha$) $a = k\pi$; then $\omega = C\cos k\pi t$, and since

$$\frac{\omega(+1)}{\omega(-1)} = \frac{y(-1)}{y(+1)} = 1,$$

we have for $u_0$ the following problem (see (2.74; 0)):

$$\Delta u_0 = f, \qquad u_0^{+\prime}|_\gamma - u_0^{-\prime}|_\gamma = 0,$$
$$u_0^+|_\gamma - u_0^-|_\gamma = 0; \qquad u_0|_\Gamma = \varphi;$$

$\beta$) $a = (k\pi + \pi/2)$; then $\omega = C\sin(k\pi + \pi/2)t$; thus $\omega(+1)/\omega(-1) = y(-1)/y(+1) = -1$, and consequently we get for $u_0$ the problem (see (2.74; 0))

$$\Delta u_0 = f,$$
$$u_0^+|_\gamma + u_0^-|_\gamma = 0,$$
$$u_0^{+\prime}|_\gamma + u_0^{-\prime}|_\gamma = 0,$$
$$u_0|_\Gamma = \varphi.$$

2°. $a(+0;\tau) = a$, $a(-0;\tau) = 0$. This corresponds to a one-sided neighborhood of $\gamma$ (see [1]). From the coalescence conditions (of class $C^1$ relative to direction $\nu$) for $\omega(\nu/\epsilon;\tau)$ on $\gamma$ and also from the boundary conditions $(\partial\omega/\partial t)(\pm 1) = 0$ $(t = \nu/\epsilon)$ it follows that problem (2.59) lies on a spectrum only for $a = k\pi$, and moreover the characteristic vector

$$\omega = \begin{cases} D\cos k\pi t, & \text{if } \quad 0 < t < 1, \\ D \quad , & \text{if } \quad -1 < t < 0. \end{cases}$$

Since $\omega(+1)/\omega(-1) = y(-1)/y(+1) = (-1)^k$, for $u_0$ we have the problem

$$\Delta u_0 = f,$$
$$u_0^+|_\gamma - (-1)^k u_0^-|_\gamma = 0,$$
$$u_0^{+\prime}|_\gamma - (-1)^k u_0^{-\prime}|_\gamma = 0,$$
$$u_0|_\Gamma = \varphi.$$

Received 20 SEPT 1967

## Bibliography

[1] M. I. Višik and L. A. Ljusternik, *The asymptotic behavior of solutions of linear differential equations with large or quickly changing coefficients and boundary conditions*, Uspehi Mat. Nauk **15** (1960), no. 4 (94), 27–95 = Russian Math. Surveys **15** (1960), no. 4, 23–91. MR 23 # A1919.

[2] ———, *Regular degeneration and boundary layer for linear differential equations with small parameter*, Uspehi Mat. Nauk **12** (1957), no. 5 (77), 3–122; English transl., Amer. Math. Soc. Transl. (2) **20** (1962), 239–364. MR **20** # 2539; **25** # 322.

[3] O. A. Oleĭnik, *Solution of fundamental boundary value problems for second order equations with discontinuous coefficients*, Dokl. Akad. Nauk SSSR **124** (1959), 1219–1222. (Russian) MR **21** # 1442.

[4] L. Ja. P′janzina, *Approximation of solutions of boundary value problems for parabolic and hyperbolic equations by means of solutions of the Cauchy problem*, Dokl. Akad. Nauk SSSR **154** (1964), 1276–1279 = Soviet Math. Dokl. **5** (1964), 285–288. MR **29** # 1443.

[5] R. Courant, *Methods of mathematical physics*. Vol. II: *Partial differential equations*, Interscience, New York, 1962; Russian transl., "Mir", Moscow, 1964. MR **25** # 4216; **31** # 4968.

[6] O. A. Oleĭnik, *On properties of solutions of certain boundary problems for equations of elliptic type*, Mat. Sb. **30 (72)** (1952), 695–702. (Russian) MR **14**, 280.

[7] S. Agmon, A. Douglis and L. Nirenberg, *Estimates near the boundary for solutions of elliptic partial differential equations satisfying general boundary conditions*. I, Comm. Pure Appl. Math. **12** (1959), 623–727; Russian transl., IL, Moscow, 1962. MR **23** # A2610.

[8] A. S. Demidov, *The unique solvability of boundary value problems for a second order elliptic equation with certain conjugacy conditions on the surfaces of discontinuity of the coefficients*, Vestnik Moskov. Univ. Ser. I Mat. Meh. **24** (1969), no. 3, 30–36. (Russian) MR **40** # 7627.

[9] Z. G. Šeftel′, *Energy inequalities and general boundary problems for elliptic equations with discontinuous coefficients*, Sibirsk. Mat. Ž. **6** (1965), 636–668. (Russian) MR **31** # 6056.

[10] I. C. Gohberg and M. G. Kreĭn, *The basic propositions on defect numbers, root numbers and indices of linear operators*, Uspehi Mat. Nauk **12** (1957), no. 2 (74), 43–118; English transl., Amer. Math. Soc. Transl. (2) **13** (1960), 185–264. MR **20** # 3459; **22** # 3984.

[11] O. A. Oleĭnik, *Boundary-value problems for linear elliptic and parabolic equations with discontinuous coefficients*, Izv. Akad. Nauk SSSR Ser. Mat. **25** (1961), 3–20; English transl., Amer. Math. Soc. Transl. (2) **42** (1964), 175–194. MR **23** # A1136.

[12] V. A. Il′in, *On the solvability of the Dirichlet and Neumann problems for a linear elliptic operator with discontinuous coefficients*, Dokl. Akad. Nauk SSSR **137** (1961), 28–30 = Soviet Math. Dokl. **2** (1961), 228–230. MR **23** # A3353.

Translated by H. N. Parker

Trudy Moskov. Mat. Obšč.
Tom 23 (1970)

Trans. Moscow Math. Soc.
Vol. 23 (1970)

# ESTIMATES IN $\mathscr{L}_2$ AND THE SOLVABILITY OF GENERAL BOUNDARY VALUE PROBLEMS FOR DEGENERATE ELLIPTIC SECOND-ORDER EQUATIONS

*V. P. GLUŠKO*

**Contents**



## §1. Introduction

Interest in the study of boundary value problems for degenerate differential equations increased markedly again after the appearance of papers by Fichera [**1**] and Oleĭnik [**2**]. The fundamental results in this direction are due to M. V. Keldyš [**3**]. The results he obtained were later developed and generalized in papers by Oleĭnik [**4**], N. D. Vvedenskaja [**5**], A. M. Il′in [**6**], and others.

The study of generalized solutions of degenerate elliptic second-order equations began with papers by Mihlin [**7**] and Višik [**8**]. Following these, a number of papers appeared in which degenerate equations of second, and higher, order were studied by methods similar to Višik's. A bibliography of these papers is in [**9**]. The method of "elliptic regularization" was applied to the study of degenerate elliptic-parabolic second-order equations first by Oleĭnik [**2**] and then by Kohn and Nirenberg [**10**].

In order to describe our basic results pertaining to estimates of solutions of elliptic equations degenerating along the boundary $S$ of a region $\mathscr{D}$, let us write the differential operator under consideration in a neighborhood of a point $x^0 \in S$ in a local coordinate system $t, y$ ($y = (y_1, y_2, \cdots, y_{n-1})$ being the local coordinate system on $S$ with center at $x^0$, and $t$ the point's coordinate on the interior normal to $S$):

---

*AMS (MOS) subject classifications* (1970). Primary 35J40.

$$L_0 = \partial_t (a_{nn}(t, y)\partial_t) + \sum_{k=1}^{n-1} a_{nk}(t, y)\partial_t\partial_{y_k}$$

(1.1)

$$+ \sum_{k,l=1}^{n-1} a_{kl}(t, y)\partial_{y_k}\partial_{y_l} + a_n(t, y)\partial_t + \sum_{k=1}^{n-1} a_k(t, y)\partial_{y_k} + a_0.$$

By virtue of our conditions on $L_0$, the coefficient $a_{nn}(t, y) > 0$ for $t > 0$, and $a_{nn}(0, y) = 0$. If we denote by $K_d$ the hemisphere $t^2 + |y|^2 \leq d^2$, $t > 0$, with center at the point $x^0((t, y) = (0, 0))$ and introduce the function $\alpha(t, y) = \sqrt{a_{nn}(t, y)}$, then for any integer $m \geq 0$ we can define the norm

$$\|v\|^2_{m,\alpha} = \sum_{j+2s+|\tau| \leqslant m} \int_{K_d} |\alpha^j(t, y)\partial_t^{s+j}\partial_y^\tau v(t, y)|^2 dt dy \tag{1.2}$$

on functions $v(t, y) \in C^m(K_d)$. In this paper inequalities of two types are proved:

$$\|v\|_{m,\alpha} \leqslant C\{\|L_0 v\|_{m-2,\alpha} + \langle B_0 v\rangle_{m-m^*-1} + \|v\|_0\}; \tag{1.3}$$

$$\|v\|_{m,\alpha} \leqslant C\{\|L_0 v\|_{m-2,\alpha} + \|v\|_0\}, \tag{1.4}$$

where $v(t, y)$ is some function in $C^m(K_d)$ which vanishes near $t^2 + |y|^2 = d^2$, $B_0$ is a boundary operator of degeneracy order $m^*$ (cf. Definition 2.2), and the boundary norm is

$$\langle v\rangle^2_m = \int_{\overline{K}_d \cap (t=0)} \sum_{|\tau|=m} |\partial_y^\tau v(0, y)|^2 dy.$$

When inequality (1.3) is fulfilled, we say that the boundary condition is "retained" at the point $x^0$; but if inequality (1.4) is fulfilled, then we say that the boundary condition is "dropped" at that point. In our case a simple criterion exists for determining when the boundary condition is "retained" or "dropped" at $x^0$. Let us introduce the quantity

$$b_s = a_n(0, 0) + s\partial_t a_{nn}(t, y)|_{(t,y)=(0,0)},$$

where $s$ is a real parameter. If it turns out that $b_{(m-1)/2} < 0$ at the point $x^0$, the boundary condition is "retained" at that point, but if $b_0 > 0$, the boundary condition is "dropped."

In this paper we consider only the case when the boundary $S$ consists of a finite number of nonintersecting closed manifolds $S_q$, the "order of degenerating" varying weakly along each manifold $S_q$, and all the points of each manifold $S_q$ belonging to one type by the $b_s$ criterion. In [**2**] and [**11**] it is shown that instead of $b_s$, it is possible to introduce a different criterion which is independent of the choice of a particular coordinate system.

Besides conditions on smoothness of coefficients and the condition $b_{(m-1)/2} < 0$, we must introduce two additional necessary and sufficient conditions for the validity of inequality (1.3). These are the condition of $\alpha$-

ellipticity on the operator $L_0$ (cf. Definition 2.1) and the condition of complementarity (cf. Definition 2.3) on the boundary operator $B_0$ with respect to $L_0$ at the point $x^0 \in S$. The latter condition is the analogue of the well-known Šapiro-Lopatinskiĭ condition in the case of uniformly elliptic operators.

If the boundary condition is "dropped" at the point $x^0$, for the validity of inequality (1.4) it is necessary and sufficient that the $\alpha$-ellipticity condition be fulfilled.

Introducing the $\alpha$-ellipticity condition has permitted us to extend estimates obtained earlier for a special kind of operator $L_0$ [**12**] to a maximally broad (in a well-defined sense) class of degenerate elliptic operators $L_0$ with real coefficients.

Using very well-known "interior" estimates for elliptic operators and inequalities (1.3)-(1.4), we can obtain estimates in the whole region with the aid of a partition of unity in $\overline{\mathscr{D}}$. Taking into account that the right sides of inequalities (1.3) and (1.4) are evaluated through the constant $\|v\|_{m,\alpha}$, it is natural to call inequalities (1.3) and (1.4) coercion inequalities and the corresponding boundary value problems in $\mathscr{D}$, coercive value problems.

In an article by Kohn and Nirenberg [**11**], an estimate is obtained of the solution of the Dirichlet problem for a degenerate second-order equation. The authors consider the more general case of the behavior of $L_0$ on the boundary; hence their estimates, generally speaking, are weaker than coercion estimates.

Let us describe in general terms our method of proving inequalities (1.3) and (1.4). Instead of operators $L_0$ and $B_0$, we introduce operators $L_1$ and $B_{01}$, distinguished on subordinate (in a well-defined sense) terms, where the coefficients in $B_{01}$ are constants and the coefficients in $L_1$ may depend only on $t$. Then with the aid of the Fourier transform $F_{n-1}$ "in a tangential direction," the equation $L_1(\partial_y, \partial_t)v = f$ is reduced to the ordinary differential equation $\widetilde{L}(\xi, \partial_t)u = \tilde{f}$ on $(0, d)$ $(u(t,\xi) = F_{n-1}v;\ \tilde{f}(t,\xi) = F_{n-1}f)$, with the condition $u(d,\xi) = 0$, as well as the condition $B_{01}(\xi, \partial_t)u|_{t=0} = \tilde{g}(\xi)$ when deriving (1.3). Results of [**13**]-[**18**] permit us to smooth out (with respect to $t$) the solution $u(t,\xi)$ of the equation $\widetilde{L}u = \tilde{f}$ and to construct a model for the solution of corresponding boundary value problems for each $\xi \neq 0$.

After this we might have returned to the space $(t, y)$ with the aid of the inverse Fourier transform $F_{n-1}^{-1}$, obtained a model for the solution $v(t, y)$ of the original problem, and estimated in norms of the form (1.2) the integral operators involved in this model. Instead of this difficult (but more perspicuous) approach we prefer one based on Parseval's equation and estimates in the space $(t, \xi)$.

The proof of inequalities (1.3) and (1.4) consists of two steps. In the first step we estimate $\|v\|_{m,\alpha}$ through the norm $\|f\|_{m-2,\alpha}$ and through $\|v\|_0 = \sum_{|\tau|=m} \|\partial_y v\|_0$. This step corresponds to estimating "normal" derivatives of a solution through "tangents" when proving analogous inequalities for uniformly elliptic operators; however, here this is connected with significantly greater difficulties.

In the second step, in order to estimate $\|\partial_y v\|_0$ ($|\tau| = m$) we use a simple "power" inequality which solutions of the equation $Lu = \tilde{f}$ satisfy on $(0, d)$, with the condition $u(d, \xi) = 0$. Then we turn to the general operators $L_0$ and $B_0$ and "paste together" the estimates obtained into inequalities for functions defined in the whole region $\mathscr{D}$.

As is well known, the proof of the normal solvability of general boundary value problems is reduced, after obtaining coercion inequalities, to constructing a right regularizer in an appropriate boundary value problem. Although the general scheme of constructing a regularizer is usual for the theory of elliptic equations, when obtaining estimates in the so-called $\epsilon$-norms here (as in a number of other cases), additional difficulties arise connected with the fact that considering the equation $\widetilde{L}u = \tilde{f}$ in the finite interval $0 < t < d$ does not permit using the method of similarity transformations.

The size of this article has not permitted us to consider some other cases here for which inequalities of the type (1.4) are possible. Part of the results which bear an auxiliary character are cited without proof.

The author is deeply grateful to Professor S. G. Kreĭn for a number of extremely useful conversations on the theme of this article.

## §2. Basic assumptions and results

Let $\mathscr{D}$ be a bounded region of the Euclidean space $E_n$ and $S = \bigcup_{q=1}^{q_0} S_q$ its boundary, consisting of closed nonintersecting $(n-1)$-dimensional manifolds $S_q$ (components). The second order differential operator

$$\mathscr{L} = \sum_{i,j=1}^{n} a_{ij}(x)\,\partial_{x_i}\partial_{x_j} + \sum_{i=1}^{n} a_i(x)\,\partial_{x_i}\partial_{x_j} + a_0(x)$$

is defined in $\mathscr{D}$.

CONDITION 2.1. The coefficients $a_{ij}(x)$ and $a_i(x)$ $(1 \leqq i, j \leqq n)$ of the operator $\mathscr{L}$ are real, sufficiently smooth functions in $\mathscr{D}$. The operator $\mathscr{L}$ is elliptic at each interior point $x \in \mathscr{D}$. For the sake of definiteness we shall assume that the characteristic polynomial $\omega_1(x, \lambda) = \sum_{i,j=1}^{n} a_{ij}(x)\,\lambda_i\lambda_j > 0$ for $x \in \mathscr{D}$ and all real $\lambda = (\lambda_1, \cdots, \lambda_n) \neq 0$.

The boundary manifolds $S_q$ are characteristic for $\mathscr{L}$, i.e. $\omega_1(x^0, \sigma) = 0$ for $x^0 \in S$, and $\sigma = (\sigma_1, \cdots, \sigma_n)$ is the interior normal to $S$ at the point $x^0$.

CONDITION 2.2. Each component $S_q$ of the boundary $S$ belongs to class $C^m$ $(m \geqq 2)$. More precisely, for any point $x^0 \in S$ there is a neighborhood $\mathscr{A}_{x^0}(\delta)$ possessing the following properties:

a) There exists a one-to-one mapping

$$t = t(x_1, x_2, \ldots, x_n); \quad y_k = y_k(x_1, x_2, \ldots, x_n) \quad (1 \leqslant k \leqslant n-1) \tag{2.1}$$

of the region $\mathscr{A}_{x^0}(\delta) \cap \mathscr{D}$ into the hemisphere $K$: $t^2 + |y|^2 \leqq \delta^2$, $t > 0$, $|y|^2 = y_1^2 + y_2^2 + \cdots + y_{n-1}^2$ in the space $E_n$.

b) The set $S_q \cap \mathscr{A}_{x^0}(\delta)$ turns into part of the hyperplane $t=0$.

c) The mapping (2.1) and its inverse belong to $C^m$.

d)
$$\left.\frac{\partial t(x)}{\partial x_i}\right|_{x=x^0} = \sigma_i \quad (1 \leqslant i \leqslant n),$$
where the $\sigma_i$ are the direction cosines of the interior normal at the point $x^0 \in S_q$.

e) The coordinates $t$ of the point $x \in \overline{\mathscr{D}}$ in two local systems of coordinates $(t^1, y^1)$ and $(t^0, y^0)$ corresponding to the two neighborhoods $\mathscr{A}_{x^1}(\delta_1)$ and $\mathscr{A}_{x^0}(\delta_0)$ $(x^1, x^0 \in S_q)$ coincide in the intersection of these neighborhoods.

f) An $m$-times differentiable transformation of the variables $y^1 = (y_1^1, y_2^1, \cdots, y_{n-1}^1)$ into the variables $y^0 = (y_1^0, y_2^0, \cdots, y_{n-1}^0)$ exists on the intersection of the two neighborhoods $\mathscr{A}_{x^1}(\delta_1)$ and $\mathscr{A}_{x^0}(\delta_0)$.

Let us denote a mapping (2.1) satisfying conditions a)–f) by $T_{x^0}(x) = (t, y)$, assuming that $T_{x^0}(x^0) = (0,0)$.

Property e) and the compactness of boundary $S$ imply the existence of a "boundary layer" $\mathscr{U}_d$ (the intersection of $\mathscr{D}$ with a $d$-neighborhood of $S$) in which the function $t = t(x)$ in (2.1) is defined and has continuous derivatives up to order $m$. Together with the function $t = t(x)$, the function

$$a(x) = \sum_{i,j=1}^{n} a_{ij}(x)\, \partial_{x_i} t(x)\, \partial_{x_j} t(x) \tag{2.2}$$

is defined in $\mathscr{U}_d$, the $a_{ij}(x)$ being coefficients for higher derivatives of operator $\mathscr{L}$.

Considering $\mathscr{L}$ in $\mathscr{A}_{x^0}(\delta) \cap \mathscr{D}$, after the transformation $T_{x^0}(x) = (t, y)$ we can write $\mathscr{L}$ in the form

$$\begin{aligned} L_0 &= \partial_t (a_{nn}^0(t, y)\, \partial_t) + \sum_{k=1}^{n-1} a_{nk}^0(t, y)\, \partial_t \partial_{y_k} \\ &+ \sum_{k,l=1}^{n-1} a_{kl}^0(t, y)\, \partial_{y_k} \partial_{y_l} + a_n^0(t, y)\, \partial_t + \sum_{k=1}^{n-1} a_k^0(t, y)\, \partial_{y_k} + a_0^0(t, y), \end{aligned} \tag{2.3}$$

where $\partial_t = \partial/\partial t$ and $\partial_{y_k} = \partial/\partial y_k$. Obviously $a_{nn}^0(t,y) \equiv a(T_{x^0}^{-1}(t,y))$. Condition 2.1 implies that $a_{nn}^0(t,y) > 0$, $t > 0$, and that $a_{nn}^0(0,y) = 0$.

Condition 2.3. For each point $x^0 \in S$ there exists $\alpha_{x^0}(t)$, positive for $t > 0$, equal to zero for $t = 0$, and such that the limits

$$\lim_{(t,\, y) \to (0,\, 0)} \frac{a_{nn}^0(t, y)}{\alpha_{x^0}^2(t)} = 1, \tag{2.4}$$

$$\lim_{(t,\, y) \to (0,\, 0)} \frac{a_{nk}^0(t, y)}{\alpha_{x^0}(t)} = \gamma_k \quad (1 \leqslant k \leqslant n-1), \tag{2.5}$$

$$\lim_{(t,\, y) \to (0,\, 0)} \frac{|\operatorname{grad}_y a_{nn}^0(t, y)|}{\alpha_{x^0}(t)} = 0 \tag{2.6}$$

exist.

We shall take

$$\alpha_{x^0}(t) \equiv \sqrt{a^0_{nn}(t, 0)},$$

since fulfillment of Condition 2.3 by some function $\alpha_{x^0}(t)$ implies fulfillment also of the latter condition.

Conditions (2.4) and (2.6) mean that the "order of degenerating" along each component $S_q$ does not vary substantially. As to condition (2.5), it should be noted here that ellipticity of $L_0$ for $t > 0$ implies, as is well known, the validity of the estimate

$$|a^0_{nk}(t, y)| \leqslant 2\sqrt{a^0_{nn}(t, y)\, a^0_{kk}(t, y)}$$

and, by virtue of (2.4), the boundedness of the function $|a^0_{nk}(t, y)\alpha^{-1}_{x^0}(t)|$ in a neighborhood of the point $(t, y) = (+0, 0)$. In addition, condition (2.5) ensures the existence of this function's limit as $(t, y) \to (+0, 0)$.

Introducing the notation $\alpha\partial_t = \alpha_{x^0}(t)\,\partial_t$, let us write $L_0$ in the form $L_0 = L_0' + L_0''$, where

$$(2.7) \qquad L_0' = \frac{a^0_{nn}(t, y)}{\alpha^2_{x^0}(t)}(\alpha\partial_t)^2 + \sum_{k=1}^{n-1} \frac{a^0_{nk}(t, y)}{\alpha_{x^0}(t)}\alpha\partial_t\partial_{y_k} + \sum_{k,l=1}^{n-1} a^0_{kl}\partial_{y_k}\partial_{y_l}.$$

**Definition** 2.1. We say that the operator $\mathscr{L}$ is *$\alpha$-elliptic* in the region $\overline{\mathscr{D}}$ if it is elliptic in $\mathscr{D}$ and the quadratic form in $\eta$, $\xi = (\xi_1, \cdots, \xi_{n-1})$

$$(2.8) \qquad \zeta(\eta, \xi) = \eta^2 + \sum_{k=1}^{n-1} \gamma_k \eta \xi_k + \sum_{k,l=1}^{n-1} a^0_{kl}(0, 0)\,\xi_k \xi_l \neq 0$$

does not vanish for any real $(\eta, \xi) \neq 0$ at any point $x^0$ of the boundary characteristic manifold $S_q$ when conditions (2.4), (2.5) are fulfilled.

Since we agreed to assume the characteristic polynomial $\omega_1(x, \lambda)$ positive in $\mathscr{D}$, condition (2.8) implies $\zeta(\eta, \xi) > 0$ for $(\eta, \xi) \neq 0$. As is obvious from (2.7) and (2.8), $\alpha$-ellipticity in fact means the ellipticity of $L_0'$ with respect to the derivatives $\alpha\partial_t$ and $\partial_{y_k}$ $(1 \leqq k \leqq n-1)$ at the points $x^0 \in S_q$.

Let a boundary differential operator $\mathscr{B}_q$ be defined on $S_q$ such that in a neighborhood of each point $x^0$, after the transformation $T_{x^0}$, $\mathscr{B}_q$ can be written in the form

$$(2.9) \qquad B_0(-i\partial_y, \partial_t) = \sum_{s=0}^{r} \sum_{|\tau| \leqslant \rho_s} b_{\tau, s}(y)(-i\partial_y)^\tau \partial_t^s,$$

where

$$\tau = (\tau_1, \tau_2, \ldots, \tau_{n-1}); \ |\tau| = \tau_1 + \tau_2 + \ldots + \tau_{n-1}; \ \xi^\tau = \xi_1^{\tau_1}\xi_2^{\tau_2}\ldots\xi_{n-1}^{\tau_{n-1}};$$

the $b_{\tau,s}(y)$ are sufficiently smooth, complex-valued functions.[1]

**Definition** 2.2. We say that the *degeneracy order* of the operator $\mathscr{B}(B_0)$ at the point $x^0 \in S_q$ equals $m^*$ if the degree of the polynomial in $\eta$, $\xi$

[1] As usual, $i = \sqrt{-1}$, except when $i$ is a superscript or subscript.

$$\sum_{s=0}^{r} \sum_{|\tau| \leqslant \rho_s} b_{\tau,s}(0)\, \xi^\tau \eta^{2s}$$

equals $m^*$. Obviously, $2r \leqq m^*$. If at each point $x^0 \in S_q$ the degeneracy order of $\mathscr{B}$ equals $m^*$, we say that the degeneracy order of $\mathscr{B}$ on $S_q$ equals $m^*$.

Let us call an operator of the form

$$B_1(-i\partial_y, \partial_t) = \sum_{s=0}^{r} \Lambda_s(-i\partial_y)\, \partial_t^s \tag{2.10}$$

the principal part of operator $B_0$ of degeneracy order $m^*$, where the

$$\Lambda_s(\xi) = \sum_{|\tau| = m^* - 2s} b_{\tau,s}(0)\, \xi^\tau$$

are homogeneous polynomials in $\xi$ of degree $m^* - 2s$. (In particular, possibly some $\Lambda_s(\xi) \equiv 0$.)

Let us define the functions (cf. [1], [2] and [11])

$$\hat{b}(x) = \sum_{i=1}^{n} \Big(a_i(x) - \sum_{j=1}^{n} \partial_{x_j} a_{ij}(x)\Big) \sigma_i;$$

$$\Delta_s(x) = \hat{b}^2(x)\, |\operatorname{grad} \psi|^2 \tag{2.11}$$

$$+ s \sum_{i=1}^{n} \Big(a_i(x) - \sum_{j=1}^{n} \partial_{x_j} a_{ij}(x)\Big) \partial_{x_i} \Big( \sum_{i',j'=1}^{n} a_{i'j'}(x)\, \partial_{x_{i'}} \psi \partial_{x_{j'}} \psi \Big)$$

on the boundary manifolds $S_q$, where $x \in S_q$, $\psi(x_1, \cdots, x_n) = 0$ is the surface equation of $S_q$ in a neighborhood of the point $x$ ($\psi > 0$ in $\mathscr{D}$, $\operatorname{grad} \psi \neq 0$), $\sigma = (\sigma_1, \cdots, \sigma_n)$ is the interior normal to $S_q$ at the point $x$, and $s$ is a real number.

As was shown in [2] and [10], the symbols $\hat{b}$ and $\Delta_s$ on the characteristic manifold $S_q$ does not depend on the choice of a particular coordinate system.

CONDITION 2.4. Let us assume that for a given integer $m \geqq 2$ the components $S_q$ of the boundary $S$ can be decomposed into two sets (corresponding to $q \in Q_1$ or $Q_2$) by the following rule: $q \in Q_1$ if $\hat{b}(x) > 0$ for $x \in S_q$; $q \in Q_2$ if $\hat{b}(x) < 0$ and $\Delta_{(m-1)/2}(x) > 0$ for $x \in S_q$.[2)]

Let us note that either of the sets $Q_1$, $Q_2$ may prove to be the empty set.

At each point $x^0 \in S_q$ $(q \in Q_2)$, let us define the function $b_s(t) = a_n^0(t,0) + s\partial_t a_{nn}^0(t,0)$, where $a_{nn}^0(t,y)$ and $a_n^0(t,y)$ are coefficients of (2.3), $s \geqq 0$, and $0 \leqq t \leqq d$. By definition we shall assume $b_s = b_s(+0)$ and $b = b_0(+0)$. Using the invariance of $\hat{b}(x)$ and $\Delta_s(x)$ relative to the choice of coordinate system at the point $x^0$ on $S_q$, it is easy to obtain from the definition of $Q_2$ that at the points $x^0 \in S_q$ $(q \in Q_2)$, $b_s < 0$ for $0 \leqq s \leqq (m-1)/2$.

---

[2)] We are not considering here the case of $\hat{b}(x) < 0$, $\Delta_{(m-1)/2}(x) < 0$ $(x \in S_q)$. We note only that in this case $a_{nn}^0(t,0) = O(t)$ and inequality (1.4) is valid for $v(t,y) \in H_\alpha^m(K_\delta)$.

At each point $x^0 \in S_q$ $(q \in Q_2)$ we can consider the homogeneous polynomial in $\xi$ of degree $m^*$

$$\vartheta_{x^0}(\xi) = \sum_{s=1}^{r} \frac{b^s}{b_1 b_2 \ldots b_s} \Lambda_s(\xi) \left(-\frac{c(\xi)}{b}\right)^s + \Lambda_0(\xi), \tag{2.12}$$

where

$$c(\xi) = -\sum_{k,l=1}^{n-1} a^0_{kl}(0, 0)\, \xi_k \xi_l. \tag{2.13}$$

DEFINITION 2.3. We shall say that the boundary operator $\mathscr{B}_q$ of degeneracy order $m^*$ satisfies the *complementarity condition* with respect to $\mathscr{L}(L_0)$ at the point $x^0 \in S_q$ $(q \in Q_2)$ if for any $\xi \neq 0$ the value of $\vartheta_{x^0}(\xi) \neq 0$.

Fulfillment of the complementarity condition by $\mathscr{B}_q$ with respect to $\mathscr{L}$ on $S_q$ means that this condition is fulfilled at each point $x^0 \in S_q$. The complementarity condition is the analog of the well-known Šapiro-Lopatinskiĭ (cf. [**19**]) condition for uniformly elliptic operators. In fact, the condition $\vartheta_{x^0}(\xi) \neq 0$ for $\xi \neq 0$ can be interpreted as the condition that the polynomial in $z$

$$\sum_{s=1}^{r} \frac{b^s}{b_1 b_2 \ldots b_s} \Lambda_s(\xi) z^s + \Lambda_0(\xi)$$

is not divisible by the binomial $bz + c(\xi)$ for $\xi \neq 0$.

The following condition pertains to the smoothness of the coefficients of the operators $\mathscr{L}$ in $\overline{\mathscr{D}}$ and $\mathscr{B}_q$ on $S_q$.

CONDITION 2.5. For any integer $m \geqq 2$, let the coefficients $a_{ij}(x)$ $(1 \leqq i, j \leqq n)$ of operator $\mathscr{L}$ belong to $C^{m-1}(\overline{\mathscr{D}})$, and let the coefficients $a_i(x)$ $(0 \leqq i \leqq n)$ belong to $C^{m-2}(\overline{\mathscr{D}})$. The coefficients of the boundary operators $\mathscr{B}_q$ of degeneracy order $m_q^*$ belong to $C_q^{m-m^*}$ (wherever they are defined).

Let us denote by $c_m$ positive constants depending on the coefficients' norms in the above-mentioned spaces. When necessary, the dependence of the constant $c_m$ on other parameters is indicated in parentheses: $c_m(\delta, \phi, \cdots)$.

Let us defined the basic functional spaces and norms used in this paper. Let $K_d$ be the hemisphere $t^2 + |y|^2 \leqq d^2$, $t > 0$, in $E_n$. Let us define the norm

$$\|v\|^2_{m,\alpha} = \sum_{j+2s+|\tau| \leqslant m} \int_{K_d} |\alpha^j(t, y)\, \partial_t^{j+s} \partial_y^\tau v(t, y)|^2\, dt\, dy \tag{2.14}$$

on the set of functions $v(t,y) \in C^m(K_d)$, where $\alpha = \alpha(t,y)$ is a "weight" function, positive for $t > 0$ and equal to zero for $t = 0$.

Let us construct a finite cover of $\overline{\mathscr{D}}$ which consists of a strictly interior subregion $\mathscr{D}_{\delta_0}$ of the region $\mathscr{D}$ and a finite system of neighborhoods $\mathscr{A}_{x^k}(\delta_k)$ (cf. Condition 2.2), where $x^k \in S$, $k = 1, 2, \cdots, N$. Let $\phi_k(x)$ $(k = 0, 1, \cdots, N)$ be functions which belong to $C_0^\infty(E_n)$ and form a partition of unity corresponding to this cover:

$$\sum_{k=0}^{N} \varphi_k(x) \equiv 1 \text{ in } \overline{\mathscr{D}}; \quad \operatorname{supp} \varphi_0 \subset \mathscr{D}_{\delta_0};$$

$$\operatorname{supp} \varphi_k \subset \mathscr{A}_{x^k}(\delta_k), \quad k = 1, 2, \ldots, N.$$

DEFINITION 2.4. Let the integer $m \geqq 1$. We say that the function $v(x) \in \mathscr{L}_2(\mathscr{D})$ with generalized derivatives in $\mathscr{D}$ up to order $m$ belongs to $H_\alpha^m(\mathscr{D})$ if a cover

$$\{\mathscr{D}_{\delta_0}, \mathscr{A}_{x^k}(\delta_k) \ (x^k \in S, \ k = 1, 2, \ldots, N, \ \delta_k \ll d)\}$$

of $\mathscr{D}$ exists such that the form

$$|||v|||_{m,\alpha} = \Big\{ \sum_{|\mu| \leqslant m} \int_{\mathscr{D}} |\partial_x^\mu (\varphi_0 v)|^2 \, dx + \sum_{k=1}^{N} \|(\varphi_k v)_k\|_{m,\alpha_k}^2 \Big\}^{1/2} \tag{2.15}$$

is finite, where

$$(v)_k = v(T_{x^k}^{-1}(t, y)), \quad \alpha_k(t, y) = \sqrt{(a)_k},$$

the function $a(x)$ was introduced in (2.2), and the norm $\|\cdot\|_{m,\alpha}^2$ is defined by formula (2.14).

We can show, using corresponding results of [**20**], that the norms (2.15) generated by various covers $\{\mathscr{D}_{\delta_0}, \mathscr{A}_{x^k}(\delta_k)\}$ are equivalent. Obviously the space $H_\alpha^m(\mathscr{D})$ is contained in S. L. Sobolev's space $W_2^{[m/2]}(\mathscr{D})$, and the set of functions $v(x) \in C^m(\mathscr{D})$ is embedded compactly into the space $H_\alpha^m(\mathscr{D})$.

By definition we take $H_\alpha^0(\mathscr{D}) = \mathscr{L}_2(\mathscr{D})$ for $m = 0$, and

$$|||v|||_{0,\alpha}^2 \equiv |||v|||^2 \equiv \int_{\mathscr{D}} |v(x)|^2 \, dx.$$

Let us introduce boundary norms on the manifolds $S_q$ in the usual way. The norm on a finite function $v(y)$ $(y \in E_{n-1})$ is given by the formula

$$\langle v \rangle_m^2 = \int_{E_{n-1}} |\xi|^{2m} |F_{n-1} v(\xi)|^2 \, d\xi,$$

where $F_{n-1}v(\xi)$ is the Fourier transform in $E_{n-1}$ of the function $v(y)$.

Let the neighborhoods $\mathscr{A}_{x^k}(\delta_k)$ $(x^k \in S_q, k = 1, 2, \cdots, N_q)$ form a finite cover of $S_q$, and let the $\phi_k(x)$ form a corresponding partition of unity. Then the norm of the function $v(x)$ on the boundary $S_q$ is defined by the formula

$$\langle v \rangle_{S_q, m} = \sum_{k=1}^{N_q} \langle (\varphi_k v)_k \rangle_m. \tag{2.15'}$$

As is well known (cf. [**20**]), boundary norms corresponding to various finite covers $\{A_{x^k}(\delta_k)\}$ and introduced in this way will be equivalent. Let us denote by $H^m(S_q)$ the set of functions $v$ defined on $S_q$ with finite norm (2.15′).

Now we can formulate the basic results of the paper.

THEOREM 2.1. *For a given integer $m \geqq 2$, let the operator $\mathscr{L}$ and the region $\mathscr{D}$ with boundary $S = \bigcup_{q=1} S_q$ satisfy Conditions 2.1–2.5. On each boundary manifold $S_q$, $q \in Q_2$, a boundary operator $\mathscr{B}_q$ is defined which satisfies Condition 2.5 and has degeneracy order $m_q^* \leq 2[(m-2)/2]$ on $S_q$.*

*Then for the inequality*

$$(2.16) \qquad |||v|||_{m,\alpha} \leqslant c_m \Big\{ |||\mathscr{L} v|||_{m-2,\alpha} + \sum_{q \in Q_2} \langle \mathscr{B}_q v \rangle_{S_q, m-m_q^*-1} + |||v||| \Big\}$$

*to be valid, where $v$ is any function in $H_\alpha^m(\mathscr{D})$ and the positive constant $c_m$ does not depend on $v$, it is necessary and sufficient that operator $\mathscr{L}$ be $\alpha$-elliptic in $\overline{\mathscr{D}}$ and that the boundary operators $\mathscr{B}_q$ satisfy the complementarity condition with respect to $\mathscr{L}$ on $S_q$, $q \in Q_2$.*

With respect to the given integer $m \geqq 2$, the operator $\mathscr{L}$ defined in a region $\mathscr{D}$ with boundary $S = \bigcup_{q=1}^{q0} S_q$, and the system of boundary operators $\mathscr{B}_q$ each of which is defined on $S_q$, $q \in Q_2$, and has degeneracy order $m_q^* \leqq 2[(m-2)/2]$, we pose the following boundary value problem: find the function $v(x) \in H_\alpha^m(\mathscr{D})$ which satisfies the conditions

$$(2.17) \qquad \mathscr{L} v = f \quad \text{in } \mathscr{D};$$

$$(2.18) \qquad \mathscr{B}_q v = g_q \quad \text{on } S_q \quad \text{for } q \in Q_2,$$

where $f \in H_\alpha^{m-2}(\mathscr{D})$ and $g_q \in H^{m-m_q^*-1}(S_q)$ are given functions.

Let us construct the operator $\mathfrak{A}$ corresponding to boundary value problem (2.17)–(2.18). For this purpose we introduce the space

$$\mathscr{H}_\alpha^{m-(m^*)} = H_\alpha^{m-2}(\mathscr{D}) \times \prod_{q \in Q_2} H^{m-m_q^*-1}(S_q),$$

where $\times$ denotes a direct product of spaces, the norms of $H_\alpha^{m-2}(\mathscr{D})$ and $H^{m-m_q^*-1}(S_q)$ are defined by formulas (2.15) and (2.15′), and the norm in $\mathscr{H}_\alpha^{m-(m^*)}$ is equal to the sum of the norms of its components. Since under Conditions 2.1–2.5 the estimate

$$(2.19) \qquad |||\mathscr{L} v|||_{m-2,\alpha} + \sum_{q \in Q_2} \langle \mathscr{B}_q v \rangle_{S_q, m-m_q^*-1} \leqslant c_m |||v|||_{m,\alpha}$$

is valid, we can define $\mathfrak{A}$ as an operator mapping from $H_\alpha^m(\mathscr{D})$ into $\mathscr{H}_\alpha^{m-(m^*)}$ by the formulas

$$\mathscr{L} v = f \in H_\alpha^{m-2}(\mathscr{D});$$

$$\mathscr{B}_q v|_{S_q} = g_q \in H^{m-m_q^*-1}(S_q) \quad \text{for } q \in Q_2$$

($v \in H_\alpha^m(\mathscr{D})$). From (2.19) it follows that $\mathfrak{A}$ is a continuous operator from $H_\alpha^m(\mathscr{D})$ into $\mathscr{H}_\alpha^{m-(m^*)}$. Let us denote the kernel of $\mathfrak{A}$ in $H_\alpha^m(\mathscr{D})$ by $\mathscr{N}$ ($v \in \mathscr{N} : v \in H_\alpha^m(\mathscr{D})$, $\mathscr{L} v = 0$ in $\mathscr{D}$, $\mathscr{B}_q v|_{S_q} = 0$, $q \in Q_2$).

COROLLARY 2.1. *Let the conditions of Theorem* 2.1 *be fulfilled, ensuring the validity of* (2.16). *Then the operator* $\mathfrak{A}$ *corresponding to the boundary value problem* (2.17)–(2.18) *possesses the following properties*:

1) *The kernel* $\mathcal{N}$ *of* $\mathfrak{A}$ *is finite dimensional.*

2) *The range of* $\mathfrak{A}$ *is closed in* $\mathscr{H}_\alpha^{m-(m^*)}$.

3) *The addend* $\| |v| \|$ *in the right side of* (2.16) *can be discarded if and only if the solution of boundary value problem* (2.17)–(2.18) *is unique in* $H_\alpha^m(\mathscr{D})$.

Corollary 2.1 is deduced from inequalities (2.16) and (2.19) on the basis of well-known theorems (cf. [**20**]).

THEOREM 2.2. *Let the integer* $m \geq 2$, *and let Conditions* 2.1–2.5 *be fulfilled.* (*In Condition* 2.5 *defining smoothness of coefficients, m should be changed to* $m+1$ *for even m.*) *Suppose the operator* $\mathscr{L}$ *is* $\alpha$-*elliptic in* $\overline{\mathscr{D}}$ *and the boundary operators* $\mathscr{B}_q$ *satisfy the complementarity condition on* $S_q$, $q \in Q_2$ $(m_q^* \leqq 2[(m-2)/2])$. *Then the boundary value problem* (2.17)–(2.18) *is normally solvable in* $H_\alpha^m(\mathscr{D})$. (*The operator* $\mathfrak{A}$ *corresponding to this problem in Noetherian.*)

The proof of Theorem 2.2 is deduced in §9.

## §3. Some auxiliary inequalities

In this section we cite without proof a number of inequalities involving space norms with "weight." On the whole, we use as a "weight" function the function $\alpha(t) = \sqrt{a(t)}$, where $a(t)$ is a positive function for $t > 0$ which vanishes at $t = +0$ and belongs to at least $C^1[0,d]$.

In inequalities containing the operator $L_0$ (cf. (2.3)) or the boundary operator $B_0$ $(B_1)$ (cf. (2.9) and (2.10)) the function $\alpha(t) = \sqrt{a_{nn}^0(t,0)}$, where $a_{nn}^0(t,y)$ is the coefficient of $\partial_t^2$ in $L_0$. In some inequalities we shall also use the "weight" function $\alpha(t,y) = \sqrt{a_{nn}^0(t,y)}$.

Let us introduce functional spaces on the set of functions $v(t,y)$ $(y = (y_1, y_2, \cdots, y_{n-1}))$ defined in the strip $E_{n,d}$ $(\{(t,y) \in E_{n,d}\colon\ 0 < t < d,\ y \in E_{n-1}\})$.

DEFINITION 3.1. We denote by $H_\alpha^m(E_{n,d})$ $(m \geq 0)$ the space of functions $v(t,y)$ which have generalized derivatives up to order $m$ in $E_{n,d}$ and for which the norm

$$|v|_{m,\alpha} = \Big\{\int_0^d \int_{E_{n-1}} \sum_{j+2s+|\tau| \leqslant m} |(\alpha\partial_t)^j \partial_t^s \partial_y^\tau v(t,y)|^2\, dy\, dt\Big\}^{1/2} \tag{3.1}$$

is finite, where $\alpha = \alpha(t)$ is the "weight" function and $\alpha^2(t) \in C^{m-1}[0,d]$.

Using the Fourier transform $F_{n-1}$ in the variable $y \in E_{n-1}$, we can introduce into $H_\alpha^m$ the equivalent norm

$$|v|'_{m,\alpha} = \Big\{\sum_{p+j+2s=m} \int_{E_{n-1}} (1+|\xi|^2)^p \, \|(\alpha\partial_t)^j \partial_t^s u\|^2\, d\xi\Big\}^{1/2}, \tag{3.2}$$

where $u(t,y) = F_{n-1}v$, $\xi = (\xi_1, \cdots, \xi_{n-1})$ and

$$\|u\|^2 = \int_0^d |u(t,\xi)|^2\,dt.$$

DEFINITION 3.2. We denote by $H^m_\alpha(K_\delta)$ the linear subset of $H^m_\alpha(E_{n,d})$ which consists of all functions $v(t,y) \in H^m_\alpha(E_{n,d})$ whose carrier is enclosed in the hemisphere $K_\delta$ $(0 < \delta \leq d)$.

DEFINITION 3.3. We denote by $H^m(E_{n-1})$ the Sobolev space of functions defined in $E_{n-1}$, with norm

$$\langle\langle g\rangle\rangle_m = \Big\{ \sum_{|\tau|\leqslant m} \int_{E_{n-1}} |\partial_y^\tau g(y)|^2\,dy\Big\}^{1/2}. \tag{3.3}$$

As is well known, (3.3) is equivalent to the norm

$$\langle g\rangle_m = \Big\{ \int_{E_{n-1}} |\xi|^{2m}\,|F_{n-1}\,g(\xi)|^2\,d\xi$$

on the functions $g(y) \in H^m(E_{n-1})$ which are finite in $E_{n-1}$. Taking into account that the norm (3.2) is equivalent to (3.1), we can reduce the proof of the inequalities we need for $v(t,y) \in H^m_\alpha(E_{n,d})$ to the proof of corresponding inequalities for functions $u(t)$ $(0 < t < d)$ of one variable and containing the parameter $\xi \in E_{n-1}$.

DEFINITION 3.4. The function $u(t)$ belongs to the space $\breve{H}^m_\alpha(0,d)$ $(m \geq 0)$ if $u(t) \in \mathscr{L}_2(0,d)$ has generalized derivatives up to order $m$ on $(0,d)$ and the norm

$$\|u\|_{m,\alpha} = \Big\{ \sum_{j+2s\leqslant m} \|(\alpha\partial_t)^j\,\partial_t^s\,u\|^2\Big\}^{1/2}$$

is finite, where the "weight" function $\alpha = \alpha(t)$ was introduced above.

Let us denote by $c_m$ various constants depending on function $\alpha^2(t)$ and its derivatives up to order $m-1$ on $[0,d]$.

LEMMA 3.1. *Let the integer $m \geq 2$, let $\alpha^2(t) \in C^{m-1}[0,d]$, and let $u(t) \in \breve{H}^m_\alpha(0,d)$. Then for $p > 0$, $j \geq 0$, $1 \leq p+j+2s \leq m$, $s \leq [(m-2)/2]$, arbitrary $\epsilon > 0$ and arbitrary $\xi$ the inequality*

$$|\xi|^{2p}\,\|(\alpha\partial_t)^j\,\partial_t^s\,u\|^2 \leqslant \varepsilon^{m-2s-j} \sum_{\beta+2i=m} \|(\alpha\partial_t)^\beta\,\partial_t^i\,u\|^2$$

$$+\|u^{\left[\frac{m}{2}\right]}\|^2) + \varepsilon^{2\left[\frac{m}{2}\right]-2s-j}\,|\xi|^{\frac{2p\left(m-2\left[\frac{m}{2}\right]\right)}{m-2s-j}}\,\|u^{\left[\frac{m}{2}\right]}\|^2 \tag{3.4}$$

$$+\,c_m\big(\varepsilon^{-2s-j}\,|\xi|^{\frac{2pm}{m-2s-j}} + \varepsilon^{m-2s-j}\big)\,\|u\|^2$$

*is valid, where the constant $c_m = c_m(p,j,s,m,d) > 0$.*

Inequality (3.4) is valid also for $p = 0$ if $j + 2s \leq m-1$ and the remaining conditions of Lemma 3.1 are fulfilled.

If in Lemma 3.1 $p + j + 2s \leq m-1$, the inequality

$$
\begin{aligned}
(3.5)\qquad & |\xi|^{2p}\,\|(\alpha\partial_t)^j\,\partial_t^s u\|^2 \leqslant \varepsilon^{m-2s-j}\sum_{\beta+2i=m}\|(\alpha\partial_t)^\beta\,\partial_t^i u\|^2 \\
& + \|u^{\left[\frac{m}{2}\right]}\|^2) + \varepsilon^{2\left[\frac{m}{2}\right]-2s-j}\Big(\frac{p}{m-2s-j}\,|\xi|^{2\left(m-2\left[\frac{m}{2}\right]\right)} \\
& + \frac{m-2s-j-p}{m-2s-j}\Big)\|u^{\left[\frac{m}{2}\right]}\|^2 + c_m\Big(\frac{p\varepsilon^{m-2s-j-p}}{m-2s-j}\,|\xi|^{2m} \\
& + \frac{(m-2s-j-p)\,\varepsilon^{-2s-j-\frac{p(m-p)}{m-2s-j-p}}}{m-2s-j} + \varepsilon^{m-2s-j}\Big)\|u\|^2
\end{aligned}
$$

is easily obtained from (3.4).

**Lemma** 3.2. *Let the integer $s \geqq 1$, and let $\alpha^2(t) \geqq \kappa t$ ($\kappa =$ constant $> 0$) and $p \geqq 0$. Then for any function $u(t) \in C^{s+1}[0,d]$, any $\epsilon > 0$, and any $\xi$ the inequality*

$$
(3.6)\qquad |\xi|^{2p}\,\|u^{(s)}\|^2 \leqslant \varepsilon\,(\|\alpha u^{(s+1)}\|^2 + \|u\|^2) + c\varepsilon^{-2s}\,|\xi|^{2p(2s+1)}\,\|u\|^2
$$

*is valid, where the constant $c = c(s,p,\kappa,d) > 0$.*

When proving (3.6) an inequality proved by P. I. Lizorkin and S. V. Uspenskiĭ (cf. [**21**]) is used.

The following lemma contains the well-known estimate of the modulus of the function $u(t)$ and its derivatives at the point $t = +0$.

**Lemma** 3.3. *Let the integer $m \geqq 2$, and let $u(t) \in C^{[m/2]}[0,d]$. Then for any $\xi$ the inequality*

$$
(3.7)\qquad |\xi|^{2(m-2s-1)}\,|\partial_t^s u(+0)|^2 \leqslant c\Big\{\sum_{p+2\beta=m}|\xi|^{2p}\,\|\partial_t^\beta u\|^2 + \|u\|^2\Big\}
$$

*is valid, where $0 \leqq s \leqq [(m-2)/2]$ and the constant $c = c(s,m,d) > 0$.*

Now let us turn to the estimates in $H_\alpha^m(E_{n,d})$ containing the operator $L_0$ (cf. (2.3)) and the boundary operator $B_0$ (cf. (2.9)). Taking $d > 0$ sufficiently small, we can assert that $L_0$ ($B_0$) is defined in the hemisphere $K_d$. ($B_0$ is in the sphere $|y| \leqq d$.) Let us call the operator

$$
\begin{aligned}
(3.8)\qquad L_1 &= \partial_t(a_{nn}^0(t,0)\,\partial_t) + \sum_{k=1}^{n-1} a_{nk}^0(t,0)\,\partial_t\,\partial_{y_k} \\
&+ \sum_{k,l=1}^{n-1} a_{kl}^0(0,0)\,\partial_{y_k}\,\partial_{y_l} + a_n^0(0,0)\,\partial_t
\end{aligned}
$$

the principal part of $L_0$ in $K_d$.

Obviously $L_1$ is defined on functions $v(t,y) \in H_\alpha^m(E_{n,d})$ for $m \geqq 2$ and $\alpha^2(t) = a_{nn}^0(t,0)$, with $|L_1 v|_{m-2,\alpha} \leqq c_m |v|_{m,\alpha}$.

The operator $B_1$ (cf. (2.10)), the principal part of $B_0$, is defined on functions $v(t,y) \in H_\alpha^m(E_{n,d})$ ($m^* \leqq m-2$), and the estimate $\langle\!\langle \gamma B_1 v\rangle\!\rangle_{m-m^*-1} \leqq c_m |v|_{m,\alpha}$ is valid, where $\gamma$ is the operator of passage to the limit as $t \to +0$.

**Lemma** 3.4. *Let the integer $m \geqq 2$, let Conditions* 2.3 *and* 2.5 *be fulfilled, let $\alpha^2(t) = a^0_{nn}(t,0)$, and let the degeneracy order of the boundary operator $B_0$ equal $m^* \leqq 2[(m-2)/2]$. Then for any $\epsilon > 0$ there is a $\delta \in (0,d]$ such that for any function $v(t,y) \in H^m_\alpha(K_\delta)$ the inequalities*

$$\sum_{j+2s+|\tau|=m-2} |(\alpha\partial_t)^j \partial_t^s \partial_y^\tau L_1 v|_0 \leqslant \sum_{j+2s+|\tau|=m-2} |(\alpha\partial_t)^j \partial_t^s \partial_y^\tau L_0 v|_0 + \varepsilon |v|_{m,\alpha} + c_m(\varepsilon)|v|_0, \tag{3.9}$$

$$\langle \gamma B_1 v\rangle_{m-m^*-1} \leqslant \langle \gamma B_0 v\rangle_{m-m^*-1} + \varepsilon |v|_{m,\alpha} + c'_m(\varepsilon)|v|_0 \tag{3.10}$$

*are valid for even $m$; for odd $m$ inequality* (3.10) *is retained, but the right side of inequality* (3.9) *must be changed to*

$$\sum_{j+2s+|\tau|=m-2} |(\alpha\partial_t)^j \partial_t^s \partial_y^\tau L_0 v|_0 + \varepsilon |v|_{m,\alpha} + c_m(\varepsilon)|v|_{m-1,\alpha}. \tag{3.11}$$

*The constants $c_m(\epsilon) > 0$ and $c_m(\epsilon) > 0$ do not depend on $v$.*

We use the norm

$$\|v\|_{m,\alpha,\varepsilon} = \int_{E_{n,d}} \sum_{j+2s\leqslant m} \varepsilon^{2(j+2s)} |\alpha^j(t)\partial_t^{j+s} v(t,y)|^2\,dt\,dy + \int_{E_{n,d}} \varepsilon^{2m}\Big(\sum_{|\tau|=m-2\left[\frac{m}{2}\right]} |\partial_y^\tau \partial_t^{\left[\frac{m}{2}\right]} v(t,y)|^2 + \sum_{|\tau|=m} |\partial_y^\tau v(t,y)|^2\Big)\,dt\,dy + |v|_0^2, \tag{3.12}$$

where $0 < \epsilon \leqq 1$, along with norms (3.1) and (3.2) in $H^m_\alpha(E_{n,d})$. If $\alpha^2(t) \in C^{m-1}[0,d]$ $(m \geqq 2)$, then for every $\epsilon \in (0,1)$ the norm (3.12) is equivalent to (3.1). Norms (3.12) and (3.1) coincide for $m = 0, 1$ and $\epsilon = 1$. We use the symbol $\|\cdot\|_{m,\alpha}$ to denote the norm (3.12) for $\epsilon = 1$.

**Corollary** 3.1. *Let the integer $m \geqq 2$, let Conditions* 2.3 *and* 2.5 *be fulfilled, and let $\alpha^2(t) = a^0_{nn}(t,0)$. Then for any $\epsilon > 0$ there is a $\delta \in (0,d)$ such that for any function $v(t,y) \in H^m_\alpha(K_\delta)$ the inequalities*

$$|L_1 v|^2_{m-2,\alpha} \leqslant c_{m-2}\|L_0 v\|^2_{m-2,\alpha} + \varepsilon |v|^2_{m,\alpha} + c'_m(\varepsilon)|v|_0^2; \tag{3.13}$$

$$|L_1 v - L_0 v|^2_{m-2,\alpha} \leqslant \varepsilon |v|^2_{m,\alpha} + c_m(\varepsilon)|v|_0^2 \tag{3.13$'$}$$

*are valid for even $m$, and*

$$|L_1 v|^2_{m-2,\alpha} \leqslant c_{m-2}\|L_0 v\|^2_{m-2,\alpha} + \varepsilon |v|^2_{m,\alpha} + c_m(\varepsilon)|v|^2_{m-1,\alpha} \tag{3.14}$$

*for odd $m$. The constants $c_{m-2}$ and $c_m(\epsilon)$ do not depend on $v$.*

**Lemma** 3.5. *Let the integer $m \geqq 2$, let Conditions* 2.3 *and* 2.5 *be fulfilled, let $\alpha^2(t) = a^0_{nn}(t,0)$, and let $B$ be a boundary operator of degeneracy order $m^* \leqq m-2$. If $\phi(t,y) \in C^m(K_d)$ and vanishes identically near $t^2 + |y|^2 = d^2$, then for any function $v(t,y) \in C^m(K_d)$ the inequalities*

$$\|L_0(\varphi v) - \varphi L_0 v\|_{m-2,\alpha} \leqslant c_m(\varphi)\|v\|_{m-1,\alpha}; \tag{3.15}$$

$$\langle \gamma B(\varphi v) - \gamma(\varphi Bv)\rangle_{m-m^*-1} \leqslant c'_m(\varphi)\|v\|_{m-1,\alpha} \tag{3.16}$$

*are valid. The norms in the right sides of inequalities* (3.15) *and* (3.16) *are defined by formula* (2.14). *The constants* $c_m(\phi)$ *and* $c'_m(\phi)$ *do not depend on* $v$.

REMARK. Condition (2.3) implies that in a sufficiently small hemisphere $K_d$ the estimates

$$\frac{1}{2} \leqslant \frac{\tilde{\alpha}^{2j}(t, y)}{\alpha^{2j}(t)} \leqslant \frac{3}{2} \quad (1 \leqslant j \leqslant m) \tag{3.17}$$

are valid, where $\tilde{\alpha}^2(t, y) = a^0_{nn}(t, y)$ and $\alpha^2(t) = a^0_{nn}(t, 0)$. Therefore we can change the "weight" function $\alpha(t)$ in the norm $\|\cdot\|_{m,\alpha}$ to $\tilde{\alpha}(t,y)$. It is easy to show that the norm $\|v\|_{m,\alpha}$ is equivalent to the norm $\|v\|_{m,\tilde{\alpha}}$ on functions $v(t,y) \in H^m_\alpha(K_d)$. If $d > 0$ is sufficiently small, we can change in this way any of the norms of the form $\|\cdot\|_{m,\alpha}$ or $\|\cdot\|_{m,\alpha}$ in inequalities (3.13)–(3.16) to the norm $\|\cdot\|_{m,\tilde{\alpha}}$.

Along with $L_0$, the operators

$$\begin{aligned} L_2 &= q_n(t, y)\partial_t^2 + \sum_{k=1}^{n-1} q_k(t, y)\partial_t\partial_{y_k} + \sum_{k,l=1}^{n-1} q_{kl}(t, y)\partial_{y_k}\partial_{y_l} \\ &\quad + (q_0(t, y) + \partial_t q_n(t, y))\partial_t; \\ L_3 &= \sum_{k=1}^{n-1} a^0_k(t, y)\partial_{y_k} + a^0_0(t, y) \end{aligned} \tag{3.18}$$

are defined in $K_d$, where

$$q_n(t, y) = a^0_{nn}(t, y) - a^0_{nn}(t, 0);$$

$$q_k(t, y) = a^0_{nk}(t, y) - a^0_{nk}(t, 0) \ (1 \leqslant k \leqslant n-1);$$

$$q_{kl}(t, y) = a^0_{kl}(t, y) - a^0_{kl}(0, 0) \ (1 \leqslant k, \ l \leqslant n-1);$$

$$q_0(t, y) = a^0_n(t, y) - a^0_n(0, 0).$$

In the sphere $|y| \leqq d$ the operators

$$\begin{aligned} B_2 &= \sum_{s=0}^{r} \sum_{|\tau|=m^*-2s} (b_{\tau,s}(y) - b_{\tau,s}(0))(-i\partial_y)^\tau \partial_t^s; \\ B_3 &= \sum_{s=0}^{r} \sum_{|\tau|\leqslant m^*-2s-1} b_{\tau,s}(y)(-i\partial_y)^\tau \partial_t^s \end{aligned} \tag{3.19}$$

are defined as well. By the construction of $L_2$ and $L_3$ ($B_2$ and $B_3$) the representation $L_0 = L_1 + L_2 + L_3$ ($B_0 = B_1 + B_2 + B_3$) is valid.

Let us introduce the function $\psi_\delta(t, y) \in C^\infty(E_n)$ such that $\psi_\delta(t, y) \equiv 1$ for $t^2 + |y|^2 \leqq \delta^2$, and $\psi_\delta(t, y) \equiv 0$ for $t^2 + |y|^2 \geqq 4\delta^2$. Now we can construct operators $L_\delta$ and $B_\delta$ on functions $v(t, y) \in H^m_\alpha(E_{n,d})$ ($m \geqq 2$, $m^* \leqq m-2$)

by the following formulas:

$$L_\delta = q_{n,\delta}(t, y)\,\partial_t^2 + \sum_{k=1}^{n-1} q_{k,\delta}(t, y)\,\partial_t\partial_{y_k} + \sum_{k,l=1}^{n-1} q_{kl,\delta}(t, y)\,\partial_{y_k}\partial_{y_l} + q_{0,\delta}(t, y)\,\partial_t;$$

(3.20)

$$B_\delta = \sum_{s=0}^{r} \sum_{|\tau|=m^*-2s} \rho_{\tau,s,\delta}(y)\,\partial_y^\tau\partial_t^s,$$

where

$$q_{n,\delta}(t, y) = \psi_\delta q_n(t, y); \quad q_{k,\delta}(t, y) = \psi_\delta q_k(t, y);$$
$$q_{kl,\delta}(t, y) = \psi_\delta q_{kl}(t, y);$$
$$q_{0,\delta}(t, y) = \psi_\delta(q_0(t, y) + \partial_t q_n(t, y)); \ \rho_{\tau,s,\delta}(y) = \psi_\delta(0, y)(b_{\tau,s}(y) - b_{\tau,s}(0)).$$

Let us note that $L_\delta$ and $B_\delta$ coincide with $L_2$ and $B_2$, respectively, in $K_\delta$, and that $L_{0\delta} = L_1 + L_\delta + \psi_\delta L_3$ coincides with $L_0$ in $K_\delta$.

LEMMA 3.6. *Let the integer $m \geqq 2$, and let Conditions 2.3 and 2.5 be fulfilled. Then for any $\delta \in (0, d/2)$ and $\epsilon \in (0, 1]$, the estimate*

(3.21)
$$\| L_\delta v \|_{m-2,\alpha,\varepsilon} \leqslant (\mathcal{N}(\delta) + \varepsilon\mathcal{M}(\delta)) \Big\{ \| v \|^2_{2,\alpha} + \sum_{3 \leqslant j+2s+|\tau| \leqslant m} \int_{E_{n,d}} \varepsilon^{2(j+2s+|\tau|-2)} |\alpha^j(t)\,\partial_t^{j+s}\partial_y^\tau v|^2\,dt\,dy \Big\}^{1/2}$$

*is valid for all $v(t, y) \in H^m_\alpha(E_{n,d})$, where*

$$\alpha^2(t) = a^0_{nn}(t, 0); \ \mathcal{N}(\delta) > 0, \ \mathcal{M}(\delta) \geqslant 0, \ \mathcal{N}(\delta) \to 0 \ \text{as} \ \delta \to 0, \ \mathcal{M}(\delta) \equiv 0$$

*for $m = 2$.*

For every $\epsilon \in (0, 1]$ we shall consider the norm

$$\langle\langle g \rangle\rangle_{m,\varepsilon} = \Big\{ \int_{E_{n-1}} \sum_{|\tau| \leqslant m} |(\varepsilon\partial_y)^\tau g(y)|^2\,dy \Big\}^{1/2}$$

equivalent to (3.3) on functions $g(y) \in H^m(E_{n-1})$.

LEMMA 3.7. *Let the integer $m \geqq 2$, let Condition 2.5 be fulfilled, and let the operator $B_\delta$ have degeneracy order $m^* \leqq m - 2$. Then for any $\delta \in (0, d/2)$ and any $\epsilon \in (0, 1]$, the estimate*

(3.22)
$$\langle\langle \gamma B_\delta v \rangle\rangle_{m-m^*-1,\varepsilon} \leqslant (\mathcal{N}(\delta) + \varepsilon\mathcal{M}(\delta)) \times \Big\{ \sum_{\substack{0 \leqslant |\tau| \leqslant m-m^*-1 \\ |\mu|+2s=m^*}} \varepsilon^{2|\tau|} \int_{E_{n-1}} |\xi^{\tau+\mu}\partial_t^s u(0, \xi)|^2\,d\xi \Big\}^{1/2}$$

*is valid for all functions $v(t, y) \in H^m_\alpha(E_{n,d})$, where $u^{(s)}(0, \xi) = F_{n-1}(\partial_t^s v(t, y))|_{t=0}$, and $\mathcal{N}(\delta)$ and $\mathcal{M}(\delta)$ satisfy the same conditions as in Lemma 3.6.*

## §4. Simplification of the problem

As is obvious from (2.15) and (2.16), the proof of inequality (2.16) can be reduced to obtaining corresponding interior estimates for $\phi_0 v(x)$ in the subregion $\mathscr{D}_{\delta_0}$ of $\mathscr{D}$ ($\overline{\mathscr{D}}_{\delta_0} \subset \mathscr{D}$) and estimates of $(\phi_k v)_k$ in a neighborhood of the boundary points $x^k \in S$. Since interior estimates for elliptic operators are well known (cf. [**19**]), the problem consists of obtaining estimates of $(\phi_k v)_k$ in norms (2.14).

Let $x^0 \in S$, and let the mapping $T_{x^0}(x) = (t, y)$ take $\mathscr{A}_{x^0}(d) \cap \mathscr{D}$ into the hemisphere $K_d$: $t^2 + |y|^2 \leqq d^2$, $t > 0$. Let us write the operator $\mathscr{L}$ after the transformation $T_{x^0}(x) = (t, y)$ in the form of (2.3). Let us extract the principal part $L_1$ of this operator (cf. (3.8)). The coefficients of $L_1$ do not depend on $y$; therefore Condition 2.3 reduces to an existence condition on the limits

$$\lim_{t \to +0} \frac{a^0_{nk}(t, 0)}{\sqrt{a^0_{nn}(t, 0)}} = \gamma_k \quad (1 \leqslant k \leqslant n-1). \tag{4.1}$$

In what follows we use the notation

$$b_s(t) = a^0_n(t, 0) + s\partial_t a^0_{nn}(t, 0); \quad b_s = b_s(+0); \quad b = b_0. \tag{4.2}$$

On the hyperplane $t = 0$ we consider the boundary operator $B_0(-i\partial_y, \partial_t)$ of the form (2.9) and of degeneracy order $m^*$. Its principal part $B_1$ is defined by (2.11).

DEFINITION 4.1. We denote by $V^m(K_d)$ the set of functions $\phi(t, y) \in C^m(\overline{K}_d)$ with carrier in $\overline{K}_d$ and identically vanishing near $t^2 + |y|^2 = d^2$.

THEOREM 4.1. *Let the integer $m_1 \geqq 2$, and let the boundary operator $B_1$ have degeneracy order $m^* \leqq m_1 - 2$ and satisfy the complementarity condition at $t = 0$ with respect to the operator $L_1$, which is $\alpha$-elliptic in $K_d$ when condition* (4.1) *is fulfilled. Suppose that for some $m \geqq m_1$ the coefficients $a^0(t, 0) \in C^{m-1}[0, d]$ $(1 \leqq k \leqq n)$ and $b_{(m-1)/2} < 0$.*

*Then a $\delta$ $(0 < \delta \leqq d)$ exists such that the conditions*

$$v \in H^{m_1}_\alpha(K_d); \quad \phi \in V^m(K_\delta); \quad |L_1(\phi v)|_{m-2,\alpha} < \infty; \quad \langle \gamma B_1(\phi v) \rangle_{m-m^*-1} < \infty$$

*imply that $\phi v \in H^m_\alpha(K_\delta)$ and that the inequality*

$$|\varphi v|^2_{m,\alpha} \leqslant c_m \{ |L_1(\varphi v)|^2_{m-2,\alpha} + \langle \gamma B_1(\varphi v) \rangle^2_{m-m^*-1} + |\varphi v|^2_0 \} \tag{4.3}$$

*is valid. The constant $c_m > 0$ does not depend on $\phi$ or $v$.*

The proof of Theorem 4.1 will be deduced in §5.

The following theorem pertains to the case when the boundary conditions are "dropped" at $t = 0$.

THEOREM 4.2. *Suppose that $b > 0$, and that the coefficients $a^0_{nk}(t, 0)$ $(1 \leqq k \leqq n)$ of the operator $L_1$ belong to $C^{m-1}[0, d]^{nk}$ and fulfill condition* (4.1), *while $L_1$ itself is $\alpha$-elliptic in $K_d$. Then there exists a $\delta$ $(0 < \delta \leqq d)$ such that the conditions $v \in H^2_\alpha(K_d)$, $\phi \in V^m(K_\delta)$ and $|L_1(\phi v)|_{m-2,\alpha} < \infty$ imply that $\phi v \in H^m_\alpha(K_\delta)$ and that the inequality*

$$|\varphi v|^2_{m,\alpha} \leqslant c_m \{ |L_1(\varphi v)|^2_{m-2,\alpha} + |\varphi v|^2_0 \} \tag{4.4}$$

*is valid. The constant $c_m > 0$ does not depend on $\phi$ or $v$.*

The proof of Theorem 4.2 is deduced in §6.

The results cited in §3 permit us to change to the general operator $L_0$ and the general boundary operator $B_0$ in inequalities (4.3) and (4.4).

COROLLARY 4.1. *Let the integer* $m \geqq 2$ *and let the boundary operator* $B_0$ *of degeneracy order* $m^* \leqq 2\lfloor (m-2)/2 \rfloor$ *satisfy the complementarity condition on the hyperplane* $t=0$ *with respect to the operator* $L_0$, *which is* $\alpha$-*elliptic in* $K_d$. *Suppose that* $b_{(m-1)/2} < 0$ *and the coefficients of* $L_0$ *and* $B_0$ *satisfy Conditions 2.3 and 2.5. Then a* $\delta$ $(0 < \delta \leqq d)$ *exists such that for any functions* $\phi \in V^m(K_\delta)$ *and* $v \in H^m_\alpha(K_d)$ *the inequality*

$$|\varphi v|^2_{m,\alpha} \leqslant c_m(\varphi)\left(\|\varphi L_0 v\|^2_{m-2,\alpha} + \langle \gamma\varphi B_0 v\rangle^2_{m-m^*-1} + \|v\|^2_{m-1,\alpha}\right) \tag{4.5}$$

*is valid, where the constant* $c_m(\phi) > 0$ *does not depend on* $v$.

PROOF. Inequality (4.3) and estimates (3.13), (3.14) and (3.10) imply that a $\delta > 0$ exists such that for any $\phi(t, y) \in V^m(K_\delta)$ the inequality

$$|\varphi v|^2_{m,\alpha} \leqslant c_m\{\|L_0(\varphi v)\|^2_{m-2,\alpha} + \langle \gamma B_0(\varphi v)\rangle^2_{m-m^*-1} + |\varphi v|^2_{m-1,\alpha}\} \tag{4.6}$$

is valid. In order to obtain inequality (4.5) from (4.6), it remains to apply to (4.6) estimates (3.15) and (3.16), as well as the trivial estimate $|\phi v|_{m-1,\alpha} < c(\phi)\|v\|_{m-1,\alpha}$, where the constant $c(\phi) > 0$.

COROLLARY 4.2. *For the integer* $m \geqq 2$ *let the conditions of Theorem 4.2 be fulfilled, let the coefficients of* $L_0$ *satisfy Conditions 2.3 and 2.5, and let* $L_0$ *itself be* $\alpha$-*elliptic in* $K_d$. *Then a* $\delta$ $(0 < \delta \leqq d)$ *exists such that for any functions* $\phi \in V^m(K_\delta)$ *and* $v \in H^m_\alpha(K_d)$ *the inequality*

$$|\varphi v|^2_{m,\alpha} \leqslant c_m(\varphi)\left(\|\varphi L_0 v\|^2 + \|v\|^2_{m-1,\alpha}\right) \tag{4.7}$$

*is valid. The constant* $c_m(\phi) > 0$ *does not depend on* $v$.

The proof of Corollary 4.2 is analogous to the proof of Corollary 4.1.

Now we can deduce the

PROOF OF THEOREM 2.1. SUFFICIENCY. Let us consider an expanding system of regions $\{\mathscr{D}_\delta\}$ $(\delta \to 0)$, each of which is strictly interior with respect to $\mathscr{D}$ $(\overline{\mathscr{D}}_\delta \subset \mathscr{D})$, with $\mathscr{D}_\delta \to \mathscr{D}$ as $\delta \to 0$. By virtue of Corollaries 4.1 and 4.2, for each point $x^0 \in S$ there exists a neighborhood $\mathscr{A}_{x^0}(\delta)$ and a transformation $T_{x^0}$ taking $\mathscr{A}_{x^0}(\delta) \cap \mathscr{D}$ into $K_\delta$ such that either inequality (4.5) for $b_{(m-1)/2} < 0$ or inequality (4.7) for $b > 0$ is valid in $K_\delta$. Actually, if $x^0 \in S_q$ and $q \in Q_2$, then $\hat{b}(x^0) < 0$ by Condition 2.4, and $\Delta_{(m-1)/2} > 0$. Since the sign of $\hat{b}$ and $\Delta_s$ does not depend on the choice of coordinate system, we obtain

$$b = a^0_n(0, 0) < 0, \quad b^2 + \frac{1}{2}(m-1)\, b\partial_t a^0_{nn}(0, 0) > 0$$

in the local coordinate system $(t, y)$ $((t, y) = T_{x^0}(x))$. Hence it follows that $b_{(m-1)/2} < 0$. That the value of $b > 0$ for $x^0 \in S_q$ and $q \in Q_1$ is shown analogously.

It is easy to verify that the remaining conditions of Corollaries 4.1 and 4.2 follow immediately from the conditions of Theorem 2.1.

The system of regions $\{\mathscr{D}_\delta\}$ and neighborhoods $\{\mathscr{A}_x(\delta)\}$ $(x \in S)$ forms a cover of $\overline{\mathscr{D}}$. Since $\mathscr{D}$ is compact, we can select a finite cover of $\overline{\mathscr{D}}$ consisting of one region $\mathscr{D}_{\delta_0}$ $(\overline{\mathscr{D}}_{\delta_0} \subset \mathscr{D})$ and the system of neighborhoods $\mathscr{A}_{x^k}(\delta_k)$ $(k = 1, 2, \cdots, N)$. Let $\sum_{k=0}^N \phi_k(x) \equiv 1$ be a partition of unity in $\overline{\mathscr{D}}$ $(\operatorname{supp} \phi_0 \subset \mathscr{D}_{\delta_0}$, $\operatorname{supp} \phi_k \subset \mathscr{A}_{x^k}(\delta_k))$ corresponding to this cover.

Taking into account the Remark on p. 125, we obtain the inequalities

$$|(\varphi_k v)_k|^2_{m,\alpha_k^0} \leqslant c_m(\varphi_k)\{\|(\varphi_k \mathscr{L} v)_k\|^2_{m-2,\alpha_k} + \langle \gamma(\varphi_k \mathscr{B}_q v)_k\rangle^2_{m-m_q^*-1} + \|(v)_k\|^2_{m-1,\alpha_k}\} \tag{4.8}$$

when the conditions of Corollary 4.1 are fulfilled, and

$$|(\varphi_k v)_k|^2_{m,\alpha_k^0} \leqslant c_m(\varphi_k)\{\|(\varphi_k \mathscr{L} v)_k\|^2_{m-2,\alpha_k} + \|(v)_k\|^2_{m-1,\alpha_k}\}, \tag{4.9}$$

when the conditions of Corollary 4.2 are fulfilled, where

$$(v)_k = v(T^{-1}_{x^k}(t, y)); \ \alpha_k^2(t, y) = (a)_k; \ \alpha_k^0 = \alpha_k(t, 0);$$

the norms $\|(v)\|_{m-1,\alpha_k}$ are computed with respect to the region

$$\widetilde{\mathscr{D}}_k = T_{x^k}(\mathscr{A}_{x^k}(\delta_k) \cap \mathscr{D}) \subset K_{\delta_k}.$$

Consider the addend $\|(v)_k\|^2_{m-1,\alpha_k}$ in the right sides of (4.8) and (4.9). Let $T_{x^k}(x) = (t, y^k)$ be the transformation introduced in §2, and $J_k(x)$ the Jacobian of this transformation. We have

$$\begin{aligned} \|(v)_k\|^2_{m-1,\alpha_k} &= \sum_{j+2s+|\tau| \leqslant m-1} \int_{\widetilde{\mathscr{D}}_k} |\alpha_k^j(t, y^k)\, \partial_t^{j+s} \partial^\tau_{y^k} (v)_k|^2 \, dt dy^k \\ &= \sum_{j+2s+|\tau| \leqslant m-1} \int_{\mathscr{A}_{x^k}(\delta_k)\cap\mathscr{D}} a^j(x)\, |\partial_t^{j+s}\partial^\tau_{y^k} v(x)|^2 \, |J_k(x)| \, dx \\ &\leqslant (N+1) \sum_{\substack{0 \leqslant l \leqslant N \\ j+2s+|\tau| \leqslant m-1}} \int_{\mathscr{A}_{x^k}(\delta_k)\cap\mathscr{D}} a^j(x)\, |\partial_t^{j+s}\partial^\tau_{y^k}\varphi_l v(x)|^2 \, |J_k(x)| \, dx. \end{aligned} \tag{4.10}$$

By the construction of the transformations $T_{x^k}$, an $m$-times differentiable transformation of the variables $y^k = (y_1^k, \cdots, y_{n-1}^k)$ into the variables $y^l = (y_1^l, \cdots, y_{n-1}^l)$ exists in the nonempty intersection of any two neighborhoods $\mathscr{A}_{x^k}(\delta_k)$ and $\mathscr{A}_{x^l}(\delta_l)$. Taking this into account, we obtain easily from (4.10) the estimate

$$\|(v)_k\|^2_{m-1,\alpha_k} \leqslant c_0\Big\{\sum_{|\mu| \leqslant m-1} \int_{\mathscr{D}} |\partial_x^\mu \varphi_0 v|^2 \, dx + \sum_{l=1}^N \|(\varphi_l v)_l\|^2_{m-1,\alpha_l}\Big\}, \tag{4.11}$$

where the constant $c_0 > 0$ depends only on $\mathscr{D}$ and on $a(x)$.

Using the equivalence of the norms $\|\cdot\|_{m-1,\alpha_k}$ and $|\cdot|_{m-1,\alpha_k^0}$ (cf. the Remark on p. 125), we obtain

$$\| (v)_k \|^2_{m-1,\alpha_k} \leqslant \tilde{c}_0 \Big\{ \sum_{|\mu| \leqslant m-1} \int_{\mathscr{D}} | \partial^\mu_x \varphi_0 v |^2 \, dx + \sum_{l=1}^{N} | (\varphi_l v)_l |^2_{m-1,\alpha^0_l} \Big\} \tag{4.12}$$

from inequality (4.11).

Finally we can write down well-known interior estimates (cf. [**19**]) in the form

$$\begin{aligned} \sum_{|\mu| \leqslant m} \int_{\mathscr{D}} | \partial^\mu_x (\varphi_0 v) |^2 \, dx \leqslant c_m (\varphi_0, \delta_0, \varepsilon) \Big\{ \sum_{\mu \leqslant m-2} \int_{\mathscr{D}} | \partial^\mu_x (\varphi_0 \mathscr{L} v) |^2 \, dx \\ + \int_{\mathscr{D}} | v(x) |^2 \, dx \Big\} + \varepsilon \sum_{|\mu| \leqslant m} \int_{\mathscr{D}_{\delta_0}} | \partial^\mu_x v(x) |^2 \, dx, \end{aligned} \tag{4.13}$$

where $\epsilon$ is any positive number. The estimate

$$\sum_{|\mu| \leqslant m} \int_{\mathscr{D}_{\delta_0}} | \partial^\mu_x v(x) |^2 \, dx \leqslant c(\delta_0) \Big\{ \sum_{|\mu| \leqslant m} \int_{\mathscr{D}} | \partial^\mu_x \varphi_0 v |^2 \, dx + \sum_{k=1}^{N} | (\varphi_k v)_k |^2_{m,\alpha^0_k} \Big\} \tag{4.14}$$

is easily derived for the last sum in the right side of (4.13).

For even $m$, by virtue of inequality (3.5), for any $\epsilon > 0$ there is a $c(\epsilon) > 0$ such that

$$| (\varphi_k v)_k |^2_{m-1,\alpha^0_k} \leqslant \varepsilon | (\varphi_k v)_k |^2_{m,\alpha^0_k} + c(\varepsilon) | (\varphi_k v)_k |^2_0. \tag{4.15}$$

Moreover, the inequality

$$\sum_{|\mu| \leqslant m-1} \int_{\mathscr{D}} | \partial^\mu_x \varphi_0 v |^2 \, dx \leqslant \varepsilon \sum_{|\mu| \leqslant m} \int_{\mathscr{D}} | \partial^\mu_x \varphi_0 v |^2 \, dx + c(\varepsilon) ||| \varphi_0 v |||^2 \tag{4.16}$$

is very well known.

Choosing $\epsilon > 0$ sufficiently small in (4.13), (4.15) and (4.16), we obtain from (4.8)–(4.9) and (4.12)–(4.16) the inequality

$$\begin{aligned} \sum_{|\mu| \leqslant m} \int_{\mathscr{D}} | \partial^\mu_x \varphi_0 v |^2 \, dx + \sum_{k=1}^{N} | (\varphi_k v)_k |^2_{m,\alpha^0_k} \\ \leqslant c_m \Big\{ ||| \mathscr{L} v |||^2_{m-2,\alpha} + \sum_{q \in Q_2} \langle \mathscr{B}_q v \rangle^2_{s_q, m-m^*_q-1} + ||| v |||^2 \Big\} \end{aligned} \tag{4.17}$$

for even $m$.

In order to derive (2.16) from (4.17), we must change the norms $| \cdot |_{m,\alpha^0_k}$ in the left side of (4.17) to the equivalent $\| \cdot \|_{m,\alpha_k}$.

Turning to the proof of (2.16) for odd $m$, it should be noted that when the conditions of Theorem 2.1 are fulfilled for some odd $m$, they are fulfilled also for the even number $m-1$. After changing $m$ to $m-1$ in (4.17), we obtain

$$\begin{aligned} \| (v)_k \|^2_{m-1,\alpha_k} \leqslant c_0 \sum_{|\mu| \leqslant m-1} \int_{\mathscr{D}} | \partial^\mu_x \varphi_0 v |^2 \, dx \\ + c_m \Big\{ ||| \mathscr{L} v |||^2_{m-3,\alpha} + \sum_{q \in Q_2} \langle \mathscr{B}_q v \rangle^2_{s_q, m-m^*_q-2} + ||| v |||^2 \Big\} \end{aligned} \tag{4.18}$$

from (4.12) and (4.17). If we take into account that

$$|||\mathcal{L}v|||_{m-3,\alpha} \leqslant |||\mathcal{L}v|||_{m-2,\alpha}; \ \langle \mathcal{B}_q v\rangle_{s_q, m-m_q^*-2} \leqslant c\,\langle \mathcal{B}_q v\rangle_{s_q, m-m_q^*-1},$$

then (4.8)-(4.9), (4.12)-(4.14), (4.16) and (4.18) imply the validity of the inequality for odd $m$.

The sufficiency in Theorem 2.1 has been proved. The proof of the necessity of the $\alpha$-ellipticity condition on $\mathcal{L}$ in $\overline{\mathcal{D}}$ and of the complementarity condition on the boundary operator $\mathcal{B}_q$ with respect to operator $\mathcal{L}$ on $S_q$ $(q \in Q_2)$ will be deduced in §8.

## §5. Proof of Theorem 4.1

Since by the condition of the theorem we have

$$b_{1/2(m-1)} < 0, \quad \partial_t a^0_{nn}(t, 0)|_{t=+0} \geqslant 0,$$

and the functions $a^0_n(t,0)$ and $\partial_t a^0_{nn}(t,0)$ are continuous in the half right-neighborhood of the point $t=0$, we can find $\delta_1$ $(0<\delta_1 \leqq d)$ and $\beta_1>0$ such that the inequality

$$-\beta_1^{-1} \leqslant b_s(t) \leqslant -\beta_1 \tag{5.1}$$

will be fulfilled for the functions $b_s(t) = a^0_n(t,0) + s\partial_t a^0_{nn}(t,0)$, $t \in [0, \delta_1]$ and $0 \leqq s \leqq (m-1)/2$.

We introduce the linear functions in $\xi = (\xi_1, \cdots, \xi_{n-1})$

$$\omega(t, \xi) = \sum_{k=1}^{n-1} a^0_{nk}(t, 0)\,\xi_k; \quad \omega_0(\xi) = \sum_{k=1}^{n-1} \gamma_k \xi_k,$$

where the $\gamma_k$ are defined by formulas (4.1). Using the functions $\omega_0(\xi)$ and $c(\xi)$ (2.13), we can write the $\alpha$-ellipticity condition on $L_1$ in the form

$$\eta^2 + \omega_0(\xi)\,\eta - c(\xi) \geqslant \nu_0(\eta^2 + |\xi|^2), \tag{5.2}$$

where the constant $\nu_0 = \nu_0(x^0)$ and $|\xi|^2 = \xi_1^2 + \cdots + \xi_{n-1}^2$.

Condition (4.1) implies that a $\delta_2$ $(0<\delta_2 \leqq d)$ exists such that for all $\xi$ and $t \in [0, \delta_2]$ the estimate

$$|\omega(t, \xi) - \omega_0(\xi)\,\alpha(t)| \leqslant \frac{1}{2}\nu_0 \alpha(t)\,|\xi| \tag{5.3}$$

is valid.

If we take $\delta = \min\{\delta_1, \delta_2\}$, then we can assert that all the conditions of the theorem and conditions (5.1) and (5.3) are fulfilled in the hemisphere $K_\delta$. Without loss of generality we shall assume that $\delta = d$ in Theorem 4.1.

Let $v(t, y) \in H^m_\alpha(K_d)$ and

$$L_1 v = f_1(t, y); \quad B_1 v|_{t=+0} = g_1(y). \tag{5.4}$$

By construction the functions $f_1(t, y)$ and $g_1(y)$ are finite on each hyperplane $t = t_0$ $(0 \leqq t_0 \leqq d)$, and $f_1(t, y) \in H^{m-2}_\alpha(E_{n,d})$ and $g(y) \in H^{m-m^*-1}(E_{n-1})$.

Let us apply the Fourier transform operator $F_{n-1}$ in the variable $y=(y_1,\cdots y_{n-1})$ to identities (5.4). Since the coefficients in $L_1$ and $B_1$ do not depend on $y$, we obtain

$$L(\xi,\partial_t)u=f(t,\xi);\quad B_1(\xi,\partial_t)u|_{t=0}=g(\xi);\quad u(d,\xi)=0, \tag{5.5}$$

where

$$\begin{gathered} u(t,\xi)=F_{n-1}v;\quad f(t,\xi)=F_{n-1}f_1;\ g(\xi)=F_{n-1}g_1;\\ L(\xi,\partial_t)=\partial_t(a(t)\partial_t)+(b+i\omega(t,\xi))\partial_t+c(\xi);\\ a(t)\equiv\alpha^2(t)\equiv a^0_{nn}(t,0). \end{gathered} \tag{5.6}$$

For the remaining notation, see (2.10), (2.13) and (4.2).

In (5.5), to the first two conditions, which are obtained from (5.4), we have added the condition $u(d,\xi)=0$, which follows naturally from the fact that $v(d,y)\equiv 0$ for $y\in E_{n-1}$.

We shall analyze boundary value problem (5.5) on $[0,d]$ for an ordinary second order differential operator $L$ degenerating at $t=+0$. We shall be interested in not only questions of existence and smoothness (with respect to $t$) of solutions of problem (5.5) but also the dependence of solutions on the parameter $\xi$. When investigating the smoothness of solutions of the equation $Lu=f(t,\xi)$, we shall rely on the results of [**13**]–[**18**]. Assertions made below without proof either have been obtained in the aforementioned papers or can be proved by methods developed in them.

We consider the equation

$$(a(t)u')'+p(t,\xi)u'-b\lambda u=f(t,\xi) \tag{5.7}$$

on the segment $[0,d]$ of the real axis, where $a(t)=a^0_{nn}(t,0)>0$ for $t>0$, $a(0)=0$;

$$p(t,\xi)=b+i\omega(t,\xi),\ \mathrm{Re}\,p(t,\xi)\equiv|b|=\mathrm{const},\ \omega(t,\xi)=\sum_{k=1}^{n-1}a^0_{nk}(t,0)\xi_k;$$

$a^0_{nk}(0,0)=0$; $\lambda=-b^{-1}(c(\xi)-\nu)$; $c(\xi)$ is defined in (2.13); $\xi\in E_{n-1}$ and $\nu\geqq 0$ are parameters; a prime denotes differentiation with respect to $t$.

Any solution of equation (5.7) is a solution of the Volterra integral equation

$$\begin{aligned} u(t,\xi)=&-\lambda\int_t^d k_1(t,s,\xi)(u(s,\xi)+(b\lambda)^{-1}f(s,\xi))\,ds\\ &+a(d)\mu_1(\xi)\int_t^d e^{P(s,\xi)}\frac{ds}{a(s)}+\mu_0(\xi), \end{aligned} \tag{5.8}$$

where

$$P(s,\xi)=\int_s^d\frac{p(\tau,\xi)}{a(\tau)}\,d\tau;\qquad k_1(t,s,\xi)=-b\int_t^s e^{P(\tau,\xi)-P(s,\xi)}\frac{d\tau}{a(\tau)}; \tag{5.9}$$

$\mu_0(\xi)$ and $\mu_1(\xi)$ are some constants.

The following assertions are valid for $b > 0$:

5.1°. The kernel $k_1(t, s, \xi)$ of the Volterra integral operator (5.8) is bounded; consequently for any fixed $\xi$ and $\nu$ the series

$$W_\nu(t, s, \xi) = \sum_{\varkappa=1}^{\infty} (-\lambda)^\varkappa k_\varkappa(t, s, \xi),$$

where $\lambda = -b^{-1}(c(\xi) - \nu)$, $\nu \geqq 0$, and for $\kappa \geqq 2$ the integral

$$k_\varkappa(t, s, \xi) = \int_t^s k_{\varkappa-1}(t, \tau, \xi) k_1(\tau, s, \xi)\, d\tau,$$

converges absolutely and uniformly on $0 \leqq t \leqq s \leqq d$.

5.2°. The identities

$$\begin{aligned} W_\nu(t, s, \xi) &= -\lambda\Big[k_1(t, s, \xi) + \int_t^s W_\nu(t, \tau, \xi) k_1(\tau, s, \xi)\, d\tau\Big] \\ &= -\lambda\Big[k_1(t, s, \xi) + \int_t^s k_1(t, \tau, \xi) W_\nu(\tau, s, \xi)\, d\tau\Big] \end{aligned} \tag{5.10}$$

are valid for $0 < t \leqq s \leqq d$.

5.3°. If $f(t, \xi) \in \mathscr{L}_2(0, d)$, then any solution of equation (5.7) belongs to $C^0[0, d]$ and is representable in the form

$$\begin{aligned} u(t, \xi) &= \mu_0(\xi) + \mu_1(\xi) a(d) \int_t^d e^{P(s,\xi)} \frac{ds}{a(s)} \\ &+ \int_t^d W_\nu(t, \tau, \xi) \Big[\mu_0(\xi) + \mu_1(\xi) a(d) \int_\tau^d e^{P(s,\xi)} \frac{ds}{a(s)} + \frac{f(\tau, \xi)}{b\lambda}\Big] d\tau \end{aligned} \tag{5.11}$$

for some $\mu_0(\xi)$ and $\mu_1(\xi)$.

5.4°. If $f(t, \xi) \in \mathscr{L}_2(0, d)$, then for any solution of (5.7)

$$\lim_{t \to +0} a(t) u'(t) = 0.$$

We make the following assertions for the more general case when $\operatorname{Re} p(t, \xi) \equiv b(t)$.

5.5°. For $b(0) < a'(0)/2$ the integral operator

$$\mathscr{K} f(t, \xi) = \int_t^d e^{P(t,\xi) - P(s,\xi)} f(s, \xi) \frac{ds}{a(s)}$$

exists for every $\xi$ and is bounded in $\mathscr{L}_2(0, d)$, since

$$\|\mathscr{K} f(-, \xi)\| \leqslant k_0 \|f(-, \xi)\|,$$

the constant $k_0$ not depending on $\xi$.

5.6°. If $m \geqq 1$, $a(t) \in C^{m-1}[0, d] \cap C^1[0, d]$, $p(t, \xi) \in C^{m-1}[0, d]$, $f(t, \xi) \in C^{m-1}[0, d]$ and $b(0) < -ma'(0)$, then any solution $u(t, \xi)$ of equation (5.7) belongs to $C^m[0,d]$ and satisfies the relations

$$u^{(s)}(0,\xi)=b_s^{-1}\Big[f^{(s-1)}(0,\xi)+b\lambda u^{(s-1)}(0,\xi) \tag{5.12}$$
$$-\sum_{k=1}^{s-1}\Big(\sum_{j=1}^{k}\frac{(s-j-1)!}{(s-k-1)!\,(k-j)!}p_j^{(s-k)}(0,\xi)\Big)u^{(k)}(0,\xi)\Big],$$

where $\lambda=-b^{-1}(c(\xi)-\nu)$; $p_j(t,\xi)=b(t)+ja'(t)+i\omega(t,\xi)$; $b_s=b(0)+sa'(0)$ and $s=1,2,\cdots,m$.

5.7°. If $m\geqq 2$, $a(t)\equiv\alpha^2(t)\in C^{m-1}[0,d]$, $p(t,\xi)\in C^{m-1}[0,d]$, $f(t,\xi)\in\widetilde{H}_\alpha^{m-2}(0,d)$ (the space $\widetilde{H}_\alpha^m(0,d)$ was introduced in Definition 3.4) and $b(0)<(-(m-1)/2)a'(0)$, then any solution $u(t,\xi)$ of equation (5.7) belongs to $\widetilde{H}_\alpha^m(0,d)$.

In §2 when considering boundary operators $B_0$ of degeneracy order $m^*$, we introduced the concept of the principal part $B_1$ of this operator. Let us call boundary operators of the form (2.9) equivalent if they have identical principal parts. In the class of equivalent boundary operators we distinguish an operator $B_{01}$, called the *principal equivalent* with respect to operator $L_1-\nu$.

DEFINITION 5.1. Let us call the boundary operator $B_{01}(-i\partial_y,\partial_t)$ the *principal equivalent* of the operator $B_0(-i\partial_y,\partial_t)$ with respect to the operator $L_1-\nu$ if: 1) $B_0$ and $B_{01}$ have identical principal parts, and 2) for any solution $u(t,\xi)$ of the equation $L(\xi,\partial_t)u-\nu u=0$ (cf. (5.6)) for any $\xi\in E_{n-1}$, the equality

$$B_{01}(\xi,\partial_t)u(t,\xi)|_{t=0}=\vartheta(\xi)u(0,\xi) \tag{5.13}$$

is satisfied, where $\vartheta(\xi)$ is defined by formula (2.12).

LEMMA 5.1. *Let the integer $m\geqq 2$, let $a(t)$ and $\omega(t,\xi)$ belong to $C^{m-1}[0,d]$, and let $p(t,\xi)=b+i\omega(t,\xi)$, where $b+[(m-2)/2]a'(0)<0$. Then for any boundary operator $B_0$ of degeneracy order $m^*\leqq m-2$, its principal equivalent $B_{01}$ with respect to $L_1-\nu$ exists.*

PROOF. Applying (5.12) to any solution $u(t,\xi)$ of the equation $L(\xi,\partial_t)u-\nu u=0$, we can write down the equalities

$$u^{(s)}(0,\xi)=bb_s^{-1}\lambda_0u^{(s-1)}(0,\xi)-\sum_{j=0}^{s-1}\sigma_j^s(\xi)u^{(j)}(0,\xi), \tag{5.14}$$

where

$$s=1,2,\ldots,\Big[\frac{m-2}{2}\Big];\ \lambda_0=-b^{-1}c(\xi);\ b_s=b+sa'(0);\ \sigma_0^s(\xi)\equiv 0;$$

$$\sigma_j^s(\xi)=\sum_{i=1}^{j}\frac{(s-i-1)!}{(s-j-1)!\,(j-i)!}\,\frac{p_i^{(s-j)}(0,\xi)}{b_s}\quad(1\leqslant j\leqslant s-2);$$

$$\sigma_{s-1}^s(\xi)=-b_s^{-1}\nu+\sum_{i=1}^{s-1}b_s^{-1}p_i'(0,\xi).$$

If we take advantage of the linear dependence of $p_j(t,\xi)$ on $\xi$, then for $0 \leqq j < s$ we can construct polynomials $R_j^s(\xi)$ in $\xi$ of degree $\leqq s-j$ such that (5.14) will imply the validity of the relation

$$u^{(s)}(0,\xi) + \sum_{j=0}^{s-1} R_j^s(\xi)\, u^{(j)}(0,\xi) = \frac{b^s}{b_1 b_2 \dots b_s} \lambda_0^s u(0,\xi). \tag{5.15}$$

For $s = 1, 2$ equalities (5.15) are easily obtained immediately from (5.14). Let us prove (5.15) for the remaining $s$ by induction. From (5.14) and (5.15) we have

$$\begin{aligned} u^{(s)}(0,\xi) = \frac{b}{b_s}\lambda_0 \Big[ \frac{b^{s-1}}{b_1 b_2 \dots b_{s-1}} \lambda_0^{s-1} u(0,\xi) \\ - \sum_{j=0}^{s-2} R_j^{s-1}(\xi)\, u^{(j)}(0,\xi) \Big] - \sum_{j=0}^{s-1} \sigma_j^s(\xi)\, u^{(j)}(0,\xi). \end{aligned} \tag{5.16}$$

Moreover, from (5.14)

$$\lambda_0 u^{(j)}(0,\xi) = \frac{b_{j+1}}{b}\Big[u^{(j+1)}(0,\xi) + \sum_{i=0}^{j} \sigma_i^{j+1}(\xi)\, u^{(i)}(0,\xi)\Big] \tag{5.17}$$

for $0 \leqq j \leqq \lfloor m/2 \rfloor - 2$. From (5.16) and (5.17) we obtain

$$\begin{aligned} u^{(s)}(0,\xi) &= \frac{b^s}{b_1 b_2 \dots b_s}\lambda_0^s u(0,\xi) - \sum_{j=0}^{s-2} \frac{b}{b_s} R_j^{s-1}(\xi)\, \lambda_0 u^{(j)}(0,\xi) \\ &- \sum_{j=0}^{s-1} \sigma_j^s(\xi)\, u^{(j)}(0,\xi) = \frac{b^s}{b_1 b_2 \dots b_s}\lambda_0^s u(0,\xi) - \sum_{j=0}^{s-1} R_j^s(\xi)\, u^{(j)}(0,\xi), \end{aligned} \tag{5.18}$$

where

$$R_j^s(\xi) = \sigma_j^s(\xi) + b_j b_s^{-1} R_{j-1}^{s-1}(\xi) + \Big\{\sum_{i=j}^{s-2} b_{i+1} b_s^{-1} R_j^{s-1}(\xi)\, \sigma_j^{i+1}(\xi)\Big\} \tag{5.19}$$

for $1 \leqq j \leqq s-2$. The expression in braces in (5.19) should be discarded for $j = s-1$; the addend $b_j b_s^{-1} R_{j-1}^{s-1}(\xi)$ should be discarded for $j = 0$.

From (5.18) and (5.19) we conclude by induction that (5.15) is valid for $1 \leqq s \leqq \lfloor (m-2)/2 \rfloor$, the degree of polynomial $R_j^s(\xi)$ not exceeding $s-j$.

Let us introduce the boundary operator

$$B_{02}(-i\partial_y, \partial_t) = \sum_{s=0}^{r-1} \Big( \sum_{j=s+1}^{r} \Lambda_j(-i\partial_y)\, R_s^j(-i\partial_y)\Big)\partial_t^s \tag{5.20}$$

and assume that

$$B_{01}(-i\partial_y, \partial_t) = B_1(-i\partial_y, \partial_t) + B_{02}(-i\partial_y, \partial_t). \tag{5.21}$$

We show that $B_{01}$ is the principal equivalent of $B$ with respect to $L_1 - \nu$.

In fact, the degree of the polynomial in $\xi$

$$\sum_{j=s+1}^{r} \Lambda_j(\xi) R_s^j(\xi)$$

does not exceed $m^* - 2s - 1$; therefore the principal parts of $B_0$ and $B_{01}$ coincide. For any solution $u(t,\xi)$ of the equation $L(\xi, \partial_t)u - \nu u = 0$ we have from (5.20), (5.12) and (5.15)

$$B_{01}(\xi, \partial_t) u(t, \xi)|_{t=0} = \sum_{s=0}^{r} \Lambda_s(\xi) u^{(s)}(0, \xi)$$

$$+ \sum_{s=0}^{r-1} \Big( \sum_{j=s+1}^{r} \Lambda_j(\xi) R_s^j(\xi) \Big) u^{(s)}(0, \xi) = \Lambda_0(\xi) u(0, \xi)$$

$$+ \sum_{s=1}^{r} \Lambda_s(\xi) \Big[ u^{(s)}(0, \xi) + \sum_{j=0}^{s-1} R_j^s(\xi) u^{(j)}(0, \xi) \Big]$$

$$= \Big[ \sum_{s=1}^{r} \frac{b^s}{b_1 b_2 \ldots b_s} \Lambda_s(\xi) \lambda_0^s + \Lambda_0(\xi) \Big] u(0, \xi).$$

Recalling that $\lambda_0 = -b^{-1}c(\xi)$, we obtain (5.13) from this. The lemma is proved.

LEMMA 5.2. *Let the coefficients $a(t)$, $\omega(t,\xi)$ and $c(\xi)$ in the equation*

$$(a(t)u')' + (b(t) + i\omega(t, \xi))u' + (c(\xi) - \nu)u = f(t, \xi) \tag{5.22}$$

*satisfy conditions (5.2) and (5.3), let $b(t) \in C^1[0,d]$, let the constant $\nu \geqq 0$, and let $f(t,\xi) \in \mathcal{L}_2(0,d)$. Then for any solution of equation (5.22) which is continuous on $[0,d]$ and satisfies the conditions $u(d,\xi) = 0$ and $\alpha(t)u'(t,\xi)|_{t=0} = 0$, the estimate*

$$\nu_0 \| \alpha u' \|^2 + (\nu + \nu_0 |\xi|^2) \| u \|^2 \leqslant -b(0) |u(0, \xi)|^2 + \frac{2}{\nu + \nu_0 |\xi|^2} \| f \|^2 + \max_{0 \leqslant t \leqslant d} |b'(t)| \, \| u \|^2 \tag{5.23}$$

*is valid for any real $\xi$ ($\xi \neq 0$ if $\nu = 0$) and $\alpha(t) = \sqrt{a(t)}$.*

PROOF. We multiply (5.22) by $\bar{u}(t,\xi)$ and integrate the resulting identity with respect to $t \in (0,d)$. We have

$$\int_0^d (au')'\bar{u}\,dt + \int_0^d b(t) u'\bar{u}\,dt + i \int_0^d \omega(t, \xi) u'\bar{u}\,dt + c(\xi) \int_0^d |u|^2 dt = \int_0^d f\bar{u}\,dt + \nu \int_0^d |u|^2 dt. \tag{5.24}$$

Integrating by parts and using the conditions of the lemma, we obtain

$$\int_0^d (au')'\bar{u}dt = au'\bar{u}\Big|_{t=0}^{t=d} - \int_0^d a(t)\,|u'|^2\,dt = -\|\alpha u'\|^2;$$

(5.25) $$\operatorname{Re}\int_0^d b(t)u'\bar{u}dt = \frac{1}{2}\int_0^d b(t)\,\partial_t|u|^2\,dt = \frac{1}{2}b(t)\,|u(t,\xi)|^2\Big|_{t=0}^{t=d}$$

$$-\frac{1}{2}\int_0^d b'(t)\,|u|^2\,dt \leqslant -\frac{1}{2}b(0)\,|u(0,\xi)|^2 + \frac{1}{2}\max_{0\leqslant t\leqslant d}|b'(t)|\,\|u\|^2.$$

After simple transformations we derive

(5.27) $$\int_0^d [\,|\eta^*(t)|^2 + \omega_0(\overline{\xi^*(t)})\,\eta^*(t) - c(\xi^*(t))]\,dt$$
$$= \int_0^d [(b(t) + i\omega(t,\xi) - i\omega_0(\xi)\,\alpha(t))\,u'\bar{u} - (\nu u + f)\,\bar{u}]\,dt,$$

from (5.24) and (5.25), where

$$\eta^*(t) = -i\alpha(t)\,u'(t,\xi);\ \ \xi_k^*(t) = \xi_k u(t,\xi).$$

As is well known, condition (5.2) implies the validity of the inequality

(5.28) $$\operatorname{Re}\Big[\,|\eta^*|^2 + \omega_0(\overline{\xi^*})\,\eta^* + \sum_{k,l=1}^{n-1} a_{kl}^0(0,0)\,\xi_k^*\overline{\xi_l^*}\Big] \geqslant \nu_0[\,|\eta^*|^2 + |\xi^*|^2]$$

for any complex $\eta^*$ and $\xi^*$. From (5.3) and (5.26)-(5.28) we obtain

$$\nu_0[\,\|\alpha u'\|^2 + |\xi|^2\|u\|^2] \leqslant -\frac{1}{2}b(0)\,|u(0,\xi)|^2 + \frac{1}{2}\max_{0\leqslant t\leqslant d}|b'(t)|\,\|u\|^2$$
$$+\frac{1}{2}\nu_0|\xi|\int_0^d \alpha(t)\,|u'|\,|u|\,dt + \int_0^d |f|\,|u|\,dt - \nu\|u\|^2.$$

Using the elementary inequality

(5.29) $$2|a|\,|b| \leqslant \varepsilon|a|^2 + \frac{1}{\varepsilon}|b|^2 \quad (\varepsilon > 0),$$

we obtain (5.23) from this. The lemma is proved.

COROLLARY 5.1. *Let the coefficients* $a(t) \in C^1[0,d]$, *let* $p(t,\xi) \in C^1[0,d]$, *and let* $c(\xi) - \nu$ *and the right side of equation* (5.22) *satisfy the same conditions as in Lemma* 5.2. *If* $b(0) < 0$, *estimate* (5.23) *is valid for any solution* $u(t,\xi)$ *of equation* (5.22) *which satisfies the condition* $u(d,\xi) = 0$.

The proof of Corollary 5.1 follows immediately from Lemma 5.2 and assertions 5.3° and 5.4°.

**COROLLARY** 5.2. *If in Lemma* 5.2 $b(t) = b = \text{const} < 0$ *and the remaining conditions of the lemma are fulfilled, then the estimate*

$$(5.30)\qquad (\nu + \nu_0 |\xi|^2)^2 \|u(-, \xi)\|^2 \leqslant |b|(\nu + \nu_0 |\xi|^2)|u(0, \xi)|^2 + 2\|f(-, \xi)\|^2$$

*is valid for any solution* $u(t, \xi)$ *of equation* (5.7) *which vanishes for* $t = d$. *Hence, in particular, any solution of the equation* $L(\xi, \partial_t)u - \nu u = 0$ $(\nu \geqq 0)$ *which satisfies the conditions* $u(0, \xi) = u(d, \xi) = 0$ *is identically equal to zero on* $[0, d]$.

Instead of boundary value problem (5.5), we shall now consider the boundary value problem

$$(5.31)\qquad L(\xi, \partial_t)u - \nu u = f(t, \xi);\ B_{01}(\xi, \partial_t)u|_{t=0} = g(\xi);\ u(d, \xi) = 0,$$

as well as the boundary value problem

$$(5.32)\qquad L(\xi, \partial_t)u - \nu u = 0;\ B_{01}(\xi, \partial_t)u|_{t=0} = g(\xi);\ u(d, \xi) = 0,$$

where $\nu \geqq 0$ and the boundary operator $B_{01}(\xi, \partial_t)$ (the principal equivalent of $B_0$ with respect to $L_1 - \nu$) is defined in (5.21).

**LEMMA** 5.3. *Let the integer* $m \geqq 2$, *let* $a(t)$ *and* $\omega(t, \xi)$ *belong to* $C^{m-1}[0, d]$, *let* $p(t, \xi) = b + i\omega(t, \xi)$, *and let* $b + [(m-2)/2]a'(0) < 0$. *If the degeneracy order* $m^*$ *of* $B_{01}$ $(B_1)$ *does not exceed* $m - 2$, *then boundary value problem* (5.32) *is uniquely solvable for any* $\xi \neq 0$, $g(\xi)$ *if and only if* $\vartheta(\xi) \neq 0$ $(\vartheta(\xi)$ *being defined in* (2.12)).

**PROOF.** By virtue of 5.6°, any solution of the equation $Lu - \nu u = 0$ belongs to $C^{[(m-2)/2]}[0, d]$. Let us consider the solution of the equation $Lu - \nu u = 0$ obtained from (5.11) for $\mu_0(\xi) = 0$:

$$(5.33)\qquad u(t, \xi) = \mu_1(\xi)a(d)\left[\int_t^d e^{P(s,\xi)} \frac{ds}{a(s)} + \int_t^d W_\nu(t, s, \xi) \int_s^d e^{P(\tau,\xi)} \frac{d\tau}{a(\tau)} ds\right].$$

This solution satisfies the condition $u(d, \xi) = 0$. Making use of identity (5.10), we can write

$$W_\nu(t, d, \xi) = (\nu - c(\xi))\left[\int_t^d e^{P(s,\xi)} \frac{ds}{a(s)} + \int_t^d W_\nu(t, s, \xi) \int_s^d e^{P(\tau,\xi)} \frac{d\tau}{a(\tau)} ds\right].$$

The latter identity and (5.33) imply that

$$(5.34)\qquad u(t, \xi) = \mu(\xi) W_\nu(t, d, \xi),$$

where $\mu(\xi) = \mu_1(\xi)a(d)(\nu - c(\xi))^{-1}$. The solution (5.34) satisfies the conditions $Lu = \nu u$ and $u(d, \xi) = 0$, and it remains for us to select $\mu(\xi)$ in such a way that the boundary condition $B_{01}u|_{t=0} = g(\xi)$ is fulfilled. By a property of operator $B_{01}$ (cf. (5.13)) we have

$$(5.35)\qquad B_{01}(\xi, \partial_t)u(t, \xi)|_{t=0} = \vartheta(\xi)\mu(\xi)W_\nu(0, d, \xi).$$

Let us show that $W_\nu(0,d,\xi)\neq 0$. If $W_\nu(0,d,\xi)=0$, then the function $W_\nu(t,d,\xi)$, being the solution of the equation $LW_\nu=\nu W_\nu$ on $(0,d)$, would satisfy the conditions $W_\nu(0,d,\xi)=W_\nu(d,d,\xi)=0$. Then by Corollary 5.2 we would have $W_\nu(t,d,\xi)\equiv 0$ on $(0,d)$. On the other hand, from identity (5.10) we have

$$a(t)\,\partial_t W_\nu(t,d,\xi)=(c(\xi)-\nu)\Big[e^{P(t,\xi)}+\int_t^d e^{P(t,\xi)-P(s,\xi)}W_\nu(s,d,\xi)\,ds\Big]$$

for $0<t<d$; consequently $\lim_{t\to d}\partial_t W_\nu(t,d,\xi)=(c(\xi)-\nu)a^{-1}(d)\neq 0$. We have arrived at a contradiction which shows that our premise $W_\nu(0,d,\xi)=0$ is false.

Thus $W_\nu(0,d,\xi)\neq 0$, and by the condition of the lemma, $\vartheta(\xi)\neq 0$ for $\xi\neq 0$. Therefore, having chosen $\mu(\xi)=g(\xi)\,[\vartheta(\xi)\,W_\nu(0,d,\xi)]^{-1}$ in (5.34) and (5.35), let us construct the (unique) solution

$$u(t,\xi)=\frac{g(\xi)\,W_\nu(t,d,\xi)}{\vartheta(\xi)\,W_\nu(0,d,\xi)}$$

of problem (5.32). Introducing the function

$$h_\nu(t,\xi)=\frac{W_\nu(t;d,\xi)}{W_\nu(0,d,\xi)},\tag{5.36}$$

we can write this solution in the form

$$u(t,\xi)=\frac{g(\xi)}{\vartheta(\xi)}h_\nu(t,\xi).\tag{5.37}$$

Thus the sufficiency of the condition $\vartheta(\xi)\neq 0$ for $\xi\neq 0$ has been proved. In order to show the necessity of this condition, let us note that any solution of the equation $Lu-\nu u=0$ which satisfies the condition $u(d,\xi)=0$ is representable in the form (5.34). Then the unique solvability of problem (5.32) indicates that the equation

$$\mu(\xi)\,[\vartheta(\xi)\,W_\nu(0,d,\xi)]=g(\xi)$$

is uniquely solvable with respect to $\mu(\xi)$ for any $\xi\neq 0$ and $g(\xi)$. Since $W_\nu(0,d,\xi)\neq 0$, it follows that $\vartheta(\xi)\neq 0$ for $\xi\neq 0$. The lemma is proved.

COROLLARY 5.3. *Let the conditions of Corollary* 5.2 *be fulfilled, and let* $\vartheta(\xi)\neq 0$ *for* $\xi\neq 0$. *Then the estimate*

$$\|u(-,\xi)\|^2\leqslant c_0\,|\xi|^{-2(m^*+1)}\,|g(\xi)|^2,\tag{5.38}$$

*where the constant* $c_0=|b|\,[\nu_0\min_{|\xi|=1}|\vartheta(\xi)|^2]^{-1}$, *is valid for the solution* $u(t,\xi)$ *of problem* (5.32) *for* $\xi\neq 0$.

PROOF. As was shown in Lemma 5.3, the solution of problem (5.32) is representable in the form (5.37). Therefore

$$\|u(-,\xi)\|^2=|g(\xi)|^2\,|\vartheta(\xi)|^{-2}\,\|h_\nu(-,\xi)\|^2.\tag{5.39}$$

The function $h_\nu(t,\xi)$ is the solution of the equation $Lu-\nu u=0$ and by construction satisfies the conditions $h_\nu(0,\xi)=1$ and $h_\nu(d,\xi)=0$. Hence from (5.30) we obtain

$$|\xi|^2\,\|h_\nu(-,\xi)\|^2 \leqslant |b|\,\nu_0^{-1}. \tag{5.40}$$

Taking into account that $\vartheta(\xi)$ is a homogeneous function of $\xi$ of degree $m^*$, we obtain the required inequality (5.38) from (5.39) and (5.40).

LEMMA 5.4. *Let $a(t)$, $\omega(t,\xi)\in C^1[0,d]$, $f(t,\xi)\in\mathscr{L}_2(0,d)$, $p(t,\xi)=b+i\omega(t,\xi)$ and $b=\mathrm{const}<0$. If conditions (5.2) and (5.3) are fulfilled, then for any $\nu\geqq 0$ a solution of the problem*

$$L(\xi,\partial_t)u-\nu u=f(t,\xi);\quad u(0,\xi)=u(d,\xi)=0 \tag{5.41}$$

*exists which is continuous on $[0,d]$ and satisfies the estimate*

$$\|u(-,\xi)\|^2\leqslant 2(\nu+\nu_0|\xi|^2)^{-2}\,\|f(-,\xi)\|^2 \tag{5.42}$$

*for any $\xi$ ($\neq 0$ if $\nu=0$).*

PROOF. Consider the function

$$u^-(t,\xi)=\int_0^d \mathscr{Y}_\nu^-(t,s,\xi)\frac{f(s,\xi)}{c(\xi)-\nu}\,ds, \tag{5.43}$$

where

$$\mathscr{Y}_\nu^-(t,s,\xi)=\begin{cases}\dfrac{W_\nu(0,s,\xi)\,W_\nu(t,d,\xi)}{W_\nu(0,d,\xi)}-W_\nu(t,s,\xi) & \text{for } t<s;\\[2ex] \dfrac{W_\nu(0,s,\xi)\,W_\nu(t,d,\xi)}{W_\nu(0,d,\xi)} & \text{for } t>s.\end{cases}$$

Let us show that the function $u^-(t,\xi)$ is a solution of problem (5.41). This fact can be proved immediatey. Let us derive (5.43) from (5.11). Assuming $\mu_0(\xi)=0$, we have

$$u^-(t,\xi)=\mu_1(\xi)\,a(d)\left[\int_t^d e^{P(s,\xi)}\frac{ds}{a(s)}+\int_t^d W_\nu(t,s,\xi)\int_s^d e^{P(\tau,\xi)}\frac{d\tau}{a(\tau)}\,ds\right]$$
$$-\int_t^d W_\nu(t,s,\xi)\frac{f(s,\xi)}{c(\xi)-\nu}\,ds.$$

Using (5.20), we can write the latter in the form

$$u^-(t,\xi)=\mu(\xi)\,W_\nu(t,d,\xi)-\int_t^d W_\nu(t,s,\xi)\frac{f(s,\xi)}{c(\xi)-\nu}\,ds, \tag{5.44}$$

where

$$\mu(\xi)=-\mu_1(\xi)\,a(d)\,(c(\xi)-\nu)^{-1}.$$

Obviously $u^-(d,\xi)=0$. Now let us choose $\mu(\xi)$ in such a way that $u^-(0,\xi)=0$. For this purpose let us assume

$$\mu(\xi) = \int_0^d \frac{W_\nu(0, s, \xi)}{W_\nu(0, d, \xi)} \frac{f(s, \xi)}{c(\xi) - \nu} ds. \tag{5.45}$$

When proving Lemma 5.3, we established that $W_\nu(0,d,\xi) \neq 0$ for any $\xi$ ($\neq 0$ if $\nu = 0$).

We obtain (5.43) from (5.44) and (5.45). By virtue of 5.3° the solution $u^-(t,\xi)$ of the equation $(L-\nu)u^- = f$ is continuous on $[0,d]$, and by construction it satisfies the conditions $u^-(0,\xi) = u^-(d,\xi) = 0$. By virtue of Corollary 5.2, estimate (5.30) is valid for $u^-(t,\xi)$, which implies estimate (5.42). The lemma is proved.

LEMMA 5.5. *Let the integer $m \geqq 2$, let $a(t)$ and $\omega(t,\xi) \in C^{m-1}[0,d]$, let $f(t,\xi) \in \widetilde{H}_\alpha^{m-2}(0,d)$, let the boundary operator $B_{01}$ have degeneracy order $m^* \leqq m-2$, let conditions (5.1)–(5.3) be fulfilled, and let $\vartheta(\xi) \neq 0$ for $\xi \neq 0$. Then for $\xi \neq 0$ a unique solution $u(t,\xi) \in \widetilde{H}_\alpha^m(0,d)$ of problem (5.31) exists, representable in the form*

$$u(t, \xi) = u^-(t, \xi) + [g(\xi) - B_{01}(\xi, \partial_t) u^-|_{t=0}] \vartheta^{-1}(\xi) h_\nu(t, \xi), \tag{5.46}$$

*where the functions $u^-(t,\xi)$ and $h_\nu(t,\xi)$ are defined by formulas (5.43) and (5.36), respectively. The estimate*

$$|\xi|^{2m} \|u\|^2 \leqslant c_{1/2m} \Big\{ \sum_{p+2s=m-2} (1 + |\xi|^2)^p \|f^{(s)}\|^2 + |\xi|^{2(m-m^*-1)} |g(\xi)|^2 \Big\} \tag{5.47}$$

*is valid for this solution, where the constant $c_{m/2} > 0$ depends on $\nu_0, \beta_1, m^*, m$ and the moduli of the coefficients of operators $B_1$ and $B_{02}$.*

PROOF. First let us construct the partial solution $u^-(t,\xi)$ of the equation $Lu - \nu u = f$ by formula (5.43). By Lemma 5.4 such a solution exists and satisfies the conditions $u^-(0,\xi) = u^-(d,\xi) = 0$. By virtue of 5.7° this solution belongs to $\widetilde{H}_\alpha^m(0,d)$ for every $\xi$. Since the highest order $r$ of derivatives with respect to $t$ which enter into $B_{01}$ does not exceed $[m^*/2] \leqq [(m-2)/2]$, and since assertion 5.6° implies that $u^-(t,\xi)$ has continuous derivatives up to order $[(m-2)/2]$ with respect to $t$ on $[0,d]$ $(f(t,\xi) \in \widetilde{H}_\alpha^{m-2}(0,d) \subset C^{[m/2]-2}[0,d])$, we can calculate $g_{01}(\xi) = B_{01}(\xi, \partial_t) u^-(t,\xi)|_{t=0}$. Introducing the function $u_2(t,\xi) = u(t,\xi) - u^-(t,\xi)$, we can reduce problem (5.31) to the problem

$$Lu_2 - \nu u_2 = 0; \ B_{01}(\xi, \partial_t) u_2|_{t=0} = g(\xi) - g_{01}(\xi); \ u_2(d, \xi) = 0. \tag{5.48}$$

By Lemma 5.3 problem (5.8) is uniquely solvable for $\xi \neq 0$, and its solution is representable in the form $u_2(t,\xi) = [g(\xi) - g_{01}(\xi)]\vartheta^{-1}(\xi) h_\nu(t,\xi)$. By virtue of 5.7°, $u_2(t,\xi)$ belongs to $\widetilde{H}_\alpha^m(0,d)$ and is the unique solution of problem (5.48) in $\widetilde{H}_\alpha^m(0,d)$. Consequently the unique solution $u(t,\xi) = u^-(t,\xi) + u_2(t,\xi)$ of problem (5.31) exists, belongs to $\widetilde{H}_\alpha^m(0,d)$ and is representable in the form (5.46).

Next let us evaluate $\|u(-,\xi)\|$. Using estimates (5.42) and (5.38), we obtain

$$|\xi|^{2m}\|u\|^2 \leqslant 2(|\xi|^{2m}\|u^-\|^2 + |\xi|^{2m}\|u_2\|^2)$$
$$\leqslant 4\nu_0^{-2}|\xi|^{2(m-2)}\|f\|^2 + 2c_0|\xi|^{2(m-m^*-1)}|g(\xi)-g_{01}(\xi)|^2. \tag{5.49}$$

By the construction of $g_{01}(\xi)$ and $B_{01}$ (cf. Lemma 5.1) we have

$$|g_{01}(\xi)| \leqslant |B_1(\xi,\partial_t)u^-|_{t=0}| + |B_{02}(\xi,\partial_t)u^-|_{t=0}|$$
$$\leqslant c\left\{\sum_{s=1}^{r}|\xi|^{m^*-2s}|\partial_t^s u^-(0,\xi)| + \sum_{s=1}^{r-1}(1+|\xi|)^{m^*-2s-1}|\partial_t^s u^-(0,\xi)|\right\}, \tag{5.50}$$

where the constant $c>0$ depends only on the moduli of the coefficients of $B_1$ and $B_{02}$.

We have taken into account in this connection that $u^-(0,\xi)=0$. We obtain the estimates

$$|\partial_t^s u^-(0,\xi)| \leqslant c_{s+1}\sum_{j=0}^{s-1}(1+|\xi|)^{2s-2-2j}|f^{(j)}(0,\xi)| \tag{5.51}$$

for $1\leqq s\leqq r$ from (5.12), where the constant $c_{s+1}$ depends on the constant $\beta_1$ in condition (5.1) and on the derivatives of $a(t)$ and $a^0_{nk}(t,0)$ up to order $s$. Let us prove (5.51) by induction. For $s=1$ we obtain $|\partial_t u^-(0,\xi)|\leqq \beta_1^{-1}|f(0,\xi)|$ from (5.12); consequently inequality (5.51) is valid for $s=1$. Next by induction we derive

$$|\partial_t^s u^-(0,\xi)| \leqslant \beta_1^{-1}\Big[|f^{(s-1)}(0,\xi)|$$
$$+|c(\xi)|c_{s+1}\sum_{j=0}^{s-2}(1+|\xi|)^{2s-4-2j}|f^{(j)}(0,\xi)|$$
$$+\sum_{k=1}^{s-1}\tilde{c}_{s-k+1}c_{k+1}(1+|\xi|)\sum_{j=0}^{k-1}(1+|\xi|)^{2k-2-2j}|f^{(j)}(0,\xi)| \tag{5.52}$$

from (5.12) and (5.51). Here we have made use of the trivial estimate

$$|\sigma_k^s(\xi)| \leqslant \tilde{c}_{s-k+1}(1+|\xi|)\quad(1\leqslant k\leqslant s-1).$$

Let us make the estimate $|c(\xi)|\leqq c_0|\xi|^2$ and change the order of summation in (5.52). We have

$$|\partial_t^s u^-(0,\xi)| \leqslant \beta_1^{-1}\Big[|f^{(s-1)}(0,\xi)| + c_0c_{s+1}\sum_{j=0}^{s-2}(1+|\xi|)^{2s-2-2j}|f^{(j)}(0,\xi)|$$
$$+\sum_{j=0}^{s-2}\Big(\sum_{k=j+1}^{s-1}\tilde{c}_{s-k+1}c_{k+1}(1+|\xi|)^{2k-1-2j}\Big)|f^{(j)}(0,\xi)|\Big].$$

Very simple estimates in the latter inequality bring us to (5.51). Inequality (5.51) is proved.

Now let us apply estimates (5.51) to (5.50). We obtain

$$|g_{01}(\xi)| \leqslant c\Big\{\sum_{s=1}^{r} |\xi|^{m^*-2s} c_{s+1} \sum_{j=0}^{s-1} (1+|\xi|)^{2s-2-2j} |f^{(j)}(0,\xi)|$$

$$+ \sum_{s=1}^{r-1} (1+|\xi|)^{m^*-2s-1} c_{s+1} \sum_{j=0}^{s-1} (1+|\xi|)^{2s-2-2j} |f^{(j)}(0,\xi)|\Big\}$$

$$\leqslant \tilde{c}_{r+1}\Big\{\sum_{j=0}^{r-1} (1+|\xi|)^{m^*-2-2j} |f^{(j)}(0,\xi)|\Big\}.$$

Let us apply inequality (3.7) for an estimate of $|\xi|\,|f^{(j)}(0,\xi)|$. Taking into account that $r \leqq [(m-2)/2]$, we obtain the estimate

$$|\xi|^{m-m^*-1} |g_{01}(\xi)| \leqslant \tilde{c}_{r+1} \sum_{j=0}^{r-1} (1+|\xi|)^{m-4-2j} |\xi|\,|f^{(j)}(0,\xi)|$$

$$\leqslant \overset{*}{c}_{r+1} \sum_{j=0}^{r-1} (1+|\xi|)^{m-4-2j} (\|f^{(j+1)}\| + (1+|\xi|^2)\|f^{(j)}\|) \tag{5.53}$$

$$\leqslant c_{r+1} \sum_{j=0}^{r} (1+|\xi|)^{m-2-2j} \|f^{(j)}\|.$$

Inequality (5.47) is easily derived from estimates (5.49) and (5.53). The lemma is proved.

Lemma 5.6. *Let the integer* $m \geqq 2$, *let*

$$a(t) \equiv \alpha^2(t) \in C^{m-1}[0,\ d];\ \omega(t,\xi) \in C^{m-1}[0,d];\ f(t,\xi) \in \widetilde{H}_\alpha^{m-2}(0,d)$$

*and let conditions* (5.1) *be fulfilled. Then the estimate*

$$\sum_{p+j+2s=m} |\xi|^{2p} \|(\alpha\partial_t)^j \partial_t^s u\|^2 \tag{5.54}$$

$$\leqslant c_m\Big\{\sum_{p+j+2s=m-2} (1+|\xi|^2)^p \|(\alpha\partial_t)^j \partial_t^s f\|^2 + (1+|\xi|^2)^m \|u\|^2\Big\},$$

*where the constant* $c_m > 0$ *does not depend on* $\xi$, $f(t,\xi)$ *or* $u(t,\xi)$, *is valid for any solution* $u(t,\xi)$ *of equation* (5.7).

Lemma 5.6 will be proved in §7. Here we prove a corollary implied by Lemma 5.5 and 5.6.

Corollary 5.4. *Let the conditions of Lemma* 5.5 *be fulfilled. Then the estimate*

$$\sum_{p+j+2s=m} |\xi|^{2p} \|(\alpha\partial_t)^j \partial_t^s u\|^2$$

(5.55) $$\leqslant c_m \Big\{ \sum_{p+j+2s=m-2} (1+|\xi|^2)^p \|(\alpha\partial_t)^j \partial_t^s f\|^2 + |\xi|^{2(m-m^*-1)} |g(\xi)|^2 + \|u\|^2 \Big\},$$

*where the constant $c_m > 0$ does not depend on $\xi$, $f(t,\xi)$, $g(\xi)$ or $u(t,\xi)$, is valid for any solution $u(t,\xi) \in \widetilde{H}^m_\alpha(0,d)$ of problem* (5.5).

PROOF. A solution $u(t,\xi)$ of problem (5.5) can be considered a solution of the following problem:

(5.56) $$Lu = f(t,\xi);\ B_{01}(\xi,\partial_t)u|_{t=0} = g_2(\xi);\ u(d,\xi) = 0,$$

where

$$g_2(\xi) = g(\xi) + B_{02}(\xi,\partial_t)u|_{t=0}$$

and $B_{02}$ is defined by (5.20). By virtue of inequalities (5.54) and (5.47) the estimate

$$\sum_{p+j+2s=m} |\xi|^{2p} \|(\alpha\partial_t)^j \partial_t^s u\|^2$$

(5.57) $$\leqslant c_m \Big\{ \sum_{p+j+2s=m-2} (1+|\xi|^2)^p \|(\alpha\partial_t)^j \partial_t^s f\|^2 + |\xi|^{2(m-m^*-1)} |g_2(\xi)|^2 + \|u\|^2 \Big\}$$

is valid for a solution of problem (5.56). Since

$$|g_2(\xi)|^2 \leqslant 2|g(\xi)|^2 + 2|B_{02}(\xi,\partial_t)u|_{t=0}|^2,$$

it remains for us to estimate $|B_{02}(\xi,\partial_t)u|_{t=0}|$ in order to obtain (5.55) from (5.57). The definition (5.20) of $B_{02}$ implies the estimate

(5.58) $$|B_{02}(\xi,\partial_t)u|_{t=0}|^2 \leqslant c \sum_{s=0}^{r-1} \Big( \sum_{j=s+1}^{r} |\xi|^{2(m^*-2j)} (1+|\xi|^2)^{j-s} \Big) |u^{(s)}(0,\xi)|^2.$$

For even $m$ we have $r \leq m^*/2 \leq (m-2)/2$, and it is easy to obtain the estimate

(5.59′) $$|\xi|^{2(m-m^*-1)} |B_{02}(\xi,\partial_t)u|_{t=0}|^2 \leqslant c^* \sum_{s=0}^{r-1} (1+|\xi|^2)^{m-2s-3} |\xi|^2 |u^{(s)}(0,\xi)|^2$$

from (5.58).

For odd $m$ the value of $r \leq (m-3)/2$. In this case we derive the inequality

(5.59″) $$|\xi|^{2(m-m^*-1)} |B_{02}(\xi,\partial_t)u|_{t=0}|^2 \leqslant c^{**} \sum_{s=0}^{r-1} (1+|\xi|^2)^{m-2s-4} |\xi|^4 |u^{(s)}(0,\xi)|^2$$

from (5.58).

Making use of estimate (3.7), we obtain

(5.60′) $$|\xi|^{2(m-m^*-1)} |B_{02}(\xi,\partial_t)u|_{t=0}|^2 \leqslant c^* c \sum_{s=0}^{r} (1+|\xi|^2)^{m-1-2s} \|u^{(s)}\|^2$$

from (5.59′) and

$$(5.60'')\quad |\xi|^{2(m-m^*-1)}|B_{02}(\xi,\partial_t)u|_{t=0}|^2\leqslant c^{**}c\sum_{s=0}^{r}(1+|\xi|^2)^{m-2-2s}|\xi|^2\|u^{(s)}\|^2$$

from (5.59″). In order to estimate the right sides of (5.60′) and (5.60″), let us apply inequalities (3.4) and (3.5). From these inequalities we have

$$(5.61)\quad \begin{aligned}&\sum_{s=0}^{r}(1+|\xi|^2)^{m-1-2s}\|u^{(s)}\|^2\leqslant c(\varepsilon)\|u\|^2\\&+\varepsilon\Big\{\sum_{s=0}^{r}\|(\alpha\partial_t)^{m-2s}\partial_t^s u\|^2+\|\partial_t^{\frac{m}{2}}u\|^2+|\xi|^{2m}\|u\|^2\Big\}\end{aligned}$$

for even $m$ and any $\epsilon>0$. Applying inequality (5.61) to (5.60′), we obtain

$$(5.62)\quad |\xi|^{2(m-m^*-1)}|B_{02}(\xi,\partial_t)u|_{t=0}|^2\leqslant\varepsilon\sum_{p+j+2s=m}|\xi|^{2p}\|(\alpha\partial_t)^j\partial_t^s u\|^2+c(\varepsilon)\|u\|^2$$

for even $m$ and any $\epsilon>0$.

If $m$ is an odd number, then once again we make use of inequality (5.61) after changing $m$ to $m-1$ in it. Applying the resulting inequality to (5.60″), after simple estimates we arrive at inequality (5.62).

Choosing $\epsilon>0$ sufficiently small in inequality (5.62), we derive inequality (5.55) from (5.57) and (5.62).

PROOF OF THEOREM 4.1. Let $v(t,y)\in H_\alpha^{m_1}(K_d)$, having chosen $d>0$ so small that conditions (5.1) and (5.3) are fulfilled. The function $v(t,y)$ can be considered a solution of problem (5.4). By the condition of the theorem, $f_1(t,y)\in H_\alpha^{m-2}(K_d)$ and $g_1(y)\in H^{m-m^*-1}(E_{n-1})$. Then for every $\xi\in E_{n-1}$ the function $u(t,\xi)=F_{n-1}v$ is a solution of problem (5.5). From 5.7° it follows that $u(t,\xi)\in\widetilde{H}_\alpha^m(0,d)$, since $f(t,\xi)=F_{n-1}f_1\in\widetilde{H}_\alpha^{m-2}(0,d)$. By Corollary 5.4 estimate (5.55) is valid for $u(t,\xi)$.

Using the equivalence of norms (3.1) and (3.2), we obtain inequality (4.3) from inequality (5.55) after integrating with respect to $\xi\in E_{n-1}$. Since the magnitudes on the right side of inequality (4.3) are finite, its left side also is finite. This means that $v(t,y)\in H_\alpha^m(K_d)$. Theorem 4.1 is proved.

## §6. Proof of Theorem 4.2

As in the proof of Theorem 4.1, we shall assume without loss of generality that $\delta=d$ and that along with condition (5.3) the inequalities

$$(6.1)\quad \beta_1\leqslant b_s(t)\leqslant\beta_1^{-1};\ \beta_1\leqslant b_{s+1}(t)-\frac{1}{2}a'(0)\leqslant\beta_1^{-1}$$

are fulfilled for all $t\in[0,d]$, where

$$0\leqslant s\leqslant\left[\frac{m}{2}\right];\ a(t)=a_{nn}^0(t,0);\ b_s(t)=a_n^0(t,0)+sa'(t);$$

$\beta_1>0$ is a constant.

Denoting $u(t,\xi) = F_{n-1}v$, we consider the problem

$$L(\xi, \partial_t)u - \nu u = f(t, \xi); \; u(d, \xi) = 0, \tag{6.2}$$

where

$$v(t, y) \in H_\alpha^m(K_d); \; f(t, \xi) = F_{n-1}f_1, \; f_1(t, y) \in H_\alpha^{m-2}(K_d);$$

the operator $L(\xi, \partial_t)$ is defined in (5.6), the constant $\nu \geqq 0$.

We shall use some of the results of [**13**]-[**18**] on the existence of smooth solutions of equation (5.7). We shall cite them without proof.

The following assertions are valid for $b > 0$:

6.1°. Any solution of equation (5.7) which is continuous on $[0, d]$ is a solution of the Volterra integral equation

$$u(t, \xi) = \lambda \int_0^t l_1(t, s, \xi) \left( u(s, \xi) + \frac{f(s, \xi)}{b\lambda} \right) ds + \mu_0(\xi), \tag{6.3}$$

where

$$l_1(t, s, \xi) = b \int_s^t e^{P(\tau,\xi)-P(s,\xi)} \frac{d\tau}{a(\tau)},$$

the function $P(s,\xi)$ is defined in (5.9), and $\mu_0(\xi)$ is a constant.

6.2°. The kernel $l_1(t,s,\xi)$ of the Volterra integral operator in (6.3) is bounded; consequently for any fixed $\xi \in E_{n-1}$ the series

$$Z_\nu(t, s, \xi) = \sum_{\varkappa=1}^{\infty} \lambda^\varkappa l_\varkappa(t, s, \xi)$$

converges absolutely and uniformly on $0 \leqq s \leqq t \leqq d$, where $\lambda = b^{-1}(\nu - c(\xi))$ and for $\kappa \geqq 2$

$$l_\varkappa(t, s, \xi) = \int_s^t l_{\varkappa-1}(t, \tau, \xi)\, l_1(\tau, s, \xi)\, d\tau.$$

6.3°. The identities

$$\begin{aligned} Z_\nu(t, s, \xi) &= \lambda \left[ l_1(t, s, \xi) + \int_s^t Z_\nu(t, \tau, \xi)\, l_1(\tau, s, \xi)\, d\tau \right] \\ &= \lambda \left[ l_1(t, s, \xi) + \int_s^t l_1(t, \tau, \xi)\, Z_\nu(\tau, s, \xi)\, d\tau \right] \end{aligned} \tag{6.4}$$

are valid for $0 < s \leqq t \leqq d$.

6.4°. If $f(t,\xi) \in \mathscr{L}_2(0,d)$, then any solution of equation (5.7) which belongs to $C^0[0,d]$ is representable in the form

$$u(t, \xi) = \mu_0(\xi) \left[ 1 + \int_0^t Z_\nu(t, s, \xi)\, ds \right] + \int_0^t Z_\nu(t, s, \xi) \frac{f(s, \xi)}{b\lambda}\, ds. \tag{6.5}$$

Formula (6.5) defines a family of solutions $u(t,\xi) \in C^0[0,d]$ of equation (5.7) which depend on the constant $\mu_0$.

6.5°. If $f(t,\xi) \in \mathscr{L}_2(0,d)$, then for any solution $u(t,\xi) \in C^0[0,d]$ of equation (5.7)

$$\lim_{t \to +0} a(t)\, u'(t) = 0.$$

We formulate the following assertions for the more general case $\operatorname{Re} p(t,\xi) \equiv b(t)$.

6.6°. For $b(0) > a'(0)/2$ the integral operator

$$\mathcal{M} f(t,\xi) = \int_0^t e^{P(t,\xi)-P(s,\xi)} f(s,\xi) \frac{ds}{a(s)}$$

exists for every $\xi$ and is bounded in $\mathscr{L}_2(0,d)$, since

$$\| \mathcal{M} f(-,\xi) \| \leqslant k_0 \| f(-,\xi) \|, \tag{6.6}$$

the constant $k_0$ not depending on $\xi$.

6.7°. Let the integer $m \geqq 2$, let $a(t) \equiv \alpha^2(t) \in C^{m-1}[0,d]$, let $p(t,\xi) \in C^{m-1}[0,d]$, and let $f(t,\xi) \in \widetilde{H}^{m-2}_\alpha(0,d)$. If $b(0) > 0$, then any solution $u(t,\xi)$ of equation (5.7) which is continuous on $[0,d]$ belongs to $\widetilde{H}^m_\alpha(0,d)$.

Lemma 5.2 and 6.5° immediately imply

LEMMA 6.1. *Let*

$$a(t) \equiv \alpha^2(t) \in C^1[0,d]; \quad \omega(t,\xi) \in C^1[0,d]; \quad p(t,\xi) = b + i\omega(t,\xi),$$

*where* $b = \text{const} > 0$, $f(t,\xi) \in \mathscr{L}_2(0,d)$, *and conditions* (5.2) *and* (5.3) *are fulfilled. Then the estimate*

$$\nu_0 \| \alpha u' \|^2 + (\nu + \nu_0 |\xi|^2) \| u \|^2 \leqslant 2(\nu + \nu_0 |\xi|^2)^{-1} \| f \|^2 \tag{6.7}$$

*for all* $\xi \in E_{n-1}$ *and* $\nu \geqq 0$ ($\xi \neq 0$ *if* $\nu = 0$) *is valid for a solution* $u(t,\xi)$ *of equation* (5.7) *which belongs to* $C^0[0,d]$ *and satisfies the condition* $u(d,\xi) = 0$.

*Inequality* (6.7) *implies that this solution of problem* (6.2) *is unique in class* $C^0[0,d]$.

LEMMA 6.2. *Let the conditions of Lemma* 6.1 *be fulfilled. Then for any* $\xi \in E_{n-1}$ *and* $\nu \geqq 0$ ($\xi \neq 0$ *if* $\nu = 0$) *a unique solution of problem* (6.2) *exists which is continuous on* $[0,d]$ *and satisfies estimate* (5.42).

PROOF. For every $\mu_0(\xi)$ formula (6.5) defines a solution of equation (5.7) which is continuous on $[0,d]$. In order to obtain a solution of problem (6.2), $\mu_0(\xi)$ must be found from the condition

$$u(d,\xi) = \mu_0(\xi)\left[1 + \int_0^d Z_\nu(d,s,\xi)\, ds\right] - \int_0^d Z_\nu(d,s,\xi) \frac{f(s,\xi)}{c(\xi)-\nu}\, ds = 0. \tag{6.8}$$

Next we note that $1+\int_0^d Z_\nu(d,s,\xi)\,ds \neq 0$. If $1+\int_0^d Z_\nu(d,s,\xi)\,ds=0$ for given $\xi$ and $\nu$, then a solution $u_0(t,\xi)=1+\int_0^t Z_\nu(t,s,\xi)\,ds$ of the homogeneous equation $Lu_0-\nu u_0=0$ would exist, continuous on $[0,d]$ and satisfying the condition $u_0(d,\xi)=0$. Then by Lemma 6.1 we would have $u_0(t,\xi)\equiv 0$. On the other hand, by virtue of the boundedness of the function $Z_\nu(t,s,\xi)$ for $0\leqq s\leqq t\leqq d$, we would have $u_0(0,\xi)=1$, which is impossible in view of the continuity of $u_0(t,\xi)$ on $[0,d]$. Thus $1+\int_0^d Z_\nu(d,s,\xi)\,ds\neq 0$, and from (6.8) and (6.5) we obtain a solution of problem (6.2):

$$u(t,\xi)=\int_0^d \mathcal{Y}_\nu^+(t,s,\xi)\frac{f(s,\xi)}{c(\xi)-\nu}\,ds, \tag{6.9}$$

where

$$\mathcal{Y}_\nu^+(t,s,\xi)=\begin{cases}\dfrac{Z_\nu(d,s,\xi)\left[1+\int_0^t Z_\nu(t,\tau,\xi)\,d\tau\right]}{1+\int_0^d Z_\nu(d,\tau,\xi)\,d\tau}-Z_\nu(t,s,\xi) & \text{for } t>s;\\[2ex] \dfrac{Z_\nu(d,s,\xi)\left[1+\int_0^t Z_\nu(t,\tau,\xi)\,d\tau\right]}{1+\int_0^d Z_\nu(d,\tau,\xi)\,d\tau} & \text{for } t<s.\end{cases}$$

By Lemma 6.1 this solution is unique (in $C^0[0,d]$), estimate (6.7) is valid for it, and estimate (5.42) follows immediately from (6.7). The lemma is proved.

The following lemma will be deduced in §7.

LEMMA 6.3. *Let the integer* $m\geqq 2$, *let* $a(t)\equiv\alpha^2(t)\in C^{m-1}[0,d]$, *let* $p(t,\xi)=b+i\omega(t,\xi)\in C^{m-1}[0,d]$, *let* $f(t,\xi)\in\widetilde{H}_\alpha^{m-2}(0,d)$, *and let conditions* (6.1) *be fulfilled. If* $u(t,\xi)\in C^0[0,d]$ *is a solution of equation* (5.7), *then* $u(t,\xi)\in\widetilde{H}_\alpha^m(0,d)$, *and estimate* (5.54) *is valid.*

The following simple argument completes the

PROOF OF THEOREM 4.2. Let $v(t,y)\in H_\alpha^2(K_d)$. Then the function $u(t,\xi)=F_{n-1}v\in\widetilde{H}_\alpha^2(0,d)\subset C^0[0,d]$, $\xi\in E_{n-1}$, is a solution of problem (6.2), where $\nu=0$ and $f(t,\xi)=F_{n-1}L_1v$.

By the condition of the theorem $f(t,\xi)\in\widetilde{H}_\alpha^{m-2}(0,d)$; consequently $u(t,\xi)\in\widetilde{H}_\alpha^m(0,d)$ by Lemma 6.3, and estimate (5.54) is valid for $\xi\in E_{n-1}$. By virtue of Lemma 6.2, estimate (5.42) holds for $u(t,\xi)$ for $\xi\neq 0$. We obtain inequality (4.4) from (5.54) and (5.42), as in the proof of Theorem 4.1. Inequality (4.4) implies $v(t,y)\in H_\alpha^m(K_d)$.

## §7. Evaluation of normal derivatives

After a Fourier transformation "in a tangential direction," the problem stated in the heading of this section reduces to the proof of inequality (5.54) for a solution $u(t,\xi)$ of equation (5.7).

**Lemma** 7.1. *Let* $m \geqq 2$; *let* $a(t) \equiv \alpha^2(t) \in C^{m-1}[0,d]$; *let* $p(t,\xi) = b(t) + i\omega(t,\xi) \in C^{m-1}[0,d]$; *let* $|\omega(t,\xi)| \leqq c_0\alpha(t)|\xi|$; *let* $f(t,\xi) \in \widetilde{H}^{m-2}_\alpha[0,d)$, *and let* $u(t,\xi) \in \widetilde{H}^m_\alpha(0,d)$ *be a solution of equation* (5.22). *Then the estimates*

$$\sum_{p+j+2s=m} |\xi|^{2p}\|(\alpha\partial_t)^j\partial_t^s u\|^2 \tag{7.1}$$
$$\leqslant c_m\Big\{\sum_{p+j+2s=m-2}(1+|\xi|^2)^p\|(\alpha\partial_t)^j\partial_t^s f\|^2+\|u^{\left(\frac{m}{2}\right)}\|^2+(1+|\xi|^2)^m\|u\|^2\Big\}$$

*for even* $m$ *and*

$$\sum_{p+j+2s=m}|\xi|^{2p}\|(\alpha\partial_t)^j\partial_t^s u\|^2\leqslant c_m\Big\{\sum_{p+j+2s=m-2}(1+|\xi|^2)^p\|(\alpha\partial_t)^j\partial_t^s f\|^2 \tag{7.2}$$
$$+\|\alpha\partial_t u^{\left(\frac{m-1}{2}\right)}\|^2+(1+|\xi|^2)\|u^{\left(\frac{m-1}{2}\right)}\|^2+(1+|\xi|^2)^m\|u\|^2\Big\}$$

*for odd* $m$ *are valid for* $u(t,\xi)$. *The constant* $c_m > 0$ *does not depend on* $\xi$, $f$ *or* $u$.

**Proof.** Differentiating identity (5.22) $s$ times with respect to $t > 0$, we obtain

$$(\alpha\partial_t)^2u^{(s)}+p_{s+\frac{1}{2}}(t,\xi)u^{(s+1)}+(c(\xi)-\nu)u^{(s)}=f^{(s)}-\sum_{i=1}^{s}\psi_i^s(t,\xi)u^{(i)}, \tag{7.3}$$

where the functions

$$p_j(t,\xi)\equiv b(t)+ja'(t)+i\omega(t,\xi),$$
$$\psi_i^s(t,\xi)=\frac{s!}{(i-1)!\,(s+1-i)!}\left(p(t,\xi)+\frac{s+1}{s+2-i}a'(t)\right)^{(s+1-i)}$$

and their derivatives with respect to $t\in(0,d)$ satisfy the estimates

$$|\partial_t^s p_j(t,\xi)|\leqslant c_{s,j}(1+|\xi|);\quad |\partial_t^r\psi_i^r(t,\xi)|\leqslant c_{s,r,i}(1+|\xi|). \tag{7.4}$$

Let us apply the operator $(\alpha\partial_t)^{\beta-2}$ $(\beta\geqq 2)$ to identity (7.3). After simple transformations we obtain the identity

$$(\alpha\partial_t)^\beta u^{(s)}+p_{s+\frac{1}{2}}(t,\xi)(\alpha\partial_t)^{\beta-2}u^{(s+1)}+(c(\xi)-\nu)(a\partial_t)^{\beta-2}u^{(s)}=(\alpha\partial_t)^{\beta-2}f^{(s)}$$
$$-\sum_{j=0}^{\beta-3}\frac{(\beta-2)!}{(\beta-2-j)!\,j!}[(\alpha\partial_t)^{\beta-2-j}p_{s+\frac{1}{2}}(t,\xi)][(\alpha\partial_t)^j u^{(s+1)}] \tag{7.5}$$
$$-\sum_{i=!}^{s}\sum_{j=0}^{\beta-2}\frac{(\beta-2)!}{(\beta-2-j)!\,j!}[(\alpha\partial_t)^{\beta-2-j}\psi_i^s(t,\xi)].[(\alpha\partial_t)^j u^{(i)}].$$

We shall also use the trivial identity

$$(7.6)\quad \alpha(t)(\alpha\partial_t)^\beta u^{(s+1)} = (\alpha\partial_t)^{\beta+1}u^{(s)} - \sum_{i=0}^{\beta-1}\frac{\beta!}{i!\,(\beta-i)!}[(\alpha\partial_t)^{\beta-i}\alpha(t)][(\alpha\partial_t)^i u^{(s+1)}].$$

Let us show that (7.6) can be written in the form

$$(7.7)\qquad \alpha(t)(\alpha\partial_t)^\beta w' = \sum_{i=1}^{\beta+1} Q_i^\beta(t)(\alpha\partial_t)^i w + \sum_{j=0}^{\beta-1} S_j^\beta(t)(\alpha\partial_t)^j w',$$

where $w = u^{(s)}(t,\xi)$ and the functions $Q_i^\beta(t)$ and $S_j^\beta(t)$, depending only on $a(t)$ and the derivatives of $a(t)$ up to order $\beta$, belong to $C^{m-1-\beta}[0,d]$ ($Q_{\beta+1}^\beta(t) \equiv 1$). The validity of this assertion for $\beta = 1$ is obvious. Let us prove (7.7) by induction for the remaining $\beta > 1$. First of all we note that the function $(\alpha\partial_t)^i\alpha(t)$ can be written in the form

$$(7.8)\qquad (\alpha\partial_t)^i\alpha(t) = \begin{cases}\alpha(t)\,\theta_i(t), & \text{if } i \text{ is even;}\\ \theta_i(t), & \text{if } i \text{ is odd;}\end{cases}$$

the functions $\theta_i(t)$ are evaluated from the recursion relations

$$\theta_{2p+1}(t) = \frac{1}{2}a'(t)\,\theta_{2p}(t) + a(t)\,\theta'_{2p}(t);$$

$$\theta_{2p+2}(t) = \theta'_{2p+1}(t); \quad \theta_0(t) \equiv 1,$$

and $\theta_i(t) \in C^{m-1-i}[0,d]$. Actually, representation (7.8) is obvious for $i = 1$ ($\theta_1(t) = a'(t)/2$) and $i = 2$ ($\theta_2(t) = a''(t)/2$). Next by induction we obtain

$$(\alpha\partial_t)^{2p+1}\alpha(t) = \alpha\partial_t(\alpha(t)\,\theta_{2p}(t)) = \frac{1}{2}a'(t)\,\theta_{2p}(t) + a(t)\,\theta'_{2p}(t) = \theta_{2p+1}(t);$$

$$(\alpha\partial_t)^{2p+2}\alpha(t) = \alpha\partial_t(\theta_{2p+1}(t)) = \alpha(t)\,\theta'_{2p+1}(t) = \alpha(t)\,\theta_{2p+2}(t).$$

Representation (7.8) is proved.

Let us turn to the proof of (7.7), where we shall assume that representation (7.7) is valid for all functions $\alpha(t)(\alpha\partial_t)^s w'$, $s < \beta$ (induction hypothesis). From (7.6) we have

$$(7.9)\qquad \alpha(t)(\alpha\partial_t)^\beta w' = (\alpha\partial_t)^{\beta+1}w - \sum_{s=0}^{\beta-1}\frac{\beta!}{(\beta-s)!\,s!}\widetilde{S}_s^\beta(t)(\alpha\partial_t)^s w',$$

where $\widetilde{S}_s^\beta(t) = \alpha(t)\theta_{\beta-s}(t)$ for $(\beta - s)$ even and $\widetilde{S}_s^\beta(t) = \theta_{\beta-s}(t)$ for $(\beta - s)$ odd. As we proved above, $\theta_{\beta-s}(t) \in C^{m-1-(\beta-s)}[0,d] \subset C^{m-1-\beta}[0,d]$, and in order to obtain (7.7) from (7.9) it remains for us to consider the addend $\widetilde{S}_s^\beta(t)(\alpha\partial_t)^s w'$ in the right side of (7.9) for even $(\beta - s)$ and $s \leqq \beta - 1$. In this case by the induction hypothesis we have

$$\widetilde{S}_s^\beta(t)(\alpha\partial_t)^s w' = \theta_{\beta-s}(t)\,\alpha(t)(\alpha\partial_t)^s w'$$

$$= \theta_{\beta-s}(t)\Big\{\sum_{i=1}^{s+1} Q_i^s(t)(\alpha\partial_t)^i w + \sum_{j=0}^{s-1} S_j^s(t)(\alpha\partial_t)^j w'\Big\}.$$

Applying the latter identity to (7.9), we obtain (7.7).

Taking into account that $|\omega(t,\xi)| \leqq c_0\alpha(t)|\xi|$, we obtain

$$\|(\alpha\partial_t)^\beta\partial_t^s u\| \leqslant c_{\beta+2s}\Big\{\|(\alpha\partial_t)^{\beta-2}\partial_t^s f\| + \sum_{\substack{p+j+2i\leqslant\beta+2s\\ j\leqslant\beta-1}} |\xi|^p\|(\alpha\partial_t)^j\partial_t^i u\|\Big\} \tag{7.10}$$

from (7.5), (7.7) and (7.4). We arrive at the inequality

$$\begin{aligned}\|(\alpha\partial_t)^\beta\partial_t^s u\| \leqslant \overset{*}{c}_{\beta+2s}\Big\{&\sum_{p+j+2i=\beta+2s-2}(1+|\xi|)^p\|(\alpha\partial_t)^j\partial_t^i f\|\\ &+\sum_{p+2i\leqslant 2s+\beta}|\xi|^p\|\partial_t^i u\| + \sum_{p+2i\leqslant\beta+2s-1}|\xi|^p\|(\alpha\partial_t)\partial_t^i u\|\Big\}\end{aligned} \tag{7.11}$$

as the result of successive application of inequalities (7.10). Using estimates (3.4) and (3.5) for even $m$, we obtain (7.1) from (7.11). Applying the same inequalities to (7.11) for odd $m$ leads to (7.2). The lemma is proved.

Proof of Lemma 5.6. As we have already noted, any solution $u(t,\xi)$ of equation (5.22) is a solution of integral equation (5.8). Differentiating identity (5.8) with respect to $t>0$, we obtain

$$u'(t,\xi) = -\int_t^d e^{P_1(t,\xi)-P_1(\tau,\xi)}q(\tau,\xi)\frac{d\tau}{a(\tau)} - \mu_1(\xi)e^{P_1(t,\xi)},$$

where

$$q(t,\xi) = f(t,\xi) - (c(\xi)-\nu)u(t,\xi);\quad b_j(t) = b(t) + ja'(t);$$

$$P_j(t,\xi) = \int_t^d \frac{p_j(s,\xi)}{a(s)}ds;\quad p_j(t,\xi) = b_j(t) + i\omega(t,\xi).$$

Let us introduce the differential operator

$$D_r q(t,\xi) = \partial_t(p_r^{-1}(t,\xi)\partial_t(p_{r-1}^{-1}(t,\xi)\partial_t(\dots\partial_t(p_1^{-1}(t,\xi)q(t,\xi))\dots)))$$

and prove the validity of the identities

$$p_{s-1}^{-1}(t,\xi)D_{s-2}\partial_t^2 u(t,\xi) = \sigma_{s-1}(\xi)e^{P_s(t,\xi)} - \int_t^d e^{P_s(t,\xi)-P_s(\tau,\xi)}q_{s-1}(\tau,\xi)\frac{d\tau}{a(\tau)}, \tag{7.13}$$

where $s \geqq 2$, $q_s(t,\xi) = D_s q(t,\xi)$ and

$$\begin{aligned}\sigma_s(\xi) = (-1)^{s+1}\Big[&\frac{\mu_1(\xi)}{a^s(d)} + \frac{q(d,\xi)}{p_1(d,\xi)a^s(d)}\\ &-\frac{q_1(d,\xi)}{p_2(d,\xi)a^{s-1}(d)} + \dots + (-1)^{s+1}\frac{q_{s-1}(d,\xi)}{p_s(d,\xi)a(d)}\Big].\end{aligned} \tag{7.14}$$

We shall deduce the proof of (7.13) for $s=2$ only; it can be proved analogously by induction for $s>2$. Differentiating (7.12) with respect to $t>0$, we obtain

$$\partial_t^2 u(t,\xi) = \frac{q(t,\xi)}{a(t)} + \mu_1(\xi)\frac{p_1(t,\xi)}{a(t)} e^{P_1(t,\xi)}$$

$$+ \frac{p_1(t,\xi)}{a(t)} \int_t^d e^{P_1(t,\xi)-P_1(\tau,\xi)} q(\tau,\xi)\frac{d\tau}{a(\tau)}.$$

With the aid of integration by parts we derive from this

$$p_1^{-1}(t,\xi)\,\partial_t^2 u(t,\xi) = \frac{q(t,\xi)}{p_1(t,\xi)\,a(t)} + \frac{\mu_1(\xi)}{a(d)} e^{P_2(t,\xi)}$$

$$+ a^{-1}(t)\int_t^d [\partial_\tau e^{P_1(t,\xi)-P_1(\tau,\xi)}]\, p_1^{-1}(\tau,\xi)\, q(\tau,\xi)\, d\tau$$

$$= (a^{-1}(d)\,\mu_1(\xi) + [p_1(d,\xi)\,a(d)]^{-1} q(d,\xi))\, e^{P_2(t,\xi)}$$

$$- \int_t^d e^{P_2(t,\xi)-P_2(\tau,\xi)} \partial_\tau [p_1^{-1}(\tau,\xi)\, q(\tau,\xi)] \frac{d\tau}{a(\tau)}$$

$$= \sigma_1(\xi)\, e^{P_2(t,\xi)} - \int_t^d e^{P_2(t,\xi)-P_2(\tau,\xi)} q_1(\tau,\xi)\frac{d\tau}{a(\tau)}.$$

Before making use of identity (7.13) we shall prove the estimate

$$|q_s(t,\xi)| \leqslant c_s \sum_{i=0}^{s} (1+|\xi|)^{s-i}\, |q^{(i)}(t,\xi)|, \tag{7.15}$$

or the more general

$$|\partial_t^r q_s(t,\xi)| \leqslant c_{s+r} \sum_{i=0}^{s+r} (1+|\xi|)^{s+r-i}\, |q^{(i)}(t,\xi)|. \tag{7.15$'$}$$

Let us consider the expression

$$\partial_t^r q_s(t,\xi) = \partial_t^{r+1} [p_s^{-1}(t,\xi)\, q_{s-1}(t,\xi)] \tag{7.15$''$}$$

and show that

$$\partial_t^r q_s(t,\xi) = \sum_{0\leqslant i\leqslant s+r-1} \Phi_{s+r-i}^{r,s}(t,\xi)\, q^{(i)}(t,\xi) + \frac{q^{(r+s)}(t,\xi)}{p_1(t,\xi)\, p_2(t,\xi)\ldots p_s(t,\xi)}, \tag{7.16}$$

where $\Phi_i^{r,s}(t,\xi)$ for all $t\in[0,d]$ satisfies the inequality

$$|\Phi_i^{r,s}(t,\xi)| \leqslant c_{r,s}(1+|\xi|)^i. \tag{7.17}$$

We shall prove (7.16) and (7.17) by induction on $s$. For $s=1$ we have

$$\partial_t^r q_1(t,\xi) = \partial_t^{r+1}(p_1^{-1}(t,\xi)\, q(t,\xi)) = \sum_{i=0}^{r+1} \frac{(r+1)!}{(r+1-i)!\, i!} [\partial_t^{r+1-i} p_1^{-1}(t,\xi)]\, q^{(i)}(t,\xi). \tag{7.18}$$

Setting

$$\Phi^{r,1}_{r+1-i}(t,\xi) = \frac{(r+1)!}{(r+1-i)!\,i!}\,\partial_t^{r+1-i} p_1^{-1}(t,\xi) \tag{7.19}$$

in (7.18) and taking into account that for $|p_j(t,\xi)| \geqq |b_j(t)| \geqq \beta_1 > 0$ the trivial estimate

$$|\partial_t^i p_j^{-1}(t,\xi)| \leqslant c_{j,i}(1+|\xi|)^i \quad (0 \leqslant t \leqslant d) \tag{7.20}$$

holds, we obtain (7.16) and (7.17) for $s = 1$. Let (7.16) and (7.17) be proved for some $s - 1 \geqq 1$. Then from (7.15″) we obtain

$$\begin{aligned}
\partial_t^r q_s(t,\xi) &= \sum_{i=0}^{r+1} \frac{(r+1)!}{(r+1-i)!\,i!}\,(p_s^{-1}(t,\xi))^{(r+1-i)} \Bigg\{ \frac{q^{(s-1+i)}(t,\xi)}{p_1(t,\xi)\,p_2(t,\xi)\dots p_{s-1}(t,\xi)} \\
&\quad + \sum_{j=0}^{s+i-2} \Phi^{i,s-1}_{j+i-1-j}(t,\xi)\,q^{(j)}(t,\xi)\Bigg\} = \frac{q^{(r+s)}(t,\xi)}{p_1(t,\xi)\,p_2(t,\xi)\dots p_s(t,\xi)} \\
&\quad + \sum_{j=0}^{s-2}\Bigg(\sum_{i=0}^{r} \frac{(r+1)!}{(r+1-i)!\,i!}\,(p_s^{-1}(t,\xi))^{(r+1-i)}\,\Phi^{i,s-1}_{s+i-1-j}(t,\xi)\Bigg) q^{(j)}(t,\xi) \\
&\quad + \sum_{j=s-1}^{s+r-2}\Bigg(\sum_{i=j+2-s}^{r} \frac{(r+1)!}{(r+1-i)!\,i!}\,(p_s^{-1}(t,\xi))^{(r+1-i)}\,\Phi^{i,s-1}_{s+i-1-j}(t,\xi)\Bigg) q^{(j)}(t,\xi) \\
&\quad + \sum_{j=s-1}^{s+r-1} \frac{(r+1)!}{(j+1-s)!\,(r+s-j)!}\,(p_s^{-1}(t,\xi))^{(r+s-j)}\,\frac{q^{(j)}(t,\xi)}{p_1(t,\xi)\,p_2(t,\xi)\dots p_{s-1}(t,\xi)} \\
&\quad + \sum_{j=0}^{s+r-1} p_s^{-1}(t,\xi)\,\Phi^{r+1,s-1}_{s+r-j}(t,\xi)\,q^{(j)}(t,\xi) \\
&= \frac{q^{(r+s)}(t,\xi)}{p_1(t,\xi)\,p_2(t,\xi)\dots p_s(t,\xi)} + \sum_{j=0}^{r+s-1} \Phi^{r,s}_{r+s-j}(t,\xi)\,q^{(j)}(t,\xi).
\end{aligned} \tag{7.20′}$$

Identity (7.16) has been proved by induction. Using the estimates

$$|p_j(t,\xi)| \geqslant |b_j(t)| \geqslant \beta_1 > 0 \left(1 \leqslant j \leqslant \left[\frac{m-1}{2}\right]\right)$$

and estimates (7.17) for $\Phi_i^{r,s-1}(t,\xi)$, it is easy to establish from (7.20′) that estimates (7.17) are valid for $\Phi_i^{r,s}(t,\xi)$. The validity of (7.15) and (7.15′) for $1 \leqq s \leqq m/2$ follows immediately from (7.16) and (7.17).

Let us apply identity (7.16) to (7.13). For $2 \leqq \lfloor m/2 \rfloor$ we obtain

$$\begin{aligned}
\frac{u^{(s)}(t,\xi)}{p_1(t,\xi)\,p_2(t,\xi)\dots p_{s-1}(t,\xi)} &= -\sum_{j=2}^{s-1} p_{s-1}^{-1}(t,\xi)\,\Phi^{0,s-2}_{s-j}(t,\xi)\,u^{(j)}(t,\xi) \\
&\quad + \sigma_{s-1}(\xi)\,e^{P_s(t,\xi)} - \int_t^d e^{P_s(t,\xi)-P_s(\tau,\xi)}\,q_{s-1}(\tau,\xi)\,\frac{d\tau}{a(\tau)}.
\end{aligned} \tag{7.21}$$

Next note that condition (5.1) and the condition $|\omega(t,\xi)| \leq c_0 \alpha(t)|\xi|$ imply the estimate

$$\frac{1}{|p_j(t,\xi)|^2} = \frac{1}{|b_j(t)|^2} - \frac{|\omega(t,\xi)|^2}{|p_j(t,\xi)|^2 |b_j(t)|^2} \geqslant \beta_1^2 - \frac{c_0^2\alpha^2(t)|\xi|^2}{\beta_1^4}.$$

Consequently

$$\frac{1}{|p_j(t,\xi)|} \geqslant \beta_1 - \frac{c_0\alpha(t)|\xi|}{\beta_1^2} \quad \left(1 \leqslant j \leqslant \frac{m-1}{2}\right). \tag{7.22}$$

Let us consider and, with the aid of (5.1) and (7.22), evaluate the expression

$$\left|\frac{u^{(s)}(t,\xi)}{p_1(t,\xi)p_2(t,\xi)\dots p_{s-1}(t,\xi)}\right| \geqslant \beta_1 \left|\frac{u^{(s)}(t,\xi)}{p_1(t,\xi)p_2(t,\xi)\dots p_{s-2}(t,\xi)}\right| - \frac{c_0\alpha(t)|\xi|}{\beta_1^2}\left|\frac{|u^{(s)}(t,\xi)|}{p_1(t,\xi)p_2(t,\xi)\dots p_{s-2}(t,\xi)}\right|$$

$$\geqslant \beta_1 \left|\frac{u^{(s)}(t,\xi)}{p_1(t,\xi)p_2(t,\xi)\dots p_{s-2}(t,\xi)}\right| - \frac{c_0\alpha(t)|\xi|}{\beta_1^s}|u^{(s)}(t,\xi)|.$$

Continuing further, we arrive at the estimate

$$\left|\frac{u^{(s)}(t,\xi)}{p_1(t,\xi)p_2(t,\xi)\dots p_{s-1}(t,\xi)}\right| \geqslant \beta_1^{s-1}|u^{(s)}(t,\xi)| - c\alpha(t)|\xi|\,|u^{(s)}(t,\xi)|, \tag{7.23}$$

where the constant

$$c = c_0(\beta_1^{-s} + \beta_1^{2-s} + \dots + \beta_1^{2(s-1)-s}) > 0.$$

For $2 \leq s \leq [m/2]$ we obtain

$$\begin{aligned}\beta_1^{s-1}|u^{(s)}(t,\xi)| \leqslant c\alpha(t)|\xi|\,|u^{(s)}(t,\xi)| + \sum_{j=2}^{s-1}\frac{c_{0,s-2}}{\beta_1}(1+|\xi|)^{s-j}|u^{(j)}(t,\xi)| \\ + |\sigma_{s-1}(\xi)|\,|e^{P_s(t,\xi)}| + \left|\int_t^d e^{P_s(t,\xi)-P_s(\tau,\xi)} q_{s-1}(\tau,\xi)\frac{d\tau}{a(\tau)}\right|\end{aligned} \tag{7.24}$$

from (7.21), (7.17) and (7.23). By virtue of condition (5.1), we can make use of assertion 5.5° to estimate the $\mathscr{L}_2(0,d)$ norm of the integral operator in the right side of (7.24). In addition condition (5.1) permits us to evaluate

$$\int_0^d |e^{P_s(t,\xi)}|^2\,dt = a(d)\int_0^d \exp\left\{\int_t^d \frac{2b_{s-1/2}(\tau)}{a(\tau)}\,d\tau\right\}\frac{dt}{a(t)} \leqslant \frac{a(d)}{2\beta_1}.$$

Therefore we obtain the estimate

$$\beta_1^{s-1}\|u^{(s)}\| \leqslant c|\xi|\,\|(\alpha\partial_t)u^{(s-1)}\| + \sum_{j=2}^{s-1}\frac{c_{0,s-2}}{\beta_1}(1+|\xi|)^{s-j}\|u^{(j)}\|$$
(7.25)
$$+a(d)(2\beta_1)^{-1}|\sigma_{s-1}(\xi)| + k_0\|q_{s-1}\| \quad \left(2\leqslant s\leqslant\left[\frac{m}{2}\right]\right)$$

from (7.24). Analogously for $s=1$ we obtain

$$\|u'\| \leqslant a(d)(2\beta_1)^{-1}\,|\mu_1(\xi)| + k_0\|q\| \tag{7.26}$$

from (7.12). Using (7.15) and conditions (5.1), it is easy to obtain the estimate

$$|\sigma_{s-1}(\xi)| \leqslant c\left\{|\mu_1(\xi)| + \sum_{j=0}^{s-2}|q_j(d,\xi)|\right\}$$

from (7.14). Making use of inequality (7.15) for $t=d$, we derive from this the inequality

$$|\sigma_{s-1}(\xi)| \leqslant c^*\left\{|\mu_1(\xi)| + \sum_{i=0}^{s-2}(1+|\xi|)^{s-2-i}|q^{(i)}(d,\xi)|\right\}. \tag{7.27}$$

In addition, from (7.15) we have

$$\|q_{s-1}\| \leqslant c_{s-1}\sum_{i=0}^{s-1}(1+|\xi|)^{s-1-i}\|q^{(i)}\|. \tag{7.28}$$

Using very well-known embedding theorems, we can estimate

$$|q^{(i)}(d,\xi)| \leqslant c\{\|q^{(i+1)}\| + \|q^{(i)}\|\}. \tag{7.29}$$

From (7.25) and (7.27)–(7.29) we obtain the estimate

$$\|u^{(s)}\| \leqslant c_s\left\{|\xi|\,\|(\alpha\partial_t)u^{(s-1)}\| + \sum_{j=2}^{s-1}(1+|\xi|)^{s-1}\|u^{(j)}\|\right.$$
(7.30)
$$\left.+|\mu_1(\xi)| + \sum_{i=0}^{s-1}(1+|\xi|)^{s-1-i}\|q^{(i)}\|\right\} \quad \left(2\leqslant s\leqslant\left[\frac{m}{2}\right]\right).$$

Let us use the inequality obtained in (7.30) when estimating $\|u^{(s)}\|$ for $s\geqq 2$. First we shall obtain an estimate of $\|u'\|$ and $|\mu_1(\xi)|$.

From identity (7.12) we obtain the estimate

$$|\mu_1(\xi)|^2\int_{^1/_2 d}^{d}|e^{P_1(t,\xi)}|^2\,dt \leqslant 2\int_{^1/_2 d}^{d}|u'(t,\xi)|^2\,dt$$
(7.31)
$$+2\int_0^d\left|\int_t^d e^{P_1(t,\xi)-P_1(\tau,\xi)}q(\tau,\xi)\frac{d\tau}{a(\tau)}\right|^2 dt.$$

Let us note that in (7.31) the quantity

$$\int_{1/2d}^{d} |e^{P_1(t,\xi)}|^2\,dt = \int_{1/2d}^{d} \exp\left\{\int_t^d \frac{2b_1(\tau)}{a(\tau)}\,d\tau\right\} dt = c_1^{-2} > 0$$

does not depend on $\xi$, and make use of assertion 5.5°. Then we shall have

$$|\mu_1(\xi)|^2 \leqslant 2c_1^2\left\{\int_{1/2d}^{d} |u'(t,\xi)|^2\,dt + k_0^2\|q\|^2\right\}. \tag{7.32}$$

For any $\epsilon > 0$ we have

$$\int_{1/2d}^{d} |u'|^2\,dt \leqslant \varepsilon \int_{1/2d}^{d} |u''|^2\,dt + c(\varepsilon)\int_{1/2d}^{d} |u|^2\,dt \tag{7.33}$$

from the well-known Ehrling-Nirenberg inequality (cf. [**22**]). The trivial identity $u'' = \alpha^{-2}(t)(\alpha\partial_t)^2u - \alpha^{-1}(t)\alpha'(t)u'$ enables us to obtain the estimate

$$|u''|^2 \leqslant c_2\{|(\alpha\partial_t)^2u|^2 + |u'|^2\}.$$

for $d/2 \leqq t \leqq d$. Then for any $\epsilon \in (0, (2c_2)^{-1})$ we obtain from (7.33)

$$\int_{1/2d}^{d} |u'(t,\xi)|^2\,dt \leqslant 2c_2\varepsilon \int_{1/2d}^{d} |(\alpha\partial_t)^2u(t,\xi)|^2\,dt + 2c(\varepsilon)\int_{1/2d}^{d} |u(t,\xi)|^2\,dt. \tag{7.34}$$

Let us apply estimates (7.32) and (7.34) to (7.26). We have

$$\|u'\|^2 \leqslant 4a^2(d)\,c_1^2\beta_1^{-2}\{2c_2\varepsilon\|(\alpha\partial_t)^2u\|^2 + 2c(\varepsilon)\|u\| + k_0^2\|q\|^2\} + 2k_0^2\|q\|^2.$$

If we apply estimate (7.1) for $m = 2$ to the last inequality and choose $\epsilon$ sufficiently small, we obtain

$$\|u'\|^2 \leqslant c_2^*\{\|f\|^2 + |\xi|^4\|u\|^2 + \|u\|^2\}, \tag{7.35}$$

and (7.32) and (7.35) imply that

$$|\mu_1(\xi)| \leqslant \tilde{c}_2\{\|f\| + |\xi|^2\|u\| + \|u\|\}. \tag{7.36}$$

Let us return to the estimate of $\|u^{(s)}\|$ for $s \geqq 2$. After elementary estimates in (7.30), we have by virtue of (7.36)

$$\|u^{(s)}\| \leqslant \tilde{c}_s\left\{\sum_{i=0}^{s-1}(1+|\xi|)^{2s-2i-2}\|f^{(i)}\| + |\xi|\,\|(\alpha\partial_t)u^{(s-1)}\|\right.$$
$$\left. + \sum_{j=1}^{s-1}(1+|\xi|)^{2s-2j}\|u^{(j)}\| + (1+|\xi|)^{2s}\|u\|\right\}.$$

Applying inequalities (3.5) to the latter estimate brings us to the following inequality:

$$\|u^{(s)}\| \leqslant c_s^*\left\{\sum_{i=0}^{s-1}(1+|\xi|)^{2s-2i-2}\|f^{(i)}\| + |\xi|\,\|(\alpha\partial_t)u^{(s-1)}\| + (1+|\xi|)^{2s}\|u\|\right\}. \tag{7.37}$$

Now we can prove inequality (5.54) for even $m$. For this purpose let us evaluate $\|u^{m/2}\|^2$ in (7.1) with the aid of (7.35) if $m=2$ and with the aid of (7.37) if $m>2$. In the latter case, to obtain (5.54), we must still make use of inequality (3.5) for $s=(m-2)/2$, $j=1$ and sufficiently small $\epsilon>0$.

For odd $m$ we must evaluate $\|u^{(m-1)/2}\|$ in inequality (7.2) with the aid of (5.54) after changing $m$ in the latter to $m-1$ ($m-1$ is even!). In this way we arrive at the inequality

$$\sum_{p+\beta+2s=m} |\xi|^{2p}\|(\alpha\partial_t)^\beta\partial_t^s u\|^2 \leqslant c_m\Big\{\sum_{p+j+2i=m-2}(1+|\xi|^2)^p\|(\alpha\partial_t)^j\partial_t^i f\|^2 + \|(\alpha\partial_t)u^{\left(\frac{m-1}{2}\right)}\|^2+(1+|\xi|^2)^m\|u\|^2\Big\}. \tag{7.38}$$

It remains for us to evaluate $\|(\alpha\partial_t)u^{(m-1)/2}\|$. Let us write (7.13) in the form

$$\alpha(t)p_s^{-1}(t,\xi)D_{s-1}\partial_t^2 u(t,\xi)=\alpha(d)\sigma_s(\xi)e^{P_{s+1/2}(t,\xi)} - \int_t^d e^{P_{s+1/2}(t,\xi)-P_{s+1/2}(\tau,\xi)}\alpha(\tau)q_s(\tau,\xi)\frac{d\tau}{a(\tau)}, \tag{7.39}$$

where $s=(m-1)/2$. Repeating estimates (7.22)–(7.29) with minor changes and using (7.36) and (3.5), we obtain from (7.39) the inequality

$$\begin{aligned}\|(\alpha\partial_t)u^{(s)}\| \leqslant c_{s+1}\Big\{&|\xi|\,\|(\alpha\partial_t)^2u^{(s-1)}\|+(1+|\xi|)\|u^{(s)}\| \\ &+|\xi|^2\|(\alpha\partial_t)u^{(s-1)}\|+(1+|\xi|)^{2s+1}\|u\| \\ &+\sum_{i=0}^{s-1}(1+|\xi|)^{2s-1-2i}\|f^{(i)}\|+\|(\alpha\partial_t)f^{(s-1)}\|\Big\}.\end{aligned} \tag{7.40}$$

Let us apply inequality (5.54) in order to estimate the addends

$$\|(\alpha\partial_t)^2u^{(s-1)}\|,\ \|u^{(s)}\|,\ |\xi|\,\|(\alpha\partial_t)u^{(s-1)}\| \quad \left(s=\frac{1}{2}(m-1)\right)$$

in the right side of (7.40). We obtain

$$\|(\alpha\partial_t)u^{\left(\frac{m-1}{2}\right)}\|^2 \leqslant c_m\Big\{\sum_{p+j+2s=m-2}(1+|\xi|^2)^p\|(\alpha\partial_t)^j\partial_t^s f\|^2+(1+|\xi|^2)^m\|u\|^2\Big\} \tag{7.41}$$

The validity of (5.54) for odd $m$ follows from inequalities (7.38) and (7.41). The lemma is proved.

**Proof of lemma** 6.3. If $f(t,\xi)\in\widetilde{H}_\alpha^{m-2}(0,d)$ and conditions (6.1) are fulfilled, then by virtue of 6.7°, $u(t,\xi)$ belongs to $\widetilde{H}_\alpha^m(0,d)$. Differentiating (6.3) with respect to $t>0$, we obtain

$$u'(t,\xi)=\int_0^t e^{P_1(t,\xi)-P_1(\tau,\xi)}q(\tau,\xi)\frac{d\tau}{a(\tau)}, \tag{7.42}$$

where $q(\tau,\xi)=f(t,\xi)-(c(\xi)-\nu)u(t,\xi)$.

From (7.42) by successive differentiation and integration by parts, as in Lemma 5.6, we obtain the identity

$$p_{s-1}^{-1}(t,\xi)\,D_{s-2}\partial_t^2u(t,\xi)=\int_0^t e^{P_s(t,\xi)-P_s(\tau,\xi)}q_{s-1}(\tau,\xi)\frac{d\tau}{a(\tau)}, \tag{7.43}$$

where $0<t<d$, $2\leqq s\leqq[m/2]$ and the notation used was introduced at the beginning of the proof of Lemma 5.6. Repeating estimates (7.23)–(7.24) and making use of assertion 6.6°, from (7.43) we can obtain

$$\|u^{(s)}\|\leqslant c_s\Big\{|\xi|\,\|\alpha u^{(s)}\|+\sum_{j=2}^{s-1}(1+|\xi|)^{s-j}\|u^{(j)}\|+\|q_{s-1}\|\Big\} \tag{7.44}$$

for $s=2,3,\cdots,[m/2]$. We are taking into account in this connection that by virtue of condition (6.1), $|p_s(t,\xi)|\geqq b_s(t)\geqq\beta_1>0$.

Inequality (7.15) enables us to extend estimate (7.44) and obtain

$$\begin{aligned}\|u^{(s)}\|\leqslant c_s^*\Big\{&|\xi|\,\|\alpha u^{(s)}\|+\sum_{i=1}^{s-1}(1+|\xi|)^{2s-2i}\|u^{(i)}\|\\&+\sum_{i=0}^{s-1}(1+|\xi|)^{2s-2-2i}\|f^{(i)}\|\Big\}\quad\Big(2\leqslant s\leqslant\Big[\frac{m}{2}\Big]\Big).\end{aligned} \tag{7.45}$$

We can obtain an estimate analogously for $\|u'\|$ from identity (7.42):

$$\|u'\|\leqslant c_1\{\|f\|+|\xi|^2\|u\|\}. \tag{7.46}$$

Application of estimates (7.45), (7.46) and (3.4) to inequality (7.1) enables us to obtain the required inequality (5.54) for even $m$.

If we start from estimate (7.2) for the proof of inequality (5.54) for odd $m$, then it is necessary first of all to estimate $\|(\alpha\partial_t)u^{((m-1)/2)}\|$. To do this we must prove the identity

$$\alpha(t)\,p_{s-1}^{-1}(t,\xi)\,D_{s-2}\partial_t^2u(t,\xi)=\int_0^t e^{P_{s+1/2}(t,\xi)-P_{s+1/2}(\tau,\xi)}\alpha(\tau)\,q_{s-1}(\tau,\xi)\frac{d\tau}{a(\tau)} \tag{7.47}$$

for $s=(m-1)/2$ and $\alpha(t)=\sqrt{a(t)}$. Although the proof of (7.47) is basically analogous to the proof of (7.43), it should be noted that when integrating by parts it is impossible to guarantee the boundedness of $|f^{(s-1)}(t,\xi)|$ as $t\to+0$ (by the condition $f\in\widetilde{H}_a^{m-2}(0,d)$). Therefore it is necessary to rely on the boundedness of the quantity $|a(t)f^{(s-1)}(t,\xi)|$ as $t\to+0$, which follows from the elementary estimate

$$|a(t)f^{(j)}(t,\xi)|\leqslant|a(d)f^{(j)}(d,\xi)|+c_2(\|af^{(j+1)}\|+\|f^{(j)}\|).$$

As in the proof of Lemma 5.6, from (7.47) we easily obtain the estimate

$$\begin{aligned}(7.48)\quad &\|(\alpha\partial_t)u^{\left(\frac{m-1}{2}\right)}\| \leqslant c_m\{|\xi|\,\|(\alpha\partial_t)^2u^{\left(\frac{m-3}{2}\right)}\|+(1+|\xi|)\|u^{\left(\frac{m-1}{2}\right)}\|+|\xi|^2\|(\alpha\partial_t)u^{\left(\frac{m-3}{2}\right)}\|\\ &+\sum_{i=1}^{\frac{1}{2}(m-3)}(1+|\xi|)^{m-2i}\|u^{(i)}\|+\sum_{j=0}^{\frac{1}{2}(m-3)}(1+|\xi|)^{m-3-2j}\|(\alpha\partial_t)f^{(j)}\|\}.\end{aligned}$$

Keeping in mind that $b>0$ and the even number $m-1\geqq 2$, we can apply inequality (5.54) for the subsequent estimates. From this inequality we find

$$\begin{aligned}(7.49)\quad &\|(\alpha\partial_t)^2u^{\left(\frac{m-3}{2}\right)}\|+\|u^{\left(\frac{m-1}{2}\right)}\|+|\xi|\,\|(\alpha\partial_t)u^{\left(\frac{m-3}{2}\right)}\|\\ &\leqslant c_{m-1}\Big\{\sum_{p+j+2s=m-3}(1+|\xi|)^p\|(\alpha\partial_t)^j\partial_t^s f\|+(1+|\xi|)^{m-1}\|u\|\Big\}.\end{aligned}$$

After applying estimate (7.49), together with (3.5), to (7.48), we obtain the required estimate for $\|(\alpha\partial_t)u^{(s)}\|$, where $s=(m-1)/2$:

$$(7.50)\quad \|(\alpha\partial_t)u^{(s)}\|\leqslant c_m\Big\{\sum_{p+j+2i=m-2}(1+|\xi|)^p\|(\alpha\partial_t)^j\partial_t^i f\|+(1+|\xi|)^m\|u\|\Big\}$$

From inequalities (7.2), (7.49) and (7.50) we derive the required estimate (5.54) for odd $m$ also. The lemma is proved.

## §8. Necessity of the $\alpha$-ellipticity condition and the complementarity condition

NECESSITY OF THE $\alpha$-ELLIPTICITY CONDITION IN THEOREM 2.1. If inequality (2.16) is fulfilled for any function $v\in C^m(\mathscr{D})$, they by virtue of very well-known results the operator $\mathscr{L}$ is elliptic at the interior points of $\mathscr{D}$. Let us assume that the condition of $\alpha$-ellipticity (cf. Definition 2.1) is violated at a certain point $x^0\in S$. For definiteness let us assume that $x^0\in S_q$, $q\in Q_2$, and $m$ is an even number.

Let inequality (2.16) be fulfilled for any function $v\in H^m(\mathscr{D})$. Let us apply it to the function $\phi v$, where $\phi\in C_0^\infty(E_n)$ and is nonzero only in some sufficiently small neighborhood of the point $x^0$. Using the equivalence of norms of the form (2.15) which correspond to various covers of $\mathscr{D}$, we obtain the inequality

$$(8.1)\quad \|\varphi v\|_{m,\tilde{\alpha}}\leqslant c_m'\{\|L_0(\varphi v)\|_{m-2,\tilde{\alpha}}+\langle B_0(\varphi v)\rangle_{m-m^*-1}+|\varphi v|_0\}$$

from (2.16), where

$$\tilde{\alpha}(t,y)=\sqrt{a_{nn}^0(t,y)},\quad \varphi(t,y)\in V^m(K_\delta)$$

and $\delta > 0$ is sufficiently small.

If we make use of the equivalence of norms $\|\cdot\|_{m,\tilde{\alpha}}$ and $|\cdot|_{m,\alpha}$ (cf. the Remark on p. 125), then from inequalities (8.1) and (3.13), for sufficiently small $\delta > 0$, we obtain

(8.2) $$|\varphi v|^2_{m,\alpha} \leqslant c_m \{ |L_1(\varphi v)|^2_{m-2,\alpha} + \langle B_0(\varphi v)\rangle^2_{m-m^*-1} + |\varphi v|^2_0\},$$

where $\alpha(t) = \sqrt{a^0_{nn}(t,0)}$ and $L_1$ is the principal part (3.8) of $L_0$ in a neighborhood of $x^0$.

Let us represent $L_1$ in the form

$$L_1 = L_1' + \sum_{k=1}^{n-1} (a^0_{nk}(t, 0) - \gamma_k\alpha(t))\,\partial_t\partial_{y_k} + \left(a^0_n(0, 0) + \frac{1}{2}\partial_t a^0_{nn}(t, 0)\right)\partial_t,$$

where

$$L_1' = (\alpha\partial_t)^2 + \sum_{k=1}^{n-1} \gamma_k\alpha\partial_t\partial_{y_k} + \sum_{k,l=1}^{n-1} a^0_{kl}(0, 0)\,\partial_{y_k}\partial_{y_l}$$

and $\gamma_k$ $(1 \leqq k \leqq n-1)$ is defined in (4.1). Using condition (4.1), we can assert that for any $\epsilon > 0$ there is a $\delta_1 = \delta_1(\epsilon) > 0$ such that for $0 < \delta \leqq \delta_1$ the inequality

(8.3) $$\begin{aligned}|L_1(\varphi v)|^2_{m-2,\alpha} \leqslant 2 \sum_{\beta+|\tau|=m-2} \int_{K_\delta} |(\alpha\partial_t)^\beta\partial^\tau_y L_1'(\varphi v)|^2\,dtdy \\ + \varepsilon|\varphi v|^2_{m,\alpha} + J_{m-1}(\varphi v)\end{aligned}$$

is valid, where the functional $J_{m-1}(\phi v)$ depends on $\phi v(t, y)$ and the derivatives of $\phi v(t, y)$ with respect to $t$ and $y$ up to order $m-1$. From inequalities (8.2) and (8.3), for sufficiently small $\delta > 0$ we have

(8.4) $$\begin{aligned}|\varphi v|^2_{m,\alpha} \leqslant 4c_m \Big\{ \sum_{\beta+|\tau|=m-2} \int_{K_\delta} |(\alpha\partial_t)^\beta\partial^\tau_y L_1'(\varphi v)|^2\,dtdy \\ + \langle B_0(\varphi v)\rangle^2_{m-m^*-1} + J_{m-1}(\varphi v) + |\varphi v|^2_0\Big\}.\end{aligned}$$

Now let us choose $v(t, y)$, having assumed

$$v(t, y) = \exp\left\{i(\xi, y) - i\eta\int_t^\delta \frac{ds}{\alpha(s)} - \theta\int_t^\delta \frac{ds}{\alpha^2(r)}\right\},$$

where $\xi = (\xi_1, \cdots, \xi_{n-1})$, the $\eta$ are arbitrary real parameters, $(\xi, y) = \xi_1 y_1 + \cdots + \xi_{n-1}y_{n-1}$, and $\theta$ is a positive number. It is easy to show that for any $m \geqq 2$ we can choose $\theta = \theta(m) > 0$ so large that $v(t, y)$ will belong to $C^m(K_\delta)$. In addition, we shall have

(8.5) $$\partial^j_t\partial^\tau_y v(t, y)|_{t=+0} = 0 \quad (2j + |\tau| \leqslant m-1)$$

on the hyperplane $t = 0$.

Introducing the notation

$$w(t, y) = \exp\left\{-\theta\int_t^{\delta}\frac{ds}{\alpha^2(s)}\right\}\varphi(t, y);$$

$$\Phi(\xi, \eta, t, y) = (\xi, y) - \eta\int_t^{\delta}\frac{ds}{\alpha(s)},$$

let us represent $\phi v(t, y)$ in the form

$$\varphi v(t, y) = e^{i\Phi(\xi,\eta,t,y)}w(t, y),$$

it being assumed that $|\phi v|_0 = |w|_0 \neq 0$. The identity

$$(\alpha\partial_t)^\beta\partial_y^\tau e^{i\Phi(\xi,\eta,t,y)} = (i\eta)^\beta(i\xi)^\tau e^{i\Phi(\xi,\eta,t,y)} \tag{8.6}$$

is easily verified. Taking (8.5) into account, we obtain the inequality

$$\begin{aligned}&|e^{i\Phi(\xi,\eta,t,y)}w(t, y)|^2_{m,\alpha}\\ &\leqslant 4c_m\Big\{\sum_{\beta+|\tau|=m-2}\int_{K_\delta}|(\alpha\partial_t)^\beta\partial_y^\tau L_1'(e^{i\Phi(\xi,\eta,t,y)}w)|^2\,dtdy\\ &\quad + J_{m-1}(e^{i\Phi(\xi,\eta,t,y)}w(t, y)) + |w|_0^2\Big\}\end{aligned} \tag{8.7}$$

from (8.4). Since $L_1''$ is a differential operator with constant coefficients with respect to the derivatives $\alpha\partial_t$ and $\partial_{y_k}$, by means of simple estimates we derive the inequality

$$\begin{aligned}\sum_{\beta+|\tau|=m}\xi^{2\tau}\eta^{2\beta}|w|_0^2 &\leqslant 4c_m\Big\{\sum_{\beta+|\tau|=m-2}\xi^{2\tau}\eta^{2\beta}\Big|\eta^2+\sum_{k=1}^{n-1}\gamma_k\eta\xi_k\\ &+\sum_{k,l=1}^{n-1}a^0_{kl}\xi_k\xi_l\Big|^2|w|_0^2 + J'_{m-1}(e^{i\Phi(\xi,\eta,t,y)}, w) + J_{m-1}(e^{i\Phi(\xi,\eta,t,y)}w)\Big\}\end{aligned} \tag{8.8}$$

from (8.7) and (8.6), where $J'_{m-1}(e^{i\Phi(\xi,\eta,t,y)}, w(t, y))$ contains the derivatives of $e^{i\Phi(\xi,\eta,t,y)}$ with respect to $t$ and $y$ up to order $m-1$ and the derivatives of $w(t, y)$ with respect to $t$ and $y$ up to order $m$. After changing $\xi$ to $\lambda\xi$ and $\eta$ to $\lambda\eta$ $(\lambda > 0)$ in (8.8), we obtain

$$\begin{aligned}(\eta^2+|\xi|^2)^m|w|_0^2 &\leqslant 4c_m\Big\{(\eta^2+|\xi|^2)^{m-2}\Big|\eta^2+\sum_{k=1}^{n-1}\gamma_k\eta\xi_k\\ &+\sum_{k,l=1}^{n-1}a^0_{kl}\xi_k\xi_l\Big|^2|w|_0^2+\lambda^{-2m}[J'_{m-1}(e^{i\Phi(\lambda\xi,\lambda\eta,t,y)}, w)+J_{m-1}(e^{i\Phi(\lambda\xi,\lambda\eta,t,y)}w)]\Big\}.\end{aligned} \tag{8.9}$$

It is easy to verify that

$$\lambda^{-2m}[J'_{m-1}(e^{i\Phi(\lambda\xi,\lambda\eta,t,y)}, w) + J_{m-1}(e^{i\Phi(\lambda\xi,\lambda\eta,t,y)}w)] \to 0$$

as $\lambda\to\infty$. Therefore (8.9) necessarily implies the estimate

(8.10) $$\left|\eta^2+\sum_{k=1}^{n-1}\gamma_k\eta\xi_k+\sum_{k,l=1}^{n-1}a^0_{kl}(0,0)\,\xi_k\xi_l\right|\geqslant\nu_0(\eta^2+|\xi|^2),$$

where the constant $\nu_0>0$. Inequality (8.10) contradicts our assumption that the $\alpha$-ellipticity condition is violated at the point $x^0\in S$. The necessity of the $\alpha$-ellipticity condition in Theorem 2.1 is proved.

NECESSITY OF THE COMPLEMENTARITY CONDITION ON $S_q$ $(q\in Q_2)$ IN THEOREM 2.1. Let the condition of complementarity be violated at some point $x^0\in S_q$ $(q\in Q_2)$; that is, a vector $\xi^0=(\xi^0_1,\cdots,\xi^0_{n-1})\neq 0$ exists for which $\vartheta_{x^0}(\xi^0)=0$ (cf. Definition 2.12). If inequality (2.16) is valid in spite of this, then, as was shown at the beginning of this section, estimate (8.2), where $\phi\in V^m(K_\delta)$ and $\delta>0$ is sufficiently small, is valid. Applying estimates of the type (3.10), we obtain from (8.2) the inequality

(8.11) $$|\varphi v|^2_{m,\alpha}\leqslant c'_m\{|L_1(\varphi v)|^2_{m-2,\alpha}+\langle B_{01}(\varphi v)\rangle^2_{m-m^*-1}+|\varphi\dot v|^2_0\},$$

where boundary operator $B_{01}$ is the principal equivalent (cf. Definition 5.1) of $B_0$ with respect to $L_1$ in a neighborhood of the point $x^0$.

We are considering the point $x^0\in S_q$, $q\in Q_2$; therefore $b=a^0_n(0,0)<0$. Moreover, $\partial_t a^0_{nn}(t,0)\geqq 0$ for $t=+0$. Hence for sufficiently small $\delta>0$ a constant $\kappa>0$ exists such that for all $t\in[0,\delta]$ the inequalities

(8.12) $$\varkappa\leqslant\partial_t a^0_{nn}(t,0)-b\leqslant\varkappa^{-1}$$

are fulfilled.

Finally, we take $\delta>0$ so small that estimate (5.3) is fulfilled for $t\in[0,\delta]$. Having chosen $\delta$ in this way, we fix it.

Now, after first changing $L_0$ to $L_1$, $B$ to $B_{01}$, and the norm $\|\cdot\|_{m-2,\alpha}$ to the norm $|\cdot|_{m-2,\alpha}$ in inequalities (3.15)–(3.16), let us apply them to (8.11). We obtain the inequality

(8.13) $$|\varphi v|^2_{m,\alpha}\leqslant c_m(\varphi)\{|\varphi L_1 v|^2_{m-2,\alpha}+\langle\varphi B_{01}v\rangle^2_{m-m^*-1}+\|v\|^2_{m-1,\alpha}\},$$

where the norm $\|v\|_{m-1,\alpha}$ is computed on the hemisphere $K_\delta$.

Let us apply inequality (8.13) to the function $v_\lambda(t,y)=\exp\{i\lambda(\xi^0,y)\}u_\lambda(t)$, where $\lambda$ is a real parameter and $u_\lambda(t)$ is the solution of the ordinary differential equation

(8.14) $$L_\lambda u_\lambda\equiv(a^0_{nn}(t,0)\,u'_\lambda)'+(b+i\omega(t,\lambda\xi^0))\,u_\lambda+c(\lambda\xi^0)\,u_\lambda=0,$$

which satisfies the condition $u_\lambda(\delta)=0$.

We shall consider those solutions $u_\lambda(t)$ of equation (8.14) for which $\|u_\lambda\|\neq 0$. (It is sufficient for this purpose to require, along with the condition $u_\lambda(\delta)=0$, fulfillment of the condition $u'_\lambda(\delta)=1$.) The condition of Theorem 2.1 implies $b_{(m-1)/2}<0$; consequently, by assertion 5.7°, $u_\lambda(t)\in\widetilde{H}^m_\alpha(0,\delta)$ for any real $\lambda$. We easily compute

$$
\begin{aligned}
& L_1 v_\lambda = (L_\lambda u_\lambda) e^{i\lambda(\xi^0, y)} = 0 \quad (0 \leqslant t \leqslant \delta); \\
(8.15) \qquad & B_{01}(-i\partial_y, \partial_t) v_\lambda |_{t=0} = e^{i\lambda(\xi^0, y)} B_{01}(\lambda\xi^0, \partial_t) u_\lambda |_{t=0} \\
& = e^{i\lambda(\xi^0, y)} \vartheta_{x^0}(\lambda\xi^0) u_\lambda(0) = e^{i\lambda(\xi^0, y)} \lambda^{m^*} \vartheta_{x^0}(\xi^0) u_\lambda(0) = 0.
\end{aligned}
$$

When deriving the last equality in (8.15) we made use of property (5.13), the homogeneous function $\vartheta_{x^0}(\xi)$, and the assumption $\vartheta_{x^0}(\xi^0) = 0$.

By virtue of (8.15), inequality (8.13) can be written in the form

$$
(8.16) \qquad |\varphi v_\lambda|^2_{m,\alpha} \leqslant c_m(\varphi) \| v_\lambda \|^2_{m-1,\alpha}.
$$

Without loss of generality let us assume in what follows that $|\xi^0| = 1$. Using inequality (5.54), we obtain the estimate

$$
\begin{aligned}
& \| v_\lambda \|^2_{m-1,\alpha} \leqslant c(\delta) \sum_{\beta + 2s + |\tau| \leqslant m-1} |\lambda|^{2|\tau|} \| (\alpha\partial_t)^\beta \partial_t^s u_\lambda \|^2 \\
(8.17) \qquad & \leqslant \tilde{c}_{m-1}(\delta)(1 + |\lambda|^2)^{m-1} \| u_\lambda \|^2.
\end{aligned}
$$

If we choose $\phi(t, y) \in V^m(K_\delta)$ so that $\phi(t, y) \equiv 1$ in $K_{3\delta/4}$, we find that

$$
(8.18) \qquad |\lambda|^{2m} \int_0^{1/2\delta} |u_\lambda|^2 dt \leqslant c^*_m(\delta)(1 + |\lambda|^2)^{m-1} \int_0^\delta |u_\lambda|^2 dt
$$

for any real $\lambda$ from (8.16) and (8.17). Now if we prove that

$$
(8.19) \qquad \int_0^\delta |u_\lambda|^2 dt \leqslant \left(1 + \frac{1}{\varkappa^2}\right) \int_0^{1/2\delta} |u_\lambda|^2 dt,
$$

then from (8.18) and (8.19) we shall derive the inequality

$$
|\lambda|^{2m} \leqslant c^*_m(\delta)\left(1 + \frac{1}{\varkappa^2}\right)(1 + |\lambda|^2)^{m-1},
$$

which must be fulfilled for any $\lambda$. Since this is impossible, the contradiction obtained completes the proof of the necessity of the complementarity condition in Theorem 2.1.

Let us prove inequality (8.19). After multiplying (8.14) by $\bar{u}_\lambda(t)$ and integrating with respect to $t \in (s, s + \frac{1}{2}\delta)$, $0 < s < \frac{1}{2}\delta$, we obtain

$$
\begin{aligned}
& \int_s^{s+\frac{1}{2}\delta} \{\partial_t(\alpha^2(t)\partial_t u_\lambda)\bar{u}_\lambda + i\omega(t, \lambda\xi^0)(\partial_t u_\lambda)\bar{u}_\lambda + c(\lambda\xi^0)|u_\lambda|^2\} dt \\
& + b \int^{s+\frac{1}{2}\delta} (\partial_t u_\lambda)\bar{u}_\lambda dt = 0.
\end{aligned}
$$

With the aid of integration by parts we derive

$$
\begin{aligned}
&\frac{1}{2} b\,|u_\lambda|^2\Big|_{t=s}^{t=s+\frac{1}{2}\delta} + \frac{1}{2}\alpha^2(t)\,\partial_t |u_\lambda(t)|^2\Big|_{t=s}^{t=s+\frac{1}{2}\delta}\\
&= \operatorname{Re}\int_s^{s+\frac{1}{2}\delta} \{|-i\alpha(t)\,\partial_t u_\lambda|^2 + \omega_0(\lambda\xi^0)(-i\alpha(t)\,\partial_t u_\lambda)(\bar{u}_\lambda)\\
&\quad - c(\lambda\xi^0)\,|u_\lambda|^2 - i(\omega(t,\lambda\xi^0) - \omega_0(\lambda\xi^0)\,\alpha(t))(\partial_t u_\lambda)\,\bar{u}_\lambda\}\,dt.
\end{aligned}
\tag{8.20}
$$

After repeating in the right side of (8.20) the estimates derived in Lemma 5.2, we arrive at the inequality

$$
\begin{aligned}
&b\,|u_\lambda(t)|^2\Big|_{t=s}^{t=s+\frac{1}{2}\delta} + \alpha^2(t)\,\partial_t |u_\lambda(t)|^2\Big|_{t=s}^{t=s+\frac{1}{2}\delta}\\
&\geqslant \frac{3}{2}\nu_0 \int_s^{s+\frac{1}{2}\delta} \{\alpha^2(t)\,|\partial_t u_\lambda|^2 + |\lambda|^2\,|u_\lambda|^2\}\,dt \geqslant 0.
\end{aligned}
$$

From the latter inequality we derive the estimate

$$
(b - \partial_t a^0_{nn}(t,0))\,|u_\lambda(t)|^2\Big|_{t=s}^{t=s+\frac{1}{2}\delta} + \partial_t\{\alpha^2(t)\,|u_\lambda(t)|^2\}\Big|_{t=s}^{t=s+\frac{1}{2}\delta} \geqslant 0. \tag{8.21}
$$

If we integrate (8.21) with respect to $s \in (0, \frac{1}{2}\delta)$ and take into account that $u_\lambda(\delta) = 0$, then after simple estimates we obtain

$$
\begin{aligned}
&\int_0^{\frac{1}{2}\delta} (\partial_s a^0_{nn}(s,0) - b)\,|u_\lambda(s)|^2\,ds\\
&\geqslant \int_{\frac{1}{2}\delta}^{\delta} (\partial_s a^0_{nn}(s,0) - b)\,|u_\lambda(s)|^2\,ds.
\end{aligned}
\tag{8.22}
$$

If we make use of inequalities (8.12), then (8.22) implies the estimate

$$
\int_0^{\frac{1}{2}\delta} |u_\lambda(s)|^2\,ds \geqslant \varkappa^2 \int_{\frac{1}{2}\delta}^{\delta} |u_\lambda(s)|^2\,ds.
$$

From this we have

$$
\int_0^{\delta} |u_\lambda(s)|^2\,ds \leqslant (1 + \varkappa^{-2}) \int_0^{\frac{1}{2}\delta} |u_\lambda(s)|^2\,ds,
$$

which proves the validity of (8.19).

The proof of Theorem 2.1 is thus completed.

## §9. Proof of Theorem 2.2 (Noetherian-ness)

Since the finite dimensionality of the kernel and the closedness of the range of the operator $\mathfrak{A}$ corresponding to boundary value problem (2.17)-(2.18) has already been proved (cf. Corollary 2.1), for the proof that $\mathfrak{A}$ is Noetherian it is sufficient to construct a right regularizer $\mathscr{R}$ for $\mathfrak{A}$, i.e. an operator $\mathscr{R}$ mapping from $\mathscr{A}_{\alpha}^{m-(m')}$ into $H_{\alpha}^{m}(\mathscr{D})$ such that

$$\mathfrak{A}\mathscr{R} = I + T, \tag{9.1}$$

where $T$ is a bounded operator from $\mathscr{A}_{\alpha}^{m-(m')}$ into $\mathscr{A}_{\alpha}^{m+1-(m')}$ (a smoothing operator).

We shall effect the construction of the regularizer $\mathscr{R}$ as usual, by the method of "pasting together" local regularizers. The existence of a regularizer in any strictly interior subregion of $\mathscr{D}$ is very well known (cf. [**23**]); so it remains for us to construct local regularizers in a neighborhood of points on the boundary $S$. In this connection we shall assume that, with the aid of the transformation $T_{x^0}(x) = (t, y)$, conversion of the point $x^0 \in S$ to the local coordinates $(t, y)$ in the neighborhood $\mathscr{A}_{x^0}(\delta)$ has been accomplished. Let us note that by the condition of Theorem 2.2 only two cases are possible: $x^0 \in S_q$, $q \in Q_1$, and $x^0 \in S_q$, $q \in Q_2$.

LEMMA 9.1. *Let the integer $m \geqq 2$, let the operator $L_1$ be $\alpha$-elliptic in $E_{n,d}$, and let one of the following two conditions be fulfilled: either $b > 0$ or $b_{(m-1)/2} = b + ((m-1)/2)\partial_t a_{nn}^0(t,0)|_{t=0} > 0$. Then for some $d > 0$ the operators $R'_+$ (in the case of $b > 0$) and $R'_-$ (in the case of $b_{(m-1)/2} < 0$) exist, defined on the functions $f(t, y) \in H_{\alpha}^{m-2}(E_{n,d})$ and mapping into $H_{\alpha}^{m}(E_{n,d})$, such that*

$$\| R'_{\pm} f \|_{m,\alpha} \leqslant c_m \| f \|_{m-2,\,\alpha}; \tag{9.2}$$

$$L_1 R'_{\pm} f = f + T'_{\pm} f, \tag{9.3}$$

*where $T'_{\pm}$ is a bounded operator from $H_{\alpha}^{m-2}(E_{n,d})$ into $H_{\alpha}^{m}(E_{n,d})$ (a smoothing operator), $\alpha(t) = \sqrt{a_{nn}^0(t,0)}$, and the constant $c_m > 0$ does not depend on $f$.*

PROOF. Let us introduce the operators

$$R'_{\pm} f \equiv v^{\pm}(t, y) \equiv F_{n-1}^{-1}(u^{\pm}(t, \xi)),$$

where $u^+(t,\xi)$ is defined by formula (6.9) for $b > 0$ and $\nu = \nu_0 > 0$, and $u^-(t,\xi)$ by formula (5.43) for $b_{(m-1)/2} < 0$ and $\nu = \nu_0 > 0$.

By construction $u^{\pm}(t,\xi)$ satisfies inequality (5.42) for sufficiently small $d > 0$, which implies the validity of the estimate

$$(1 + |\xi|^2)^m \| u^{\pm} \|^2 \leqslant 2\nu_0^{-2} (1 + |\xi|^2)^{m-2} \| \tilde{f} \|^2, \tag{9.4}$$

where $m \geqq 2$ and $\tilde{f}(t,\xi) = F_{m-1} f$.

By virtue of Lemma 5.6 ($b_{(m-1)/2}<0$) and Lemma 6.3 ($b>0$), the estimate

$$\sum_{p+j+2s=m} |\xi|^{2p} \|(\alpha\partial_t)^j \partial_t^s u^\pm\|^2$$
(9.5)
$$\leqslant c_m' \Big\{ \sum_{p+j+2s=m-2} (1+|\xi|^2)^p \|(\alpha\partial_t)^j \partial_t^s \widetilde{f}\|^2 + (1+|\xi|^2)^m \|u^\pm\|^2 \Big\}$$

is valid for almost all $\xi$. Successive application of inequalities (9.5) and (9.4) brings us to the estimate

$$\sum_{p+j+2s=m} (1+|\xi|^2)^p \|(\alpha\partial_t)^j \partial_t^s u^\pm\|^2$$
$$\leqslant c_m \sum_{p+j+2s=m-2} (1+|\xi|^2)^p \|(\alpha\partial_t)^j \partial_t^s \widetilde{f}\|^2.$$

Integrating the latter inequality with respect to $\xi\in E_{n-1}$ and using the equivalence of the norms $\|\cdot\|_{m,\alpha}$ and $|\cdot|_{m,\alpha}$, we arrive, with the help of Parseval's equation, at estimate (9.2). If we now write (5.7) for $u^\pm(t,\xi)$ and $\nu=\nu_0$ and apply the inverse Fourier transform $F_{n-1}^{-1}$ to the resulting identity, then from (5.7) we obtain (9.3), where by inequality (9.2), which has been proved, the operator $T'_\pm f=\nu_0 R'_\pm f$ maps from $H_\alpha^{m-2}(E_{n,d})$ into $H_\alpha^m(E_{n,d})$ and is bounded. The lemma is proved.

REMARK. By construction the operator $R'_- f$ satisfies the condition $\gamma R'_- f=0$.

COROLLARY 9.1. *As is obvious from the proof of Lemma* 9.1, *the operator* $R'_+$ ($R'_-$) *maps from* $H_\alpha^{m-2}(E_{n,d})$ *into* $H_\alpha^m(E_{n,d})$ *and is bounded since*

(9.6)
$$\|R'_\pm f\|_{m',\alpha}\leqslant c_{m'}\|f\|_{m'-2,\alpha}$$

*for any integers* $m'$, $2\leqq m'\leqq m$.

COROLLARY 9.2. *Let the conditions of Lemma* 9.1 *be fulfilled. Then for any* $\delta\in(0,d)$ *and* $\epsilon\in(0,1]$ *the inequality*

(9.7)
$$\|L_\delta R'_\pm f\|_{m-2,\alpha,\varepsilon}\leqslant c_m(\mathcal{N}(\delta)+\varepsilon\mathcal{M}(\delta))\|f\|_{m-2,\alpha,\varepsilon}$$

*is valid for any function* $f(t,y)\in H_\alpha^{m-2}(E_{n,d})$. *The constants* $\mathcal{N}(\delta)$ *and* $\mathcal{M}(\delta)$ *are introduced in Lemma* 3.6; *the constant* $c_m>0$ *does not depend on* $f$, $\epsilon$ *or* $\delta$.

To prove (9.7) we must apply inequality (3.21) to the function $R'_\pm f$ and make use of estimates (9.6).

LEMMA 9.2. *Let the integer* $m\geqq 2$, *let the operator* $L_{0\delta}$ (*cf.* p. 126) *be* $\alpha$-*elliptic at* $(t,y)=(0,0)$, *and let Conditions* 2.3 *and* 2.5 *be fulfilled, as well as one of the following conditions: either* $b>0$ *or* $b_{(m-1)/2}<0$. *Then for some* $\delta$ $(0<\delta\leq d)$ *there exist operators* $R_+$ (*in the case of* $b>0$) *and* $R_-$ (*in the case of* $b_{(m-1)/2}<0$), *defined on the functions* $f(t,y)\in H_\alpha^{m-2}(E_{n,d})$ *and mapping into* $H_\alpha^m(E_{n,d})$, *such that*

(9.8)
$$\|R_\pm f\|_{m,\alpha}\leqslant c_m\|f\|_{m-2,\alpha};$$

(9.9)
$$L_{0\delta}R_\pm f=f-T_\pm f \text{ in } E_{n,d},$$

*where the* $T_{\pm}$ *are bounded operators from* $H_{\alpha}^{m-2}(E_{n,d})$ *into* $H_{\alpha}^{m-1}(E_{n,d})$ *(smoothing operators) and*

$$\alpha(t)=\sqrt{a_{nn}^{0}(t,0)}.$$

PROOF. Let us write $L_{0\delta}R'_{\pm}f$ in the form

$$L_{0\delta}R'_{\pm}f=L'_{1}R'_{\pm}f+L_{\delta}R'_{\pm}f+\psi_{\delta}L_{3}R'_{\pm}f$$
$$=(I+L_{\delta}R'_{\pm})f+(T'_{\pm}f+\psi_{\delta}L_{3}R'_{\pm}f) \tag{9.10}$$

and show that for sufficiently small $\delta>0$ the operator $(I+L_{\delta}R'_{\pm})$ has an inverse into $H_{\alpha}^{m-2}(E_{n,d})$. Let us introduce the norm (3.12) into $H_{\alpha}^{m-2}(E_{n,d})$ for this purpose and choose $\delta$ and $\epsilon$ so small that the series

$$\sum_{\varkappa=0}^{\infty}(-L_{\delta}R'_{\pm})^{\varkappa}=(I+L_{\delta}R'_{\pm})^{-1} \tag{9.11}$$

will converge. Actually, if we choose $\delta=\delta^{*}$ such that $c_{m}\mathcal{N}(\delta^{*})\leqq\frac{1}{4}$ and then take $\epsilon=\epsilon^{*}$ for some $c_{m}\mathcal{M}(\delta^{*})\epsilon^{*}\leqq\frac{1}{4}$, we obtain

$$\|L_{\delta^{*}}R'_{\pm}f\|_{m-2,\alpha,\varepsilon^{*}}\leqslant\frac{1}{2}\|f\|_{m-2,\alpha,\varepsilon^{*}} \tag{9.12}$$

from inequality (9.7) for any function $f(t,y)\in H_{\alpha}^{m-2}(E_{n,d})$. As is well known, (9.12) implies that series (9.11) converges (in the norm of the operators) and that the bounded operator $(I+L_{\delta^{*}}R'_{\pm})^{-1}$ exists, with

$$\|(I+L_{\delta^{*}}R'_{\pm})^{-1}f\|_{m-2,\alpha,\varepsilon^{*}}\leqslant2\|f\|_{m-2,\alpha,\varepsilon^{*}}. \tag{9.13}$$

Since norms (3.12) are equivalent for $\epsilon=\epsilon^{*}$ and $\epsilon=1$, (9.13) implies the boundedness of operator $(I+L_{\delta^{*}}R'_{\pm})$ in the norm $\|\cdot\|_{m-2,\alpha}$ also, with

$$\|(I+L_{\delta^{*}}R'_{\pm})^{-1}f\|_{m-2,\alpha}\leqslant c(\delta^{*})\|f\|_{m-2,\alpha}. \tag{9.14}$$

Now let us change $f(t,y)$ to $(I+L_{\delta^{*}}R'_{\pm})^{-1}f$ in (9.10). Then we obtain

$$L_{0\delta}R_{\pm}f=f+T_{\pm}f,$$

where

$$R_{\pm}=R'_{\pm}(I+L_{\delta^{*}}R'_{\pm})^{-1};\quad T_{\pm}=(T'_{\pm}+\psi_{\delta^{*}}R'_{\pm})(I+L_{\delta^{*}}R'_{\pm})^{-1}.$$

Inequalities (9.2) and (9.14) imply the validity of the estimates

$$\|R_{\pm}f\|_{m,\alpha}\leqslant c_{m}\|(I+L_{\delta^{*}}R'_{\pm})^{-1}f\|_{m-2,\alpha}\leqslant c_{m}c(\delta^{*})\|f\|_{m-2,\alpha};$$
$$\|T_{\pm}f\|_{m-1,\alpha}\leqslant\nu_{0}c_{m}c(\delta^{*})\|f\|_{m-2,\alpha}+c'(\delta^{*})\|R_{\pm}f\|_{m,\alpha}$$
$$\leqslant(\nu_{0}c(\delta^{*})c_{m}+c'(\delta^{*})c(\delta^{*})c_{m})\|f\|_{m-2,\alpha}.$$

Consequently $T_{\pm}$ is a smoothing operator, and (9.8) and (9.9) are valid. The lemma is proved.

LEMMA 9.3. *Let the integer* $m\geqq2$, *and let the boundary operator* $B_{01}$ *(cf. Lemma 5.1) have degeneracy order* $m^{*}\leqq m-2$ *and satisfy the complementarity* *condition at $t=0$ with respect to the $\alpha$-elliptic operator $L_1$. Assume that $b_{(m-1)/2}<0$ and $a^0_{nn}(t,0)\in C^{m-1}$ $(1\leqq k\leqq n)$. Then for some $d>0$ there exists an operator $R''_\lambda$, defined on the functions $g(y)\in H^{m-m^*-1}(E_{n-1})$ and mapping into $H^m_\alpha(E_{n,d})$, such that*

$$\| R''_\lambda g\|_{m,\alpha}\leqslant c_m\,\langle\langle g\rangle\rangle_{m-m^*-1}; \tag{9.15}$$

$$L_1R''_\lambda g=0 \text{ in } E_{n,d}; \tag{9.16}$$

$$\gamma B_{01}R''_\lambda g=g+T''_\lambda g \text{ in } E_{n-1}, \tag{9.17}$$

*where $T''_\lambda$ is a bounded operator mapping from $H^{m-m^*-1}(E_{n-1})$ into $H^m(E_{n-1})$ (a smoothing operator), $\lambda\geqq 1$ is a parameter, the constant $c_m>0$ does not depend on $g(y)$ or $\lambda$, and $\alpha(t)=\sqrt{a^0_{nn}(t,0)}$.*

PROOF. Define $R''_\lambda$ by the formula

$$R''_\lambda g\equiv v_\lambda(t,y)\equiv F^{-1}_{n-1}(\widetilde{R}''_\lambda(t,\xi)\,\widetilde{g}(\xi)), \tag{9.18}$$

where $\tilde g(\xi)$ is the Fourier transform $F_{n-1}$ of the function $g(y)$,

$$\widetilde{R}''_\lambda(t,\xi)=\frac{h_0(t,\xi)}{\vartheta(\xi)}\;\frac{|\xi|^{m^*+1}}{|\xi|^{m^*+1}+\lambda},$$

the functions $h_0(t,\xi)$ and $\vartheta(\xi)$ are defined in Lemma 5.3, $0\leqq t\leqq d$, $y\in E_{n-1}$, and $\lambda\geqq 1$.

The function $v_\lambda(t,y)$ has generalized derivatives up to order $m$ in $E_{n,d}$. In fact, with the help of inequality (5.38) we obtain

$$\int_0^d\int_{E_{n-1}}(1+|\xi|^2)^m\,|F_{n-1}v_\lambda|^2\,d\xi\,dt\leqslant c_0\,\langle\langle g\rangle\rangle^2_{m-m^*-1}, \tag{9.19}$$

where the constant $c_0>0$ does not depend on $\lambda$.

By construction the function $\widetilde{R}''_\lambda(t,\xi)$ (like $h_0(t,\xi)$) is a solution of the equation $L\widetilde{R}''_\lambda=0$ on $(0,d)$. Hence inequality (5.54) implies the validity of the estimate

$$\sum_{p+j+2s=m}|\xi|^{2p}\,\|(\alpha\partial_t)^j\,\partial_t^s\,(F_{n-1}v_\lambda)\|^2\leqslant c_m\,(1+|\xi|^2)^m\,\|F_{n-1}v_\lambda\|^2. \tag{9.20}$$

The constant $c_m>0$ depends only on the coefficients of $L_1$.

Using the equivalence of norms (3.2) and (3.12), we obtain inequality (9.15) easily from (9.19) and (9.20). Simultaneously we have proved (9.16), since

$$F_{n-1}L_1v_\lambda=LF_{n-1}v_\lambda=L\widetilde{R}''_\lambda(t,\xi)\,\widetilde{g}(\xi)=0.$$

By the construction of $\widetilde{R}''_\lambda(t,\xi)$ (cf. Lemma 5.3) we have

$$\begin{aligned}B_{01}(\xi,\partial_t)\,\widetilde{R}''_\lambda(t,\xi)\,\widetilde{g}(\xi)|_{t=0}&=\frac{|\xi|^{m^*+1}}{|\xi|^{m^*+1}+\lambda}\,\widetilde{g}(\xi)\\&=\widetilde{g}(\xi)-\frac{\lambda}{|\xi|^{m^*+1}+\lambda}\,\widetilde{g}(\xi).\end{aligned}$$

Hence

$$\gamma B_{01}(-i\partial_y, \partial_t) v_\lambda = g(y) + T''_\lambda g(y),$$

where

$$T''_\lambda g = -F^{-1}_{n-1}(\lambda(\lambda + |\xi|^{m^*+1})^{-1} \tilde{g}(\xi)).$$

It is obvious that $T''_\lambda$ maps from $H^{m-m^*-1}(E_{n-1})$ into $H^m(E_{n-1})$ for each $\lambda \geqq 1$, and that

(9.21) $$\langle\langle T''_\lambda g\rangle\rangle_m \ll \lambda \langle\langle g\rangle\rangle_{m-m^*-1}.$$

Equality (9.17) is proved. With this the proof of the lemma is completed.

**Lemma** 9.4. *Let the integer* $m \geqq 2$, *and let the boundary operator* $B_{0\delta} \equiv B_{01} + B_\delta + \psi_\delta B_{03}$, *where* $B_{03} = B_3 - B_{02}$ (*cf. pp.* 125 *and* 135), *have degeneracy order* $m^* \leqq m-2$ *and satisfy the complementarity condition at* $(t, y) = (0,0)$ *with respect to the* $\alpha$-*elliptic operator* $L_1$. *The coefficients of* $L_1$ *and* $B_{0\delta}$ *satisfy Conditions* 2.3 *and* 2.5, *and* $b_{(m-1)/2} < 0$. *Then for some* $\delta$ $(0 < \delta \leqq d)$ *there exists an operator* $R_{2,\lambda}$, *defined on functions* $g(y) \in H^{m-m^*-1}(E_{n-1})$ *and mapping into* $H^m_\alpha(E_{n,d})$, *such that*

(9.22) $$\| R_{2,\lambda} g \|_{m,\alpha} \ll c_m \langle\langle g\rangle\rangle_{m-m^*-1};$$

(9.23) $$L_1 R_{2,\lambda} g = 0 \text{ in } E_{n,d};$$

(9.24) $$\gamma B_{0\delta} R_{2,\lambda} g = g + T_{2,\lambda} g,$$

*where* $T_{2,\lambda}$ *is a bounded operator mapping from* $H^{m-m^*-1}(E_{n-1})$ *into* $H^{m-m^*}(E_{n-1})$ (*a smoothing operator*), $\lambda \geqq 1$ *is a parameter, and the constant* $c_m > 0$ *does not depend on* $g(y)$ *or* $\lambda$.

**Proof.** By Lemma 9.3 the operator $R''_\lambda$ exists, satisfying conditions (9.15)–(9.17). Let us write $\gamma B_{0\delta} R''_\lambda g$ in the form

$$\begin{aligned}\gamma B_{0\delta} R''_\lambda g &= \gamma B_{01} R''_\lambda g + \gamma B_\delta R''_\lambda g + \gamma \psi_\delta B_{03} R''_\lambda g \\ &= (I + \gamma B_\delta R''_\lambda) g + (T''_\lambda + \gamma\psi_\delta B_{03} R''_\lambda) g\end{aligned}$$

and show that for sufficiently small $\delta > 0$ the inverse operator $(I + \gamma B_\delta R''_\lambda)^{-1}$ exists in $H^{m-m^*-1}(E_{n-1})$. For this purpose we introduce a new norm $\langle\langle g\rangle\rangle_{m-m^*-1,\epsilon}$ into $H^{m-m^*-1}(E_{n-1})$ and show that for sufficiently small $\delta$ and $\epsilon$ the norm of operator $\gamma B_\delta R''_\lambda$ does not exceed $\frac{1}{2}$:

(9.25) $$\langle\langle \gamma B_\delta R''_\lambda g\rangle\rangle_{m-m^*-1,\varepsilon} \ll \frac{1}{2} \langle\langle g\rangle\rangle_{m-m^*-1,\varepsilon}.$$

Let us apply inequality (3.22) to the function $v(t, y) = R''_\lambda g \in H^m_\alpha(E_{n,d})$. Let us estimate the addends of the form $|\xi|^{2|\tau|} |u^{(s)}(0,\xi)|^2$ in the right side of this inequality $(u(t,\xi) = F_{n-1}v)$. If $|\tau| \geqq 1$, with the aid of inequality (3.7) we have

(9.26) $$|\xi|^{2|\tau|} | u^{(s)}(0,\xi)|^2 \ll c_0\{|\xi|^{2(|\tau|-1)} \| u^{(s+1)}\|^2 + (1 + |\xi|^2)^{|\tau|+1} \| u^{(s)}\|^2\}.$$

Since $v(t, y) = R''_\lambda g$ is a solution of equation (9.16) and consequently $u(t,\xi)$, $0 < t < d$, is a solution of the equation $L(\xi, \partial_t)u = 0$, we can apply equations (5.14) in order to evaluate $|u^{(s)}(0,\xi)|$, $s = [m^*/2]$. Since

$$s = \left[\frac{m^*}{2}\right] \leqslant \left[\frac{m-2}{2}\right] \leqslant \frac{m-1}{2},$$

we have $b_s < 0$, and from (5.14) we have the estimate

$$|u^{(s)}(0, \xi)|^2 \leqslant c'_s \left\{ |\xi|^4 |u^{(s-1)}(0, \xi)|^2 + \sum_{i=1}^{s-1} (1 + |\xi|^2) |u^{(i)}(0, \xi)|^2 \right\}, \tag{9.27}$$

the constant $c'_s > 0$ depending only on the coefficients of $L_1$. Moreover, from Lemma 5.6 we have

$$\|u^{(s)}\|^2 \leqslant c''_s (1 + |\xi|^2)^{2s} \|u\|^2 \tag{9.28}$$

for $s = 1, 2, \cdots, [m/2]$. From inequalities (9.26)–(9.28) we derive the estimates

$$|\xi|^{2|\tau+\mu|} |u^{(s)}(0, \xi)|^2 \leqslant c_s (1 + |\xi|^2)^{m^*+1+|\tau|} \|u\|^2 \tag{9.29}$$

for $|\tau + \mu| \geqq 1$ and $|\mu| + 2s = m^*$, and

$$|u^{(s)}(0, \xi)|^2 \leqslant c_s (1 + |\xi|^2)^{m^*+1} \|u\|^2 \tag{9.30}$$

for $2s = m^*$. By definition $u(t, \xi) = \widetilde{R}''_\lambda(t, \xi)\tilde{g}(\xi)$, and by virtue of inequality (5.40) the estimate

$$\|u\|^2 = \left(\frac{|\xi|^{m^*+1}}{|\xi|^{m^*+1} + \lambda}\right)^2 \frac{|\tilde{g}(\xi)|^2}{|\vartheta(\xi)|^2} \|h_0\|^2 \leqslant c \frac{|\tilde{g}(\xi)|^2}{(1 + |\xi|^2)^{m^*+1}} \tag{9.31}$$

is valid, where the constant $c > 0$ does not depend on $\lambda \geqq 1$ or $\xi$.

From inequalities (3.22) and (9.29)–(9.31) we derive the estimate

$$\langle\langle \gamma B_\delta R''_\lambda g \rangle\rangle_{m-m^*-1, \varepsilon} \leqslant c_m (\mathcal{N}(\delta) + \varepsilon \mathcal{M}(\delta)) \langle\langle g \rangle\rangle_{m-m^*-1, \varepsilon}, \tag{9.32}$$

where the constant $c_m > 0$ does not depend on $\epsilon$, $\delta$ or $\lambda$.

After repeating the argument developed in Lemma 9.2, we obtain from (9.32), for some $\epsilon > 0$ and $\delta > 0$, inequality (9.25), which implies the existence of $(I + \xi B_\delta R''_\lambda)^{-1}$ in $H^{m-m^*-1}(E_{n-1})$.

Assuming $R_{2,\lambda} = R''_\lambda(I + \gamma B_\delta R''_\lambda)^{-1}$, we obtain equations (9.23) and (9.24) from (9.16) and (9.17), where

$$T_{2,\lambda} = (T'_\lambda + \gamma \psi_\delta B_{03} R''_\lambda)(I + \gamma B_\delta R''_\lambda)^{-1}$$

is a smoothing operator. It is easy to show that $T_{2,\lambda}$ is a bounded operator from $H^{m-m^*-1}(E_{n-1})$ into $H^{m-m^*}(E_{n-1})$. Inequality (9.22) is derived from (9.15) and (9.25). The lemma is proved.

COROLLARY 9.3. *As is obvious from the proof of Lemma 9.4, the operator* $(I + \gamma B_\delta R''_\lambda)^{-1}$, *for sufficiently small* $\delta > 0$, *maps into* $H^{m'-m^*-1}(E_{n-1})$ *and is bounded since*

(9.33)
$$\langle\langle (I+\gamma B_\delta R_\lambda)^{-1} g\rangle\rangle_{m'-m^*-1} \leqslant c(\delta)\,\langle\langle g\rangle\rangle_{m'-m^*-1}$$

*for some integer $m'$ such that $m^* \neq 2 \leqq m' \leqq m$. The constant $c_m(\delta)$ does not depend on $g(y)$ or $\lambda \geqq 1$.*

We shall consider pairs of functions $\mathscr{F} = \{f(t, y), g(y)\}$, where $f(t, y) \in H_\alpha^{m-2}(E_{n,d})$ and $g(y) \in H^{m-m^*-1}(E_{n-1})$. Let us denote by $W_\alpha^{m-(m^*)}$ the corresponding space $H_\alpha^{m-2}(E_{n,d}) \times H^{m-m^*-1}(E_{n-1})$. Let us take the sum of the norms in $H_\alpha^{m-2}(E_{n,d})$ and $H^{m-m^*-1}(E_{n-1})$ as the norm of $W_\alpha^{m-(m^*)}$.

**Lemma** 9.5. *Let the integer $m \geqq 2$, and let the boundary operator $B_{0\delta}$ (cf. Lemma 9.4) have degeneracy order $m^* \leqq m-2$ and satisfy the complementarity condition at $(t, y) = (0,0)$ with respect to the $\alpha$-elliptic operator $L_{0\delta}$ (cf. p. 126). Let the coefficients of $L_{0\delta}$ and $B_{0\delta}$ satisfy Conditions 2.3 and 2.5, and $b_{(n-1)/2} < 0$. Then for some $\delta_1$ and $\delta_2$ $(0 < \delta_1, \delta_2 \leqq d)$ there exists an operator $R$, defined on the pairs of functions*

$$\mathscr{F} = \{f(t, y),\ g(y)\} \in W_\alpha^{m-(m^*)},$$

*mapping into $H_\alpha^m(E_{n,d})$, and such that*

(9.34)
$$\|R\mathscr{F}\|_{m,\alpha} \leqslant c_m\{\|f\|_{m-2,\alpha} + \langle\langle g\rangle\rangle_{m-m^*-1}\};$$

(9.35)
$$L_{0\delta_1} R\mathscr{F} = f + T_1\mathscr{F} \text{ in } E_{n,d};$$

(9.36)
$$\gamma B_{0\delta_2} R\mathscr{F} = g + T_2\mathscr{F} \text{ in } E_{n-1},$$

*where $T\mathscr{F} = \{T_1\mathscr{F}, T_2\mathscr{F}\}$ is a bounded operator mapping from $W^{m-(m^*)}$ into $W_\alpha^{m+1-(m^*)}$ (a smoothing operator).*

**Proof.** By Lemma 9.4, for $\delta = \delta_2$ $(0 < \delta_2 \leqq d)$ and $\lambda = 1$ an operator $R_2$ exists, satisfying the conditions

(9.37)
$$L_1 R_2 g = 0 \text{ in } E_{n,d};$$

(9.38)
$$\gamma B_{0\delta_2} R_2 g = g + T_2' g \text{ in } E_{n-1}$$

and estimate (9.22).

Let us construct an operator $R_1$ mapping from $H_\alpha^{m-2}(E_{n,d})$ into $H_\alpha^m(E_{n,d})$ and satisfying the conditions

(9.39)
$$\|R_1 f\|_{m,\alpha} \leqslant c_m \|f\|_{m-2,\alpha};$$

(9.40)
$$L_{0\delta_1} R_1 f = f + T_1' f;$$

(9.41)
$$\gamma B_{0\delta_2} R_1 f = 0 + T_1'' f,$$

where $T_1'$ and $T_2''$ are smoothing operators. For this purpose we consider the operator

$$R_0 f = R_-' f - R_{2,\lambda}(\gamma B_{0\delta_2} R_-' f),$$

where $R_-'$ is the operator introduced in Lemma 9.1, and $R_{2,\lambda}$ is the operator introduced in Lemma 9.4 which satisfies conditions (9.22)–(9.24) for $\delta = \delta_2$ and depends on the parameter $\lambda \geqq 1$. For some $\delta \in (0, d)$ we have

$$(9.40') \qquad L_{0\delta} R_0 f = f + L_\delta R'_- f - L_\delta R_{2,\lambda} \gamma B_{0\delta_2} R'_- f + T_3 f;$$

$$(9.41') \qquad \gamma B_{0\delta_2} R_0 f = 0 + T'_3 f,$$

where $T_3 f = \psi_\delta L_3 R_0 f + T' f$ and $T'_3 f = -T_{2,\lambda} \gamma B_{0\delta_2} R'_- f$ are smoothing operators.

In order to obtain (9.40) and (9.41) from (9.40′) and (9.41′), it is sufficient to show that $(I + L_\delta R'_- - L_\delta R_{2,\lambda} \gamma B_{0\delta_2} R'_-)$ has a bounded inverse operator in $H_\alpha^{m-2}(E_{n,d})$.

As we noted in Lemma 9.2, it is sufficient for this purpose to establish that for some $\delta$, $\epsilon$ and $\lambda$ the inequality

$$(9.42) \qquad \| (L_\delta R'_- - L_\delta R_{2,\lambda} \gamma B_{0\delta_2} R'_-) f \|_{m-2,\alpha,\varepsilon} \leqslant \frac{1}{2} \| f \|_{m-2,\alpha,\varepsilon}$$

is valid. We have already obtained estimate (9.7) for $L_\delta R'_-$ in Corollary 9.2. Let us establish an analogous estimate for $L_\delta R_{2,\lambda} \gamma B_{0\delta_2} R'_-$.

Having denoted $R_{2,\lambda} \gamma B_{0\delta_2} R'_-$ by $v(t, y)$, let us apply inequality (3.21) to estimate $L_\delta v(t, y)$. Since $v(t, y)$ is a solution of the homogeneous equation $L_1 v = 0$, inequalities analogous to (9.20) and (9.19), as well as inequality (5.38) $(u(T, \xi) = F_{n-1} v)$, are valid for $v(t, y)$. In this way, using the equivalence of norms $\| \cdot \|_{m,\alpha}$ and $| \cdot |_{m,\alpha}$, we obtain from (3.21)

$$(9.43) \qquad \begin{aligned} \| L_\delta v \|^2_{m-2,\alpha,\varepsilon} \leqslant 4 (\mathcal{N}^0(\delta) + \varepsilon \mathcal{M}(\delta))^2 \int_{E_{n-1}} \Big\{ \sum_{3 \leqslant p \leqslant m} \frac{\varepsilon^{2(p-2)} (1 + |\xi|^2)^p}{(\lambda + |\xi|^{m^*+1})^2} \\ + \frac{(1 + |\xi|^2)^2}{(\lambda + |\xi|^{m^*+1})^2} \Big\} | w_1(\xi) |^2 \, d\xi, \end{aligned}$$

where

$$w_1(\xi) = F_{n-1} (I + \gamma B_{\delta_2} R''_\lambda)^{-1} \gamma B_{0\delta_2} R'_- f.$$

The trivial inequality

$$(9.44) \qquad (\lambda + |\xi|^{m^*+1})^{2\theta} \geqslant c_\theta^{-1} \max \{ \lambda^{2\theta}, (1 + |\xi|^2)^{(m^*+1)\theta} \},$$

where the constant $c_\theta > 0$ does not depend on $\lambda$ or $\xi$, is valid for any $\lambda \geqq 1$ and $0 < \theta \leqq 1$.

Now let us choose $\lambda = \epsilon^{-m^*-1}$ $(0 < \epsilon \leqq 1)$. Then with the aid of (9.44) we easily obtain

$$(9.45) \qquad \begin{aligned} \frac{\varepsilon^{2(p-2)} (1 + |\xi|^2)^p}{(\lambda + |\xi|^{m^*+1})^2} &= \frac{\varepsilon^{2(p-2)} (1 + |\xi|^2)^p}{(\lambda + |\xi|^{m^*+1})^{2(1-\theta)} (\lambda + |\xi|^{m^*+1})^{2\theta}} \\ &\leqslant c_\theta \, \varepsilon^{2m^*} (1 + |\xi|^2) \left( \theta = \frac{p-1}{m^*+1} \right) \end{aligned}$$

for $2 \leqq p \leqq m^* + 2$ and

$$\frac{\varepsilon^{2(p-2)}(1+|\xi|^2)^p}{(\lambda+|\xi|^{m^*+1})^2} \leqslant c_1 \frac{\varepsilon^{2(p-2)}(1+|\xi|^2)^p}{(1+|\xi|^2)^{m^*+1}} = c_1\varepsilon^{2(p-2)}(1+|\xi|^2)^{p-m^*-1} \tag{9.46}$$

for $m^*+3 \leqq p \leqq m$. Inequalities (9.43), (9.45) and (9.46) imply

$$\|L_\delta v\|^2_{m-2,\alpha,\varepsilon} \leqslant c\,(\mathcal{N}(\delta)+\varepsilon\,\mathcal{M}(\delta))^2 \int\limits_{E_{n-1}} \sum_{m^*+2\leqslant p\leqslant m} (1+|\xi|^2)^{p-m^*-1} \times \varepsilon^{2(p-2)}\,|w_1(\xi)|^2\,d\xi. \tag{9.47}$$

Let us apply estimates (9.33) with $m'=p$ to the right side of the last inequality (first it is necessary to set $\delta=\delta_2$ in (9.33)). Then we obtain

$$\|L_\delta v\|^2_{m-2,\alpha,\varepsilon} \leqslant c'(\delta_2)\,(\mathcal{N}(\delta)+\varepsilon\,\mathcal{M}(\delta))^2 \int\limits_{E_{n-1}} \sum_{m^*+2\leqslant p\leqslant m} \varepsilon^{2(p-2)}(1+|\xi|^2)^{p-m^*-1} \times |F_{n-1}w(0,\xi)|^2\,d\xi, \tag{9.48}$$

where $w(t,y)=B_{0\delta_2}R'_-f$. After applying inequality (3.7), we obtain the estimate

$$\int\limits_{E_{n-1}} |\xi|^{2(p-m^*-1)}\,|F_{n-1}w(0,\xi)|^2\,d\xi \leqslant c \int\limits_{E_{n,d}} \Big(\sum_{|\tau|+2s=p-m^*} |\partial_t^s\partial_y^\tau w(t,y)|^2+|w(t,y)|^2\Big)\,dt\,dy.$$

If we make use of this estimate and the very well-known estimate

$$\int\limits_{E_{n-1}} |w(0,y)|^2\,dy \leqslant c \int\limits_{E_{n,d}} (|\partial_t w(t,y)|^2+|w(t,y)|^2)\,dt\,dy,$$

then from (9.48) we obtain

$$\|L_\delta v\|^2_{m-2,\alpha,\varepsilon} \leqslant c''(\mathcal{N}(\delta)+\varepsilon\,\mathcal{M}(\delta))^2 \sum_{m^*+2\leqslant p\leqslant m} \varepsilon^{2(p-2)}\,\|w\|^2_{p-m^*,\alpha}. \tag{9.49}$$

Since $w(t,y)$ can be represented in the form

$$w(t,y)=\sum_{|\tau|+2s\leqslant m} b^0_{\tau,s}(\delta_2,y)\,\partial_y^\tau\partial_t^s v^-,$$

where $v^-(t,y)=R'_-f$, from (9.49) and the well-known estimate

$$\|w\|_{p-m^*,\alpha} \leqslant c_p(\delta_2)\,\|v^-\|_{p,\alpha}$$

we obtain

$$\|L_\delta v\|^2_{m-2,\alpha,\varepsilon} \leqslant c'_m(\mathcal{N}(\delta)+\varepsilon\,\mathcal{M}(\delta))^2 \sum_{m^*+2\leqslant p\leqslant m} \varepsilon^{2(p-2)}\,\|R'_-f\|^2_{p,\alpha}.$$

After applying inequality (9.6) we arrive at the required estimate

$$(9.49)\qquad \| L_\delta R_{2,\lambda}\gamma B_{0\delta_2} R'_- f\|^2_{m-2,\alpha,\varepsilon} \leqslant c_m (\mathcal{N}(\delta) + \varepsilon \mathcal{M}(\delta))^2 \| f\|^2_{m-2,\alpha,\varepsilon},$$

where $\lambda = \epsilon^{-m^*-1}$ and the constant $c_m$ does not depend on $\delta$ or $\epsilon$.

On choosing sufficiently small $\delta = \delta_1$ and $\epsilon = \epsilon_1(\delta_1)$, we obtain (9.42) from (9.7) and (9.50). Thus the operator

$$(I + L_{\delta_1} R'_- - L_{\delta_1} R_{2,\lambda_1}\gamma B_{0\delta_2} R'_-)^{-1} \quad (\lambda_1 = \varepsilon_1^{-m^*-1})$$

exists, and the estimate

$$(9.51)\qquad \|(I + L_{\delta_1} R'_- - L_{\delta_1} R_{2,\lambda_1}\gamma B_{0\delta_2} R'_-)^{-1} f\|_{m-2,\alpha} \leqslant c(\delta_1)\|f\|_{m-2,\alpha}$$

is valid for it. As usual, from (9.40′) and (9.41′) we obtain (9.40) and (9.41), where

$$R_1 = R_0\Gamma;\ T'_1 = T_3\Gamma;\ T''_1 = T'_3\Gamma$$

and

$$\Gamma = (I + L_{\delta_1} R'_- - L_{\delta_1} R_{2,\lambda_1}\gamma B_{0\delta_2} R'_-)^{-1}.$$

If we make use of inequalities (9.2) and (9.22), it is easy to derive the estimate

$$\| R_0 f\|_{m,\alpha} \leqslant c_m \| f\|_{m-2,\alpha},$$

which, together with (9.51), ensures the validity of (9.39).

It remains for us to construct an operator $R$ satisfying conditions (9.34)-(9.36). Let us assume

$$R\mathcal{F} = R_1(f - L_{\delta_1} R_2 g) + R_2 g,$$

where $R_2$ satisfies conditions (9.37) and (9.38). From (9.37), (9.38), (9.40) and (9.41) we obtain (9.35) and (9.36), where

$$T_1\mathcal{F} = \psi_\delta L_3 R_2 g + T'_1(f - L_{\delta_1} R_2 g);$$

$$T_2\mathcal{F} = T'_2 g + T''_1(f - L_{\delta_1} R_2 g).$$

Inequality (9.34) is derived from (9.39) and (9.22). It is easy to show that $T_1$ and $T_2$ are smoothing. The lemma is proved.

The results obtained in Lemmas 9.2 and 9.5) enable us to construct a regularizer in the original bounded region $\mathcal{D} \subset E_n$ and to complete the

PROOF OF THEOREM 2.2. Let us construct a system of regions $\{\mathcal{D}_\delta\}$ and neighborhoods $\{\mathcal{A}_x(\delta_x)\}$ $(x \in S)$ which forms a cover of $\overline{\mathcal{D}}$ such that a local regularizer of the problem under consideration exists in $\mathcal{A}_x(\delta_x) \cap \overline{\mathcal{D}}$ for each $x \in S$. This regularizer equals $R_+ f$ (cf. Lemma 9.2) if $x \in S_q$, $q \in Q_1$, and it equals $R\mathcal{F} = R\{f, g\}$ (cf. Lemma 9.5) if $x \in S_q$, $q \in Q_2$. (For simplicity of argument, we shall assume that $\delta_1 = \delta_2$ in Lemma 9.5.) Let us select a finite cover of $\overline{\mathcal{D}}$ from the system $\{\mathcal{D}_\delta, \mathcal{A}_x(\delta_x)\}$. This cover will consist of one region $\mathcal{D}_{\delta_0}$ $(\overline{\mathcal{D}}_{\delta_0} \subset \mathcal{D})$ and a system of neighborhoods $\mathcal{A}_{x^k}(\delta_k)$ $(k = 1, 2, \cdots, N)$. Let the functions $\{\phi_k(x)\}_{k=0}^N$ form a partition of unity in $\overline{\mathscr{D}}$ which is subordinated to this cover (cf. p. 118). Let us define yet another function $\psi_k(x)$ $(k = 1, 2, \cdots, N)$ with the following properties: $\psi_k(x) \equiv 1$ for $x \in \mathscr{A}_{x^k}(\delta_k)$, and $\psi_k(x) \in C_0^\infty(\mathscr{A}_{x^k}(2\delta_k))$. Let us define analogously the function $\psi_0(x) \equiv 1$ in $\mathscr{D}_{\delta_0}$, $\psi_0(x) \in C_0^\infty(\mathscr{D})$. Let us define for the points $x^k \in S_q$, $q \in Q_2$, the operators

$$\mathfrak{A}_k = \{L_{0\delta_k}, B_{0\delta_k}\}_{x^k}; \quad \mathscr{R}_k\mathscr{F} = (R\mathscr{F})_{x^k} = (R\{f, g\})_{x^k},$$

where $L_{0\delta}$ and $B_{0\delta}$ are defined on pp. 126 and 169, and $R$ is constructed in Lemma 9.5.

Analogously, for the points $x^k \in S_q$, $q \in Q_1$, let us set

$$\mathfrak{A}_k = \{L_{0\delta_k}\}_{x^k}; \quad \mathscr{R}_k f = (R_+ f)_{x^k},$$

where $R_+$ is defined in Lemma 9.2.

Finally, we shall denote by $\mathfrak{A}_0$ the operator coinciding with $\mathscr{L}$ in $\mathscr{D}$. As is well known (cf. [**23**]), an operator $\mathscr{R}_0$ (an interior regularizer) exists such that

$$\mathfrak{A}_0\mathscr{R}_0 f = f + T_0 f \tag{9.52}$$

in a strictly interior subregion of $\mathscr{D}$. The estimates

$$|||\mathscr{R}_0\psi_0 f|||_m \leq c_m |||\psi_0 f|||_{m-2}; \quad |||T_0\psi_0 f|||_{m-1} \leq c_m |||\psi_0 f|||_{m-2} \tag{9.53}$$

hold. Here $|||\cdot|||_m$ is the norm in the Sobolev space $W_2^m(\mathscr{D})$.

Now let us define the right regularizer for problem (2.17)-(2.18) by the formula

$$\mathscr{R} = \sum_{k=0}^{N} \varphi_k R_k \psi_k.$$

We verify by the standard method that

$$\mathfrak{A}\mathscr{R}\mathscr{F} = \mathscr{F} + T\mathscr{F} \tag{9.54}$$

and that the inequalities

$$|||\mathscr{R}\mathscr{F}|||_{m,\alpha}^{\mathscr{D}} \leq c_m |[\mathscr{F}]|_{m-(m^*),\alpha}, \tag{9.55}$$

$$|[T\mathscr{F}]|_{m+1-(m^*),\alpha} \leq c_m |[\mathscr{F}]|_{m-(m^*),\alpha} \tag{9.56}$$

are valid, where $\mathscr{F} \in \mathscr{A}_\alpha^{m-(m^*)}$ (cf. p. 118) and

$$|[\mathscr{F}]|_{m-(m^*),\alpha} = |||f|||_{m-2,\alpha} + \sum_{q \in Q_2} \langle g_q \rangle_{S_q, m-m_1^*-1}$$

In proving (9.54)-(9.56) we depend on Lemmas 9.2 and 9.5, as well as on (9.52)-(9.53).

On the basis of (9.54)-(9.56) we assert that $\mathscr{R}$ is the regularizer for problem (2.17)-(2.18) and $T$ is a bounded operator from $\mathscr{A}_\alpha^{m-(m^*)}$ into $\mathscr{A}_\alpha^{m+1-(m^*)}$ (a smoothing operator). If the space $\mathscr{A}_\alpha^{m+1-(m^*)}$ is compactly embedded into $\mathscr{A}_\alpha^{m-(m^*)}$, then (9.54) implies that the range of $\mathfrak{A}$ has finite codimension in $\mathscr{A}_\alpha^{m-(m^*)}$. Since $H^{m-m_q^*}(S_q)$ is compactly embedded into $H^{m+1-m_q^*}(S_q)$, it remains for us to analyze the question of the compactness of the embedding of $H_\alpha^{m-1}(\mathscr{D})$ into $H_\alpha^{m-2}(\mathscr{D})$.

Since the operator embedding $W_2^{m-1}(\mathscr{D}_\delta)$ into $W_2^{m-2}(\mathscr{D}_\delta)$ is completely continuous for any strictly interior subregion $\mathscr{D}_\delta$ of $\mathscr{D}$, the question of the compactness of the embedding of $H_\alpha^{m-1}(\mathscr{D})$ into $H_\alpha^{m-2}(\mathscr{D})$, as it is easy to see, leads to the establishment of the complete continuity of the operator embedding $H_\alpha^{m-1}(\mathscr{D})$ into $\mathscr{L}_2(\mathscr{D})$ and to the proof of the inequality

$$|v|_{m-2,\alpha} \leqslant \varepsilon |v|_{m-1,\alpha} + c(\varepsilon)|v|_0, \tag{9.57}$$

where $v(t, y) \in H_\alpha^{m-1}(E_{n,d})$, the norm $|\cdot|_{m,\alpha}$ is defined by formula (3.1), $m \geqq 2$, and $\epsilon$ is an arbitrary positive number.

First let us consider odd $m \geqq 3$. In order to obtain inequality (9.57) in this case, we must apply inequality (3.5) to the function $u(t, \xi) = F_{n-1}v$. (We change $m$ to $m-1$ in (3.5).) After integrating the resulting inequality with respect to $\xi \in R_{n-1}$ and applying Parseval's equation, we obtain inequality (9.57) for odd $m$. The space $H_\alpha^{m-1}(\mathscr{D})$ is contained in $W_2^{[(m-1)/2]}(\mathscr{D})$ which is compactly embedded into $\mathscr{L}_2(\mathscr{D})$ for $m \geqq 3$. Thus $T$ is completely continuous as an operator mapping into $\mathscr{A}_\alpha^{m-(m^*)}$ for odd $m \geqq 3$. This means that the range of $\mathfrak{A}$ (like that of $I+T$) has finite codimension in $\mathscr{A}_\alpha^{m-(m^*)}$. Theorem 2.2 is proved for odd $m \geqq 3$.

Let us turn to the consideration of even $m \geqq 2$. First we make a supplementary assumption: $\Delta_{m/2}(x) > 0$ for $x \in S_q$, $q \in Q_2$. (We recall that only $\Delta_{(m-1)/2}(x) > 0$ is guaranteed on $S_q$ $(q \in Q_2)$.) With this supplementary condition we must change the number $m$ in Lemma 9.5 to $m+1$ and show that the operator $T$ introduced in that lemma maps from $W_\alpha^{m+1-(m^*)}$ into $W_\alpha^{m+2-(m^*)}$. (Here we are taking into account that according to the premise of Theorem 2.2, the even number $m$ must also be changed to $m+1$ in Condition 2.5 defining smoothness of coefficients.) If the regularizer $R$ constructed in Lemma 9.5 is now changed to $\hat{R} = R(I-T)$, the new regularizer $\hat{R}$, as it is easy to see, will satisfy the conditions

$$\|\hat{R}\mathscr{F}\|_{m,\alpha} \leqslant \hat{c}_m\{\|f\|_{m-2,\alpha} + \langle\!\langle g \rangle\!\rangle_{m-m^*-1}\};$$

$$L_{0\delta_1}\hat{R}\mathscr{F} = f + \hat{T}_1\mathscr{F};$$

$$\gamma B_{0\delta}\hat{R}\mathscr{F} = g + \hat{T}_2\mathscr{F},$$

where $\hat{T} = \{\hat{T}_1, \hat{T}_2\} = -T^2$ is a bounded operator mapping from $W_\alpha^{m-(m^*)}$ onto $W_\alpha^{m+2-(m^*)}$.

Assuming $\hat{R}_+ = R_+(I+T_+)$, we can carry out the same changes in $R_+$ in Lemma 9.2. Then we obtain

$$\|\hat{R}_+ f\|_{m,\alpha} \leqslant \hat{c}_m \|f\|_{m-2,\alpha};$$

$$L_{0\delta}\hat{R}_+ f = f + \hat{T}_+ f$$

from (9.8)–(9.9), where $\hat{T}_+ = -T_+^2$ is a bounded operator from $H_\alpha^{m-2}(E_{n,d})$ into $H_\alpha^m(E_{n,d})$.

If we now construct, as above, the regularizer $\mathscr{R}$ for problem (2.17) – (2.18), changing $R_+$ to $\hat{R}_+$ and $R$ to $\hat{R}$, we can easily prove the complete continuity of the new operator $T$ (cf. (9.54)) in $\mathscr{A}_\alpha^{m-(m^*)}$. Actually, in this case the compactness of the embedding of $H_\alpha^m(\mathscr{D})$ into $H_\alpha^{m-2}(\mathscr{D})$ is implied by the compactness of the embedding of $H_\alpha^m(\mathscr{D})$ into $\mathscr{L}_2(\mathscr{D})$ and by the inequality

$$|v|_{m-2,\alpha} \leqslant c\,|v|_{m-1,\alpha} \leqslant \varepsilon\,|v|_{m,\alpha} + c_m(\varepsilon)\,|v|,$$

by which inequality (9.57) is replaced.

Now let us get rid of the supplementary condition $\Delta_{m/2}(x) > 0$. For this purpose, note that this condition can fail to be fulfilled (i.e. $\Delta_{m/2}(x^0) \leqq 0$) only at the points $x^0$ of the boundary manifolds $S_q$ ($q \in Q_2$) in a neighborhood of which the order of degeneracy equals unity ($\alpha^2(t) \geqq \beta_1 t$, $\beta_1 = \text{const} > 0$). But in that case, as is obvious from inequalities (3.5) and (3.6), estimate (9.57) is valid not only for odd, but also for even, $m$. In addition, by virtue of the Lizorkin-Uspenskiĭ inequality (cf. [21]), some set $\{v\}$ bounded in the norm $\|v\|_{m-1,\alpha}$, and consequently bounded in the norm of the Sobolev-Slobodeckiĭ space $W_2^{(m-1)/2}(K_d)$, is compact in $\mathscr{L}_2(K_d)$ for $m \geqq 2$. All of this ensures the complete continuity of the operator $T$ in (9.54), without the additional constructions in a neighborhood of the point $x^0$ which we carried out above. Theorem 2.2 has been proved.

Received 10 Feb 1969

## Bibliography

[1] G. Fichera, *On a unified theory of boundary value problems for elliptic-parabolic equations of second order*, Boundary Problems in Differential Equations, Univ. of Wisconsin Press, Madison, Wis., 1960, pp. 97-120; Russian transl., Matematika **7** (1963), no. 6, 99-121. MR **22** # 2789.

[2] O. A. Oleĭnik, *Linear equations of second order with nonnegative characteristic form*, Mat. Sb. **69** (**111**) (1966), 111–140; English transl., Amer. Math. Soc. Transl. (2) **65** (1967), 167–199. MR **33** # 1603.

[3] M. V. Keldyš, *On certain cases of degeneration of equations of elliptic type on the boundary of a domain*, Dokl. Akad. Nauk SSSR **77** (1951), 181–183. (Russian) MR **13**, 41.

[4] O. A. Oleĭnik, *On equations of elliptic type degenerating on the boundary of a region*, Dokl. Akad. Nauk SSSR **87** (1952), 885–888. (Russian) MR **16**, 366.

[5] N. D. Vvedenskaja, *On a boundary problem for equations of elliptic type degenerating on the boundary of a region*, Dokl. Akad. Nauk SSSR **91** (1953), 711–714. (Russian) MR **15**, 711.

[6] A. M. Il′in, *Degenerate elliptic and parabolic equations*, Mat. Sb. **50** (**92**) (1960), 433–498. (Russian) MR **22** # 2788.

[7] S. G. Mihlin, *Degenerate elliptic equations*, Vestnik Leningrad. Univ. **9** (1954), no. 8, 19-48. (Russian) MR **17**, 493.

[8] M. I. Višik, *Boundary-value problems for elliptic equations degenerating on the boundary of a region*, Mat. Sb. **35** (**77**) (1954), 513-568; English transl., Amer. Math. Soc. Transl. (2) **35** (1964), 15–78. MR **16**, 927.

[9] M. M. Smirnov, *Degenerating elliptic and hyperbolic equations*, "Nauka", Moscow, 1966. (Russian) MR **36** # 1850.

[10] J. J. Kohn and L. Nirenberg, *Non-coercive boundary value problems*, Comm. Pure Appl. Math. **18** (1965), 433–492; Russian transl. in *Pseudodifferential operators*, "Mir", Moscow, 1967, pp. 88–165. MR **31** # 6041.

[11] ———, *Degenerate elliptic-parabolic equations of second order,* Comm. Pure Appl. Math. **20** (1967), 797–872. MR **38** # 2437.

[12] V. P. Gluško, *Coerciveness in $L_2$ of general boundary value problems for a degenerate second order elliptic equation,* Funkcional. Anal. i Priložen. **2** (1968), no. 3, 87-88 = Functional Anal. Appl. 2 (1968), 261–262. MR **38** # 3585.

[13] ———, *A one-dimensional analogue of an elliptic equation of the second order degenerating at the boundary,* Dokl. Akad. Nauk SSSR **174** (1967), 1014–1017 = Soviet Math. Dokl. **8** (1967), 704–707. MR **35** # 6888.

[14] V. P. Gluško and S. G. Kreĭn, *Degenerating linear differential equations in Banach space,* Dokl. Akad. Nauk SSSR **181** (1968), 784–787 = Soviet Math. Dokl. **9** (1968), 919–922. MR **38** # 393.

[15] V. P. Gluško, *Degenerate linear differential equations.* I, Differencial′nye Uravnenija **4** (1968), 1584–1597. (Russian) MR **38** # 3507.

[16] ———, *Degenerate linear differential equations.* II, Differencial′nye Uravnenija **4** (1968), 1956–1966. (Russian) MR **38** # 6148.

[17] ———, *Degenerate linear differential equations.* III, Differencial′nye Uravnenija **5** (1969), 443–455. (Russian) MR **39** # 5866.

[18] ———, *Degenerate linear differential equations.* IV, Differencial′nye Uravnenija **5** (1969), 599–611. (Russian) MR **39** # 5867.

[19] S. Agmon, A. Douglis and L. Nirenberg, *Estimates near the boundary for solutions of elliptic partial differential equations satisfying general boundary conditions.* I, Comm. Pure Appl. Math. **12** (1959), 623-727; Russian transl., IL, Moscow, 1962. MR **23** # A2610.

[20] L. Hörmander, *Linear partial differential operators,* Die Grundlehren der math. Wissenschaften, Band 116, Academic Press, New York; Springer-Verlag, Berlin, 1963; Russian transl., "Mir", Moscow, 1965. MR **28** # 4221; **37** # 6595.

[21] S. V. Uspenskiĭ, *Properties of $W_p^r$ classes with a fractional derivative on differentiable manifolds,* Dokl. Akad. Nauk SSSR **132** (1960), 60–62 = Soviet Math. Dokl. **1** (1960), 495–497. MR **23** A # 2042b.

[22] K. Maurin, *Methods of Hilbert space,* Monografie Mat., Tom 36, PWN, Warsaw, 1959; English transl., Monografie Mat., vol. 45, PWN, Warsaw, 1967; Russian transl., "Mir", Moscow, 1965. MR **24** # A1002; **34** # 581; **36** # 6957.

[23] L. R. Volevič, *Solvability of boundary problems for general elliptic systems,* Mat. Sb. **68 (110)** (1965), 373-416; English transl., Amer. Math. Soc. Transl. (2) **67** (1968), 182-225. MR **33** # 418.

Translated by:
F. M. Goldware

Trudy Moskov. Mat. Obšč.
Tom 23 (1970)

Trans. Moscow Math. Soc.
Vol. 23 (1970)

# INVESTIGATION OF THE GREEN MATRIX FOR A HOMOGENEOUS PARABOLIC BOUNDARY VALUE PROBLEM

*S. D. ÈĬDEL'MAN and S. D. IVASIŠEN*

## Contents



## Introduction

In this paper we present the results of an investigation of the Green matrix of a homogeneous parabolic boundary value problem in a cylinder whose axis is parallel to the time axis $t$, and whose base is a finite or infinite domain in the space $E_n$. The results of this investigation, carried out in the years 1965–1966, were presented at the International Congress of Mathematicians in Moscow in August 1966 [**1**] (see also [**2**], [**3**]). The Green matrix is constructed by means of a regularizer, which for the case of parabolic systems has been described in detail by V. A. Solonnikov [**4**]. Subsequently, S. D. Ivasišen [**5**], using the method of Ju. P. Krasovskiĭ [**6**], [**7**], constructed for the case of an infinite domain the Green matrix for the nonhomogeneous boundary value problem, and obtained its detailed structure in the neighborhood of the singular point. It should be noted that in all these works exact estimates valid right up to the boundary were obtained for all derivatives of the Green matrix.

For second order equations of divergent type, P. E. Sobolevskiĭ [**9**] has obtained estimates of the Green functions of the fundamental boundary value problems and has applied these estimates in the integral representation

---

*AMS (MOS) subject classifications* (1970). Primary 35K50; Secondary 35J55.

and in the investigation of partial powers of second order divergent elliptic operators ([10] Chapter 4, §16). Using the apparatus of partial powers of the heat conduction operator on a boundaryless manifold, R. Arima [11] has investigated the Green function of the parabolic boundary value problem in a bounded domain for the case of a single equation and has obtained estimates valid up to the boundary for the derivatives to an order less than the order of the equation. On the basis of Arima's results, S. Mizohata and Arima [12], using the parabolic equation method (Minakshisundaram's method), found the asymptotic behavior of the function $N(\lambda)$, which is the number of eigenvalues smaller than $\lambda$ of the selfadjoint elliptic boundary value problem with normal boundary conditions. In a subsequent work, Mizohata [13], using results obtained for the Green function, investigated the distribution of the eigenvalues of nonselfadjoint elliptic boundary value problems.

The results for the Green function obtained by the authors [3] have been used by V. E. Šatalov and I. A. Šišmarev [14] in the investigation of the Dirichlet series $\sum_{i=1}^{\infty} 1/\lambda_i^z$ of the general homogeneous selfadjoint boundary value problem for elliptic systems in a bounded domain. These results were also used by V. P. Lavrenčuk [15] to establish the solvability of homogeneous boundary value problems for parabolic systems whose coefficients increase with increasing space coordinates. Recently, S. D. Šmulevič [17] used the authors' results, together with A. G. Kostjučenko's method [16], to find the asymptotic behavior of the function $N(\lambda)$ for the case of elliptic problems with increasing coefficients in infinite domains.

In §1 of this work we derive the Green matrix of the homogeneous parabolic boundary value problem and prove a basic theorem (Theorem 1) on the existence and on exact estimates right up to the boundary of the Green matrix and its derivatives in a cylinder of arbitrary finite height. The Green matrix is constructed in the form $G(t, x; \tau, \xi) = Z(t, \tau, x, \xi) - v(t, x; \tau, \xi)$, where $Z$ is the fundamental matrix of the solutions of the parabolic system and the matrix $v$ may have singularities only on the boundary of the domain. Subsequently, in the same section we formulate Theorem 2, which states that the operator inverse to the operator of the homogeneous boundary value problem is an integral operator whose kernel is the Green matrix (the Green operator). The proof of Theorem 1 begins in §1 and is concluded in §§3 and 4. The lemmas about integral operators which are required in this proof are given §2. The proofs of the most important of these (Lemmas 1 and 2) are given in the Appendix.

In §4 we show that in the case of stationary parabolic boundary value problems (i.e. problems in which the coefficients of the system of equations and boundary conditions are not functions of $t$), the estimates of the Green matrix are obtained in a cylinder semi-infinite in $t$. On the basis of these estimates, it is shown in §5 that by integrating over $t$ the Green matrix of the stationary parabolic boundary value problem one obtains the Green matrix of the corresponding elliptic problem (with shifted spectrum). In §5 we also obtain estimates for the derivatives of the Green matrix of the elliptic problem, as well as for the kernels (Green matrices) of negative fractional powers of the corresponding elliptic operator.

In §6 we prove certain theorems regarding the existence in Hölder spaces of Green operators of the parabolic and elliptic boundary value problems. In that section we also give the proof of Theorem 2.

The Green functions of elliptic boundary value problems have been investigated in [**18**]–[**20**] and [**6**]–[**8**].

The extension of the results of the present work to boundary value problems for systems with discontinuous coefficients is given in [**21**], and for the case of noncylindrical domains it is given in [**22**] and [**23**].

**Note added in proof.** The authors have recently become acquainted with V. A. Solonnikov's paper [**33**], in which the author demonstrates a different method of constructing and obtaining the estimates of Green matrices of homogeneous parabolic boundary value problems. That method is presented for systems parabolic in the sense of Petrovskiĭ, of arbitrary order in $t$ for the case of noncylindrical domains. It should be noted that all results of the present paper hold for systems parabolic in the sense of Petrovskiĭ of any order. This was demonstrated by the authors in [**3**].

## §1. The Green matrix. Its existence and estimates

**1. Definition of the Green matrix. Formulation of the fundamental result.** Let $Q$ be the cylindrical domain $[0, T]\times\Omega$, where $T$ is a finite positive number and $\Omega$ is a finite or infinite domain with boundary $S$ in the space $E_n$.

We consider in the domain $Q$ the system of $N$ equations for $N$ unknown functions

$$\begin{aligned} L(t, x, D_t, D_x)\,u &\equiv D_t u - A(t, x, D_x)\,u = f(t, x), \\ A(t, x, D_x) &\equiv A_0(t, x, D_x) + A_1(t, x, D_x) \equiv \sum_{|k|=2b} a_k(t, x)\,D_x^k \\ &\quad + \sum_{|k|<2b} a_k(t, x)\,D_x^k, \end{aligned} \tag{1.1}$$

with the initial condition

$$u|_{t=0} = \varphi(x) \tag{2.1}$$

and the boundary conditions

$$B_j(t, x, D_x) u|_\Gamma = 0,$$

$$B_j(t, x, D_x) \equiv B_{j0}(t, x, D_x) + B_{j1}(t, x, D_x) \equiv \sum_{|k|=r_j} b_{jk}(t, x) D_x^k \tag{3.1}$$
$$+ \sum_{|k|<r_j} b_{jk}(t, x) D_x^k,$$

where $\Gamma = [0, T] \times S$ and $r_j \leqq 2b - 1$ $(j = 1, \cdots, bN)$.

We assume that the operators $L$ and $B_j$ satisfy the following conditions in the domain $\Omega$:

CONDITION A. The problem (1.1)–(3.1) is parabolic, i.e. the system (1.1) is (uniformly) parabolic in the sense of Petrovskiĭ, and the operators $B_j$ are related to the system (1.1) by the condition of (uniform) complementation [**25**], [**24**], [**4**]. Let $\delta$ denote the constant in the parabolicity condition, and let $\delta'$ denote the constant in the complementation condition.

CONDITION $B_l$. The coefficients $a_k(t, x)$ belong to the space $C_{x,t}^{l+\alpha,(l+\alpha)/2b}(Q)$, and the coefficients $b_{jk}(t, x)$ belong to the space $C_{x,t}^{2b-r_j+l+\alpha,(2b-r_j+l+\alpha)/2b}(\Gamma)$ (these are defined in [**4**]). The boundary $S$ belongs to $C^{2b+l+\alpha}$ in the sense of [**4**]. The constant $M_l$ is a bound of the corresponding norms of all coefficients $a_k$, $b_{jk}$ and the surface $S$; $l \geqq 0$ is an integer, and $0 < \alpha < 1$.

As has been shown in [**4**], the conditions A and $B_l$ are sufficient for the existence of a unique solution $u$ of the problem (1.1)–(1.3), belonging to the space $C_{x,t}^{2b+l+\alpha,(2b+l+\alpha)/2b}(Q)$ if $f \in C_{x,t}^{l+\alpha,(l+\alpha)/2b}(Q)$, $\phi \in C^{2b+l+\alpha}(\Omega)$ and $f$ and $\phi$ satisfy the corresponding compatibility conditions. The norm is then estimated by the norms of $f$ and $\phi$. We introduce the operator $\mathscr{L}$ defined in the domain

$$\mathscr{D}(\mathscr{L}) = \{u;\ u \in C_{x,t}^{2b+\alpha,(2b+\alpha)/2b}(Q),\ u|_{t=0} = 0,$$
$$B_j(t, x, D_x) u|_\Gamma = 0\ (j = 1, \ldots, bN)\}$$

which operates on $u$ according to the relation $\mathscr{L} u = Lu$. We denote its range of values by $\mathscr{R}(\mathscr{L})$. Clearly, $\mathscr{R}(\mathscr{L}) \subseteq C_{x,t}^{\alpha,\alpha/2b}(Q)$. It follows from the result cited above that the operator $\mathscr{L}$ has an inverse $\mathscr{L}^{-1}$ bounded from $\mathscr{R}(\mathscr{L})$ into $\mathscr{D}(\mathscr{L})$. It can be shown that this operator is an integral operator, whose kernel is the Green matrix of the problem (1.1)–(3.1).

We now proceed to define and construct the Green matrix. We shall use the following constructive definition:

DEFINITION. *The Green matrix of the problem* (1.1)–(3.1) *is the matrix* $G(t, x; \tau, \xi)$, *defined and continuous for* $0 \leqq \tau < t \leqq T$ *and* $x, \xi \in \overline{\Omega}$, *which has a first-order derivative in t and derivatives up to the order* $2b$ *in x, and which is of the form*

$$G(t, x; \tau, \xi) = Z(t, \tau, x, \xi) - v(t, x; \tau, \xi), \tag{4.1}$$

*where $Z$ is the fundamental solution matrix (f.s.m.) of the system* (1.1) [**25**], *and $v$ satisfies the following conditions*:

1) $L(t, x, D_t, D_x)v = 0$ *for* $t > \tau$;
2) $v|_{t=\tau} = 0$, *where at least one of the points $x$ and $\xi$ lies inside the domain $\Omega$*;
3)

$$B_j(t, x, D_x)v|_\Gamma = B_j(t, x, D_x)Z(t, \tau, x, \xi)|_\Gamma. \tag{5.1}$$

The following fundamental theorem then holds.

THEOREM 1.1) *Let $l$ be a nonnegative integer. When Conditions* A *and* $\mathrm{B}_l$ *are satisfied, there exists a Green matrix $G(t, x; \tau, \xi)$ of the problem* (1.1)-(3.1) *with derivatives of the form $D_t^{k_0}D_x^kG$, $2bk_0 + |k| \leq 2b + l$, for which the following estimates hold*:

$$|D_t^{k_0}D_x^kG(t, x; \tau, \xi)| \leqslant C(t-\tau)^{-\frac{n+2bk_0+|k|}{2b}} \exp\left\{-c\frac{|x-\xi|^q}{(t-\tau)^{1/(2b-1)}}\right\};$$

$$|D_t^{k_0}D_x^kG(t, x; \tau, \xi) - D_t^{k_0}D_{x_0}^kG(t, x_0; \tau, \xi)| \leqslant C|x - x_0|^\alpha$$
$$\times (t-\tau)^{-\frac{n+2b+l+\alpha}{2b}} \exp\left\{-c\frac{|x^*-\xi|^q}{(t-\tau)^{1/(2b-1)}}\right\};$$

$$(2bk_0 + |k| = 2b + l, \ |x^* - \xi| = \min(|x - \xi|, |x_0 - \xi|)); \tag{6.1}$$

$$|D_t^{k_0}D_x^kG(t, x; \tau\, \xi) - D_{t_0}^{k_0}D_x^kG(t_0, x; \tau, \xi)| \leqslant C(t - t_0)^{\frac{2b(1-k_0)+l-|k|+\alpha}{2b}}$$
$$\times (t_0 - \tau)^{-\frac{n+2b+l+\alpha}{2b}} \exp\left\{-c\frac{|x-\xi|^q}{(t-\tau)^{1/(2b-1)}}\right\};$$
$$(l < 2bk_0 + |k| \leqslant 2b + l, \ \tau < t_0 < t),$$

*where $q = 2b/(2b-1)$. The matrix $v(t, x; \tau, \xi)$ in* (4.1) *then satisfies estimates which are derived from the estimates* (6.1) *by substituting for $|x - \xi|$ and $|x^* - \xi|$ the expressions $|x - \xi| + d(\xi, S)$ and $|x^* - \xi| + d(\xi, S)$, where $d(\xi, S)$ is the distance of the point $\xi$ from the boundary $S$. The constants $C$ and $c$ in the inequalities* (6.1) *are functions of the constants $\delta$ and $\delta'$ in Condition* A, *the constant $M_l$ in Condition* $\mathrm{B}_l$ *and the numbers $n$, $N$, $b$, $r_j$, $l$, $\alpha$ and $T$.*

2) *If $\phi(x) \in C^0(\Omega)$ and $f(t, x) \in C_{x,t}^{\alpha,0}(Q)$, then the vector-function*

$$u(t_0, x) = \int_\Omega G(t, x; 0, \xi)\varphi(\xi)\,d\xi + \int_0^t d\tau \int_\Omega G(t, x; \tau, \xi)f(\tau, \xi)\,d\xi \tag{7.1}$$

*is a solution of the problem* (1.1)-(3.1).

We shall designate by the name *Green operator* the integral operator whose kernel is the Green matrix. We denote by $C_0^\alpha(Q)$ the class of vector functions $f(t,x)$ from the space $C_{x,t}^{\alpha,\alpha/2b}(Q)$ which satisfy the condition $f(t,x)|_{t=0,x\in S}=0$. The following theorem then holds.

THEOREM 2. *If* $\mathscr{R}(\mathscr{L})$ *coincides with* $C_0^\alpha(Q)$ *or if* $\mathscr{R}(\mathscr{L})\subset L_2(Q)$, *then the operator* $\mathscr{L}^{-1}$ *is a Green operator, i.e. the following relation holds for all* $f\in\mathscr{R}(\mathscr{L})$:

$$\mathscr{L}^{-1}f=\int_0^t d\tau\int_\Omega G(t,x;\tau,\xi)f(\tau,\xi)\,d\xi. \tag{8.1}$$

The proof of Theorem 1 will begin in the following subsection, and will be concluded in §§3 and 4; that of Theorem 2 will be given in §6.

**2. Beginning of the proof of Theorem 1.** Note that the proof of the second part of Theorem 1 follows directly from the definition of the Green matrix and from the estimates given in the first part. Therefore it suffices to prove the first part of the theorem only.

Since the f.s.m. $Z$ is sufficiently well known [**25**], [**26**], in view of (4.1) in order to obtain and investigate the Green matrix $G$ of the problem (1.1)–(3.1) it suffices to construct and investigate the matrix $v(t,x;\tau,\xi)$ which is a solution of the following problem:

$$\begin{gathered}L(t,x,D_t,D_x)v=0,\quad v|_{t=\tau}=0,\\ B_j(t,x,D_x)v|_{\Gamma_h}=B_j(t,x,D_x)Z(t,\tau,x,\xi)|_{\Gamma_h}\quad(j=1,\dots,bN).\end{gathered} \tag{9.1}$$

We consider the point $(\tau,\xi)$ as being arbitrarily fixed inside the domain $Q$, and we seek the solution of the problem (9.1) in the cylinder $Q_h=[\tau,\tau+h]\times\Omega$, $\Gamma_h=[\tau,\tau+h]\times S$, $h\leqq T-\tau$. When Conditions A and $B_l$ are satisfied, it follows from the results of [**4**] that there exists a unique solution of the problem (9.1), belonging, as a function of $(t,x)$ for fixed $(\tau,\xi)$, to the space

$$C_{x,t}^{2b+l+\alpha,(2b+l+\alpha)/2b}(Q_h).$$

In general, the norm of the solution $v$ increases then without bounds as the point $\xi$ approaches the boundary $S$ of the domain $\Omega$. The proof of Theorem 1 is thus reduced to obtaining exact pointwise estimates of the matrix $v$ and its derivatives.

The method of the proof consists in a direct solution of the problem (9.1) by means of the regularizer $R$: $v$ is sought in the form

$$v(t,x;\tau,\xi)=Rw,\quad w=(w_0,w_1,\dots,w_{bN}) \tag{10.1}$$

and for $w(t,x;\tau,\xi)$ we obtain the equation

$$w = Pw + g, \; g = (0, B_1 Z|_{\Gamma_h}, \dots, B_{bN} Z|_{\Gamma_h}), \tag{11.1}$$

where $P$ is a bounded operator in the space

$$C_{x,t}^{l+\alpha,(l+\alpha)/2b}(Q_h) \prod_{j=1}^{bN} C_{x,t}^{2b-r_j+l+\alpha,(2b-r_j+l+\alpha)/2b}(\Gamma_h)$$

with norm smaller than 1 if the cylinder $Q_h$ in which the solution is sought is short. The basic difficulty is how to preserve the required exact estimates (in particular, the exponentially decreasing factor in the estimates) in the course of the iteration process, in which all estimates are of the same order of singularity. The transition to a cylinder of arbitrary fixed height does not present special new difficulties.

Recall the definition of the regularizer of the problem (9.1). Let $\lambda > 0$ be a small number and let $\{\omega^{(k)}\}$ and $\{\Omega^{(k)}\}$ be two systems of sets which cover $\Omega$ and have the properties listed in [**4**]. The sets $\omega^{(k)}$ and $\Omega^{(k)}$ can be either interior or adjacent to the boundary. The interior sets are $n$-dimensional cubes with edges parallel to the coordinate axes, $\omega^{(k)}$ and $\Omega^{(k)}$ have the common center $\xi^{(k)}$, and the lengths of their edges are $\lambda/2$ and $\lambda$, respectively. When $\omega^{(k)}$ and $\Omega^{(k)}$ are near the boundary, they can be defined in a local system of coordinates $\hat{x}_1, \cdots, \hat{x}_n$ with center at the point $\xi^{(k)} \in S$ (i.e. in a rectangular system whose $\hat{x}_n$ axis is directed along the inner normal to $S$ at the point $\xi^{(k)}$, and whose other axes lie in the tangent plane) by the inequalities

$$-\frac{1}{2} a\lambda \leqslant \hat{x}_i \leqslant \frac{1}{2} a\lambda \;\; (i = 1, \dots, n-1), \;\; 0 \leqslant \hat{x}_n - F(\hat{x}') \leqslant a\lambda$$

and

$$-a\lambda \leqslant \hat{x}_i \leqslant a\lambda \;\; (i = 1, \dots, n-1), \;\; 0 \leqslant \hat{x}_n - F(\hat{x}') \leqslant 2a\lambda,$$

where $\hat{x}' = (\hat{x}_1, \cdots, \hat{x}_{n-1})$, $F(\hat{x}')$ is a function which defines the surface $S$ in the neighborhood of the point $\xi^{(k)}$, and $a$ is a positive constant, independent of $\lambda$. It should be noted that the sets $\Omega^{(k)}$ have the important property that the multiplicity of their covering of the domain $\Omega$ is finite and independent of $\lambda$. We shall denote by $k'$ the number of interior $\Omega^{(k)}$, and by $k''$ the number of those adjoining the boundary.

Consider the change of variables

$$\bar{x}_i = \hat{x}_i \;\; (i < n), \;\; \bar{x}_n = \hat{x}_n - F(\hat{x}'). \tag{12.1}$$

This transforms the set $\Omega^{(k'')}$ into the standard cube $K = \{ |\hat{x}_i| \leqq a\lambda \;\; (i < n), 0 \leqq \bar{x}_n \leqq 2a\lambda \}$, whose edge $\{ |\bar{x}_i| \leqq a\lambda, \bar{x}_n = 0 \}$ we denote by $K'$.

Moreover, let $\zeta^{(k)}(x)$ and $\eta^{(k)}(x)$ be smooth functions with the properties

$$0 \leqslant \zeta^{(k)}(x) \leqslant 1, \ \zeta^{(k)}(x) = \begin{cases} 1, & x \in \omega^{(k)}, \\ 0, & x \in \Omega \setminus \Omega^{(k)}, \end{cases}$$

(13.1)
$$\eta^{(k)} = 0 \text{ for } x \in \Omega \setminus \Omega^{(k)}, \ \sum_k \eta^{(k)}(x)\zeta^{(k)}(x) \equiv 1,$$

$$|D_x^m \zeta^{(k)}| \leqslant \frac{C_m}{\lambda^{|m|}}, \ |D_x^m \eta^{(k)}| \leqslant \frac{C_m}{\lambda^{|m|}}.$$

Everywhere in the following $\hat{x} = (\hat{x}_1, \cdots, \hat{x}_n)$ shall denote the coordinates of the point $x$ in a local system of coordinates with center at the point $\xi^{(k'')} \in S$ ($k''$-local coordinates), and $\bar{x} = (\bar{x}_1, \cdots, \bar{x}_n)$ shall denote $k''$-rectifying coordinates of the point $x$, i.e. coordinates related to the $k''$-local ones by the formulas (12.1). We shall denote by $\Pi_x^{\hat{x}}$ and $\Pi_x^{\bar{x}}$ the operators transforming from $x$ to $\hat{x}$ and from $x$ to $\bar{x}$, respectively, and we shall denote by $\Pi_{\hat{x}}^x$ and $\Pi_{\bar{x}}^x$ the inverse operators. We introduce the notation

$$\bar{f}(\bar{x}) = \Pi_x^{\bar{x}} f(x), \ \hat{f}(\hat{x}) = \Pi_x^{\hat{x}} f(x), \ \bar{f}(\bar{x}, \hat{\xi}) = \Pi_{x,\xi}^{\bar{x},\hat{\xi}} f(x, \xi),$$
$$\hat{f}(\hat{x}, \hat{\xi}) = \Pi_{x,\xi}^{\hat{x},\hat{\xi}} f(x, \xi), \ \Pi_{x,\xi}^{\bar{x},\hat{\xi}} = \Pi_x^{\bar{x}} \Pi_\xi^{\hat{\xi}}.$$

Clearly $\bar{f}(\bar{x}) = \hat{f}(\bar{x}', \bar{x}_n + F(\bar{x}'))$, where $\bar{x}' = (\bar{x}_1, \cdots, \bar{x}_{n-1})$.

The regularizer of the problem (9.1) is defined by the relations

$$v(t, x) = R(f, g_1, \dots, g_{bN}) \equiv \sum_k \eta^{(k)}(x) v^{(k)}(t, x),$$

$$v^{(k')}(t, x) = R^{(k')} f \equiv \int_\tau^t d\beta \int_{\Omega^{(k')}} Z_0^{(k')}(t - \beta, x - y) \zeta^{(k')}(y) f(\beta, y)\, dy,$$

(14.1)
$$v^{(k'')}(t, x) = \Pi_{\bar{x}}^x \bar{v}^{(k'')}(t, \bar{x}), \ \bar{v}^{(k'')}(t, \bar{x}) = \bar{R}^{(k'')} f + \sum_{j=1}^{bN} \bar{R}_j^{(k'')} g_j,$$

$$\bar{R}^{(k'')} f \equiv \int_\tau^t d\beta \int_K G_0^{(k'')}(t - \beta, \bar{x}, \bar{y}) \bar{\zeta}^{(k'')}(\bar{y}) \bar{f}(\beta, \bar{y})\, d\bar{y},$$

$$\bar{R}_j^{(k'')} g_j \equiv \int_\tau^t d\beta \int_{K'} G_j^{(k'')}(t - \beta, \bar{x} - \bar{y}') \bar{\zeta}^{(k'')}(\bar{y}') \bar{g}_j(\beta, \bar{y}')\, d\bar{y}',$$

where $Z_0^{(k')}(t - \beta, x - y)$ is the f.s.m. of the system

(15.1)
$$D_t u - A_0(\tau, \xi^{(k')}, D_x) u = 0,$$

$G_0^{(k'')}(t - \beta, \bar{x}, \bar{y})$ is the Green matrix [2], and $G_j^{(k'')}(t - \beta, \bar{x} - \bar{y}')$ are the Poisson kernels [4], [25] of the problem

(16.1) $$D_t\bar{u}(t,\bar{x}) - A_0^{(k'')}(\tau,\xi^{(k'')},D_{\bar{x}})\bar{u}(t,\bar{x}) = \bar{f}(t,\bar{x}),$$

(17.1) $$\bar{u}(t,\bar{x})|_{t=\tau} = 0,$$

(18.1) $$B_{j0}^{(k'')}(\tau,\xi^{(k'')},D_{\bar{x}})\bar{u}(t,\bar{x})|_{\bar{x}_n=0} = \bar{g}_j(t,\bar{x}') \quad (j=1,\ldots,bN),$$

where $A_0^{(k'')}(\tau,\xi^{(k'')},D_{\bar{x}})$ and $B_{j0}^{(k'')}(\tau,\xi^{(k'')},D_{\bar{x}})$ are operators obtained from $A_0(\tau,\xi^{(k'')},D_x)$ and $B_{j0}(\tau,\xi^{(k'')},D_x)$ by introducing the local coordinates $\hat{x}$ with center at the point $\xi^{(k'')}\in S$ and subsequent automatic substitution of $D_{\bar{x}}$ for $D_{\hat{x}}$.

We shall now write out the equation (11.1) for $w$ in more detail. We have (cf. [**4**])

(11.1) $$\begin{gathered} w_0(t,x;\tau,\xi) = P_0 w, \\ w_j(t,x;\tau,\xi) = P_j w + B_j(t,x,D_x)Z(t,\tau,x,\xi)|_{\Gamma_h} \qquad (j=1,\ldots,bN), \end{gathered}$$

where

$$\begin{aligned} P_0 w = {} & A_1(t,x,D_x)\sum_k \eta^{(k)}(x)W_k(t,x;\tau,\xi) \\ & + \sum_{k''}\eta^{(k'')}(x)\Pi_{\bar{x}\hat{\xi}}^{x,\xi}[A_0(\tau,\xi^{(k'')},D_{\bar{x}} - \operatorname{grad} F\cdot D_{\bar{x}_n}) - A_0^{(k'')}(\tau,\xi^{(k'')},D_{\bar{x}})] \\ & \times \overline{W}_{k''}(t,\bar{x};\tau,\hat{\xi}) + \sum_k \eta^{(k)}(x)[A_0(t,x,D_x) - A_0(\tau,\xi^{(k)},D_x)]W_k(t,x;\tau,\xi) \\ & + \sum_k [A_0(t,x,D_x)\eta^{(k)}(x) - \eta^{(k)}(x)A_0(t,x,D_x)]W_k(t,x;\tau,\xi), \end{aligned}$$

(19.1) $$\begin{aligned} P_j w = {} & -B_{j1}(t,x,D_x)\sum_k \eta^{(k)}(x)W_k(t,x;\tau,\xi)|_{\Gamma_h} \\ & + \sum_{k''}\eta^{(k'')}(x)[B_{j0}(\tau,\xi^{(k'')},D_x) - B_{j0}(t,x,D_x)]W_{k''}(t,x;\tau,\xi)|_{\Gamma_h} \\ & + \sum_{k''}\eta^{(k'')}(x)\Pi_{\bar{x},\hat{\xi}}^{x,\xi}[B_{j0}^{(k'')}(\tau,\xi^{(k'')},D_{\bar{x}}) - B_{j0}^{(k'')}(\tau,\xi^{(k'')},D_{\bar{x}} - \operatorname{grad} F\cdot D_{\bar{x}_n})] \\ & \times \overline{W}_{k''}(t,\bar{x};\tau,\hat{\xi})|_{\bar{x}_n=0} + \sum_{k''}[\eta^{(k'')}(x)B_{j0}(t,x,D_x) \\ & - B_{j0}(t,x,D_x)\eta^{(k'')}(x)]W_{k''}(t,x;\tau,\xi)|_{\Gamma_h}, \end{aligned}$$

$$W_{k'}(t,x;\tau,\xi) = R^{(k')}w_0, \quad W_{k''}(t,x,\tau,\xi) = \Pi_{\bar{x},\hat{\xi}}^{x,\xi}\overline{W}_{k''}(t,\bar{x};\tau,\hat{\xi}),$$

$$\overline{W}_{k''}(t,\bar{x};\tau,\hat{\xi}) = \overline{R}^{(k'')}w_0 + \sum_{j=1}^{bN}\overline{R}_j^{(k'')}w_j.$$

If the height $h$ of the cylinder $Q_h$ is such that

(20.1) $$h = \chi\lambda^{2b},$$

where $\chi \leqq 1$ and $\lambda$ are sufficiently small, then according to [4] the solution $w$ of the system (11.1) can be represented in the form of a series

(22.1) $$w = \sum_{\nu=0}^{\infty} w^{(\nu)},$$

in which

(22.2) $$w^{(\nu)} = Pw^{(\nu-1)}.$$

In §3 we shall obtain detailed estimates for the terms of this series, and subsequently for the sum.

NOTE. In equations (14.1), which define the regularizer, we used the f.s.m., the Green matrix and the Poisson kernels for the system (15.1) and the problem (16.1)–(18.1) with coefficients fixed at the point $(\tau, \xi^{(k')})$. If $t_1 \leqq \tau < t \leqq t_1 + h$, $0 \leqq t_1 \leqq T - h$, then all following considerations also hold in the case when the coefficients in (15.1), (16.1), and (18.1) are fixed at the point $(t_1, \xi^{(k)})$.

**3. Notation.** We set

1) $Q_{\tau,\tau_1} = (\tau, \tau_1] \times \Omega$, $Q^{(k)}_{\tau,\tau_1} = (\tau, \tau_1] \times \Omega^{(k)}$, $\Gamma_{\tau,\tau_1} = [\tau, \tau_1] \times S$.

2) $D^m_{t,x} = D_t^{m_0} D_x^{m'}$, $2bm_0 + |m'| = |m|$, $|m'| = m_1 + \ldots + m_n$.

3) $d(t, x; \beta, y) = (|t - \beta|^{1/b} + |x - y|^2)^{\frac{1}{2}}$.

4) $\gamma = \dfrac{1}{2b}$, $q = \dfrac{2b}{2b-1}$.

5) $d(\xi, S) = \inf\limits_{\eta \in S} d(\xi, \eta)$.

6) $\Psi_c(t, x; \tau, \xi) = \exp\left\{-c\left[\dfrac{|x-\xi|}{(t-\tau)^\gamma}\right]^q\right\}$.

7) $\rho_c(t, \tau, \xi) = \exp\left\{-c\left[\dfrac{d(\xi, S)}{(t-\tau)^\gamma}\right]^q\right\}$.

8) $\Phi_c(t, x; \tau, \xi) = \Psi_c(t, x; \tau, \xi)\rho_c(t, \tau, \xi)$.

9) $\Delta^{t_0,x_0}_{t,x} f(t, x) = f(t, x) - f(t_0, x_0)$; $\Delta^{t,x_0}_{t,x} = \Delta^{x_0}_x$; $\Delta^{t_0,x}_{t,x} = \Delta^{t_0}_t$.

10) $\Delta_l = \delta + \delta' + \dot{M}_l + n + N + b + l + \alpha + T$,

(23.1)

where $\delta$ and $\delta'$ are the constants in Condition A, $M_l$ and $\alpha$ are the constants in Condition $B_l$, $n$ is the dimension of $\Omega$, $N$ is the number of equations in the system (1.1), $b$ is half the order of the system, and $T$ is the height of the cylinder $Q$.

Note the following elementary inequalities, which we shall need in the following:

$$\Psi_c(t, x; \beta, y)\,\Psi_c(\beta, y; \tau, \xi) \leqslant \Psi_c(t, x; \tau, \xi) \tag{24.1}$$

for $\tau < \beta < t$; $x, y, \xi \in \Omega$; and

$$\begin{gathered} d(t, \bar{x}; \beta, \bar{y}) \geqslant \frac{1}{\sqrt{2}} d(t, \hat{x}; \beta, \hat{y}) = \frac{1}{\sqrt{2}} d(t, x; \beta, y), \\ \Psi_{c_0}(t, \bar{x}; \beta, \bar{y}) \leqslant \Psi_c(t, \hat{x}; \beta, \hat{y}) = \Psi_c(t, x; \beta, y), \end{gathered} \tag{25.1}$$

$$c = \left(\frac{1}{\sqrt{2}}\right)^q c_0 \tag{26.1}$$

for $\bar{x}, \bar{y}, \in K$ with sufficiently small $\lambda$.

We shall prove only inequality (25.1). We have

$$\begin{aligned} |\bar{x} - \bar{y}|^2 &\geqslant |\bar{x}' - \bar{y}'|^2 + \frac{1}{2}|\bar{x}_n + F(\bar{x}') - \bar{y}_n - F(\bar{y}')|^2 - |F(\bar{y}') - F(\bar{x}')|^2 \\ &\geqslant |\bar{x}' - \bar{y}'|^2 + \frac{1}{2}|\bar{x}_n + F(\bar{x}') - \bar{y}_n - F(\bar{y}')|^2 \\ &\quad - C|\operatorname{grad} F(\bar{y}' + \theta(\bar{x}' - \bar{y}'))|^2 |\bar{x}' - \bar{y}'|^2, \end{aligned}$$

but, in view of the properties of the boundary $S$, $|\operatorname{grad} F(\bar{y})| \leqq a_1 |\bar{y}^1|^\alpha$, where $a_1$ is a constant, fixed on the given surface, independent of the point on $S$ which is chosen as the origin of the local system of coordinates. Therefore

$$|\bar{x} - \bar{y}|^2 \geqslant (1 - C_0\lambda^{2\alpha})|\bar{x}' - \bar{y}'|^2 + \frac{1}{2}|\bar{x}_n + F(\bar{x}') - \bar{y}_n - F(\bar{y}')|^2.$$

Choosing $\lambda$ such that $1 - C_0\lambda^{2\alpha} \geqq \frac{1}{2}$, we obtain

$$\begin{aligned} |\bar{x} - \bar{y}|^2 &\geqslant \frac{1}{2}(|\bar{x}' - \bar{y}'|^2 + |\bar{x}_n + F(\bar{x}') - \bar{y}_n - F(\bar{y}')|^2) \\ &= \frac{1}{2}|\hat{x} - \hat{y}|^2 = \frac{1}{2}|x - y|^2, \end{aligned}$$

which leads to (25.1).

**4. Classes of functions.** In the following we consider the classes of functions (vector functions, matrix functions) of the type of the f.s.m., the halfspace Green matrix, and the Poisson kernels, their derivatives, and matrices which have singularities only in the case when the parametric point falls on the boundary of the domain. In the following $l$, $p$, $r$ and $s$ are nonnegative numbers, and $l$ is an integer.

1) The class $U^{l+\alpha,r,s}_{C,c}(Q_{\tau,\tau_1}, Q)$ consists of matrix functions $u(t,x;\tau,\xi)$ defined in $Q_{\tau,\tau_1}\times Q$ which are continuous and have derivatives $D^m_{t,x}u$ ($|m|\leq l$) satisfying the following estimates:

$$(27.1)\quad |D^m_{t,x}u| \leqslant C(t-\tau)^{\nu s} d^{-n-r-|m|-s}(t,x;\tau,\xi)\,\Psi_c(t,x;\tau,\xi) \quad (|m|\leqslant l),$$

$$(28.1)\quad |\Delta^{x_0}_x D^m_{t,x}u| \leqslant C|x-x_0|^\alpha (t-\tau)^{\nu s} d^{-n-r-|m|-s-\alpha}(t,x^*;\tau,\xi)\Psi_c(t,x^*;\tau;\xi)$$
$$(|m|\leqslant l,\ |x^*-\xi| = \min(|x-\xi|,\ |x_0-\xi|)),$$

$$|\Delta^{t_0}_t D^m_{t,x}u| \leqslant C|(t-t_0)^{\nu(l-|m|+\alpha)}(t-\tau)^{\nu s} d^{-n-r-|m|-s-\alpha}(t_0,x;\tau,\xi)\,\Psi_c(t,x;\tau,\xi)$$
$$(29.1)\quad (l-2b<|m|\leqslant l,\ \tau<t_0<t).$$

2) The classes $H^{l+\alpha}_{p,c}(\overline{Q}_{\tau,\tau_1}, Q)$ and $H^{l+\alpha}_{p,c}(\Gamma_{\tau,\tau_1}, Q)$ consist of matrices $h(t,x;\tau,\xi)$ defined in $\overline{Q}_{\tau,\tau_1}\times Q$ and in $\Gamma_{\tau,\tau_1}\times Q$, respectively, which are continuous and have derivatives $D^m_{t,x}h$ ($|m|\leq l$) satisfying the inequalities:

$$|D^m_{t,x}h| \leqslant C(t-\tau)^{\nu(l-|m|+\alpha)} d^{-n-p-l-\alpha}(t,x;\tau,\xi)\,\Phi_c(t,x;\tau,\xi) \quad (|m|\leqslant l),$$
$$(30.1)$$

$$|\Delta^{x_0}_x D^m_{t,x}h| \leqslant C|x-x_0|^\alpha (t-\tau)^{\nu(l-|m|)} d^{-n-p-l-\alpha}(t,x^*;\tau,\xi)\,\Phi_c(t,x^*;\tau,\xi)$$
$$(31.1)\quad (|m|\leqslant l),$$

$$|\Delta^{t_0}_t D^m_{t,x}h| \leqslant C(t-t_0)^{\nu(l-|m|+\alpha)} d^{-n-p-l-\alpha}(t_0,x;\tau,\xi)\,\Phi_c(t,x;\tau,\xi)$$
$$(32.1)\quad (l-2b<|m|\leqslant l),$$

where $D^m_{t,x}h$ denotes, for $x\in S$, the derivative of order $|m|$ with respect to $t$, $\bar{x}'$. Here $\bar{x}'=(\bar{x}_1,\cdots,\bar{x}_{n-1})$ are the $k''$-rectifying coordinates of the point $x$ on $S$. Moreover, in the case of the classes $H^{l+\alpha}_{2b,c}(\overline{Q}_{\tau,\tau_1}, Q)$ and $H^{l+\alpha}_{2b-1,c}(\Gamma_{\tau,\tau_1}, Q)$ we assume that the matrices $h$ satisfy, respectively, the condition

$$(33.1)\quad |I^{(k)}(h)| = \left|\iint_{Q_1^{(k)}} \zeta^{(k)}(y)\,h(\beta,y;\tau,\xi)\,d\beta dy\right| \leqslant C\rho_c(t,\tau,\xi)$$

for all $k$, where

$$Q_1^{(k)} = \overline{Q}^{(k)}_{\tau,t} \cap \left\{d(\beta,y;\tau,\xi) \leqslant \frac{1}{2}(t-\tau)^\nu\right\},$$

and

$$(34.1) \qquad |J^{(k'')}(h)| = \Big| \iint\limits_{\mathcal{K}_1'} \bar{\zeta}^{(k'')}(\bar{y}')\, \bar{h}^{(k'')}(\beta, \bar{y}'; \tau, \hat{\xi})\, d\beta d\bar{y}' \Big| \leqslant C\rho_c(t, \tau, \xi)$$

for all $k''$, where

$$\bar{h}^{(k'')}(\beta, \bar{y}'; \tau, \hat{\xi}) = \Pi_{y,\xi}^{\bar{y},\hat{\xi}} h(t, x, \tau, \xi)$$

and

$$\mathcal{K}_1' = [\tau, t] \times K' \cap \{(\beta, \bar{y}'),\ d(\beta, \bar{y}', F(\bar{y}'); \tau, \hat{\xi}) \leqslant \frac{1}{2}(t-\tau)^\gamma\}.$$

3) The classes $\overset{\circ}{H}{}_c^{l+\alpha}(Q_{\tau_1,\tau_2}, Q)$ and $\overset{\circ}{H}{}_c^{l+\alpha}(\Gamma_{\tau_1,\tau_2}, Q)$ are classes of matrix functions $h(t, x; \tau, \xi)$, defined in $\overline{Q}_{\tau_1,\tau_2} \times Q$ and in $\Gamma_{\tau_1,\tau_2} \times Q$, $\tau_1 > \tau$, respectively, which are continuous and have derivatives $D_{l,x}^m h$ ($|m| \leqq l$) satisfying inequalities which differ from the inequalities (30.1)–(32.1) in that they do not involve powers of $d$, and the powers of $t - \tau$ are replaced by corresponding powers of $t - \tau_1$. Further, we shall denote by $H_c^{l+\alpha}(\overline{Q}_{\tau_1,\tau_2}, Q)$ and $H_c^{l+\alpha}(\Gamma_{\tau_1,\tau_2}, Q)$ the corresponding classes which do not involve even the powers of $t - \tau_1$.

The norm of the matrix $h$ in the classes 2 and 3 is defined as the greatest of the exact lower bounds of the constants $C$ for which all inequalities appearing in the definition of the class are satisfied, and we shall denote these norms by $\|h\|$ (in case of possible ambiguity, we shall add the appropriate subscript).

It should be noted that when $l' \leqq l$ and $\alpha' \leqq \alpha$ we have

$$H_{p,c}^{l+\alpha} \subset H_{p,c}^{l'+\alpha'}, \quad U_{C,c}^{l+\alpha,r,s} \subset U_{C',c}^{l'+\alpha',r,s}$$

where $C'$ is defined by the constants $C, l, l', \alpha, \alpha'$ and $\gamma$.

We shall now define some other classes of functions, which are independent of the choice of the parametric points.

4) The classes $\underset{\circ}{C}{}^{l+\alpha}(\overline{Q}_{\tau,\tau_1})$ and $\underset{\circ}{C}{}^{l+\alpha}(\Gamma_{\tau,\tau_1})$ shall denote the vector functions $f(t, x)$ which belong to the spaces $C_{x,t}^{l+\alpha,(l+\alpha)\gamma}(\overline{Q}_{\tau,\tau_1})$ and $C_{x,t}^{l+\alpha,(l+\alpha)\gamma}(\Gamma_{\tau,\tau_1})$, respectively, and satisfy the conditions

$$D_t^k f(t, x)\big|_{t=\tau} = 0 \quad (k = 0, 1, \ldots, [l\gamma]).$$

The set of the vector functions from $C_{x,t}^{l+\alpha,(l+\alpha)\gamma}(\overline{Q}_{\tau,\tau_1})$ which satisfy the conditions

$$D_t^k f(t, x)\big|_{t=\tau, x\in S} = 0 \quad (k = 0, 1, \ldots, [l\gamma]),$$

shall be called the class $C_0^{l+\alpha}(\overline{Q}_{\tau,\tau_1})$.

## §2. Lemmas concerning integral operators

Here we shall present some lemmas, fundamental for the subsequent derivations, concerning integral operators with kernels which are f.s.m., halfspace Green matrices, and Poisson kernels, multiplied by cutoff functions, described in §1. These operators appear in the definition of the regularizer (14.1). The proof of Lemmas 1 and 2 which we shall now formulate is quite tedious and will be given in the Appendix.

Let $Z_0(t-\beta, x-y)$ be the f.s.m. of the system (15.1), let $G_0(t-\beta, \bar{x}, \bar{y})$ be the Green matrix and let $G_j(t-\beta, \bar{x}-\bar{y}')$ be the Poisson kernels of the problem (16.1)-(18.1). We consider the operators

$$R^{(k)}h = \int_\tau^t d\beta \int_{\Omega^{(k)}} R^{(k)}(t, x; \beta, y)\zeta(y)h(\beta, y; \tau, \xi)\,dy, \tag{1.2}$$

$$R_j h = \Pi^{x,\xi}_{\bar{x},\bar{\xi}} \bar{R}_j h$$

$$= \Pi^{x,\xi}_{\bar{x},\bar{\xi}} \int_\tau^t d\beta \int_{K'} G_j(t-\beta, \bar{x}-\bar{y}')\bar{\zeta}(\bar{y}')\bar{h}(\beta, \bar{y}'; \tau, \hat{\xi})\,d\bar{y}' \tag{2.2}$$

$$(j = 1, \dots, bN),$$

where

$$R^{(k)}(t, x; \beta, y) = \begin{cases} Z_0(t-\beta, x-y) & \text{for } k = k', \\ \Pi^{x,y}_{\bar{x},\bar{y}} G_0(t-\beta, \bar{x}, \bar{y}) & \text{for } k = k'', \end{cases} \tag{3.2}$$

and $\zeta(y)$ is a cutoff function of the type of the function $\zeta^{(k)}$ defined in §1, with support in $\Omega^{(k)}$.

From known estimates for $Z_0$, $G_0$ and $G_j$ (cf. [**25**], [**2**] and [**4**]) one can easily obtain that

$$\begin{gathered} Z_0(t-\beta, x-y) \in U^{l_0+\alpha_0,0,s}_{C_0,c_0}((\beta, T]\times E_n, [0, T)\times E_n), \\ G_0(t-\beta, \bar{x}, \bar{y}) \in U^{l_0+\alpha_0,0,s}_{C_0,c_0}((\beta, T]\times E_n^+, [0, T)\times E_n^+), \\ G_j(t-\beta, \bar{x}-\bar{y}') \in U^{l_0+\alpha_0,2b-r_j-1,s}_{C_0,c_0}((\beta, T]\times E_n^+, [0, T)\times E_{n-1}) \end{gathered} \tag{4.2}$$

for arbitrary $l_0 \geqq 0$ (integer), $0 \leqq \alpha_0 \leqq 1$, $s \geqq 0$ and $T > 0$ where $E_n^+ = \{\bar{x}_n \geqq 0\}$. It should be noted that the constants $C_0$ and $c_0$ depend on $l_0$, $\alpha_0$ and $s$, but in the following we shall use (4.2) for a finite number of values of $l_0$, $\alpha_0$ and $s$. Therefore by $C_0$ and $c_0$ we shall mean constants for which (4.2) hold for all these values of $l_0, \alpha_0$ and $s$.

The following lemmas are valid.

LEMMA 1. *Assume that*

$$S \in C^{2b+l+\alpha}, \quad \tau_1 - \tau = \chi\lambda^{2b}, \tag{5.2}$$

*where* $\chi \leqq 1$ *and* $\lambda$ *is sufficiently small so that* (25.1) *is satisfied. If* $h(t,x;\tau,\xi) \in H^{l+\alpha}_{2b,c_1}(\overline{Q}_{\tau_1,\tau}, Q)$, *then*

$$u_k(t, x; \tau, \xi) = R^{(k)}h \in H^{2b+l+\alpha}_{0,c}(\overline{Q}^{(k)}_{\tau,\tau_1}, Q) \text{ and } \|R^{(k)}h\| \leqslant C\|h\|.$$

*Here* $c = \min(c_1, 2^{-q/2}c_0)$, *where* $c_0$ *is given in* (4.2). *The constant* $C$ *depends only on* $\Delta_l$ (23.1).

LEMMA 2. *If* $h(t,x;\tau,\xi) \in H^{l+\alpha}_{r_j,c_1}(\Gamma_{\tau,\tau_1}, Q)$, *then when condition* (5.2) *is satisfied,*

$$v_j(t, x; \tau, \xi) = R_j h \in H^{l+r_j+\alpha}_{0,c}(\overline{Q}^{(k'')}_{\tau,\tau_1}, Q) \text{ and } \|R_j h\| \leqslant C\|h\|,$$

*where* $c$ *and* $C$ *are the same as in Lemma* 1.

LEMMA 3. *Let* $S$ *satisfy the condition in* (5.2) *and let* $\tau_2 - \tau_1 = \chi\lambda^{2b}$, *where* $\lambda$ *is the same as in Lemma* 1. *Then the operators* $R^{(k)}h$ *and* $R_j h$, *defined by equations* (1.2) *and* (2.2) *with the integration with respect to* $\beta$ *over* $(\tau_1, \tau)$ *instead of* $(\tau, t)$, *are bounded from* $\mathring{H}^{l+\alpha}_{c_1}(\overline{Q}_{\tau_1,\tau_2}, Q)$ *and* $\mathring{H}^{l+\alpha}_{c_1}(\Gamma_{\tau_1,\tau_2}, Q)$ *into* $\mathring{H}^{2b+l+\alpha}_{c}(\overline{Q}^{(k)}_{\tau_1,\tau_2}, Q)$ *and* $\mathring{H}^{l+r_j+\alpha}_{c}(\overline{Q}^{(k'')}_{\tau_1,\tau_2}, Q)$, *respectively, where* $c$ *is the same as in Lemma* 1.

The proof of Lemma 3 is similar to that of Lemmas 1 and 2, but is much simpler, since in this case the functions $h$ have no singularities.

We shall now state a lemma concerning the operation of the operators $R^{(k)}$ and $R_j$ on classes of functions which do not depend on a parametric point.

LEMMA 4. *When the conditions* (5.2) *are satisfied, the operators* $R^{(k)}$ *and* $R_j$ *are bounded from* $\mathring{C}^{l+\alpha}(\overline{Q}_{\tau,\tau_1})$ *and* $\mathring{C}^{l+\alpha}(\Gamma_{\tau,\tau_1})$ *into* $\mathring{C}^{2b+l+\alpha}(\overline{Q}^{(k)}_{\tau,\tau_1})$ *and* $\mathring{C}^{r_j+l+\alpha}(\overline{Q}^{(k'')}_{\tau,\tau_1})$, *respectively.*

The proof of this lemma is analogous to that of the preceding lemmas.

## §3. Estimates of the Green matrix and its derivatives in a cylinder of short height

In this section we shall derive estimates for the Green matrix in a cylinder $Q_h$ of short height, making systematic use of Lemmas 1 and 2 and of the properties of the f.s.m., the halfspace Green matrix, and the Poisson kernels.

**1. Estimates of the solution of system (11.1).** As we noted in §1, the construction of the Green matrix in the cylinder $Q_h$ is reduced to the solution of the system (11.1). For sufficiently small $h$ the solution of this system is given by the series (21.1). Here we shall estimate the terms of the series (21.1), and shall use these to find the properties of the sum.

In the following we shall make a detailed investigation of the first terms of the series (21.1) $w^{(0)} = (0, \beta_j Z|_{\Gamma_h})$, and we shall establish that $w^{(0)}_j$ belongs to the class $H^{2b-r_j+l+\alpha}_{r_j,c_2}(\Gamma_h, Q)$, where the estimates (30.1)–(32.1) directly follow from well-known properties of the f.s.m., with only the condition (34.1) requiring verification. Subsequently, using Lemmas 1 and 2, we estimate the second term of the series (21.1), and again an essential step is the proof of the validity of inequalities (33.1) and (34.1). In the investigation of $w^{(1)}$ we use only the properties of operators which represent $w^{(1)}$ in terms of $w^{(0)}$. Therefore all considerations can be extended to the investigation of $w^{(\nu)}$ when $w^{(\nu-1)}$ is already known.

In the following we denote by $C$ and $c$ constants which can depend only on $\Delta_l$ (23.1). When their magnitudes are significant, we denote such constants by $C_i$ and $c_i$ $(i=1,2,\cdots)$.

It should be noted that under the assumptions of Theorem 1, and on the basis of the results of [**26**], we have for the f.s.m. $Z$ of the system (1.1)

$$Z(t, \tau, x, \xi) \in U^{2b+l+\alpha,0,s}_{C_1,c_1}(Q_{\tau,T}, Q) \tag{1.3}$$

with arbitrary $s \geqq 0$. In the following we shall use (1.3) with a finite number of values of $s$, depending on $b, r_j$, $l$ and $\alpha$.

We shall estimate $w^{(\nu)}$ $(\nu=1,2,\cdots)$, defined by (22.1), taking $w_0^{(0)}=0$ and $w_j^{(0)}=\beta_j Z\,|_{\Gamma_h}$ $(j=1,2,\cdots,bN)$. We assume that the height $h$ of the cylinder $Q_h$ satisfies condition (20.1), where for the time being $\lambda$ is the same as in §2.

From Condition $\mathrm{B}_l$ and (1.3) we obtain that

$$w_j^{(0)} \in H^{2b-r_j+l+\alpha}_{r_j,c_2}(\Gamma_h, Q), \ \|w_j^{(0)}\| = C_2 \ (j=1, \dots, bN), \tag{2.3}$$

where $c_2 = \frac{1}{2}c_1$. A specific proof is required only for the inequality (34.1) for $w_j^{(0)}$ in the case $r_j = 2b-1$, i.e. the inequality

$$|J^{(k'')}(w_j^{(0)})| \leqslant C_2\rho_{c_2}(t, \tau, \xi). \tag{3.3}$$

It is known [**25**] that $Z(t,\tau,x,\xi) = Z_0(t-\tau, x-\xi; \tau,\xi) + W(t,\tau,x,\xi)$, where the singularity of $W$ is less than that of $Z_0$. Therefore

$$w_j^{(0)} = B_{j0}Z_0\,|_{\Gamma_h} + (B_{j1}Z_0 + B_jW)\,|_{\Gamma_h} = w_{j1}^{(0)} + w_{j2}^{(0)}, \ w_{j2}^{(0)} \in H^{l+1+\alpha}_{2b-1-\alpha,c_2}(\Gamma_h, Q)$$

and

$$|J^{(k'')}(w_{j2}^{(0)})| \leqslant C \iint_{\mathcal{K}_1'} d^{-n-2b+1+\alpha}(\beta, y; \tau, \xi)\,\Phi_{c_2}(\beta, y; \tau, \xi)\, d\beta\, d\overline{y'} \leqslant C_2\rho_{c_2}(t, \tau, \xi).$$

Let us estimate $J^{(k'')}(w_{j1}^{(0)})$. We use the representation

$$\bar{w}_{j1}^{(0)}(\beta, \bar{y}'; \tau, \hat{\xi}) = \Pi_{y,\xi}^{\bar{y},\hat{\xi}}[(B_{j0}Z_0)|_{\Gamma_h}] = \sum_{|\nu|\leqslant 2b-1} c_{j\nu}(\beta, \bar{y}')D_{\bar{y}}^{\nu}\bar{Z}_0(\beta, \bar{y}; \tau, \hat{\xi})|_{\bar{y}_n=0}$$

$$(4.3)\quad = \sum_{|\nu|\leqslant 2b-1} [c_{j\nu}(\beta, \bar{y}') - c_{j\nu}(\tau, \bar{\xi}')] D_{\bar{y}}^{\nu}\hat{Z}_0(\beta - \tau, \bar{y}' - \bar{\xi}', \bar{y}_n + F(\bar{y}') - \hat{\xi}_n; \tau, \hat{\xi})|_{\bar{y}_n=0}$$

$$+ \sum_{|\nu|\leqslant 2b-1} c_{j\nu}(\tau, \bar{\xi}')D_{\bar{y}}^{\nu}\hat{Z}_0(\beta - \tau, \bar{y}' - \bar{\xi}', \bar{y}_n + F(\bar{y}') - \hat{\xi}_n; \tau, \hat{\xi})|_{\bar{y}_n=0},$$

where the $c_{j\nu}$ depend on the coefficients of the operator $B_{j0}$ and on the derivatives of $F$ up to the order $2b-1$. In view of (4.3) it suffices to estimate $J^{(k'')}(D_{\bar{y}}^{\nu}\hat{Z}_0|_{\bar{y}_n=0})$ for $|\nu| = 2b-1$, as the integrals $J^{(k'')}$ of the other components of (4.3) can be estimated in the same way as $J^{(k'')}(w_{j2}^{(0)})$, if we use the properties of the coefficients of $B_{j0}$ and the function $F$. In the case $|\nu| = 2b-1$ and $\nu_n < 2b-1$, one can drop in the integral $J^{(k'')}(D_{\bar{y}}^{\nu}\hat{Z}_0|_{\bar{y}_n=0})$ one derivative of $\hat{Z}_0$, and the result can be estimated in a manner analogous to that used in the estimate of the corresponding integrals in the proof of Lemma 2 (cf. the Appendix). Now let $\nu_n = 2b-1$, and let us write

$$J^{(k'')}(D_{\bar{y}_n}^{2b-1}\hat{Z}_0|_{\bar{y}_n=0}) = \iint_{\mathcal{K}_1'} \bar{\zeta}^{(k'')}(\bar{y}') \int_{-\infty}^{0} D_{\bar{y}_n}^{2b}\hat{Z}_0(\beta - \tau, \bar{y}' - \bar{\xi}', \bar{y}_n$$

$$+ F(\bar{y}') - \hat{\xi}_n; \tau, \hat{\xi})\, d\beta d\bar{y} = \iint_{\mathcal{K}_1^{(1)}} \bar{\zeta}^{(k'')}D_{\bar{y}_n}^{2b}\hat{Z}_0 d\beta d\bar{y} + \iint_{\mathcal{K}_1^{(2)}} \bar{\zeta}^{(k'')}D_{\bar{y}_n}^{2b}\hat{Z}_0 d\beta d\bar{y} = J_1 + J_2,$$

where

$$\mathcal{K}_1^{(1)} = \{(\beta, \bar{y}), (\beta, \bar{y}') \in \mathcal{K}_1', \ -\infty < \bar{y}_n \leqslant 0\} \cap \left\{d(\beta, y; \tau, \xi) \geqslant \frac{1}{2}(t-\tau)^{\gamma}\right\},$$

$$\mathcal{K}_1^{(2)} = \{(\beta, \bar{y}), (\beta, \bar{y}') \in \mathcal{K}_1', \ -\infty < y_n \leqslant 0\} \setminus \mathcal{K}_1^{(1)}.$$

Using the estimates of $Z_0$ and the fact that the point with local coordinates $(\bar{y}', \bar{y}_n + F(\bar{y}'))$, where $\bar{y}' \in K'$, $-\infty < y_n < 0$, lies outside the domain $\Omega$, we have

$$|J_1| \leqslant C \iint_{\mathcal{K}_1^{(1)}} (\beta - \tau)^{\gamma} d^{-n-2b-1}(\beta, y; \tau, \xi)\Psi_{c_1}(\beta, \bar{y}', \bar{y}_n + F(\bar{y}'); \tau, \bar{\xi}', \hat{\xi}_n)\, d\beta d\bar{y}$$

$$\leqslant C\rho_{c_2}(t, \tau, \xi)(t-\tau)^{\gamma} \iint_{\frac{1}{2}(t-\tau)^{\gamma} \leqslant |z|} |z|^{-n-2} dz = C\rho_{c_2}(t, \tau, \xi).$$

To estimate $J_2$ we note that $\hat{Z}_0(\beta-\tau,\bar{y}'-\bar{\xi}',\bar{y}_n+F(\bar{y}')-\hat{\xi}_n;\tau,\xi)$, as a function of $\beta$ and $\bar{y}$, is a solution of a parabolic system (for small $\lambda$), so that the derivative $D_{\bar{y}_n}^{2b}\hat{Z}_0$ can be represented as a linear function of the derivatives $D_\beta\hat{Z}_0$ and $D_{\bar{y}}^m\hat{Z}_0$ ($|m|\leqq 2b, m_n<2b$). In the case of integrals $J_2$ which contain, instead of $D_{\bar{y}_n}^{2b}\hat{Z}_0$, the lowest derivatives, one can easily obtain the estimates by integrating by parts (cf. the proof of Lemma 1 in the Appendix).

We shall now estimate $w_j^{(1)}$ ($j=0,\cdots,bN$). Using (22.1), (19.1), (13.1), Condition $B_l$, (2.3), and Lemma 2, we have

$$\begin{aligned}|w_0^{(1)}| &= |P_0w^{(0)}| \leqslant Cd^{-n-2b-l-\alpha}(t,x;\tau,\xi)\Phi_{c_3}(t,x;\tau,\xi)\\ &\times\Big[\sum_{|m|\leqslant 2b-1}\lambda^{|m|-2b+1}(t-\tau)^{\nu(2b+l-|m|+\alpha)}\\ &+\sum_{k''}|\eta^{(k'')}(x)|\Big(|\Pi_x^{\bar{x}}|\operatorname{grad}F(\bar{x}')|(t-\tau)^{\nu(l+\alpha)}\\ &+\sum_{|m|\leqslant 2b-1}(t-\tau)^{\nu(2b+l-|m|+\alpha)}\Big)+\sum_{k''}|\eta^{(k'')}(x)|\\ &\times\sum_{|m|=2b}|a_m(t,x)-a_m(\tau,\xi^{(k'')})|(t-\tau)^{\nu(l+\alpha)}\\ &+\sum_{|m|\leqslant 2b-1}\lambda^{|m|-2b}(t-\tau)^{\nu(2b+l-|m|+\alpha)}\Big],\end{aligned}$$

where $c_3=\min(c_2,2^{-q/2}c_0)$ and $c_0$ is the same as in §2. Since the conditions imposed on the boundary $S$ imply $|\operatorname{grad}F(\bar{x}')|\leqq a_1|\bar{x}'|^\alpha\leqq C\lambda^2$ and since $t-\tau\leqq\chi\lambda^{2b}$ and

$$|a_m(t,x)-a_m(\tau,\xi^{(k'')})|\leqslant C[(t-\tau)^{\nu\alpha}+|x-\xi^{(k'')}|^\alpha]\leqslant C\lambda^\alpha$$

for $x\in\Omega^{(k'')}$, and the multiplicity of the covering of $\Omega$ by the sets $\Omega^{(k)}$ is finite, it follows from the last inequality that

$$|w_0^{(1)}|\leqslant C\Lambda(t-\tau)^{\nu(l+\alpha)}d^{-n-2b-l-\alpha'}(t,x;\tau,\xi)\Phi_{c_3}(t,x;\tau,\xi),$$

where $\Lambda=\lambda^\alpha+\lambda^\gamma$. Using Condition $B_l$, and following analogous considerations, we obtain for the derivatives $D_{t,x}^m w^{(1)}$ ($|m|\leqq l$) the estimates (30.1)-(32.1), in which instead of $C$, $c$ and $p$ we have $C\Lambda$, $c_3$ and $2b$, respectively. If, moreover, we prove the validity of the inequality

$$|I^{(k)}(w_0^{(1)})|\leqslant C\Lambda\rho_{c_3}(t,\tau,\xi), \tag{5.3}$$

it will follow that $w_0^{(1)}\in H_{2b,c_3}^{l+\alpha}(Q_h,Q)$.

Let us turn now to the proof of (5.3). Using the representation (22.1) for $w_0^{(1)}$, Lemma 2, and the fact that the coefficients of system (1.1) satisfy Hölder's condition, we obtain the following representation:

$$w_0^{(1)}(t, x; \tau, \xi) = \widetilde{w}_0^{(1)}(t, x; \tau, \xi)$$

$$(6.3) \qquad + \sum_{l''} \eta^{(l'')}(x) \sum_{|m|=2b} \{\Pi^{x,\xi}_{\overline{\overline{x}}, \hat{\hat{\xi}}} [\widetilde{a}_m(\tau, \xi^{(l'')}, \overline{\overline{x}}') D^m_{\overline{\overline{x}}} \overline{\overline{W}}{}^{(0)}_{l''}(t, \overline{\overline{x}}; \tau, \hat{\hat{\xi}})]$$

$$+ [a_m(\tau, \xi) - a_m(\tau, \xi^{(l'')})] D^m_x W^{(0)}_{l''}(t, x; \tau, \xi)\},$$

where

$$|\widetilde{w}_0^{(1)}| \leqslant C\left(1 + \frac{\chi^{\nu(1-\alpha)}}{\lambda^\alpha}\right) d^{-n-2b+\alpha}(t, x; \tau, \xi)\, \Phi_{c_3}(t, x; \tau, \xi),$$

$$(7.3) \qquad \widetilde{a}_m(\tau, \xi^{(l'')}, \overline{\overline{x}}') = \sum_{0<|\nu|\leqslant 2b} \widetilde{a}_{m\nu}(\tau, \xi^{(l'')}) (D_{\overline{\overline{x}}_1} F(\overline{\overline{x}}'))^{\nu_1} \ldots (D_{\overline{\overline{x}}_{n-1}} F(\overline{\overline{x}}'))^{\nu_{n-1}},$$

$\tilde{a}_{m\nu}(\tau, \xi^{(l'')})$ are determined by the coefficients of system (1.1), and

$$W^{(0)}_{l''} = \Pi^{x,\xi}_{\overline{\overline{x}}, \hat{\hat{\xi}}} \overline{\overline{W}}{}^{(0)}_{l''} = \sum_{j=1}^{bN} \overline{\overline{R}}{}^{(l'')}_j (w_j^{(0)}),$$

Here $\overline{\overline{x}}$ denotes the $l''$-rectifying coordinates of the point $x$, and $\hat{\hat{\xi}}$ denotes the $l''$-local coordinates of $\xi$.

In view of (6.3) and (7.3) we have

$$|\dot{I}^{(k)}(w_0^{(1)})| \leqslant C\left(1 + \frac{\chi^{\nu(1-\alpha)}}{\lambda^\alpha}\right) \rho_{c_3}(t, \tau, \xi) \iint_{Q_1^{(k)}} d^{-n-2b+\alpha}(\beta, y; \tau, \xi)\, d\beta dy$$

$$(8.3) \quad + \sum_{l''} \sum_{|m|=2b} \left| \iint_{Q_1^{(k)}} \zeta^{(k)}(y) \eta^{(l'')}(y) \{\Pi^{y,\xi}_{\overline{\overline{y}}, \hat{\hat{\xi}}} [\widetilde{a}_m(\tau, \xi^{(l'')}, \overline{\overline{y}}') D^m_{\overline{\overline{y}}} \overline{\overline{W}}{}^{(0)}_{l''}(\beta, \overline{\overline{y}}; \tau, \hat{\hat{\xi}})]\right.$$

$$\left. + [a_m(\tau, \xi) - a_m(\tau, \xi^{(l'')})] D^m_y W^{(0)}_{l''}(\beta, y; \tau, \xi)\}\, d\beta dy \right|.$$

The first term on the right side of (8.3) can be estimated, clearly, as

$$C\left(1 + \frac{\chi^{\nu(1-\alpha)}}{\lambda^\alpha}\right)(t-\tau)^{\nu\alpha} \rho_{c_3}(t, \tau, \xi) \leqslant \dot{C}\Lambda \rho_{c_3}(t, \tau, \xi).$$

We shall now estimate the integral of the second term in (8.3). Consider the most difficult case $k = k''$. Denoting the integral in this case by $I''$, we have

$$I'' = \iint_{\mathcal{K}_1} \bar{\zeta}^{(k'')}(\bar{y})\, \bar{\eta}^{(l'')}(\bar{y}) \{\Pi^{\bar{y},\xi}_{\overline{\overline{y}}, \hat{\hat{\xi}}} [\widetilde{a}_m(\tau, \xi^{(l'')}, \overline{\overline{y}}') D^m_{\overline{\overline{y}}} \overline{\overline{W}}{}^{(0)}_{l''}(\beta, \overline{\overline{y}}; \tau, \hat{\hat{\xi}})]$$

$$(9.3) \qquad + [a_m(\tau, \xi) - a_m(\tau, \xi^{(l'')})]\, \Pi^{\bar{y}}_y D^m_y W^{(0)}_{l''}(\beta, y; \tau, \xi)\}\, d\beta d\bar{y},$$

where $\bar{y}$ are the $k''$-rectifying coordinates of the point $y$, and

$$\mathcal{K}_1 = \left\{(\beta, \bar{y}) \in [\tau, t] \times K, d(\beta, y; \tau, \xi) \leqslant \frac{(t-\tau)^{\gamma}}{2}\right\}$$

is the image of $Q_1^{(k'')}$ resulting from the transformation from $y$ to $\bar{y}$. Note that the coordinates $\bar{y}$, $\bar{\bar{y}}$ of the point $y \in \Omega^{(k'')} \cap \Omega^{(l'')}$ are related by the functions $\bar{\bar{y}} = h_1(\bar{y})$ and $\bar{y} = h_2(\bar{\bar{y}})$ which have the same smoothness as the functions $F$, and in view of the properties of $F$ their norms are bounded by constants independent of $k''$ and $l''$.

Let us transform the integrand in (9.3). We have

$$\text{(10.3)}\quad \Pi^{\bar{y},\xi}_{\bar{\bar{y}},\hat{\hat{\xi}}} [D^m_{\bar{\bar{y}}} \overline{\overline{W}}{}^{(0)}_{l''} (\beta, \bar{\bar{y}}; \tau, \hat{\hat{\xi}})] = \Pi^{\xi}_{\hat{\xi}} \sum_{|\nu| \leqslant |m|} \tilde{b}_{m\nu}(\bar{y}) D^{\nu}_{\bar{y}} \overline{\overline{W}}{}^{(0)}_{l''} (\beta, h_1(\bar{y}); \tau, \hat{h}_1(\hat{\xi})),$$

where $\tilde{b}_{m\nu}(\bar{y})$ depend on the functions $h_1$ and on the derivatives of the functions $h_2$, and $\hat{\hat{\xi}} = \hat{h}_1(\hat{\xi})$. Further, in an analogous manner we obtain

$$\text{(11.3)}\quad \begin{aligned} \Pi^{\bar{y}}_{y} D^m_{\bar{y}} W^{(0)}_{l''} (\beta, y; \tau, \xi) &= \Pi^{\bar{y}}_{y} D^m_{\bar{y}} \Pi^{y,\xi}_{\bar{\bar{y}},\hat{\hat{\xi}}} \overline{\overline{W}}{}^{(0)}_{l''} (\beta, \bar{\bar{y}}; \tau, \hat{\hat{\xi}}) \\ &= \Pi^{\xi}_{\hat{\xi}} \sum_{|\nu| \leqslant |m|} \tilde{c}_{m\nu}(\bar{y}) D^{\nu}_{\bar{y}} \overline{\overline{W}}{}^{(0)}_{l''} (\beta, h_1(\bar{y}); \tau, \hat{h}_1(\hat{\xi})), \end{aligned}$$

where $\tilde{c}_{m\nu}(\bar{y})$ depend on the functions $F^{(k'')}$ and $h_1$ and on the derivatives of the functions $F^{(l'')}$ and $h_2$ (here $F^{(k'')}$ is the function $F$ used to represent the equation of the surface $S$ in $k''$-local coordinates). Using (9.3)–(11.3), we obtain

$$\text{(12.3)}\quad \begin{aligned} I'' = \sum_{|\nu| \leqslant |m|} \Pi^{\xi}_{\hat{\xi}} \iint_{\mathcal{K}_1} \bar{\zeta}^{(k'')}(\bar{y}) \bar{\eta}^{(l'')}(\bar{y}) \{\tilde{a}_m(\tau, \xi^{(l'')}, h_1'(\bar{y})) \tilde{b}_{m\nu}(\bar{y}) \\ + [a_m(\tau, \xi) - a_m(\tau, \xi^{(l'')})] \tilde{c}_{m\nu}(\bar{y})\} D^{\nu}_{\bar{y}} \overline{\overline{W}}{}^{(0)}_{l''} (\beta, h_1(\bar{y}); \tau, \hat{h}_1(\hat{\xi}))\, d\beta d\bar{y}. \end{aligned}$$

Note that in view of the definition of the functions $\tilde{a}_m$, $\tilde{b}_{m\nu}$ and $\tilde{c}_{m\nu}$ the properties of the functions $F$ and the inqualities

$$|\operatorname{grad} F(\bar{y}')| \leqslant C\lambda^{\alpha},$$

$$|a_m(\tau, \xi) - a_m(\tau, \xi^{(l'')})| \leqslant C|\xi - \xi^{(l'')}|^{\alpha} \leqslant C(|\xi - y|^{\alpha} + |y - \xi^{(l'')}|^{\alpha}) \leqslant C\lambda^{\alpha}$$

for $(\tau, y) \in Q_1^{(k'')}$, the expression in the braces in (12.3) is estimated by $C\lambda^{\alpha}$ and the first-order derivatives of this expression are, in general, bounded by the constant $C$. The terms of (12.3) for which $|\nu| \leqq 2b - 1$ are bounded by the right side of (5.3), if one uses for $D^{\nu}_{\bar{y}} \overline{\overline{W}}{}^{(0)}_{l''}$ the estimate which follows from Lemma 2. If $|\nu| = 2b$ and $\nu_n < 2b$, then one differentiation with respect to $\bar{y}'$ should be shifted from $\overline{\overline{W}}{}^{(0)}_{l''}$ to the other factors, and the estimate for the result can then be found by analogy with the preceding case. To estimate the integrals in (12.3) in the case $\nu_n = 2b$ we use the fact that since $\overline{\overline{W}}{}^{(0)}_{l''}(t, \overline{\overline{x}}; \tau, \hat{\overline{\xi}})$, as a function of $t$ and $\overline{\overline{x}}$, is a solution of a parabolic system, therefore $\overline{\overline{W}}{}^{(0)}_{l''}(t, h_1(\bar{x}); \tau, \hat{h}_1(\hat{\xi}))$, as a function of $t$ and $\bar{x}$, is also the solution of some parabolic system, whence it follows that $D^{2b}_{\overline{\overline{y}}} \overline{\overline{W}}{}^{(0)}_{l''}$ can be expressed as a linear combination of the $t$-derivative, the $\bar{y}$-derivatives of order $2b$ which include at least one differentiation with respect to $\bar{y}'$, and derivatives of lower order. Estimating the integrals thus obtained, we obtain the required estimates of the terms in (12.3). This concludes the proof of (5.3).

The estimates of $w^{(1)}_j$ $(j \geqq 1)$ are obtained in the same manner as those of $w^{(0)}_0$. Carrying out the necessary steps, we obtain $w^{(1)}_j \in H^{2b - r_j + l + \alpha}_{r_j, c_3}(\Gamma_h, Q)$. We shall now dwell briefly on the proof of the inequality

$$(13.3) \qquad |J^{(k'')}(w^{(1)}_j)| \leqslant C\Lambda \rho_{c_3}(t, \tau, \xi), \quad r_j = 2b - 1,$$

in a manner analogous to (3.3) and (5.3). Just as for $w^{(1)}_0$, the representation

$$w^{(1)}_j(t, x; \tau, \xi) = \tilde{w}^{(1)}_j(t, x; \tau, \xi) + \sum_{l''} \eta^{(l'')}(x)$$

$$(14.3) \qquad \times \sum_{|m| = r_j = 2b-1} \{\Pi^{x,\xi}_{\overline{\overline{x}}, \hat{\overline{\xi}}} [\tilde{b}_{jm}(\tau, \xi^{(l'')}, \bar{x}') D^m_{\overline{\overline{x}}} \overline{\overline{W}}{}^{(0)}_{l''}(t, \overline{\overline{x}}; \tau, \hat{\overline{\xi}}) |_{\overline{\overline{x}}_n = 0}]$$

$$+ [b_{jm}(\tau, \xi^{(l'')}) - b_{jm}(\tau, \tilde{\xi})] D^m_x W^{(0)}_{l''}(t, x; \tau, \xi) |_{\Gamma_h}\}$$

is valid, where

$$(15.3) \qquad |\bar{\tilde{w}}^{(1)}_j(t, \bar{x}'; \tau, \hat{\xi})| \leqslant C\left(1 + \frac{\chi^{\gamma(1-\alpha)}}{\lambda^{\alpha}}\right) d^{-n-2b+1+\alpha}(t, x; \tau, \xi) \Phi_{c_3}(t, x; \tau, \xi),$$

the form of $\tilde{b}_{jm}$ is analogous to that of $\tilde{a}_m$ in (6.3), and $\tilde{\xi}$ is the point of $S \cap \bar{\Omega}^{(l'')}$ nearest to $\xi$. Using (14.3), (15.3) and following the same steps as in the derivation of the estimate (5.3), we obtain the inequality

$$|J^{(k'')}(w^{(1)}_j)| \leqslant C\Lambda \rho_{c_3}(t, \tau, \xi)$$

$$(16.3) \qquad + \sum_{l''} \sum_{|\nu| \leqslant 2b-1} |\Pi^{\xi}_{\hat{\xi}} \iint_{\mathcal{K}'_1} \bar{\zeta}^{(k'')}(\bar{y}) \bar{\eta}^{(l'')}(\bar{y}) \{\tilde{b}_{jm}(\tau, \xi^{(l'')}, h'_1(\bar{y})) \tilde{b}_{m\nu}(\bar{y})$$

$$+ [b_{jm}(\tau, \xi^{(l'')}) - b_{jm}(\tau, \tilde{\xi})] \tilde{c}_{m\nu}(\bar{y})\} D^{\nu}_{\bar{y}} \overline{\overline{W}}{}^{(0)}_{l''}(\beta, h_1(\bar{y}); \tau, \hat{h}_1(\hat{\xi}))\}|_{\bar{y}_n = 0} d\beta d\bar{y}'|,$$

from which we obtain the estimate (13.3) as follows. In the cases $|\nu| \leqq 2b - 2$ and $|\nu| \leqq 2b - 1$, but $|\nu''| \neq 0$, the integral in (16.3) can be estimated in a manner similar to the proof of (5.3). In the case $|\nu| = \nu_n = 2b - 1$ it can be represented in the form

$$\iint\limits_{\mathcal{K}_1'} \eta^{(l'')}(\bar{y})\{\tilde{b}_{jm}(\tau, \xi^{(l'')}, h_1'(\bar{y}))\tilde{b}_{m\nu}(\bar{y}) + [b_{jm}(\tau, \xi^{(l'')})$$

$$-b_{jm}(\tau, \bar{\xi})]\tilde{c}_{m\nu}(\bar{y})\}|_{\bar{y}_n=0}\int_{2a\lambda}^{0} D_{\bar{y}_n}[\bar{\zeta}^{(k'')}(\bar{y})\, D_{\bar{y}_n}^{2b-1}\overline{\overline{W}}_{l''}^{(0)}(\beta, h_1(\bar{y}); \tau, \hat{h}_1(\hat{\xi}))]\, d\bar{y}_n d\beta d\bar{y}',$$

since $\bar{\zeta}^{(k'')}|_{\bar{y}_n=2a\lambda}=0$. The term in the last expression which contains $D_{\bar{y}_n}^{2b-1}\overline{\overline{W}}_{l''}^{(0)}$ can be immediately estimated by means of the estimates of $\overline{\overline{W}}_{l''}^{(0)}$, since the integral over $\bar{y}$ is $n$-dimensional. We split the term which contains $D_{\bar{y}_n}^{2b}\overline{\overline{W}}_{l''}^{(0)}$, as in the proof of (3.3), into a sum of integrals over

$$\mathcal{K}_1^{(1)} = \{(\beta, \bar{y}), (\beta, \bar{y}') \in \mathcal{K}_1', \; 0 < \bar{y}_n < 2a\lambda\} \cap \left\{d(\beta, y; \tau, \xi) > \frac{1}{2}(t-\tau)^{\nu}\right\}$$

and over

$$\mathcal{K}_1^{(2)} = \{(\beta, \bar{y}), (\beta, \bar{y}') \in \mathcal{K}_1', \; 0 < \bar{y}_n < 2a\lambda\} \setminus \mathcal{K}_1^{(1)}.$$

The first of these integrals can be immediately estimated by means of the estimates of $\overline{\overline{W}}_{l''}^{(0)}$, and in the second of these the derivative $D_{\bar{y}_n}^{2b}\overline{\overline{W}}_{l''}^{(0)}$ can be expressed in terms of $D_\beta\overline{\overline{W}}_{l''}^{(0)}$ and $D_{\bar{y}}^{\mu}\overline{\overline{W}}_{l''}^{(0)}$, where $|\mu| = 2b$ and $\mu_n \leqq 2b-1$ or $|\mu| \leqq 2b-1$, and the integrals with such derivatives can then be estimated quite easily (cf. the proof of Lemma 1).

Thus we have established that

$$w_0^{(1)} \in H_{2b,c_3}^{l+\alpha}(Q_h, Q), \quad w_j^{(1)} \in H_{r_j,c_3}^{2b-r_j+l+\alpha}(\Gamma_h, Q) \quad (j=1,\ldots,bN),$$

$$\max_{j=0,1,\ldots,bN} \|w_j^{(1)}\| = C_3\Lambda,$$

where $C_3$ is a constant.

Consecutively deriving the estimates for $w_j^{(\nu)}$ $(j=0,1,\cdots,bN)$ for $\nu=2,3,\cdots$ by analogy with the case $\nu=1$, we see that

$$w_0^{(\nu)} \in H_{2b,c_3}^{l+\alpha}(Q_h, Q), \quad w_j^{(\nu)} \in H_{r_j,c_3}^{2b-r_j+l+\alpha}(\Gamma_h, Q) \quad (j=1,\ldots,bN),$$

and there exists a constant $C_4$, independent of $\nu$, such that

$$\|w_j^{(\nu)}\| < C_3 C_4^{\nu-1}\Lambda^{\nu} \quad (j=0,1,\ldots,bN) \tag{17.3}$$

for $\nu=1,2,\cdots$.

If we choose $\lambda$, $\chi$, and consequently $h$ sufficiently small, then it follows from (17.3) that the series

$$w_0 = \sum_{\nu=0}^{\infty} w_0^{(\nu)}, \quad w_j = \sum_{\nu=0}^{\infty} w_j^{(\nu)} \quad (j=1,\ldots,bN)$$

are uniformly convergent in $Q_h$ and $\Gamma_h$, respectively, and their sums satisfy

$$(18.3) \qquad w_0 \in H^{l+\alpha}_{2b,c_3}(Q_h, Q), \quad w_j \in H^{2b-r_j+l+\alpha}_{r_j,c_3}(\Gamma_h, Q) \quad (j = 1, \ldots, bN).$$

**2. Concluding estimate.** Now we can establish the estimates of the matrix $v(t, x; \tau, \xi)$ which is the solution of problem (9.1) in the cylinder $Q_h$ with small $h$. Since $v$ can be represented in the form (10.1), where $w$ is the solution of problem (11.1), therefore, using the properties of the functions $\eta^{(k)}(x)$ and the boundary $S$, and also (18.3) and Lemmas 1 and 2, we obtain

$$(19.3) \qquad v \in H^{2b+l+\alpha}_{0,c_3}(Q_h, Q).$$

Taking account of the structure (4.1) of the Green matrix $G$ and the estimates (1.3), (19.3), we finally have

$$(20.3) \qquad G(t, x; \tau, \xi) \in U^{2b+l+\alpha,0,0}_{C,c}(Q_{\tau,\tau+h}, Q).$$

This concludes the proof of Theorem 1 for a cylinder of short height.

## §4. Estimates of the Green matrix and its derivatives in a cylinder of arbitrary height. Stationary parabolic boundary value problems

**1. Proof of Theorem 1 for a cylinder of arbitrary height.** In the preceding section we established the estimates for the Green matrix in a cylinder $Q_h$ of short height $h$. The quantity $h$ does not depend, as one can easily see, on the constants in the estimates of f.s.m. $Z$. We shall now use this fact to obtain estimates of the Green matrix for $\tau < t \leqq T$. To achieve this it suffices to obtain estimates for $t > \tau + h$ for the matrix $v$ which is the solution of (9.1). Note that, as we have already pointed out in §1, the solution of the problem (9.1) at any fixed point $(\tau_1 \xi) \in Q$ exists, is unique, and is defined for $\tau \leqq t \leqq T$. Our problem consists in obtaining exact pointwise estimates for $v$ for $t > \tau + h$.

Let $\tau_1 = \tau + h/2$. We consider a smooth function $\chi_h(t)$ such that

$$\chi_h(t) = \begin{cases} 0, & t \leqslant \dfrac{h}{6}, \\ 1, & t \geqslant \dfrac{h}{3}. \end{cases}$$

Introduce the auxiliary matrix $\omega(t, x; \tau, \xi; \tau_1) = \chi_h(t - \tau_1) v(t, x; \tau, \xi)$. This matrix satisfies the following conditions:

$$L(t, x, D_t, D_x)\,\omega = \chi_h'(t - \tau_1)\, v(t, x; \tau, \xi) \equiv g_0(t, x; \tau, \xi; \tau_1),$$

$$\omega|_{t=\tau_1}=0,$$
$$(1.4)\qquad B_j(t,x,D_x)\,\omega|_\Gamma=\chi_h(t-\tau_1)B_j(t,x,D_x)Z|_\Gamma\equiv g_j(t,x;\tau,\xi;\tau_1)$$
$$(j=1,2,\ldots,bN).$$

Therefore on the basis of the theorem about the uniqueness of the solution of the boundary value problem in the class $C^{2b+\alpha,(2b+\alpha)\gamma}_{x,t}(\bar{Q}_{\tau_1,T})$ the matrix $\omega$ can be obtained as the solution of problem (1.4). Since in view of the properties of $\chi_h$ the matrix $g_0$ is equal to zero for $t<\tau_1+h/6$ and $t>\tau_1+h/3$, on the basis of the estimates of $v$ for $\tau\leqq t\leqq\tau+h$ obtained in §3 we find that $g_0\in\overset{\circ}{H}{}^{l+\alpha}_{c_3}(\bar{Q}_{\tau_1,T},Q)$. From the properties of $\chi_h$, Condition $\text{B}_l$ and (1.3) it also follows that $g_j\in\overset{\circ}{H}{}^{2b-r_j+l+\alpha}_{c_3}(\Gamma_{\tau_1,T},Q)$. Now in problem (1.4) we can follow the same arguments which we made in §§1 and 3 concerning problem (9.1) in the domain $Q_h$, but here instead of Lemmas 1 and 2 we must use Lemma 3 and we must take $w_j^{(0)}=g_j$ $(j=0,1,\cdots,bN)$. As a result of this we find that there exists a unique solution $\omega(t,x;\tau,\xi;\tau_1)$ of the problem (1.4), which is defined in $\bar{Q}_{\tau_1,\tau_1+h_0}$ with a certain $h_0>0$ and which belongs to $\overset{\circ}{H}{}^{2b+l+\alpha}_{c_3}(\bar{Q}_{\tau_1,\tau_1+h_0},Q)$. Since $\omega=v$ for $t>\tau_1+h/3$ it follows from this and from (19.3) that $v(t,x;\tau,\xi)\in H^{2b+l+\alpha}_{c_3}(\bar{Q}_{\tau_1,\tau_1+h_0},Q)$. It should be noted that in view of Conditions A and $\text{B}_l$ the choice of $h_0$ does not depend on $\tau_1$.

Considering now the matrix

$$\omega(t,x;\tau,\xi;\tau_2)=\chi_{h_0}(t-\tau_2)\,v(t,x;\tau,\xi),\quad \tau_2=\tau_1+\frac{h_0}{2},$$

and solving for it a problem analogous to (1.4), we find that

$$v\in H^{2b+l+\alpha}_{c_3}(\bar{Q}_{\tau_1,\tau_2+h_0},Q).$$

Following a similar line of argument, we reach $T$ after a finite number of steps and find that

$$v\in H^{2b+l+\alpha}_{c_3}(\bar{Q}_{\tau_1,T},Q).$$

From this and (19.3) it follows that

$$v\in H^{2b+l+\alpha}_{0,c}(\bar{Q}_{\tau,T},Q),\quad c<c_3,$$

and, taking account of (1.3), we finally obtain

$$G\in U^{2b+l+\alpha,0,0}_{C,c}(Q_{\tau,T},Q).$$

Hence, using the definition of the classes, we obtain the required estimates (6.1) for the Green matrix $G$ and the corresponding estimates for $v$. Thus Theorem 1 is proved.

**2. Estimates of the Green matrix of stationary parabolic boundary value problems in a semi-infinite cylinder.** Consider the stationary case of problem (1.1)–(3.1), i.e. the problem

$$L(x, D_t, D_x)u = f(t, x), \tag{2.4}$$

$$u|_{t=0} = \varphi(x), \tag{3.4}$$

$$B_j(x, D_x)u|_\Gamma = 0 \quad (j = 1, 2, \ldots, bN). \tag{4.4}$$

Estimates for the Green matrix of this problem can be obtained in the semi-infinite cylinder $Q_\infty = [0 \times \infty) \times \Omega$. To obtain these, we must find the dependence of the constants in the estimates of the Green matrix in a cylinder $Q$ on the height of the cylinder.

From the results of [**25**], Chapter 1, §4, it follows that for all $T > \tau$ the f.s.m. of the system (2.4) satisfies

$$Z(t-\tau, x, \xi) \in U^{2b+l+\alpha,0,s}_{C_1 e^{A_1(T-\tau)},c_1}(Q_{\tau,T}, Q_\infty), \tag{5.4}$$

where $C_1$, $c_1$ and $A_1$ are positive constants independent of $T$. Using this result, we retrace the derivation of the estimates of the matrix $v(t,x;\tau,\xi) \equiv v(t-\tau,x,\xi)$, checking the dependence of the constants in these estimates on $T$. In the following, $C_i$, $c_i$ and $A_i$ denote positive constants independent of $T$, and $C_i > 1$.

Clearly from (5.4) and Lemmas 1 and 2 it follows that

$$\| v(t, x; \tau, \xi)\|_{H^{2b+l+\alpha}_{0,c_3}(\overline{Q}_{\tau,\tau+h_0}, Q)} \leqslant C_2 e^{A_1(T-\tau)}. \tag{6.4}$$

For the same reason it follows that

$$\| v(t, x; \tau, \xi)\|_{H^{2b+l+\alpha}_{c_3}(\overline{Q}_{\tau_1,\tau_1+h_0}, Q)} \leqslant C_3 e^{A_1(T-\tau)} = M. \tag{7.4}$$

Using (5.4) and (7.4), we have

$$\| g_0(t, x; \tau, \xi; \tau_2)\|_{\overset{\circ}{H}{}^{l+\alpha}_{c_3}(\overline{Q}_{\tau_2,T}, Q)} \leqslant C_4 M,$$

$$\| g_j(t, x; \tau, \xi; \tau_2)\|_{\overset{\circ}{H}{}^{2b-r_j+l+\alpha}_{c_3}(\Gamma_{\tau_2,T}, Q)} \leqslant C_4 M,$$

and hence

$$\| \omega(t, x; \tau, \xi, \tau_2)\|_{H^{2b+l+\alpha}_{c_3}(\overline{Q}_{\tau_2,\tau_2+h_0}, Q)} \leqslant C_5 C_4 M,$$

and this, together with (7.4), implies that

$$\| v(t, x; \tau, \xi)\|_{H^{2b+l+\alpha}_{c_3}(\overline{Q}_{\tau_1,\tau_2+h_0}, Q)} \leqslant C_6 M, \quad C_6 = C_4 C_5.$$

Following the same line of reasoning, we obtain

$$\|v(t, x; \tau, \xi)\|_{H^{2b+l+\alpha}_{C_3}(\overline{Q}_{\tau_1, \tau_k+h_0}, Q)} \leqslant C_6^{k-1} M, \tag{8.4}$$

where $\tau_k = \tau_1 + (k-1) h_0/2$. Clearly from (8.4) it follows that

$$\|v(t, x; \tau, \xi)\|_{H^{2b+l+\alpha}_{C_3}(\overline{Q}_{\tau_1, T}, Q)} \leqslant C_6^m M \leqslant C_7 e^{A_2(T-\tau)}, \tag{9.4}$$

where

$$m = \left[\frac{2(T-\tau_1)}{h_0}\right] + 1, \ C_7 = C_3 C_6, \ A_2 = A_1 + \frac{2}{h_0} \ln C_6.$$

Using (5.4), (6.4), (9.4) and the definitions of the corresponding classes, we obtain the proof of the following theorem.

THEOREM 3. *Let the problem* (2.4)–(4.4) *satisfy Conditions* A *and* $C_l$:

$$a_k(x) \in C^{l+\alpha}(\Omega), \quad b_{jk}(x) \in C^{2b-r_j+l+\alpha}(S), \ S \in C^{2b+l+\alpha}.$$

*Then the Green matrix* $G(t-\tau, x, \xi)$ *of the problem* (2.4)–(4.4) *for* $0 \leqq \tau < t < \infty$ *has the following estimates*:

$$|D^m_{t,x} G| \leqslant C(t-\tau)^{-(n+|m|)\nu} \exp\left\{A(t-\tau) - c\left[\frac{|x-\xi|}{(t-\gamma)^\nu}\right]^q\right\}$$

$$(|m| \leqslant 2b + l),$$

$$|\Delta^{x_0}_x D^m_{t,x} G| \leqslant C|x - x_0|^\alpha (t-\tau)^{-(n+2b+l+\alpha)\nu} \exp\left\{A(t-\tau) - c\left[\frac{|x^*-\xi|}{(t-\tau)^\nu}\right]^q\right\}$$

$$(|m| = 2b + l, \ |x^* - \xi| = \min(|x-\xi|, |x_0 - \xi|)), \tag{10.4}$$

$$|\Delta^{t_0}_t D^m_{t,x} G| \leqslant C(t-t_0)^{(2b+l-|m|+\alpha)\nu} (t_0 - \tau)^{-(n+2b+l+\alpha)\nu}$$

$$\times \exp\left\{A(t-\tau) - c\left[\frac{|x-\xi|}{(t-\tau)^\nu}\right]^q\right\}$$

$$(l < |m| \leqslant 2b + l, \ \tau < t_0 < t),$$

*where* $C$, $A$ *and* $c$ *are positive constants. Here the matrix* $v(t-\tau, x, \xi)$ *has the estimates obtained from* (10.4) *by substituting* $|x-\xi| + d(\xi, S)$ *and* $|x^*-\xi| + d(\xi, S)$ *for* $|x-\xi|$ *and* $|x^*-\xi|$ *respectively.*

## §5. The Green matrix of a homogeneous elliptic boundary value problem generated by a parabolic problem. Estimates of the kernels of fractional powers of the elliptic operator

**1. Structure and estimates of the Green matrix of the elliptic problem.** Consider the following parabolic boundary value problem in the cylinder $Q_\infty = [0, \infty) \times \Omega$:

(1.5) $$D_t u + A(x, D_x) u + \lambda u = f(x),$$

(2.5) $$u|_{t=0} = 0,$$

(3.5) $$B_j(x, D_x) u|_\Gamma = 0 \quad (j = 1, \ldots, bN),$$

where the number $\lambda$ is such that $\operatorname{Re}\lambda > A$, and $A$ is the constant in the estimates (10.4) of the Green matrix $G(t-\tau, x, \xi)$ of problem (2.5), (3.5) for the system

(4.5) $$D_t u + A(x. D_x) u = f(x).$$

Let $G(t-\tau, x, \xi; \lambda)$ be the Green matrix of problem (1.5)–(3.5). It can be easily seen that

$$G(t-\tau, x, \xi; \lambda) = e^{-\lambda(t-\tau)} G(t-\tau, x, \xi).$$

The solution of problem (1.5)–(3.5) for any smooth bounded vector function $f(x)$ is

(5.5) $$u(t, x) = \int_0^t d\tau \int_\Omega e^{-\lambda(t-\tau)} G(t-\tau, x, \xi) f(\xi)\, d\xi.$$

We introduce in (5.5) the substitution $t - \tau = \beta$ and rewrite $u(t, x)$ in the form

(5′.5) $$u(t, x) = \int_\Omega \Big[\int_0^t e^{-\lambda\beta} G(\beta, x, \xi)\, d\beta\Big] f(\xi)\, d\xi.$$

It is natural to expect that when $t \to \infty$, $u(t, x)$ will converge to a vector function $u(x)$ which is the solution of the elliptic problem[1]

(6.5) $$A(x, D_x) u(x) + \lambda u(x) = f(x),$$

(7.5) $$B_j(x, D_x) u(x)|_S = 0 \quad (j = 1, \ldots . bN).$$

If we can now pass to the limit as $t \to \infty$ under the integral sign in (5′.5), then

$$u(x) = \int_\Omega \Big[\int_0^\infty e^{-\lambda\beta} G(\beta, x, \xi)\, d\beta\Big] f(\xi)\, d\xi$$

and the matrix

(8.5) $$\Phi(x, \xi; \lambda) = \int_0^\infty e^{-\lambda\beta} G(\beta, x, \xi)\, d\beta$$

---

[1] Note that this problem satisfies the complementation condition in the strong sense.

is the Green matrix of problem (6.5), (7.5).

Note that this method was used in the book [**25**] to construct the f.s.m. of the elliptic system from a known f.s.m. of the corresponding parabolic system.

The following theorem holds:

THEOREM 4. *Let the problem* (1.5)-(3.5) *satisfy Conditions* A *and* $C_l$ *of Theorem* 3, *where* $l$ *is a nonnegative integer. Then the following assertions hold*:

1) *The Green matrix of the elliptic problem* (6.5)-(7.5) *is given by* (8.5), *and the following representation is valid*:

$$\Phi(x, \xi; \lambda) = \varphi(x, \xi; \lambda) - w(x, \xi; \lambda), \tag{9.5}$$

*where* $\phi(x,\xi;\lambda)$ *is the f.s.m. of the system* (6.5).

2) *The derivatives of the matrix* $\phi(x,\xi;\lambda)$ *have the estimates*

$$|D_x^m \Phi| \leqslant C \exp\{-c_0 \delta^\nu |x-\xi|\} \times \begin{cases} 1, & n+|m|<2b, \\ 1+|\ln|x-\xi||, & n+|m|=2b, \\ |x-\xi|^{-n-|m|+2b}, & n+|m|>2b \end{cases} \tag{10.5}$$

$$(|m| \leqslant 2b+l),$$

$$|\Delta_x^{x_0} D_x^m \Phi| \leqslant C|x-x_0|^\alpha |x^*-\xi|^{-n-l-\alpha} \exp\{-c_0\delta^\nu |x^*-\xi|\} \tag{11.5}$$

$$(|m| = 2b+l),$$

*where* $l_0>0$ *and* $\delta = \operatorname{Re}\lambda - A > 0$, *and the matrix* $w(x,\xi;\lambda)$ *has the estimates obtained from* (10.5) *and* (11.5) *by substituting* $|x-\xi| + d(\xi, S)$ *and* $|x^* - \xi| + d(\xi,S)$ *for* $|x-\xi|$ *and* $|x^*-\xi|$, *respectively.*

3) *If* $f(x) \in C^\alpha(\Omega)$, *then the vector function*

$$u(x) = \int_\Omega \Phi(x, \xi; \lambda) f(\xi)\, d\xi \tag{12.5}$$

*is a solution of the problem* (6.5), (7.5). *Here* $u(t,x) \to u(u)$ *for* $t \to \infty$ *uniformly in* $x \in \Omega$, *where* $u(t,x)$ *is defined by* (5.5).

PROOF. Using the construction of the matrix $G$, we have

$$\Phi(x, \xi; \lambda) = \int_0^\infty e^{-\lambda\beta} Z(\beta, x, \xi)\, d\beta - \int_0^\infty e^{-\lambda\beta} v(\beta, x, \xi)\, d\beta$$

$$\equiv \varphi(x, \xi; \lambda) - w(x, \xi; \lambda),$$

where $\phi$, as has been shown in [**25**], Chapter 1, §7, is the f.s.m. of the system (6.5). Therefore to prove the first part of the theorem it suffices to prove the equalities

$$[A(x, D_x) + \lambda]\, w(x, \xi; \lambda) = 0, \tag{13.5}$$

$$B_j(x, D_x)\, \Phi(x, \xi; \lambda)|_S = 0, \quad \xi \in \Omega. \tag{14.5}$$

We shall first obtain estimates for $\Phi$ and $w$, and subsequently we shall use these to prove (13.5) and (14.5).

Assume $|x - \xi| = r$. From the estimates (10.4) we have

$$|D_x^m \Phi| \leqslant C \int_0^\infty \exp\left\{-c\left(\frac{r}{\beta^\nu}\right)^q - \delta\beta\right\} \beta^{-(n+|m|)\nu}\, d\beta$$

$$= C\left(\int_0^1 \ldots d\beta + \int_1^\infty \ldots d\beta\right) = C(I_1 + I_2).$$

Consider first the case $r \leqq 1$. Clearly we have

$$I_2 \leqslant \int_1^\infty \exp\{-\delta\beta\}\, d\beta \leqslant C.$$

Applying Lemma 7.1 of [**25**] to $I_1$, we obtain

$$I_1 \leqslant \int_0^1 \beta^{-(n+|m|)\nu} \exp\left\{-c\left(\frac{r}{\beta^\nu}\right)^q\right\} d\beta$$

$$\leqslant \begin{cases} C, & n + |m| < 2b, \\ C(1 + |\ln r|), & n + |m| = 2\beta, \\ Cr^{-n-|m|+2b}, & n + |m| > 2b. \end{cases}$$

In the case $r > 1$, using the fact that the function $g(\beta) = -\delta\beta - c(r\beta^{-\gamma})^q$, $0 \leqq \beta < \infty$, has for $\beta = [c/(2b-1)\delta]^{1/q} r$ a maximum equal to $-2b\delta^\gamma (c/(2b-1))^{1/q} r$, we obtain

$$I_1 \leqslant e^{-c_0\delta^\nu r} \int_0^1 e^{-\varepsilon\beta^{-\nu q}}\, \beta^{-(n+|m|)\nu}\, d\beta \leqslant Ce^{-c_0\delta^\nu r},$$

$$I_2 \leqslant e^{-c_0\delta^\nu r} \int_0^\infty e^{-\varepsilon\delta\beta}\, d\beta \leqslant Ce^{-c_0\delta^\nu r}.$$

Thus we have obtained the estimates (10.5). The estimate (11.5) can be proved in an analogous manner. Clearly, then, we can prove also the estimates for $w$ given in the theorem. To prove these, we assume $r = |x - \xi| + d(\xi, S)$ in all the derivations above.

We shall now prove (13.5). In view of the properties of $v$ we have

$$[D_t + A(x, D_x) + \lambda] e^{-\lambda t} v(t, x, \xi) = 0, \ x \in \bar{\Omega}, \ \xi \in \Omega, \ 0 < t < \infty.$$

Therefore

$$0 = \int_\varepsilon^M [D_t + A(x, D_x) + \lambda] e^{-\lambda t} v(t, x, \xi)\, dt = e^{-\lambda t} v(t, x, \xi) \Big|_{t=\varepsilon}^{t=M}$$

$$+ [A(x, D_x) + \lambda] \int_\varepsilon^M e^{-\lambda t} v(t, x, \xi)\, dt$$

$$\xrightarrow[M \to \infty]{\varepsilon \to 0} [A(x, D_x) + \lambda] \int_0^\infty e^{-\lambda t} v(t, x, \xi)\, dt = [A(x, D_x) + \lambda]\, w(x, \xi; \lambda).$$

Equation (14.5) can also be easily proved by means of the relation

$$B_j(x, D_x) G(t, x, \xi) |_S = 0, \ \xi \in \Omega, \ 0 < t < \infty.$$

The fact that the vector function (12.5) is a solution of problem (6.5), (7.5) follows directly from the properties of the matrix $\Phi$, and the uniform approach of $u(t,x)$ to $u(x)$ follows from the estimate

$$|u(x) - u(t, x)| = \Big| \int_\Omega \Big[ \int_t^\infty e^{-\lambda\beta} G(\beta, x, \xi)\, d\beta \Big] f(\xi)\, d\xi \Big|$$

$$\leqslant C \| f \|_0^\Omega e^{-\varepsilon t} \int_\Omega e^{-c|x-\xi|}\, d\xi \leqslant C \| f \|_0^\Omega e^{-\varepsilon t}, \quad t > 1.$$

**2. Estimates for the kernels of fractional negative powers of the elliptic operator.** Using the methods which were used in the proof of Theorem 4, we also obtain exact estimates for the Green kernels $\Phi_\beta(x, \xi; \lambda)$ of fractional negative powers of the elliptic operator corresponding to problem (6.5), (7.5), defined as

$$\Phi_\beta(x, \xi; \lambda) = \frac{1}{\Gamma(\beta)} \int_0^\infty t^{\beta-1} G(t, x, \xi; \lambda)\, dt, \ 0 < \beta < 1. \tag{15.5}$$

THEOREM 5. *If the conditions of Theorem 4 are satisfied, then the derivatives* $D_x^m \Phi_\beta$ $(|m| \leqq 2b + l)$ *satisfy the inequalities*

$$|D_x^m \Phi_\beta| \leqslant C \exp\{-c_0 \delta^\nu |x - \xi|\} \times \begin{cases} 1, & n + |m| < 2b\beta, \\ 1 + |\ln|x - \xi||, & n + |m| = 2b\beta, \\ |x - \xi|^{-n-|m|+2b}, & n + |m| > 2b\beta, \end{cases}$$

(16.5)

$$|\Delta_x^{x_0} D_x^k \Phi_\beta| \leqslant C |x - x_0|^\alpha \exp\{-c_0 \delta^\nu |x - \xi|\} |x^* - \xi|^{-n-l-\alpha}.$$

$$(|k| = 2b + l).$$

## §6. On the Green operators

In the preceding sections we have constructed the Green matrix of parabolic boundary value problems, and of elliptic problems generated by parabolic problems, and we have obtained exact estimates for its derivatives. Here we shall use these results to derive theorems concerning the effect of integral operators whose kernels are Green matrices of the corresponding problems, and in particular we shall prove Theorem 2, formulated in §1.

**1. The parabolic Green operator.** Consider the operator

$$u(t, x) = Gf = \int_0^t d\tau \int_\Omega G(t, x; \tau, \xi) f(\tau, \xi)\, d\xi. \tag{1.6}$$

The following theorem is valid.

THEOREM 6. *The Green operator* (1.6) *is bounded from* $C_0^\alpha(Q)$ *into* $C_0^{2b+\alpha}(Q)$.

PROOF. Since $u = Gf$ is a solution of problem (1.1)–(3.1) with $L\phi \equiv 0$, therefore $u|_{t=0} = 0$ and $D_t u|_{t=0} = f(0, x)$ if $u \in C_{x,t}^{2b+\alpha,(2b+\alpha)\gamma}(Q)$. Therefore it suffices to prove the latter.

Let $\delta$ and $h$ be sufficiently small positive numbers. We introduce the notation

$$U_\delta(x^0) = \Omega \cap \{|x - x^0| \leqslant \delta\},\ Q_{h\delta}(t^0, x^0) = [t^0 + h,\ t^0 + 2h] \times U_\delta(x^0).$$

To prove the theorem it clearly suffices to show that if $f \in C_0^\alpha(Q)$ then for any point $(t_0, x_0) \in [-h, T - 2h] \times \overline{\Omega}$ we have

$$u(t, x) \in C_{x,t}^{2b+\alpha,\ (2b+\alpha)\gamma}(Q_{h\delta}(t^0, x^0))$$

and

$$\| u \|_{2b+\alpha}^{Q_{h\delta}(t^0, x^0)} \leqslant C \| f \|_\alpha^Q,$$

where $C$ is independent of $(t^0, x^0)$.

We introduce the smooth functions $\chi_0(\tau)$ and $\chi(\xi)$, defined in $E_1$ and in $E_n$, respectively, which satisfy the conditions

$$\chi_0(\tau) = \begin{cases} 1, & \tau \geqslant \frac{2}{3} h, \\ 0, & \tau \leqslant \frac{1}{2} h, \end{cases} \qquad \chi(\xi) = \begin{cases} 1, & |\xi| \leqslant \frac{3}{2}\delta. \\ 0, & |\xi| \geqslant 2\delta, \end{cases} \tag{2.6}$$

Let us represent $u(t, x)$ in the form

$$u(t, x) = \int_0^t d\tau \int_\Omega G f \chi_1(\tau, \xi)\, d\xi$$

$$+ \int_0^t d\tau \int_\Omega Gf [1 - \chi_1(\tau, \xi)]\, d\xi = u_1 + u_2,$$

where $\chi_1(\tau, \xi) = \chi_0(\tau - t^0)\chi(\xi - x^0)$. In view of the properties of the functions $\chi_1$ and the estimates of the matrix $G$, we obtain that

$$\| u_2 \|_{2b+\alpha}^{Q_{h\delta}(t^0, x^0)} \leqslant C \| f \|_\alpha^Q,$$

where $C$ is independent of $(t^0, x^0)$. We shall prove that a similar estimate holds for $u_1$. The most difficult case is the case when the set $U_\delta(x^0)$ adjoins the boundary $S$ of the domain $\Omega$ and $t^0$ is such that the cutoff function $\chi_0(\tau - t^0)$ is different from zero in the neighborhood of the point $t = 0$. We shall consider here only the most difficult case: $t^0 = -h$, $x^0 \in S$. Let $U_{2\delta}(x^0) = U^+$ and $Q_{h\delta}(-h, x^0) = Q_0$. Using the construction of the matrix $G$, we write $u_1$ in the form

$$u_1 = \int_0^t d\tau \int_{U^+} Zf\chi(\xi - x^0)\, d\xi - \int_0^t d\tau \int_{U^+} v f \chi(\xi - x^0)\, d\xi = v_1 - v_2. \tag{3.6}$$

The proof of the theorem is thus reduced to the proof of the inequalities

$$\| v_i \|_{2b+\alpha}^{Q_0} \leqslant C \| f \|_\alpha^Q \quad (i = 1, 2). \tag{4.6}$$

Let $f_1(t, x)$ be the extension of $f$ from $Q$ into the strip $\Pi = [0, T] \times E_n$ such that $f_1 \in C_{x,t}^{\alpha, \alpha\gamma}(\Pi)$ and

$$\| f_1 \|_\alpha^\Pi \leqslant C \| f \|_\alpha^Q. \tag{5.6}$$

We represent $v_1$ in the form

$$v_1 = \int_0^t d\tau \int_{E_n} Zf_1\chi\, d\xi - \int_0^t d\tau \int_{U^-} Zf_1\chi\, d\xi = v_1' - v_1'', \tag{6.6}$$

where

$$U^- = \{E_n \setminus \Omega\} \cap \{|\xi - x^0| \leqslant 2\delta\}.$$

It is known (cf. [**25**], Chapter 3, §2.4) that

$$\| v_1' \|_{2b+\alpha}^\Pi \leqslant C \| f_1 \chi \|_\alpha^\Pi \leqslant C \| f_1 \|_\alpha^\Pi. \tag{7.6}$$

If $v_1''$ satisfies the estimate

$$\| v_1'' \|_{2b+\alpha}^{Q_0} \leqslant C \| f_1 \|_\alpha^\Pi, \tag{8.6}$$

then it follows from (7.6) and (8.6), in view of (5.6), that $v_1$ satisfies inequality (4.6).

The proof of inequality (8.6) is based on the methods developed in [**25**], [**26**], and [28], using parabolic volume potentials, with integration over the whole space $E_n$. Therefore we shall discuss this proof very briefly. Let $x^*$ be the point of $S$ nearest to $x$ and let $r(t,x) = d(t,x;0,x^*)$. Using the estimates for $Z$ and the fact that $f_1(0,x^*) = 0$, we obtain

$$\begin{aligned} | D_x^k v_1'' | &\leqslant \int_0^t d\tau \int_{U^-} | D_x^k Z | \, | \chi | \, | \Delta_{\tau,\xi}^{t,x} f_1 | \, d\xi + | \Delta_{t,x}^{0,x^*} f_1 | \, | I_k | \\ &\leqslant C t^{(2b-|k|)\nu} r^\alpha (t, x) \| f_1 \|_\alpha^\Pi (1 + t^{(|k|-2b)\nu} | I_k |), \quad (|k| \leqslant 2b), \end{aligned} \tag{9.6}$$

where

$$I_k = \int_0^t d\tau \int_{U^-} D_x^k Z \chi \, d\xi.$$

For $|k| < 2b$ we obtain directly from the estimates of $Z$ that

$$| I_k | \leqslant C t^{(2b-|k|)\nu}. \tag{10.6}$$

Let us prove estimate (10.6) for $|k| = 2b$. Since [**25**] gives

$$\begin{aligned} Z(t, \tau, x, \xi) &= G_0 (t - \tau, x - \xi; \tau, \xi) - W(t, \tau, x, \xi), \\ W(t, \tau, x, \xi) &= \int_\tau^t d\beta \int_{E_n} G_0 (t - \beta, x - y; \beta, y) \varphi(\beta, y; \tau, \xi) \, dy, \end{aligned} \tag{11.6}$$

it follows, in view of the estimates of $W$ and the fact that $G_0$ satisfies Hölder's condition in the two last arguments, that it suffices to estimate the integral

$$I^0 = \int_0^t d\tau \int_{U^-} D_\xi^k G_0 (t - \tau, x - \xi; t, x) \chi (\xi - x^0) \, d\xi \quad (|k| = 2b).$$

In $I^0$ we change the coordinates from $x, \xi$ to the coordinates $\bar{x}, \bar{\xi}$ which are rectifying in the neighborhood of $x^0$. Then $I^0$ is represented in the form

$$I^0 = \sum_{0 < |m| \leqslant 2b} \int_0^t d\tau \int_{K^-} D_{\bar{\xi}}^m \overline{G}_0 (t - \tau, \bar{x}, \bar{\xi}, t) \, c_m (\bar{\xi}) \bar{\chi} (\bar{\xi}) \, d\bar{\xi}, \tag{12.6}$$

where $K^-$ is the image of $U^-$ under the transformation $\xi \to \bar{\xi}$, and $c_m(\bar{\xi})$ are sufficiently smooth functions which depend on this transformation. If $|m| < 2b$, then from the estimates of $G_0$ there follows directly the inequality

$$\left| \int_0^t d\tau \int_{K^-} D_{\bar{\xi}}^m \bar{G}_0 c_m \bar{\chi}\, d\bar{\xi} \right| \leqslant C t^{(2b-|m|)\nu}.$$

When $|m| = 2b$ and $m_n < 2b$, one derivative with respect to $\bar{\xi}'$ is shifted from $\bar{G}_0$ to the other factors, and the integral over the boundary of $K^-$ is equal to zero in view of the properties of $\bar{\chi}(\bar{\xi})$. In the case $m_n = 2b$ we use the fact that $\bar{G}_0(t, \tau, \bar{x}, \bar{\xi}, t)$, as a function of $\tau$ and $\bar{\xi}$, is a solution of the parabolic system, and therefore

$$D_{\bar{\xi}_n}^{2b} \bar{G}_0(t-\tau, \bar{x}, \bar{\xi}, t) = \sum_{\substack{0<2bk_0+|k|\leqslant 2b \\ (k_n<2b)}} \bar{a}_{k_0 k}(t, \bar{x}, \bar{\xi}) D_\tau^{k_0} D_{\bar{\xi}}^k \bar{G}_0(t-\tau, \bar{x}, \bar{\xi}, t). \tag{13.6}$$

Integrals in which instead of $D_{\bar{\xi}_n}^{2b} \bar{G}_0$ we have the terms of the right side of (13.6) can be easily estimated.

From (9.6) and (10.6) we obtain the inequality

$$|D_x^k v_1''| \leqslant C r(t, x)^{2b-|k|+\alpha} \|f_1\|_\alpha^{\mathrm{II}} \quad (|k| \leqslant 2b). \tag{14.6}$$

We shall now estimate the increment $\Delta_t^{t_0} D_x^k v_1''(t, x)$ $(0 < |k| \leqq 2b)$, taking $t_0 < t$. If $t - t_0 \geqq r^{2b}(t_0, x)$, it follows from (14.6) that

$$|\Delta_t^{t_0} D_x^k v_1''| \leqslant C(t-t_0)^{(2b-|k|+\alpha)\nu} \|f_1\|_\alpha^{\mathrm{II}}. \tag{15.6}$$

We shall now prove estimate (15.6) for the case when $t - t_0 < r^{2b}(t_0, x)$. We represent

$$\begin{aligned} \Delta_t^{t_0} D_x^k u_1 &= \int_{t_1}^t d\tau \int_{U^-} D_x^k Z \chi \Delta_{\tau,\xi}^{t,x} f_1\, d\xi - \int_{t_1}^{t_0} d\tau \int_{U^-} D_x^k Z \chi \Delta_{\tau,\xi}^{t_0,x} f_1\, d\xi \\ &+ \int_0^{t_1} d\tau \int_{U^-} \Delta_t^{t_0} D_x^k Z \chi \Delta_{\tau,\xi}^{t_0,x} f_1\, d\xi + \Delta_{t,x}^{t_0,x} f_1 \int_{t_1}^t d\tau \int_{U^-} D_x^k Z \chi\, d\xi \\ &+ \Delta_{t_0,x}^{0,x^*} f_1 \Delta_t^{t_0} \int_0^t d\tau \int_{U^-} D_x^k Z \chi\, d\xi, \end{aligned} \tag{16.6}$$

where $t_1 = \max(2t_0 - t, 0)$. The estimate (15.6) for the first three terms of (16.6) can be obtained directly from the estimates of $Z$ of [**26**], and the estimates for the fourth and fifth terms follow from the inequalities

(17.6) $$\left|\int_{t_1}^{t} d\tau \int_{U^-} D_x^k Z\chi\, d\xi\right| \leqslant C(t-t_1)^{(2b-|k|)\gamma},$$

(18.6) $$|J| = \left|\Delta_t^{t_0}\int_0^t d\tau \int_{U^-} D_x^k Z\chi\, d\xi\right| \leqslant C(t-t_0)^{(2b-|k|+\alpha)\gamma} r^{-\alpha}(t_0, x).$$

Inequality (17.6) can be proved in the same manner as (10.6). We now prove (18.6).

From (11.6) we have

$$J = \Delta_t^{t_0}\int_0^t d\tau \int_{U^-} D_x^k G_0(t-\tau, x-\xi; \tau, \xi)\chi\, d\xi + \Delta_t^{t_0}\int_0^t d\tau \int_{U^-} D_x^k W\chi\, d\xi = J_1 + J_2.$$

(19.6)

To estimate $J_1$ we write

$$J_1 = \int_{t_1}^{t} d\tau \int_{U^-} [D_x^k G_0(t-\tau, x-\xi; \tau, \xi) - D_x^k G_0(t-\tau, x-\xi; t, z)|_{z=x}]\chi\, d\xi$$

$$-\int_{t_1}^{t} d\tau \int_{U^-} [D_x^k G_0(t_0-\tau, x-\xi; \tau, \xi) - D_x^k G_0(t_0-\tau, x-\xi; t, z)|_{z=x}]\chi\, d\xi$$

(20.6) $$+\int_0^{t_1} d\tau \int_{U^-} \Delta_t^{t_0}[D_x^k G_0(t-\tau, x-\xi; \tau, \xi) - D_x^k G_0(t-\tau, x-\xi; t_0, z)|_{z=x}]\chi\, d\xi$$

$$+\int_{t_1}^{t} d\tau \int_{U^-} [D_x^k G_0(t-\tau, x-\xi; t, z) - D_x^k G_0(t-\tau, x-\xi; t_0, z)]|_{z=x}\chi\, d\xi$$

$$+\Delta_t^{t_0}\int_0^t d\tau \int_{U^-} D_x^k G_0(t-\tau, x-\xi; t_0, x)\chi\, d\xi.$$

Starting from the estimates of $G_0$, one can easily prove that the first four integrals of (20.6) are estimated by $C(t-t_0)^{(2b-|k|+\alpha)\gamma}$. Here to obtain the estimate of the fourth integral we must introduce new variables $\bar{x}$ and $\bar{\xi}$, use (13.6) to shift one derivative from $G_0$, and obtain estimates of the integrals thus obtained, using the fact that $G_0$ is Hölder in the last two arguments. To estimate the fifth integral in the case $|k| = 2b$, we represent it, using (12.6), in the form

$$\sum_{0<|m|<2b}\left[\int_{t_0}^{t} d\tau \int_{K^-} G_{\bar{\xi}}^{m}\bar{G}_0(t-\tau, \bar{x}, \bar{\xi}, t_0)\, c_m(\bar{\xi})\,\bar{\chi}(\bar{\xi})\, d\bar{\xi} + \int_0^{t_0} d\tau \int_{K^-} \Delta_t^{t_0} D_{\bar{\xi}}^{m}\bar{G}_0 c_m \bar{\chi}\, d\bar{\xi}\right]$$

$$+\sum_{|m|=2b}\Delta_t^{t_0}\int_0^t d\tau \int_{K^-} D_{\bar{\xi}}^{m}\bar{G}_0(t-\tau, \bar{x}, \bar{\xi}, t_0)\, c_m \bar{\chi}\, d\bar{\xi}.$$

The terms of the first sum can be easily estimated by $C(t-t_0)^{\alpha\gamma}$, directly using the estimates of $G_0$. In the integrals of the second sum we use (13.6) to shift one derivative from $G_0$, and as a result we obtain integrals analogous to terms of the first sum, and there appears the difference

$$\begin{aligned} J_1' &= \Delta_t^{t_0} \int_0^t d\tau \int_{K^-} D_\tau \overline{G}_0 (t-\tau, \overline{x}, \overline{\xi}, t_0) \overline{a}_{10} (t_0, \overline{x}, \overline{\xi}) \overline{\chi} (\overline{\xi}) \, d\overline{\xi} \\ &= \int_{K^-} \Delta_t^{t_0} \overline{G}_0 (t, \overline{x}, \overline{\xi}, t_0) \overline{a}_{10} (t_0, \overline{x}, \overline{\xi}) \overline{\chi} (\overline{\xi}) \, d\overline{\xi} \\ &= \int_{U^-} \Delta_t^{t_0} G_0 (t, x-\xi; t_0, x) a_{10} (t_0, x, \xi) \chi (\xi - x^0) \, d\xi, \end{aligned}$$

which can be estimated in the following manner:

$$\begin{aligned} |J_1'| &\leqslant C(t-t_0)^{\alpha\nu} \int_{U^-} [t^{-(n+\alpha)\nu} \Psi_c (t, x; 0, \xi) + t_0^{-(n+\alpha)\nu} \Psi_c (t_0, x; 0, \xi)] \, d\xi \\ &\leqslant C(t-t_0)^{\alpha\nu} \int_{U^-} [t^{-n\nu} \Psi_{c'} (t, x; 0, \xi) + t_0^{-n\nu} \Psi_{c'} (t_0, x; 0, \xi)] \, d^{-\alpha} (t_0, x; 0, \xi) \, d\xi. \end{aligned}$$

Since $\xi \in U^-$ and $x \in U_\delta(x^0)$, therefore $d(t_0, x; 0, \xi) \geqq r(t_0, x)$, and

$$|J_1'| \leqslant C(t-t_0)^{\alpha\nu} r^{-\alpha} (t_0, x)$$

and the estimate (18.6) for $J_1$ is proved.

To estimate $J_2$ in (19.6) we write, using (11.6),

$$\begin{aligned} J_2 &= \Delta_t^{t_0} \int_0^t d\tau \int_{U^-} D_x^k \Big[ \int_\tau^t d\beta \int_{E_n} G_0 (t-\beta, x-y; \beta, y) \varphi (\beta, y; \tau, \xi) \, dy \Big] \chi \, d\xi \\ &= \Delta_t^{t_0} D_x^k \int_0^t d\beta \int_{E_n} G_0 (t-\beta, x-y; \beta, y) \psi (\beta, y) \, dy, \end{aligned}$$

where

$$\psi (\beta, y) = \int_0^\beta d\tau \int_{U^-} \varphi (\beta, y; \tau, \xi) \chi (\xi - x^0) \, d\xi.$$

If $\psi \in C_{\beta, y}^{\alpha, \alpha\gamma} (\Pi)$, it follows (see (7.6)) that

$$|J_2| \leqslant C(t-t_0)^{(2b-|k|+\alpha)\nu}.$$

The fact that $\psi \in C_{\beta, y}^{\alpha, \alpha\gamma} (\Pi)$ can be proved on the basis of the properties of $\phi$ by means of considerations analogous to those used above (cf. also [**28**], p. 414 (transl. p. 364)). Thus the estimate (15.6) is proved.

In an analogous manner we can prove the estimate for the increment in the derivatives of $v_1''$ with respect to $x$:

$$|\Delta_x^{x_0} D_x^k v_1''| \leqslant C|x-x^0|^\alpha \|f_1\|_\alpha^{\Pi} \quad (|k|=2b). \tag{21.6}$$

From the estimates (14.6), (15.6) and (21.6) there follows the estimate (8.6), and consequently the inequality (4.6) for $v_1$.

Consider now the estimate of $v_2$ from (3.6). Let $\delta$ be sufficiently small so that $U^+ \cap \Omega^{(k')} = 0$ for all $k'$, and let $h$ be the same as in §3. Then, in view of the results of §1.2 (see the Note at the end of that subsection), the matrix $v(t,x;\tau,\xi)$ for $(t,x)\in Q_0$ can be represented in the form

$$\begin{aligned} v(t,x;\tau,\xi) = \sum_{k''} \eta^{(k'')}(x)\,\Pi_{\bar x}^{x}\Big[\int_\tau^t d\beta \int_K G_0^{(k'')}(t-\beta,\bar x,\bar y)\,\bar\zeta^{(k'')}(\bar y)\,\bar w_0(\beta,\bar y;\tau,\xi)\,d\bar y \\ + \sum_{j=1}^{bN}\int_\tau^t d\beta \int_{K'} G_j^{(k'')}(t-\beta,\bar x-\bar y')\,\bar\zeta^{(k'')}(\bar y')\,\bar w_j(\beta,\bar y';\tau,\xi)\,d\bar y'\Big], \end{aligned} \tag{22.6}$$

where $G_0^{(k'')}$ is the Green matrix and $G_j^{(k'')}$ are the Poisson kernels of the problem (16.1)–(18.1) with coefficients fixed at the point $(0,\xi^{(k'')})$.

Using the representation (22.6), we have

$$\begin{aligned} v_2(t,x) = \sum_{k''} \eta^{(k'')}(x)\,\Pi_{\bar x}^{x}\Big\{\int_0^t d\beta \int_K G_0^{(k'')}(t-\beta,\bar x,\bar y)\,\bar\zeta^{(k'')}(\bar y)\,(\overline{W}_0 f)(\beta,\bar y)\,d\bar y \\ + \sum_{j=1}^{bN}\int_0^t d\beta \int_{K'} G_j^{(k'')}(t-\beta,\bar x-\bar y')\,\bar\zeta^{(k'')}(\bar y')\,(\overline{W}_j f)(\beta,\bar y')\,d\bar y'\Big\}, \end{aligned} \tag{23.6}$$

where

$$\begin{gathered} \overline{W}_j f = \Pi_y^{\bar y} W_j f, \\ W_j f = \int_0^\beta d\tau \int_{U^+} w_j(\beta,y;\tau,\xi)\,f(\tau,\xi)\,\chi(\xi-x^0)\,d\xi \qquad (j=0,1,\dots,bN). \end{gathered} \tag{24.6}$$

The following lemma holds for the operators $W_j$.

**Lemma** 5. *If $f\in \overset{\circ}{C}{}^\alpha(Q)$, then $W_j f$ belongs to the class $\overset{\circ}{C}{}^\alpha(\overline{Q}_{0,h})$ for $j>0$ and*

$$\|W_0 f\|_\alpha^{\overline{Q}_{0,h}} \leqslant C\|f\|_\alpha^Q, \quad \|W_j f\|_{2b-r_j+\alpha}^{\Gamma_{0,h}} \leqslant C\|f\|_\alpha^Q \quad (j>0). \tag{25.6}$$

Using this lemma, it follows from (23.6) in view of Lemma 4 (§2) that $v_2$ belongs to the class $\underset{\circ}{C}{}^{2b+\alpha}(Q_0)$ and satisfies the estimate (4.6). Thus to complete the proof of Theorem 6 it suffices to prove Lemma 5.

**Proof of Lemma** 5. Recall (see §1) that the $w_j$ are given by the following uniformly converging series:

$$w_j(\beta, y; \tau, \xi) = \sum_{\nu=0}^{\infty} w_j^{(\nu)}(\beta, y; \tau, \xi) \quad (j = 0, 1, \dots, bN),$$

where $w_j^{(\nu)}$ are defined in (22.1), and $w_0^{(0)} = 0$ and $w_j^{(0)} = B_j(\beta, y, D_y) Z|_{\Gamma_{0,h}}$. Therefore

$$W f = \sum_{\nu=0}^{\infty} W_j^{(\nu)} f \qquad (j = 0, 1, \dots, bN), \tag{26.6}$$

where

$$W_j^{(\nu)} f = \int_0^{\beta} d\tau \int_{U+} w_j^{(\nu)}(\beta, y; \tau, \xi) f(\tau, \xi) \chi(\xi - x^0)\, d\xi.$$

Let us now study the operators $W_j^{(\nu)} f$. From the estimate (4.6) and the properties of the f.s.m. $Z$ one can easily obtain that the vector function $v_1(\beta, y)$ defined in (3.6) belongs to the class $C_0^{2b+\alpha}(\overline{Q}_{0,h})$. Hence in view of the properties of the coefficients $B_j$ it follows that

$$W_j^{(0)} f \in \underset{\circ}{C}^{2b-r_j+\alpha}(\Gamma_{0,h}), \; \| W_j^{(0)} f \|_{2b-r_j+\alpha}^{\Gamma_{0,h}} \leqslant C_0 \| f \|_{\alpha}^{Q} \quad (j > 0). \tag{27.6}$$

Further, let us consider $W_0^{(1)} f$. Using the expression for $w_0^{(1)}$, we find

$$W_0^{(1)} f = \sum_{k''} \sum_{j=1}^{bN} \{ A_1(\beta, y, D_y) \eta^{(k'')}(y) \Pi_{y}^{\bar{y}}$$

$$+ \eta^{(k'')}(y) \Pi_{y}^{\bar{y}} [A_0^{(k'')}(0, \xi^{(k'')}, \quad D_{\bar{y}} - \operatorname{grad} F \cdot D_{\bar{y}_n}) - A_0^{(k'')}(0, \xi^{(k'')}, D_{\bar{y}})]$$

$$+ \eta^{(k'')}(y) [A_0(\beta, y, D_y) - A_0(0, \xi^{(k'')}, D_y)] \Pi_{y}^{\bar{y}} + [A_0(\beta, y, D_y) \eta^{(k'')}(y)$$

$$- \eta^{(k'')}(y) A_0(\beta, y, D_y)] \Pi_{y}^{\bar{y}} \} \int_0^{\beta} d\gamma \int_{K'} G_j^{(k'')}(\beta - \gamma, \bar{y} - \bar{z}') \bar{\zeta}^{(k'')}(\bar{z}')$$

$$\times \Pi_{z}^{\bar{z}} (W_j^{(0)} f)(\gamma, z)\, d\bar{z}'.$$

This together with Lemma 4 and (27.6) implies that

$$W_0^{(1)} f \in \underset{\circ}{C}^{\alpha}(\overline{Q}_{0,h}), \; \| W_0^{(1)} f \|_{\alpha}^{\overline{Q}_{0,h}} \leqslant C_1 \Lambda \| f \|_{\alpha}^{Q},$$

where $\Lambda$ is defined in §3.

In an analogous manner we prove that

$$W_j^{(1)}f \in \overset{\circ}{C}{}^{2b-r_j+\alpha}(\Gamma_{0,h}), \quad \|W_j^{(1)}f\|_{2b-r_j+\alpha}^{\Gamma_{0,h}} \leqslant C_1\Lambda\|f\|_\alpha^Q.$$

Extending the same considerations to $W_j^{(\nu)}f$ for $\nu = 2, 3, \cdots$, we obtain that

$$\begin{aligned} &W_0^{(\nu)}f \in \overset{\circ}{C}{}^{\alpha}(\overline{Q}_{0,h}), \quad \|W_0^{(\nu)}f\|_\alpha^{\overline{Q}_{0,h}} \leqslant C_1C_2^{\nu-1}\Lambda^\nu\|f\|_\alpha^Q, \\ &W_j^{(\nu)}f \in \overset{\circ}{C}{}^{2b-r_j+\alpha}(\Gamma_{0,h}), \quad \|W_j^{(\nu)}f\|_{2b-r_j+\alpha}^{\Gamma_{0,h}} \leqslant C_1C_2^{\nu-1}\Lambda^\nu\|f\|_\alpha^Q, \end{aligned} \tag{28.6}$$

where $C_1$ and $C_2$ are independent of $\nu$.

From (26.6) and (28.6) follows the proof of Lemma 5. Thus Theorem 6 is fully proved.

We proceed now to the proof of Theorem 2, formulated in §1.

**Proof of Theorem** 2. From Theorem 6 and from the uniqueness of the solution of the boundary value problem in the class $C_{x,t}^{2b+\alpha,(2b+\alpha)\gamma}(Q)$ it follows that the operator $\mathscr{L}^{-1}$ coincides with the operator $G$ on the set $\overset{\circ}{C}{}_0^\alpha(Q)$. Therefore, if the domain of definition $\mathscr{R}(\mathscr{L})$ of the operator $\mathscr{L}^{-1}$ coincides with $\overset{\circ}{C}{}_0^\alpha(Q)$, the assertion of Theorem 2 is proved for this case.

We consider now the case when $\mathscr{R}(\mathscr{L})$ does not coincide with $\overset{\circ}{C}{}_0^\alpha(Q)$, but is included in the space $L_2(Q)$. Let $M$ be the set of smooth vector functions which are equal to zero for small $t$ and also for large $|x|$, if the domain $\Omega$ is infinite. The assertion of the theorem in this case easily follows from the fact that the set $M$ is compact in the space $L_2(Q)$, and that the operators $\mathscr{L}_2(Q)$, and that the operators $\mathscr{L}^{-1}$ and $G$ coincide on the set, as stated above. In fact, we first note that the operator $G$ is bounded from $L_2(Q)$ into $L_2(Q)$. Indeed, using the estimate (6.1) for the Green matrix $G$, we have

$$\begin{aligned} |Gf| &\leqslant C\int_0^t d\tau\int_\Omega (t-\tau)^{-n\nu}\Psi_c(t, x; \tau, \xi)|f(\tau, \xi)|\,d\xi \\ &\leqslant C\Big[\int_0^t d\tau\int_\Omega (t-\tau)^{-n\nu}\Psi_c(t, x; \tau, \xi)|f(\tau, \xi)|^2\,d\xi\Big]^{\frac{1}{2}} \\ &\quad\times\Big[\int_0^t d\tau\int_\Omega (t-\tau)^{-n\nu}\Psi_c(t, x; \tau, \xi)\,d\xi\Big]^{\frac{1}{2}} \\ &\leqslant C\Big[\int_0^t d\tau\int_\Omega (t-\tau)^{-n\nu}\Psi_c(t, x; \tau, \xi)|f(\tau, \xi)|^2\,d\xi\Big]^{\frac{1}{2}}, \end{aligned}$$

from which we get

$$\|Gf\|_{L_2(Q)} \leqslant C\Big\{\int_0^T d\tau \int_\Omega |f(\tau,\xi)|^2 \Big[\int_\tau^T dt \int_\Omega (t-\tau)^{-n\nu}$$

$$\times \Psi_c(t,x;\tau,\xi)\,dx\Big]\,d\xi\Big\}^{\frac{1}{2}} \leqslant C\|f\|_{L_2(Q)}.$$

It is known (cf. [**27**], [**4**] and [**30**]) that the operator $\mathscr{L}^{-1}$ admits an extension to $L_2(Q)$ which is a bounded operator from $L_2(Q)$ into $L_2(Q)$. Since the operators $G$ and $\mathscr{L}^{-1}$ coincide on the dense set $M$, they coincide on the whole $L_2(Q)$, whence it follows in particular that $\mathscr{L}^{-1}f = Gf$ for $f \in \mathscr{R}(\mathscr{L})$.

**2. The elliptic Green operator and fractional powers.** We consider the operator

$$\Phi f = \int_\Omega \Phi(x,\xi;\lambda) f(\xi)\,d\xi, \tag{29.6}$$

where $\Phi(x,\xi;\lambda)$ is the Green matrix of the problem (6.5), (7.5).

THEOREM 7. *The operator $\Phi$ is bounded from the space $C^\alpha(\Omega)$ into the space $C^{2b+\alpha}(\Omega)$.*

PROOF. Let $t_0$ be an arbitrary fixed positive number and let $\chi(\tau)$ be a smooth function satisfying the condition

$$\chi(\tau) = \begin{cases} 0, & \tau \leqslant 0, \\ 1, & \tau \geqslant \frac{1}{2} t_0. \end{cases}$$

Using (8.5), we write

$$\Phi f = \int_\Omega \Big[\int_{-\infty}^{t_0} G(t_0-\tau,x,\xi;\lambda)\,d\tau\Big] f(\xi)\,d\xi$$

$$= \int_\Omega \Big[\int_0^{t_0} G(t_0-\tau,x,\xi;\lambda)\chi(\tau)\,d\tau\Big] f(\xi)\,d\xi$$

$$+ \int_\Omega \Big[\int_{-\infty}^{\frac{t_0}{2}} G(t_0-\tau,x,\xi;\lambda)[1-\chi(\tau)]\,d\tau\Big] f(\xi)\,d\xi = u_1(x) + u_2(x).$$

In view of the estimates of the Green matrix $G(t_0-\tau,x,\xi;\lambda)$, the vector function $u_2(x)$ belongs to $C^{2b+\alpha}(\Omega)$ and

$$\|u_2\|^\Omega_{2b+\alpha} \leqslant C\|f\|^\Omega_\alpha.$$

The vector function $u_1(x)$ represents the value for $t=t_0$ of the vector function

$$u_1(t, x) = \int_0^t d\tau \int_\Omega G(t-\tau, x, \xi; \lambda)\chi(\tau) f(\xi)\, d\xi.$$

But since $\chi(\tau) f(\xi) \in \overset{\circ}{C}{}^\alpha(\overline{Q}_{0,t_0})$, we have, in view of Theorem 6,

$$\| u_1(t, x)\|_{2b+\alpha}^{\overline{Q}_{0,t_0}} \leqslant C \| \chi(\tau) f(\xi)\|_\alpha^{\overline{Q}_{0,t_0}} \leqslant C \| f \|_\alpha^\Omega,$$

which implies the inequality

$$\| u_1(x) \|_{2b+\alpha}^\Omega \leqslant C \| f \|_\alpha^\Omega$$

which proves the theorem.

The estimates obtained in §5 for the kernels of fractional powers of the elliptic operator make it possible to obtain various theorems concerning these powers. Here we shall state only one theorem concerning the operator

$$\Phi_\beta f = \int_\Omega \Phi_\beta(x, \xi; \lambda) f(\xi)\, d\xi \quad (0 < \beta < 1),$$

where $\Phi_\beta(x, \xi; \lambda)$ is defined by formula (15.5). Its proof is easily obtained with the help of the technique used above.

THEOREM 8. *If the number $2b\beta + \alpha$ is not an integer, then the operator $\Phi_\beta$ is bounded from the class $C_0^\alpha(\Omega)$ into $C^{2b\beta+\alpha}(\Omega)$. Here $C_0^\alpha(\Omega)$ is the class of vector functions in the space $C^\alpha(\Omega)$ for which the quantity*

$$\| f(x)[1 + d^{-\alpha}(x, S)]\|_0^\Omega,$$

*is finite, where $d(x, S)$ is the distance of the point $x$ from the boundary $s$.*

## Appendix
## Proof of Lemmas 1 and 2

We shall use the following notation:

1) $(t, x) = P,\ (\beta, y) = Q,\ (\tau, \xi) = M;$

2) $(t, \bar{x}) = \overline{P},\ (\beta, \bar{y}) = \overline{Q},\ (\tau, \hat{\xi}) = \hat{M};$

3) $(t, \bar{x}') = \overline{P}',\ (\beta, \bar{y}') = \overline{Q}',\ (\tau, \bar{\xi}') = \overline{M}';$

4) $d\beta\, dy = dQ,\ d\beta\, d\bar{y}' = d\overline{Q}';$

5) $d_0 = d(P, M) = ((t-\tau)^{\frac{1}{b}} + |x - \xi|^2)^{\frac{1}{2}}.$

With this notation, formulas (1.2) and (2.2) for the operators $R^{(k)}h$ and $\overline{R}_j h$ take on the form

(1) $$u_k(P, M) = R^{(k)}h = \int_\tau^t d\beta \int_{\Omega^{(k)}} R^{(k)}(P, Q)\,\zeta(y)\,h(Q, M)\,dy,$$

(2) $$\bar v_j(\bar P, \hat M) = \bar R_j h = \int_\tau^t d\beta \int_{K'} G_j(\bar P - \bar Q')\,\bar\zeta(\bar y')\,\bar h(\bar Q'; \hat M)\,d\bar y'.$$

**1. Proof of Lemma 1.** Note that

(2) $$R^{(k'')}h = \Pi_{x,\xi}^{\bar x,\bar\xi} \bar R^{(k'')}h, \quad \bar R^{(k'')}h = \int_\tau^t d\beta \int_{K'} G_0(t-\beta, \bar x, \bar y)\,\bar\zeta(\bar y)\,\bar h(\bar Q; \hat M)\,d\bar y,$$

where $K$ is the standard cube, which is the image of $\Omega^{(k'')}$ under the transformation (12.1).

Since $R^{(k')}h$ and $\bar R^{(k'')}h$ satisfy the nonhomogeneous systems (15.1) and (16.1), the derivative $D_t u_k$ can be expressed in terms of the $x$ derivatives of $u_k$ and $\zeta h$. Therefore it suffices to estimate only the derivatives

$$D_x^m u_k \quad (|m| \leqslant 2b + l).$$

From (4.2), (5.2), (25.1) and (26.1) it follows that

(3) $$R^{(k)}(P, Q) \in U_{C,c_2}^{2b+l+\alpha,0,s}(Q_{\beta,\tau_1}^{(k)}, \tilde Q_{\tau,\tau_1}^{(k)}), \quad c_2 = 2^{-q/2}c_0,$$

where $\tilde Q_{\tau,\tau_1}^{(k)} = [\tau, \tau_1] \times \tilde\Omega^{(k)}$ and $\tilde\Omega^{(k)}$ is a neighborhood of $\Omega^{(k)}$ arbitrary for $k = k'$, and for $k = k''$ is such that it lies in a Ljapunov sphere with center at $\xi^{(k'')} \in S$. Note that the derivatives of $R^{(k)}(P, Q)$ with respect to $Q$ satisfy the same estimates as the derivatives with respect to $P$. For $k = k'$ this is obvious, and for the case $k = k''$ this can be easily obtained by the considerations of [2]. Therefore in view of (3) we have

(4) $$|\Delta_Q^M D_x^m R^{(k)}(P, Q)| \leqslant C d^\alpha(Q, M)(t-\tau)^{\nu s}[d^{-n-|m|-s-\alpha}(P, Q)\,\Psi_{c_2}(P, Q) + d^{-n-|m|-s-\alpha}(P;\beta,\xi)\,\Psi_{c_2}(P, M)] \quad (|m| \leqslant 2b + l)$$

for $P \in Q_{\beta,\tau_1}^{(k)}$ and $Q, M \in \tilde Q_{\tau,\tau_1}^{(k)}$.

We first estimate the derivatives $D_x^m u_k$ ($|m| \leq 2b$), taking into account that $P \in \bar Q_{\tau,\tau_1}^{(k)}$. We split the range of integration $\bar Q_{\tau,t}^{(k)}$ in (1) into three parts $\bar Q_{\tau,t}^{(k)} = Q_1^{(k)} \cup Q_2^{(k)} \cup Q_3^{(k)}$, where

$$Q_1^{(k)} = \bar Q_{\tau,t}^{(k)} \cap \left\{d(Q, M) \leqslant \frac{1}{2}(t-\tau)^\nu\right\}, \quad Q_2^{(k)} = \bar Q_{\tau,t}^{(k)} \cap \left\{d(P, Q) \leqslant \frac{1}{2}(t-\tau)^\nu\right\}.$$

The singularities of the kernel and of the function $h$ will be separate, and a sufficiently exact estimate will be obtained. We note that it suffices to carry out the derivations for the case $\xi \in \widetilde{\Omega}^{(k)}$. When $\xi \in \widetilde{\Omega}^{(k)}$, by choosing sufficiently small $\chi$ and using the fact that $t-\tau \leqq \tau_1 - \tau = \chi\lambda^{2b}$ we can ensure that $Q_1^{(k)}$ will be empty, and the estimate becomes simpler.

We have

$$
\begin{aligned}
D_x^m u_k(P, M) &= \iint\limits_{Q_1^{(k)}} \Delta_Q^M D_x^m R^{(k)}(P, Q)\,\zeta(y)\,h(Q, M)\,dQ \\
&+ D_x^m R^{(k)}(P, M) \iint\limits_{Q_1^{(k)}} \zeta(y)\,h(Q, M)\,dQ + \iint\limits_{Q_2^{(k)}} D_x^m R^{(k)}(P, Q)\,\zeta(y)\,\Delta_Q^P h(Q, M)\,dQ \\
&+ \iint\limits_{Q_2^{(k)}} D_x^m R^{(k)}(P, Q)\,\zeta(y)\,dQ\,h(P, M) \\
&+ \iint\limits_{Q_3^{(k)}} D_x^m R^{(k)}(P, Q)\,\zeta(y)\,h(Q, M)\,dQ = \sum_{i=1}^{5} E_i.
\end{aligned} \tag{5}
$$

Note that here the differentiation uses formulas which define the higher-order derivatives of integrals of volume potential type, which are singular integrals.

Using (3), (4) with corresponding $s$, and the fact that

$$h \in H_{2b, c_1}^{l+\alpha}(\overline{Q}_{\tau,\tau_1}, Q),$$

we have

$$
\begin{aligned}
\sum_{i=1}^{3} |E_i| &\leqslant C\|h\|(t-\tau)^{\nu(2b+l-|m|+\alpha)} \Big\{ \iint\limits_{Q_1^{(k)}} [d^{-n-2b-l-2\alpha}(P, Q)\,\Psi_{c_2}(P, Q) \\
&+ d^{-n-2b-l-2\alpha}(P; \beta, \xi)\,\Psi_{c_2}(P, M)]\, d^{-n-2b+\alpha}(Q, M)\,\Phi_{c_1}(Q, M)\,dQ \\
&+ d^{-n-2b-l-\alpha}\Psi_{c_2}(P, M)\,\rho_{c_1}(t, \tau, \xi) + (t-\tau)^{-\nu\alpha} \iint\limits_{Q_2^{(k)}} d^{-n-2b+\alpha}(P, Q)\,\Psi_{c_2}(P, Q) \\
&\times [d^{-n-2b-l-\alpha}(Q, M)\,\Phi_{c_1}(Q, M) + d^{-n-2b-l-\alpha}(\beta, x; M)\,\Phi_{c_1}(P, M)]\,dQ\Big\}.
\end{aligned} \tag{6}
$$

For $Q \in Q_1^{(k)}$

$$d(P, Q) \geqslant d_0 - d(Q, M) \geqslant d_0 - \frac{1}{2}(t-\tau)^\nu \geqslant \frac{1}{2} d_0,$$

$$d(P; \beta, \xi) \geqslant d_0 - (\beta-\tau)^\nu \geqslant \frac{1}{2} d_0.$$

In an analogous manner we show that for $Q \in Q_2^{(k)}$

$$d(Q, M) \geqslant \frac{1}{2} d_0, \quad d(\beta, x; M) \geqslant \frac{1}{2} d_0.$$

From these inequalities and (6) we obtain

$$\sum_{i=1}^{3} |E_i| \leqslant C\|h\|(t-\tau)^{\nu(2b+l+\alpha-|m|)} d_0^{-n-2b-l-\alpha}\Phi_c(P, M)$$

$$\times \left[d_0^{-\alpha} \iint_{Q_1^{(k)}} d^{-n-2b+\alpha}(Q, M)\, dQ + 1 + (t-\tau)^{-\nu\alpha} \iint_{Q_2^{(k)}} d(P, Q)\, dQ\right],$$

which implies the estimate

$$\text{(7)} \qquad \sum_{i=1}^{3} |E_i| \leqslant C\|h\|(t-\tau)^{\nu(2b+l-|m|+\alpha)} d_0^{-n-2b-l-\alpha}\Phi_c(P, M),$$

since the last integrals are estimated by $C(t-\tau)^{\gamma\alpha}$. We estimate, as an example, the first of these. Introducing the following change of the variables of integration:

$$y_i - \xi_i = z_i \; (i = 1, \ldots, n), \quad (\beta - \tau)^{\nu} = z_0, \quad |z|^2 = \sum_{i=0}^{n} z_i^2$$

and then transforming to spherical coordinates, we obtain

$$\iint_{Q_1^{(k)}} d^{-n-2b+\alpha}(Q, M)\, dQ \leqslant C \iint_{|z| \leqslant \frac{1}{2}(t-\tau)^{\nu}} |z|^{-n-2b+\alpha} |z_0|^{2b-1}\, dz$$

$$\leqslant C \int_0^{\frac{1}{2}(t-\tau)^{\nu}} \rho^{-1+\alpha}\, d\rho = C(t-\tau)^{\nu\alpha}.$$

To prove the estimate (7) for $E_4$ it suffices to prove the inequality

$$\text{(8)} \qquad E_4^0 = \left|\iint_{Q_2^{(k)}} D_x^m R^{(k)}(P, Q)\, \zeta(y)\, dQ\right| \leqslant C(t-\tau)^{\nu(2b-|m|)}.$$

In the case $|m| < 2b$ the estimate (8) follows directly from (3) ($s = 0$). Let $|m| = 2b$. For $k = k'$ in $E_4^0$ we integrate by parts to shift one derivative with respect to $y$ from $Z_0$ to $\zeta$, and we estimate the result using (3) and (13.1):

$$E_4^0 = \Big| -\iint\limits_{Q_2^{(k')}} D_y D_x^{m-1} Z_0 (P-Q)\,\zeta(y)\,dQ \Big|$$

$$= \Big| \iint\limits_{Q_2^{(k')}} D_x^{m-1} Z_0 D_y \zeta\, dQ - \int\limits_{\Gamma'} D_x^{m-1} Z_0 \zeta \cos(\vec{n}, y_i)\, d\Gamma' \Big|$$

$$\leqslant C \Big[ \frac{(t-\tau)^{\nu}}{\lambda} + (t-\tau)^{-\nu(n+2b-1)} \int\limits_{\Gamma'} |\cos(\vec{n}, y_i)|\, d\Gamma' \Big]$$

$$\leqslant C \Big[ \frac{(\tau_1-\tau)^{\nu}}{\lambda} + (t-\tau)^{-\nu(n+2b-1)} \int\limits_{d^{(i)} \leqslant \frac{1}{2}(t-\tau)^{\nu}} d\beta\, dy^{(i)} \Big] \leqslant C(\chi^{\nu}+1) \leqslant C.$$

Here $\Gamma'$ is the curved part of the boundary of $Q_2^{(k')}$, i.e. the set of all points $Q \in \overline{Q}_{\tau,t}^{(k')}$ for which $d(P,Q) = \frac{1}{2}(t-\tau)^{\nu}$, $n$ is the outer normal to $\Gamma'$; $d^{(i)} = d(t, x^{(i)}; \beta, y^{(i)})$ and $x^{(i)} = (x_1, \cdots, x_{i-1}, x_{i+1}, \cdots, x_n)$. To estimate $E_4^0$ in the case $k = k''$ it clearly suffices to show that

$$\overline{E}_4^0 = \Big| \iint\limits_{\mathcal{K}_2} D_{\bar{x}}^m G_0(t-\beta, \bar{x}, \bar{y})\,\bar{\zeta}(\bar{y})\, d\overline{Q} \Big| \leqslant C(t-\tau)^{\nu(2b-|m|)}, \tag{9}$$

where

$$\mathcal{K}_2 = \Big\{ \overline{Q} \in [\tau,t] \times K,\ d(t, \bar{x}', \bar{x}_n + F(\bar{x}'); \beta, \bar{y}', \bar{y}_n + F(\bar{y}')) \leqslant \frac{(t-\tau)^{\nu}}{2} \Big\}$$

is the image of $Q_2^{(k'')}$ under the transformation from $y$ to the $k''$-rectifying coordinates $\bar{y}$. For $|m| < 2b$, $E_4^0$ can be estimated by means of (4.2). If $|m| = 2b$ and $m_n < 2b$, then by virtue of the fact that

$$G_0(t-\beta, \bar{x}, \bar{y}) = G_0(t-\beta, \bar{x}' - \bar{y}', \bar{x}_n, \bar{y}_n), \tag{10}$$

[2] and the properties of $\zeta$, one derivative with respect to $\bar{x}'$ can be shifted from $G_0$ to $\bar{\zeta}$, and then, using (4.2) and (13.1), we can obtain the estimate (9) in the same manner as in the case $k = k'$. When $|m| = m_n = 2b$, then, using the system (16.1) ($\bar{f} \equiv 0$), which has the solution $G_0$ for $t > \beta$, the derivative $D_{\bar{x}_n}^{2b} G_0$ can be expressed in terms of $D_t G_0$ and $D_{\bar{x}}^m G_0$, where $D_{\bar{x}}^m$ contains at least one differentiation with respect to $\bar{x}'$, and integrals with such kernels are easy to estimate.

Finally, the estimate (7) for $E_5$ can be obtained by means of (3), the condition of the lemma, and (24.1), in the following manner:

$$|E_5| \leqslant C\|h\|(t-\tau)^{\nu(2b+2l-|m|+2\alpha)} \iint\limits_{Q_3^{(k)}} d^{-n-2b-l-\alpha}(P,Q)$$

$$\times \Psi_{c_2}(P,Q)\, d^{-n-2b-l-\alpha}(Q,M)\, \Phi_{c_1}(Q,M)\, dQ$$

$$\leqslant C\|h\|(t-\tau)^{\nu(2b+2l+2\alpha-|m|)} \Phi_c(P,M) \iint\limits_{Q_3^{(k)}} d^{-n-2b-l-\alpha}(P,Q)\, d^{-n-2b-l-\alpha}(Q,M)\, dQ,$$

but since

$$\iint\limits_{Q_3^{(k)}} d^{-n-2b-l-\alpha}(P,Q)d^{-n-2b-l-\alpha}(Q,M)\,dQ = \iint\limits_{Q_3^{(k)}\cap\{d(P,Q)\leqslant 1/2d_0\}} \ldots dQ$$

$$+ \iint\limits_{Q_3^{(k)}\cap\{d(P,Q)>\frac{1}{2}d_0\}} \ldots dQ \leqslant C\Big[d_0^{-n-2b-l-\alpha} \iint\limits_{\frac{1}{2}(t-\tau)^\nu\leqslant|z|\leqslant\frac{1}{2}d_0} |z|^{-n-l-1-\alpha}\,dz$$

$$+ d_0^{-n-2b-l-\alpha} \iint\limits_{|z|>\frac{1}{2}(t-\tau)^\nu} |z|^{-n-l-1-\alpha}\,dz\Big] \leqslant Cd_0^{-n-2b-l-\alpha}(t-\tau)^{-\nu(l+\alpha)},$$

this leads to the inequality (7) for $E_5$.

Thus we have obtained the estimates

$$|D_x^m u_k(P,M)| \leqslant C\|h\|(t-\tau)^{\nu(2b+l-|m|+\alpha)}d_0^{-n-2b-l-\alpha}\Phi_c(P,M) \qquad (|m|\leqslant 2b). \tag{11}$$

We now estimate the increments in the derivatives $D_x^m u_k$ ($|m|\leqq 2b$) with respect to $x$. Let $x$ and $x_0$ be arbitrary points $\Omega^{(k)}$, and let $d=|x-x_0|$. If $d\geqq\frac{1}{4}(t-\tau)^\gamma$, then from (11) there directly follows the estimate

$$|\Delta_x^{x_0}D_x^m u_k| \leqslant C\|h\|d^\alpha(t-\tau)^{\nu(2b+l-|m|)}d_0^{-n-2b-l-\alpha}\Phi_c(P,M). \tag{12}$$

For the sake of definiteness we assume that $|x-\xi|\leqq|x_0-\xi|$.

The estimate (12) is valid also in the case $d<\frac{1}{4}(t-\tau)^\gamma$. To show this we represent the range of integration in the form $\overline{Q}_{\tau,t}^{(k)}=Q_1^{(k)}\cup Q_2^{(k)}\cup Q_3^{(k)}$, and then we isolate from $Q_1^{(k)}$ and $Q_2^{(k)}$ the parts

$$Q_{1d}^{(k)}=Q_1^{(k)}\cap\{d(Q,M)\leqslant 2d\} \text{ and } Q_{2d}^{(k)}=Q_2^{(k)}\cap\{d(P,Q)\leqslant 2d\}$$

respectively. Thus we write

$$\Delta_x^{x_0}D_x^m u_k = \iint\limits_{Q_1^{(k)}} \Delta_x^{x_0}D_x^m R^{(k)}(P,Q)\zeta(y)h(Q,M)\,dQ$$

$$+\Big[\iint\limits_{Q_2^{(k)}} D_x^m R^{(k)}(P,Q)\zeta(y)h(Q,M)\,dQ - \iint\limits_{Q_2^{(k)}} D_{x_0}^m R^{(k)}(P_0,Q)\zeta(y)h(Q,M)\,dQ\Big] \tag{13}$$

$$+\iint\limits_{Q_3^{(k)}} \Delta_x^{x_0}R^{(k)}(P,Q)\zeta(y)h(Q,M)\,dQ = I_1+I_2+I_3,$$

where $P_0=(t,x_0)$. We now write $I_1$ in the form

$$I_1=\iint\limits_{Q_{1d}^{(k)}} \Delta_Q^M D_x^m R^{(k)}(P,Q)\zeta h\,dQ - \iint\limits_{Q_{1d}^{(k)}} \Delta_Q^M D_{x_0}^m R^{(k)}(P_0,Q)\zeta h\,dQ$$

$$+ \iint\limits_{Q_1^{(k)} \setminus Q_{1d}^{(k)}} \Delta_x^{x_0} D_x^m R^{(k)} \zeta h dQ + \Delta_x^{x_0} D_x^m R^{(k)} (P, M) \iint\limits_{Q_{1d}^{(k)}} \zeta h dQ = \sum_{i=1}^{4} I_1^{(i)}.$$

Estimating in the same manner as $E_1$, we obtain

$$| I_1^{(1)} | \leqslant C \| h \| (t - \tau)^{\nu(2b+l-|m|+\alpha)} d_0^{-n-2b-l-2\alpha} \Phi_c (P, M) \iint\limits_{Q_{1d}^{(k)}} d^{-n-2b+\alpha} (Q, M)\, dQ,$$

from which there follows the estimate (12) for $I_1^{(1)}$, since

$$\iint\limits_{Q_{1d}^{(k)}} d^{-n-2b+\alpha} (Q, M) dQ \leqslant C \int_0^{2d} \rho^{-1+\alpha} d\rho = C d^{\alpha}.$$

We can estimate $I_1^{(2)}$ in an analogous manner, noting that for $Q \in Q_1^{(k)}$ we have

$$d (P_0, Q) \geqslant d (P, Q) - | x - x_0 | \geqslant \frac{1}{4} d_0.$$

Further, in the case $k = k'$, using for $Z_0$ (4.2) with $\alpha_0 > \alpha$ and the inequality (26.1), we obtain

$$| I_1^{(3)} | \leqslant C \| h \| (t - \tau)^{\nu(2b+l-|m|)} d^{\alpha_0} \iint\limits_{Q_1^{(k')} \setminus Q_{1d}^{(k')}} [d^{-n-2b-l-\alpha_0} (P, Q) \Psi_{c_0} (P, Q)$$

$$+ d^{-n-2b-l-\alpha_0} (P_0, Q) \Psi_{c_0} (P_0, Q)] \, d^{-n-2b} (Q, M) \Phi_{c_1} (Q, M) \, dQ$$

$$\leqslant C \| h \| d^{\alpha_0} (t - \tau)^{\nu(2b+l-|m|)} d_0^{-n-2b-l-\alpha_0} \Phi_c (P, M) \iint\limits_{Q_1^{(k')} \setminus Q_{1d}^{(k')}} d^{-n-2b} (Q, M) \, dQ.$$

Since the last integral is estimated by $C(d_0/d)^{\alpha_0 - \alpha}$, we obtain the estimate (12) for $I_1^{(3)}$. The estimate of $I_1^{(4)}$ in the case $k = k'$ can be obtained immediately by applying to $Z_0$ (4.2) with $\alpha_0 > \alpha$ and the inequality

$$\left| \iint\limits_{Q_{1d}^{(k)}} \zeta (y) h (Q, M) \, dQ \right| \leqslant C \| h \| \left( \frac{d_0}{d} \right)^{\alpha_0 - \alpha} \rho_{c_1} (t, \tau, \xi), \tag{14}$$

to prove which we write

$$\iint\limits_{Q_{1d}^{(k)}} = \iint\limits_{Q_1^{(k)}} - \iint\limits_{Q_1^{(k)} \setminus Q_{1d}^{(k)}} .$$

The first integral satisfies the estimate (33.1), and for the second integral, in view of (30.1) ($p = 2b$), we have

$$\Bigl|\iint\limits_{Q_1^{(k)}\setminus Q_{1d}^{(k)}}\Bigr| \leqslant C\|h\|\rho_{c_1}(t,\tau,\xi)\int\limits_{2d}^{\frac{1}{2}d_0}\frac{d\rho}{\rho} \leqslant C\|h\|\Bigl(\frac{d_0}{d}\Bigr)^{\alpha_0-\alpha}\rho_{c_1}(t,\tau,\xi).$$

To obtain the estimates of $I_1^{(3)}$ and $I_1^{(4)}$ in the case $k=k''$ we note that

$$D_x^m R^{(k'')}(P,Q)=\sum_{|\nu|\leqslant|m|} f_\nu(x)K_\nu(P,Q), \tag{15}$$

where the $f_\nu(x)$ depend on the derivatives of the functions $F$ up to order $|m|$, are bounded and satisfy Hölder's condition with index $\alpha$, and

$$K_\nu(P,Q)=\Pi_{\bar{x},\bar{y}}^{x,y} D_{\bar{x}}^{\nu} G_0(\bar{P},\bar{Q})$$

and in view of (4.2) it belongs to the class $U_{C,c_2}^{\alpha_0,|\nu|,s}(Q_{\beta,\tau_1}^{(k'')},\widetilde{Q}_{\tau,\tau_1}^{(k'')})$ with arbitrary $0<\alpha_0\leqq 1$, $s\geqq 0$. Using (15), we write

$$I_1^{(3)}+I_1^{(4)}=\sum_{|\nu|\leqslant|m|}\Bigl[\Delta_x^{x_0} f_\nu(x)\iint\limits_{Q_1^{(k'')}\setminus Q_{1d}^{(k'')}}\Delta_Q^M K_\nu(P,Q)\zeta h\,dQ$$

$$+f_\nu(x_0)\iint\limits_{Q_1^{(k'')}\setminus Q_{1d}^{(k'')}}\Delta_x^{x_0}K_\nu(P,Q)\zeta h\,dQ+\Delta_x^{x_0}f_\nu(x)K_\nu(P,M)$$

$$\times\iint\limits_{Q_1^{(k'')}}\zeta h\,dQ+f_\nu(x_0)\Delta_x^{x_0}K_\nu(P,M)\iint\limits_{Q_{1d}^{(k'')}}\zeta h\,dQ.$$

The first term under the summation sign can be estimated in the same manner as $I_1^{(4)}$; the second and the fourth, as $I_1^{(3)}$ and $I_1^{(4)}$ in the case $k=k'$; and the third, as $E_2$. Thus we obtain the estimate (12) for $I_1$.

We now estimate $I_2$ from (13). We represent this term in the form

$$I_2=\iint\limits_{Q_{2d}^{(k)}} D_x^m R^{(k)}(P,Q)\zeta(y)\Delta_Q^P h(Q,M)\,dQ$$

$$-\iint\limits_{Q_{2d}^{(k)}} D_{x_0}^m R^{(k)}(P_0,Q)\zeta(y)\Delta_Q^{P_0}h(Q,M)\,dQ$$

$$+\iint\limits_{Q_2^{(k)}\setminus Q_{2d}^{(k)}}\Delta_x^{x_0}D_x^m R^{(k)}(P,Q)\zeta(y)\Delta_Q^P h(Q,M)\,dQ$$

$$+\iint\limits_{Q_2^{(k)}}\Delta_x^{x_0}D_x^m R^{(k)}(P,Q)\zeta(y)\,dQ\,h(P,M)$$

$$+\iint\limits_{Q_{2d}^{(k)}} D_{x_0}^m R^{(k)}(P_0,Q)\zeta(y)\,dQ\,\Delta_x^{x_0}h(P,M)=\sum_{i=1}^{5} I_2^{(i)}.$$

The integrals $I_2^{(1)}$ and $I_2^{(2)}$ can be estimated in the same manner as $E_3$, but one must take into account that for $Q \in Q_{2d}^{(k)}$,

$$d(P_0, Q) \leqslant d(P, Q) + |x - x_0| \leqslant 3d \quad \text{and} \quad \iint_{Q_{2d}^{(k)}} d^{-n-2b+\alpha}(P, Q)\, dQ \leqslant Cd^{\alpha}.$$

In a manner analogous to the estimate of $E_3$ and $I_1^{(3)}$, in the case $k = k'$ we have

$$|I_2^{(3)}| \leqslant C \|h\| (t-\tau)^{\nu(2b+l-|m|)} d_0^{-n-2b-l-\alpha} \Phi_c(P, M)\, d^{\alpha_0}$$

$$\times \iint_{Q_2^{(k')} \setminus Q_{2d}^{(k')}} [d^{-n-2b-\alpha_0+\alpha}(P, Q) + d^{-n-2b-\alpha_0+\alpha}(P_0, Q) + d^{\alpha} d^{-n-2b-\alpha_0}(P_0, Q)]\, dQ,$$

from which we obtain estimate (12), since the last integral is estimated by

$$C\left(\int_{2d}^{\frac{1}{2}d_0} \rho^{-1-\alpha_0+\alpha}\, d\rho + \int_{d}^{\frac{3}{4}d_0} \rho^{-1-\alpha_0+\alpha} d\rho + d^{\alpha} \int_{d}^{\frac{3}{4}d_0} \rho^{-1-\alpha_0} d\rho\right) \leqslant Cd^{-\alpha_0+\alpha}.$$

If $k = k''$, then, using the representation (15), we reduce the estimate of $I_2^{(3)}$ to the estimate of an integral containing $\Delta_x^{x_0} K_\nu$, which can be estimated in the same way as $I_2^{(3)}$ in the preceding case, and an integral containing $\Delta_x^{x_0} f_\nu$, which can be estimated as $E_3$.

To estimate $I_2^{(4)}$ and $I_2^{(5)}$ it suffices to verify the inequalities

$$\left| \iint_{Q_2^{(k)}} \Delta_x^{x_0} D_x^m R^{(k)} \zeta\, dQ \right| \leqslant Cd^{\alpha} (t - \tau)^{\nu(2b-|m|-\alpha)},$$

$$\left| \iint_{Q_{2d}^{(k)}} D_{x_0}^m R^{(k)} \zeta\, dQ \right| \leqslant C (t-\tau)^{\nu(2b-|m|)}.$$

The proof of these inequalities is analogous to the proof of (8): for $|m| < 2b$ these inequalities follow directly from (3), and in the case $|m| = 2b$ one must first shift the derivative from $R^{(k)}$ to $\zeta$ and then use (3).

Finally, $I_3$ in (13) can be estimated in a manner analogous to $E_5$.

Thus the estimate (12) has been proved. The estimate of the increments in $t$ of the derivatives $D_x^m u_k$ can be obtained in an analogous manner.

We shall now estimate the derivatives $D_x^m u_k$ $(2b < |m| \leqq 2b + l)$. We shall first obtain a special representation of these derivatives. Let $(t, x)$ be an arbitrary fixed point in $Q_{\tau,\tau_1}^{(k)}$. Consider $u_k(t, z; M)$ for $z$ in a neighborhood of $x$ which lies inside $\Omega^{(k)}$ and for which $|z - x| \leqq \frac{1}{4}(t-\tau)^{\gamma}$. We represent it in the form

$$u_k(t, z; M) = \iint\limits_{Q_1^{(k)}} \Delta_Q^M R^{(k)}(t, z; Q)\,\zeta h\,dQ + R^{(k)}(t, z; M) \iint\limits_{Q_1^{(k)}} \zeta h\,dQ$$
$$+ \iint\limits_{Q_2^{(k)}} R^{(k)}(t, z; Q)\,\zeta h\,dQ + \iint\limits_{Q_3^{(k)}} R^{(k)}(t, z; Q)\,\zeta h\,dQ, \tag{16}$$

where $Q_i^{(k)}$ $(i = 1, 2, 3)$ are the same as in (5). The first, second and fourth terms admit direct differentiation with respect to $z$ even at $z = x$, and the integrals thus obtained are then estimated as the corresponding terms in (5). The greatest difficulty is presented by the function

$$D_z^m v_k(t, z; M)\,|_{z=x} = D_z^m \iint\limits_{Q_2^{(k)}} R^{(k)}(t, z; Q)\,\zeta h\,dQ\,|_{z=x}.$$

In the differentiation $v_k$ in the case $k = k'$ one can shift $|m| - 2b$ derivatives from $R^{(k)}$ to $\zeta h$ and the result can be written in the form

$$D_z^m v_{k'}\,|_{z=x} = \iint\limits_{Q_2^{(k')}} D_x^{2b} Z_0(P - Q) \sum_{\mu+\nu=|m|-2b} C_{\mu\nu} D_y^{\mu} \zeta(y)\, \Delta_Q^P D_y^{\nu} h(Q, M)\,dQ$$
$$+ \sum_{\mu+\nu=|m|-2b} C_{\mu\nu} \iint\limits_{Q_2^{(k')}} D_x^{2b} Z_0(P - Q)\, D_y^{\mu} \zeta\,dQ\, D_x^{\nu} h(P, M) \tag{17}$$
$$- \sum_{\mu+\nu=|m|-2b-1} \int\limits_{\Gamma'} D_x^{2b+\mu} Z_0(P - Q)\, D_y^{\nu}(\zeta h) \cos(\vec{n}, y)\,d\Gamma',$$

where in the case when $r$ is a number, not a vector, $D_r x$ denotes a derivative or order $r$. The first two terms in (17) can be estimated as $E_3$ and $E_4$, and the estimate of the last term is obvious.

In the case $k = k''$ it is considerably more difficult to obtain an expression for $D_z^m v_{k''}\,|_{z=x}$ that is convenient for further estimates, because one cannot, generally speaking, directly shift derivatives from $R^{(k'')}$ to $\zeta h$. Note that since

$$v_{k''}(t, z; M) = \Pi_{z,\xi}^{\bar{z},\bar{\xi}}\, \bar{v}(t, \bar{z}; M),$$

where

$$\bar{v}(t, \bar{z}; \hat{M}) = \iint\limits_{\mathcal{K}_2} G_0(t - \mathfrak{p}, \bar{z}, \bar{y})\, \bar{\zeta}(\bar{y})\, \bar{h}(\bar{Q}, \hat{M})\,d\bar{Q}$$

($\mathcal{K}_2$ has already appeared in (9)), to estimate $D_z^m v_{k''}\,|_{z=x}$ $(2b < |m| \leq 2b + l)$ it suffices to estimate the derivatives $D_{\bar{z}}^m \bar{v}(t, \bar{z}; M)\,|_{\bar{z}=\bar{x}}$ for $|m| > 2b$. To obtain expressions for these derivatives, the functions $D_{\bar{z}}^m v$ are approximated by a special sequence so as to preserve the type of the corresponding exponentials in the estimates.

We introduce the sequence

$$\bar{v}^{(r)}(t, \bar{z}; \hat{M}) = \sum_{i=0}^{r} \iint_{\mathscr{K}_{2i}} G_0(t-\beta, \bar{z}, \bar{y}) \bar{\zeta}(\bar{y}) \bar{h}(\bar{Q}, \hat{M})\, d\bar{Q} = \iint_{\mathscr{K}_2^{(r)}} G_0 \bar{\zeta} \bar{h}\, d\bar{Q}, \tag{18}$$

where $\mathscr{K}_{2i} = (t_i, t_{i+1}) \cap \mathscr{K}_2$, $t_i = t - (t - t_0)/2^i$, $t_0$ is the smallest value of the first coordinate of the point $\bar{Q} \in \mathscr{K}_2$, and $\mathscr{K}_2^{(r)} = \bigcup_{i=0}^{r} \mathscr{K}_{2i}$. Clearly $\bar{v}^{(r)} \to \bar{v}$ for $r \to \infty$, and

$$D_{\bar{z}}^{m} \bar{v}^{(r)} = \iint_{\mathscr{K}_2^{(r)}} D_{\bar{z}}^{m} G_0 \bar{\zeta} \bar{h}\, d\bar{Q}.$$

Using the system (16.1) ($\bar{f} \equiv 0$) and (10), we obtain the expression

$$\begin{aligned} D_{\bar{z}}^{m} G_0(t-\beta, \bar{z}, \bar{y}) &= \sum_{\substack{|\nu|=|m| \\ (\nu_n<2b)}} a_{m\nu} D_{t,\bar{z}}^{\nu} G_0(t-\beta, \bar{z}, \bar{y}) \\ &= \sum_{\substack{|\nu|=|m| \\ (\nu_n<2b)}} (-1)^{\nu_0+\cdots+\nu_{n-1}} a_{m\nu} D_{\beta, \bar{y}'}^{\nu'} D_{z_n}^{\nu_n} G_0, \end{aligned} \tag{19}$$

where $\nu = (\nu_0, \cdots, \nu_n)$, $\nu' = (\nu_0, \cdots, \nu_{n-1}, 0)$, and the $a_{m_\nu}$ are defined by the coefficients of the system (16.1).

Let $\mu_s \leqq \nu_s$ $(s = 0, 1, \cdots, n-1)$ be such that the greatest value of $|\mu'| = 2b\mu_0 + \mu_1 + \cdots + \mu_{n-1}$ does not exceed $|m| - 2b$. Clearly $|m| - 4b < |\mu'| \leqq |m| - 2b$. Using (19) and taking account of the properties of $\zeta$, we obtain by integration by parts

$$\begin{aligned} D_{\bar{z}}^{m} \bar{v}^{(r)} = \sum_{|\nu|=|m|} a_{m\nu} \Big[ &\iint_{\mathscr{K}_2^{(r)}} D_{t,\bar{z}}^{\nu-\mu'} G_0 D_{\beta,\bar{y}'}^{\mu'} (\bar{\zeta} \bar{h})\, d\bar{Q} \\ &- \sum_{|\eta| \leqslant |\mu''|-1} \int_{\Gamma_2^{(r)}} D_{t,\bar{z}}^{\nu-(\eta+1)} G_0^{(k'')} D_{\bar{y}'}^{\eta} (\bar{\zeta}^{(k'')} \bar{h}) \cos(\vec{n}, \bar{y}')\, d\Gamma_2^{(r)} \\ &- \sum_{\eta_0 \leqslant \mu_0 - 1} \int_{\Gamma_2^{(r)}} D_{t,\bar{z}}^{\nu-(\eta_0+1)} G_0 D_{\beta}^{\eta_0} (\bar{\zeta} \bar{h}) \cos(\vec{n}, \beta)\, d\Gamma_2^{(r)} \\ &- \sum_{\eta_0 \leqslant \mu_0 - 1} \int_{K_{t_{r+1}}} D_{t,\bar{z}}^{\nu-(\eta_0+1)} G_0 D_{\beta}^{\eta_0} (\bar{\zeta} \bar{h})\, d\bar{y} \Big], \end{aligned} \tag{20}$$

where $\Gamma_2^{(r)}$ is the "curved" part of the boundary $\mathscr{K}_2^{(1)}$, $K_{t_{r+1}}$ is the part of the boundary $\mathscr{K}_2^{(1)}$ which belongs to the hyperplane $t = t_{r+1}$, and $\mu'' = (0, \mu_1, \cdots, \mu_{n-1}, 0)$. We transform the first term in the square brackets, and denote it by $J_\nu^{(r)}$. For $\nu$ such that $|\mu'| = |m| - 2b$ we write

$$
J_\nu^{(r)} = \sum_{|\eta|\leqslant|\mu''|} c_{\mu'',\eta}\Big[\iint_{\mathcal{K}_2^{(r)}} D_{t,\bar z}^{\nu-\mu'} G_0 D_{\bar y'}^{\eta} \bar\zeta \Delta_{\bar Q}^{t,\bar z} D_\beta^{\mu_0} D_{\bar y'}^{\mu''-\eta} \bar h\, d\bar Q
+ D_t^{\mu_0} D_{\bar z'}^{\mu''-\eta} \bar h \iint_{\mathcal{K}_2^{(r)}} D_{t,\bar z}^{\nu-\mu'} G_0 D_{\bar y'}^{\eta} \bar\zeta\, d\bar Q\Big], \tag{21}
$$

and for $\nu$ such that $|\mu'| < |m| - 2b$ we write

$$
J_\nu^{(r)} = \sum_{i=0}^{r} \sum_{|\eta|\leqslant|\mu''|} c_{\mu''\eta}\Big[\iint_{\mathcal{K}_{2i}} D_{t,\bar z}^{\nu-\mu'} G_0 D_{\bar y'}^{\eta} \bar\zeta \Delta_{\beta_i}^{t_i} D_\beta^{\mu_0} D_{\bar y'}^{\mu''-\eta} \bar h\, d\bar Q
+ \iint_{\mathcal{K}_{2i}} D_{t,\bar z}^{\nu-\mu'} G_0 D_{\bar y'}^{\eta} \bar\zeta D_{t_i}^{\mu_0} D_{\bar y'}^{\mu''-\eta} \bar h(t_i, \bar y; \hat M)\, d\bar Q\Big]. \tag{22}
$$

Further, we transform the last integrals in (21) and (22), and denote these by $\tilde J_\nu^{(r)}$ and $\tilde J_{\nu i}^{(r)}$, respectively. Since in (21) $D_{t,\bar z}^{\nu-\mu'}$ is equal either to $D_t$ or to $D_{\bar z'}$, $D_z^{2b-1}$, we can shift one differentiation, either in $t$ or in $\bar z'$, from $G_0$ to $\tilde J_\nu^{(r)}$. Assuming the second alternative, we obtain

$$
\tilde J_\nu^{(r)} = \iint_{\mathcal{K}_2^{(r)}} D_z^{2b-1} G_0 D_{\bar y'}^{\eta+1} \bar\zeta\, d\bar Q - \int_{\Gamma_2^{(r)}} D_z^{2b-1} G_0 D_{\bar y'}^{\eta} \bar\zeta \cos(\vec n, \bar y')\, d\Gamma_2^{(r)}. \tag{23}
$$

In $\tilde J_{\nu i}^{(r)}$ from (22) we get $D_{t,\bar z}^{\nu-\mu'} = D_t D_{\bar z_n}^{\nu_n}$, so that

$$
\tilde J_{\nu i}^{(r)} = -\int_{\Gamma_{2i}} D_{\bar z_n}^{\nu_n} G_0 D_{\bar y'}^{\eta} \bar\zeta D_{t_i}^{\mu_0} D_{\bar y'}^{\mu''-\eta} \bar h \cos(\vec n, \beta)\, d\Gamma_{2i}
+ \int_{K_{t_i}} D_{\bar z_n}^{\nu_n} G_0 D_{\bar y'}^{\eta} \bar\zeta D_{t_i}^{\mu_0} D_{\bar y'}^{\mu''-\eta} \bar h\, d\bar y - \int_{K_{t_{i+1}}} D_{\bar z_n}^{\nu_n} G_0 D_{\bar y'}^{\eta} \bar\zeta D_{t_i}^{\mu_0} D_{\bar y'}^{\mu''-\eta} \bar h\, d\bar y, \tag{24}
$$

where $\Gamma_{2i}$ is the "curved" part of the boundary, and $K_{t_i}$ is that part of the boundary $\mathcal{K}_{2i}$ which lies in the hyperplane $t = t_i$.

Substituting (23) and (24) into (21) and (22), respectively, and then substituting the results into (20), we take the limit $r \to \infty$. The right part of (20) converges to a limit uniformly in $\bar z$ in a certain neighborhood of the point $\bar x$. Consequently this limit coincides with $D_{\bar z}^m \bar v$. Taking this limit for $\bar z = \bar x$, we obtain the following expression:

$$
D_{\bar z}^m \bar v\,|_{\bar z=\bar x} = \sum_{|\nu|=|m|} a_{m\nu}\Big[J_\nu - \sum_{|\eta|\leqslant|\mu''|-1} \int_{\Gamma_2} D_{t,\bar x}^{\nu-(\eta+1)} G_0 D_{\bar y'}^{\eta} (\bar\zeta \bar h) \cos(\vec n, \bar y')\, d\Gamma_2
- \sum_{\eta_0\leqslant\mu_0-1} \int_{\Gamma_2} D_{t,\bar x}^{\nu-(\eta_0+1)} G_0 D_\beta^{\eta_0} (\bar\zeta \bar h) \cos(\vec n, \beta)\, d\Gamma_2 - \sum_{|\eta|=|\nu|-2b} a'_{\nu\eta} D_{t,\bar x}^{\eta} (\bar\zeta \bar h)\Big], \tag{25}
$$

where $J_\nu = \lim_{r\to\infty} J_\nu^{(r)}$ and in the case (21)

$$J_\nu = \sum_{|\eta|\leqslant|\mu''|} c_{\mu''\nu}\Big[\iint\limits_{\mathcal{K}_2} D_{t,\bar{x}}^{\nu-\mu'} G_0 D_{\bar{y}'}^{\eta}\bar{\zeta}\Delta_{\bar{Q}}^{\bar{P}} D_\beta^{\mu_0} D_{\bar{y}'}^{\mu''-\eta}\bar{h}\,d\bar{Q} \tag{26}$$
$$+ D_t^{\mu_0} D_{\bar{x}'}^{\mu''-\eta}\bar{h}\Big(\iint\limits_{\mathcal{K}_2} D_{\bar{x}}^{2b-1} G_0 D_{\bar{y}'}^{\eta+1}\bar{\zeta}\,d\bar{Q} - \int\limits_{\Gamma_2} D_{\bar{x}}^{2b-1} G_0 D_{\bar{y}'}^{\eta}\bar{\zeta}\cos(\vec{n}, \bar{y}')\,d\Gamma_2\Big)\Big],$$

or in the case (22)

$$J_\nu = \sum_{|\eta|\leqslant|\mu''|} c_{\mu''\nu}\Big[\sum_{i=0}^{\infty}\iint\limits_{\mathcal{K}_{2i}} D_{t,\bar{x}}^{\nu-\mu'} G_0 D_{\bar{y}'}^{\eta}\bar{\zeta}\Delta_\beta^{t_i} D_\beta^{\mu_0} D_{\bar{y}'}^{\mu''-\eta}\bar{h}\,d\bar{Q}$$
$$-\sum_{i=1}^{\infty}\int\limits_{\Gamma_{2i}} D_{\bar{x}}^{\nu_n} G_0 D_{\bar{y}'}^{\eta}\zeta D_{t_i}^{\mu_0} D_{\bar{y}'}^{\mu''-\eta}\bar{h}\cos(\vec{n}, \beta)\,d\Gamma_{2i} \tag{27}$$
$$+\sum_{i=1}^{\infty}\int\limits_{K_{t_i}} D_{\bar{x}_n}^{\nu_n} G_0 D_{\bar{y}'}^{\eta}\bar{\zeta}\Delta_{t_i}^{t_{i-1}} D_{t_i}^{\mu_0} D_{\bar{y}'}^{\mu''-\eta}\bar{h}\,d\bar{y}\Big] - D_{t,\bar{x}'}^{\mu'} D_{x_n}^{\nu_n}(\bar{\zeta}^{(k'')}\bar{h})$$
$$= J_\nu' - D_{t,\bar{x}'}^{\mu'} D_{\bar{x}_n}^{\nu_n}(\bar{\zeta}\bar{h}).$$

Let us see how the last terms appear in (25) and (27). The last term in (25) is the value for $\bar{z} = \bar{x}$ of the limit of the last term in (20). As was shown in [**2**], the Green matrix

$$\mathcal{G}_0(t-\beta, \bar{z}, \bar{y}) = Z_0(t-\beta, \bar{z}-\bar{y}) - v_0(t-\beta, \bar{z}'-\bar{y}', z_n, y_n),$$

where $Z_0$ is the f.s.m. of system (16.1) and the derivatives of the matrix $v_0$ have estimates which contain $\exp\{-c[|z_n|/(t-\beta)^\gamma]^q\}$. Thus for the points $\bar{z}$ in some neighborhood $V_\delta$ of the point $\bar{x}$ of radius $\delta > 0$ (we take $\delta$ sufficiently small so that $V_{3\delta}$ will lie inside $K$) we obtain

$$\int\limits_{K_{t_{r+1}}} D_{t,\bar{z}}^{\nu-(\eta_0+1)} v_0 D_\beta^{\eta_0}(\bar{\zeta}\bar{h})\,d\bar{y} \to 0, \quad r\to\infty.$$

Further, let $\chi'(\bar{y})$ be a smooth function with support in $V_{3\delta}$, which is equal to 1 in $V_{2\delta}$. Using the system (16.1) ($\bar{f}\equiv 0$), we write

$$\sum_{\eta_0\leqslant\mu_0-1}\int\limits_{K_{t_{r+1}}} D_{t,\bar{z}}^{\nu-(\eta_0+1)} Z_0\bar{\zeta} D_\beta^{\eta_0}\bar{h}\,d\bar{y} = \sum_{\eta_0\leqslant\mu_0-1}\ \sum_{|\eta'|=|\nu|-2b(\eta_0+1)} (-1)^{|\eta'|} a_{\nu\eta'}$$
$$\times\Big[\int\limits_{V_{3\delta}} D_{\bar{y}}^{\eta'} Z_0(t-t_{r+1}, \bar{z}-\bar{y})\chi'(\bar{y})\bar{\zeta}(\bar{y}) D_\beta^{\eta_0}\bar{h}(t_{r+1}, \bar{y}; \hat{M})\,d\bar{y}$$
$$+\int\limits_{K_{t_{r+1}}\setminus V_{2\delta}} D_{\bar{y}}^{\eta'} Z_0(1-\chi'(\bar{y}))\bar{\zeta} D_\beta^{\eta_0}\bar{h}\,d\bar{y}\Big].$$

For $\bar{z} \in V_\delta$ the limit of the last term for $r \to \infty$ clearly equals zero. In the first term we shift all differentiations from $Z_0$ by integration by parts (integrals over the boundary of $V_{3\delta}$ are zero in view of the properties of $\chi'$) and we take the limit $r \to \infty$, taking $\bar{z} \in V_\delta$. By the properties of the f.s.m., in the limit we obtain an expression whose value for $\bar{z} = \bar{x}$ coincides with the last term in (25). The last term in (27) appears in an analogous manner.

We shall now derive the estimates for the expressions in (25)-(27). This can be done using the methods which we used above in the case $|m| \leq 2b$. We shall describe briefly only the derivation of the estimates for $J_\nu$ in (27). Since $h \in H^{l+\alpha}_{2b,c_1}(\bar{Q}_{\tau,\tau_1}, Q)$, we have

$$|D^{\mu'}_{t,\bar{x}'} D^{\nu_n}_{\bar{x}_n} (\zeta \bar{h})| \leqslant C\|h\|(t-\tau)^{\nu(2b+l-|m|+\alpha)} d_0^{-n-2b-l-\alpha} \Phi_{c_1}(P, M).$$

Using (4.2) for $G_0$, the estimates for $h$ and (13.1), we obtain

$$|J'_\nu| \leqslant C\|h\| \Big[\sum_{i=0}^{\infty} \iint_{\mathcal{K}_{2i}} d^{-n-|m|+|\mu'|}(\bar{P}, \bar{Q}) \Psi_{c_0}(\bar{P}, \bar{Q})(\beta - t_i)^{\nu(l-|\mu'|+\alpha)}$$

$$\times d^{-n-2b-l-\alpha}(t_i, y; M)\Phi_{c_1}(Q, M)\, d\bar{Q} + \sum_{i=0}^{\infty} \int_{\Gamma_{2i}} d^{-n-\nu_n}(\bar{P}, \bar{Q}) \Psi_{c_0}(\bar{P}, \bar{Q})$$

$$\times (t_i - \tau)^{\nu(l-|\mu'|+\alpha)} d^{-n-2b-l-\alpha}(t_i, y; M)\Phi_{c_1}(t_i, y; M)\,|\cos(\vec{n}, \beta)|\, d\Gamma_{2i}$$

$$+ \sum_{i=1}^{\infty} \int_{K_{t_i}} d^{-n-\nu_n}(\bar{P}, t_i, \bar{y}) \Psi_{c_0}(\bar{P}; t_i; \bar{y})(t_i - t_{i-1})^{\nu(l-|\mu'|+\alpha)} d^{-n-2b-l-\alpha}(t_{i-1}, y; M)$$

Hence in view of (24.1)-(26.1) and the fact that for $\beta \in (t_i, t)$ and $\beta - t_i \leq t - \beta \leq d^{2b}(P, Q)$ we have

$$t - \beta \leqslant t - t_i,\ t_i - t_{i-1} = t - t_i,\ \sum_{i=0}^{\infty} \iint_{\mathcal{K}_{2i}} \dots d\bar{Q} = \iint_{\mathcal{K}_2} \dots d\bar{Q},$$

$$\sum_{i=0}^{\infty} \int_{\Gamma_{2i}} \dots d\Gamma_{2i} = \int_{\Gamma_2} \dots d\Gamma_2,$$

it follows that

$$|J'_\nu| \leqslant C\|h\|(t-\tau)^{\nu(2b+l-|m|)} d_0^{-n-2h-l-\alpha} \Phi_c(P, M) \Big[\iint_{\mathcal{K}_2} d^{-n-2b+\alpha}(P, Q)\, d\bar{Q}$$

$$+ (t-\tau)^{\nu(|m|-2b-|\mu'|+\alpha)} \int_{\Gamma_2} d^{-n-\nu_n}(P, Q)\,|\cos(\vec{n}, \beta)|\, d\Gamma_2 + \sum_{i=1}^{\infty} (t - t_i)^{\nu\varepsilon}$$

$$\times \int_{K_{t_i}} d^{-n+\alpha-\varepsilon}(\bar{P}; t_i, \bar{y})\, d\bar{y}\Big],$$

where $0 < \epsilon < \alpha$. From this inequality we obtain the estimate (28) for $J'_\nu$, since $Q \in \Gamma_2, d(P,Q) = \frac{1}{2}(t-\tau)^\gamma$ and

$$\iint_{\mathcal{K}_2} d^{-n-2b+\alpha}(P,Q)\,d\bar{Q} = \iint_{Q_2^{(k'')}} d^{-n-2b+\alpha}(P,Q)\,dQ \leqslant C(t-\tau)^{\nu\alpha},$$

(29)
$$\int_{\Gamma_2} |\cos(\vec{n},\beta)|\,d\Gamma_2 \leqslant \int_{|\bar{x}-\bar{y}|\leqslant C(t-\tau)^\nu} d\bar{y} = C(t-\tau)^{\nu n},$$

$$\sum_{i=1}^{\infty}(t-t_i)^{\nu\varepsilon}\int_{K_{t_i}} d^{-n+\alpha-\varepsilon}(\bar{P};t_i,\bar{y})\,d\bar{y} \leqslant$$

$$\leqslant \sum_{i=1}^{\infty} 2^{-\nu\varepsilon i}(t-t_0)^{\nu\varepsilon}\int_{|\bar{x}-\bar{y}|\leqslant C(t-\tau)^\nu} |\bar{x}-\bar{y}|^{-n+\alpha-\varepsilon}\,d\bar{y} \leqslant C(t-\tau)^{\nu\alpha}.$$

To complete the proof of Lemma 1 we must also estimate the increments in the derivatives $D_x^m u_k$ ($|m| > 2b$). For this we use (17) and (25)–(27) in the same manner as in the case $|m| \leqq 2b$ to obtain representations of these increments. The estimates for the terms of this representation are analogous to the estimates of the corresponding terms in the case $|m| \leqq 2b$ and to the estimates of $J_\nu$ given above.

**2. Proof of Lemma 2.** This proof is in many ways similar to that of Lemma 1. Here it also suffices to estimate only the $x$ derivatives of $v_j$. From the definition $v_j = R_j h$ (2.2) and the condition (5.2) on the boundary $S$ it suffices to estimate the derivatives of $\bar{v}_j = \bar{R}_j h$ with respect to $\bar{x}$, taking $\bar{x} \in K$.

The domain of integration in (2), i.e. the cylinder $[\tau, t] \times K'$, where $K'$ is the standard cube of dimension $n-1$, which lies in the hyperplane $\bar{x}_n = 0$, shall be represented as in the preceding case in the form $[r,t] \times K' = \mathcal{K}'_1 \cup \mathcal{K}'_2 \cup \mathcal{K}'_3$, where

$$\mathcal{K}'_1 = [\tau, t] \times K' \cap \left\{(\beta, \bar{y}'), d(\beta, \bar{y}', F(\bar{y}'); \hat{M}) = d(Q, M) \leqslant \frac{1}{2}(t-\tau)^\nu\right\},$$

$$\mathcal{K}'_2 = [\tau, t] \times K' \cap \left\{(\beta, \bar{y}'), d(t, \bar{x}', \bar{x}_n + F(\bar{x}'); \beta, \bar{y}', F(\bar{y}')) = d(P,Q) \leqslant \frac{1}{2}(t-\tau)^\nu\right\}.$$

Taking $\bar{x} \in K$ to be such that $\bar{x}_n > 0$, we write

(30)
$$D_{\bar{x}}^m \bar{v}_j(\bar{P}, \hat{M}) = \iint_{\mathcal{K}'_1} [D_{\bar{x}}^m G_j(\bar{P} - \bar{Q}') - D_{\bar{x}}^m G_j(\bar{P}'' - \bar{M}')]\,\zeta(\bar{y}')\,\bar{h}(\bar{Q}', \hat{M})\,d\bar{Q}'$$
$$+ D_{\bar{x}}^m G_j(\bar{P}'' - \bar{M}')\iint_{\mathcal{K}'_1} \zeta(\bar{y}')\,\bar{h}(\bar{Q}', \hat{M})\,d\bar{Q}' +$$

$$+\iint\limits_{\mathcal{K}_2'} D_x^m G_j(\bar{P}-\bar{Q}')\,\bar{\zeta}(\bar{y}')\,\bar{h}(\bar{Q}',\hat{M})\,d\bar{Q}'$$

$$+\iint\limits_{\mathcal{K}_3'} D_x^m G_j(\bar{P}-\bar{Q}')\;\bar{\zeta}(\bar{y}')\;\bar{h}(\bar{Q}';\hat{M})\,d\bar{Q}'=\sum_{i=1}^{4} L_i,$$

where $\bar{P}''=(t,\bar{x}',|\bar{x}_n-\bar{\xi}_n|)$, $|m|\leqq l+r_j$ and here, as in (5), we assume $\xi\in\widetilde{\Omega}^{(k'')}$.

Using (4.2) for $G_j$ and the fact that $h$ belongs to the class $H^{l+\alpha}_{r_j,c_1}(\Gamma_{\tau,\tau_1}Q)$, we estimate $L_i$ $(i=1,2,3,4)$:

$$|L_1|\leqslant C\|h\|\iint\limits_{\mathcal{K}_1'}(t-\tau)^{\nu(l+r_j-|m|)}\,[d^{-n-2b+1-l-\alpha}(\bar{P},\bar{Q}')\,\Psi_{c_0}(\bar{P},\bar{Q}')$$

$$+d^{-n-2b+1-l-\alpha}(\bar{P}'',\bar{M}')\,\Psi_{c_0}(\bar{P}'',\bar{M}')$$

$$\times d^{\alpha}(\bar{Q}';\tau,\bar{\xi}',\bar{x}_n-|\bar{x}_n-\bar{\xi}_n|\,d^{-n-r_j}(Q,M)\,\Phi_{c_1}(Q,M)\,d\bar{Q}'.$$

Noting that

$$d(\bar{P}'',\bar{M}')=d(\bar{P},\bar{M})\ \text{и}\ d(\bar{Q}';\tau,\bar{\xi}',\bar{x}_n-|\bar{x}_n-\bar{\xi}_n|)\leqslant d(\bar{Q}',\bar{M})\leqslant Cd(Q,M),$$

we obtain from the last inequality, taking account of (24.1)–(26.1) in a manner analogous to (6),

$$|L_1|\leqslant C\|h\|\,(t-\tau)^{\nu(l+r_j-|m|)}\,d_0^{-n+1-2b-l-\alpha}\,\Phi_c(P,M)\iint\limits_{\mathcal{K}_1'}d^{-n-r_j+\alpha}(Q,M)\,d\bar{Q}'.$$

To estimate the last integral we use the inequality $d(Q,M)\geqq d(\bar{Q}',\bar{M}')$, introduce the change of variables $(\beta-\tau)^{\gamma}=z_0$, $\bar{y}'-\bar{\xi}'=z'$, and change to spherical coordinates, to obtain

$$\iint\limits_{\mathcal{K}_1'}d^{-n-r_j+\alpha}(Q,M)\,d\bar{Q}'\leqslant C\iint\limits_{|z|\leqslant\frac{1}{2}(t-\tau)^{\nu}}|z|^{-n-r_j+\alpha}\,|z_0|^{2b-1}\,dz$$

$$\leqslant C\int_0^{\frac{1}{2}(t-\tau)^{\nu}}\rho^{-1+2b-1-r_j+\alpha}\,d\rho\leqslant C\,(t-\tau)^{\nu(2b-1-r_j+\alpha)}$$

Therefore we have the following estimate for $L_1$:

$$|L_1|\leqslant C\|h\|\,(t-\tau)^{\nu(l+r_j-|m|+\alpha)}\,d_0^{-n-l-r_j-\alpha}\,\Phi_c(P,M). \tag{31}$$

One can easily verify that a similar estimate is valid for $L_2$. For $r_j<2b-1$ this can be established as above, and for $r_j=2b-1$ one must use (34.1).

The estimate for $L_4$ is obtained in a manner analogous to the estimate of $E_5$ in Lemma 1. Using the estimates for $G_j$ and $\bar{h}$ and the inequalities (24.1)-(26.1), we have

$$|L_4| \leqslant C\|h\|(t-\tau)^{\nu(2b-r_j-|m|+2\alpha)}\Phi_c(P,M)L_0,$$

$$L_0 = \iint_{\mathcal{K}_3'} d^{-n-2b+1-l-\alpha}(P,Q)\,d^{-n-l-r_j-\alpha}(Q,M)\,d\bar{Q}' \tag{32}$$

$$= \iint_{\mathcal{K}_3'\cap\left\{d(P,Q)\leqslant\frac{1}{2}d_0\right\}} \ldots d\bar{Q}' + \iint_{\mathcal{K}_3'\cap\left\{d(P,Q)>\frac{1}{2}d_0\right\}} \ldots d\bar{Q}' = L_0' + L_0''.$$

Since for $\bar{Q}' \in K_3' \cap \{d(P,Q) \leqq \frac{1}{2}d_0\}$ we have

$$0 \leqq d(\bar{P}',\bar{Q}') \leqq \tfrac{1}{2}d_0 \quad\text{and}\quad d(Q,M) \geqq \tfrac{1}{2}d_0,$$

we obtain

$$L_0' \leqslant Cd_0^{-n-l-r_j-\alpha}\Big[(t-\tau)^{-\nu(l+2\alpha)} \int_{d(\bar{P}',\bar{Q}')\leqslant\frac{1}{2}(t-\tau)^{\nu}} d^{-n-2b+1+\alpha}(\bar{P}',\bar{Q}')\,d\bar{Q}'$$

$$+ \int_{\frac{1}{2}(t-\tau)^{\nu}\leqslant d(\bar{P}',\bar{Q}')\leqslant\frac{d_0}{2}} d^{-n-2b+1-l-\alpha}(\bar{P}',\bar{Q}')\,d\bar{Q}'\Big] \tag{33}$$

$$\leqslant Cd_0^{-n-l-r_j-\alpha}\left[(t-\tau)^{-\nu(l+2\alpha)}(t-\tau)^{\nu\alpha} + (t-\tau)^{-\nu(l+\alpha)}\right]$$

$$= Cd_0^{-n-l-r_j-\alpha}(t-\tau)^{-\nu(l+\alpha)}.$$

Since $d(Q,M) \geqq \frac{1}{2}(t-\tau)^{\nu}$ in $L_0''$, in the case $2b-l-r_j-1-\alpha<0$ we have

$$L_0'' \leqslant Cd_0^{-n-2b-l+1-\alpha}\Big[(t-\tau)^{\nu(2b-l-r_j-1-2\alpha)} \int_{d(\bar{Q}',\bar{M}')\leqslant\frac{1}{2}(t-\tau)^{\nu}} d^{-n-2b+1+\alpha}(\bar{Q}',\bar{M}')\,d\bar{Q}'$$

$$+ \int_{d(\bar{Q}',\bar{M}')>\frac{1}{2}(t-\tau)^{\nu}} d^{-n-l-r_j-\alpha}(\bar{Q}',\bar{M}')\,d\bar{Q}'\Big] \tag{34}$$

$$\leqslant Cd_0^{-n-2b-l+1-\alpha}(t-\tau)^{\nu(2b-l-r_j-1-\alpha)} \leqslant Cd_0^{-n-l-r_j-\alpha}(t-\tau)^{-\nu(l+\alpha)}.$$

In the case $2b-l-r_j-1-\alpha>0$ we obtain

$$L_0'' \leqslant Cd_0^{-n-2b-l+1-\alpha} \int_{d(\bar{Q}',\bar{M}')\leqslant 2d_0} d^{-n-l-r_j-\alpha}(\bar{Q}',\bar{M}')\,d\bar{Q}'$$

$$+ \int_{d(\bar{Q}',\bar{M}')>2d_0} d^{-n-2b+1-l-\alpha}(\bar{P}',\bar{Q}')\,d^{-n-l-r_j-\alpha}(\bar{Q}',\bar{M}')\,d\bar{Q}' \leqq \tag{35}$$

$$\leqslant C d_0^{-n-2l-r_j-2\alpha}\Big[1 + d_0^{l+\alpha} \int\limits_{d(\overline{P}',\overline{Q}')\geqslant d_0} d^{-n-2b+1-l-\alpha}(\overline{P}', \overline{Q}')\, d\overline{Q}'\Big]$$

$$\leqslant C d_0^{-n-2l-r_j-2\alpha} \leqslant C(t-\tau)^{-\nu(l+\alpha)}\, d_0^{-n-l-r_j-\alpha}.$$

The estimate (31) for $L_4$ follows from (32)–(35).

We now estimate $L_3$. We shall use the following representation (cf. [4]):

$$G_j(t, \overline{x}) = [D_t + (-1)^b(1+\delta_1)\Delta^b_{\overline{x}'}]^\theta G_j^{(\theta)}(t, \overline{x}) \equiv \sum_{|\nu'|=2b\theta} a_{\nu'} D^{\nu'}_{t,\overline{x}'} G_j^{(\theta)}(t, \overline{x}),$$

(36)
$$\Delta_{\overline{x}'} = \sum_{i=1}^{n-1} \frac{\partial^2}{\partial \overline{x}_i^2},$$

where

(37)
$$G_j^{(\theta)}(\overline{P}-\overline{Q}') \in U^{l_0+\alpha_0, 2b(1-\theta)-r_j-1, s}_{C_0, c_0}((\beta, T] \times E_n^+, [0, T) \times E_{n-1})$$

with arbitrary $l_0 \geqq 0$, $0 < \alpha_0 \leqq 1$, $s \geqq 0$ and $T > 0$, where $\theta$ is an arbitrary nonnegative integer. The constants $C_0$ and $c_0$ depend in general on $\theta$, but since we shall use (37) only for a finite number of values of $\theta$, which depend on $l$, we shall disregard this dependence of $C_0$ and $c_0$ on $\theta$, assuming that (37) has the same constants as (4.2).

Assume in (36) that $\theta = [p]+1$, where $p = (|m| - r_j)/2b$. Let $\mu_s \leqq \nu_s$ $(s = 0, 1, \cdots, n-1)$ be such that the greatest value of $|\mu'|$ does not exceed $|m| - r_j$. Note that $|\mu'| = 2b[p]$ or $|\mu'| = 2bp$. Using (36) in the same manner as in (20), we obtain

(38)
$$\begin{aligned} L_3 = \sum_{|\nu'|=2b\theta} a_{\nu'} \Big[ \iint\limits_{\mathcal{K}_2'} & D^{\nu'-\mu'}_{t,\overline{x}'} D^m_{\overline{x}} G_j^{(\theta)}(\overline{P}-\overline{Q}')\, D^{\mu'}_{\beta,\overline{y}'}[\zeta(\overline{y}')\overline{h}(\overline{Q}', \widetilde{M})]\, d\overline{Q}' \\ & - \sum_{|\eta|\leqslant|\mu''|-1} \int\limits_{\Gamma_2'} D^{\nu'-(\eta+1)}_{t,\overline{x}'} D^m_{\overline{x}} G_j^{(\theta)} D^\eta_{\overline{y}'}(\zeta\overline{h}) \cos(\vec{n}, \overline{y}')\, d\Gamma_2' \\ & - \sum_{\eta_0 \leqslant \mu_0 - 1} \int\limits_{\Gamma_2'} D^{\nu'-(\eta_0+1)}_{t,\overline{x}'} D^m_{\overline{x}} G_j^{(\theta)} D^{\eta_0}_\beta(\zeta\overline{h}) \cos(\vec{n}, \beta)\, d\Gamma_2' \Big] \\ & = \sum_{|\nu'|=2b\theta} a_{\nu'}[N_1 + N_2 + N_3]. \end{aligned}$$

where $\Gamma_2'$ is the "curved" part of the boundary of $\mathcal{K}_2'$. We shall first derive the estimates for $N_2$ and $N_3$. Using (37), the estimates of $\overline{h}$, the properties of $\overline{\zeta}$ and the inequalities (24.1)–(26.1), we obtain

$$|N_2|+|N_3|\leqslant C\|h\|(t-\tau)^{\nu(l+r_j-|m|+\alpha)}d_0^{-n-l-r_j-\alpha}\Big[(t-\tau)^{-\nu(n+2b-2)}$$
$$\times\int_{\Gamma_2'}|\cos(\vec{n},\bar{y}')|\,d\Gamma_2'+(t-\tau)^{-\nu(n-1)}\int_{\Gamma_2'}|\cos(\vec{n},\beta)|\,d\Gamma_2'\Big].$$

But since

$$\int_{\Gamma_2'}|\cos(\vec{n},\bar{y}_i)|\,d\Gamma_2'\leqslant\int_{d(\bar{P}^{(i)},\bar{Q}^{(i)})\leqslant C(t-\tau)^{\nu}}d\bar{Q}^{(i)}\leqslant C(t-\tau)^{\nu(n+2b-2)},$$
$$\int_{\Gamma_2'}|\cos(\vec{n},\beta)|\,d\Gamma_2'\leqslant\int_{|\bar{x}'-\bar{y}'|\leqslant C(t-\tau)^{\nu}}d\bar{y}'\leqslant C(t-\tau)^{\nu(n-1)},\tag{39}$$

where

$$\bar{P}^{(i)}=(t,\bar{x}_1,\ldots,\bar{x}_{i-1},\bar{x}_{i+1},\ldots,\bar{x}_{n-1}),\quad \bar{Q}^{(i)}=(\beta,\bar{y}_1,\ldots,\bar{y}_{i-1},\bar{y}_{i+1},\ldots,\bar{y}_{n-1}),$$

we obtain the estimate (31) for $N_2$ and $N_3$.

To estimate $N_1$, we obtain for it representations analogous to (26) and (27). Clearly we have

$$N_1=\sum_{i=0}^{\infty}\sum_{|\eta|\leqslant|\mu''|}c_{\mu''\eta}\iint_{\mathscr{K}_{2i}'}D_{t,\bar{x}'}^{\nu'-\mu'}D_{\bar{x}}^{m}G_j^{(\theta)}D_{\bar{y}'}^{\eta}\bar{\zeta}\Delta_{\beta}^{t_i}D_{\beta}^{\mu_0}D_{\bar{y}'}^{\mu''-\eta}\bar{h}\,d\bar{Q}'$$
$$+\sum_{i=0}^{\infty}\sum_{|\eta|\leqslant|\mu''|}c_{\mu''\eta}\iint_{\mathscr{K}_{2i}'}D_{t,\bar{x}'}^{\nu'-\mu'}D_{\bar{x}}^{m}G_j^{(\theta)}D_{\bar{y}'}^{\eta}\bar{\zeta}D_{t_i}^{\mu_0}D_{\bar{y}'}^{\mu''-\eta}\bar{h}\,d\bar{Q}'=N_1'+N_1'',\tag{40}$$

where $\mathscr{K}_{2i}'=(t_i,t_{i+1})\cap\mathscr{K}_2'$ and $t_i$ is defined in the same way as in Lemma 1. Further, we write down a representation for $N_1''$. First let $|\mu'|=2bp$, so that $|\nu'|-|\mu'|=2b(1-\{p\})$, where $\{p\}$ is the fractional part of $p$. Forming under the integral sign a special increment in the derivatives of $\bar{h}$, and transforming the integral in the auxiliary term by integration by parts, we obtain

$$N_1''=\sum_{|\eta|\leqslant|\mu''|}c_{\mu''\eta}\sum_{i=0}^{\infty}\Big[\iint_{\mathscr{K}_{2i}'}D_{t,\bar{x}'}^{\nu'-\mu'}D_{\bar{x}}^{m}G_j^{(\theta)}D_{\bar{y}'}^{\eta}\bar{\zeta}\Delta_{\bar{y}'}^{\bar{z}'}D_{t_i}^{\mu_0}D_{\bar{y}'}^{\mu''-\eta}\bar{h}\,d\bar{Q}'$$
$$+D_{t_i}^{\mu_0}D_{\bar{z}'}^{\mu''-\eta}\bar{h}\Big(\iint_{\mathscr{K}_{2i}'}D_{t,\bar{x}'}^{\nu'-\mu'-1}D_{\bar{x}}^{m}G_j^{(\theta)}D_{\bar{y}'}^{\eta+1}\bar{\zeta}\,d\bar{Q}'$$
$$-\int_{\Gamma_{2i}'}D_{t,\bar{x}'}^{\nu'-\mu'-1}D_{\bar{x}}^{m}G_j^{(\theta)}D_{\bar{y}'}^{\eta}\bar{\zeta}\cos(\vec{n},\bar{y}')\,d\Gamma_{2i}'\Big)\Big],\tag{41}$$

where $(\bar{z}',0)$ are the $k''$-rectifying coordinates of the point $z \in S \cap \Omega^{(k'')}$ nearest to $x$. Let now $|\mu'| = 2b[p]$, so that $|\nu'| - |\mu'| = 2b$. If $D_{t,\bar{x}'}^{\nu'-\mu'} = D_t$, then by integration by parts we obtain the following representation, analogous to (27):

$$N_1'' = \sum_{|\eta| \leqslant |\mu''|} c_{\mu''\eta} \Big[ \sum_{i=1}^{\infty} \int_{K_{t_i}'} D_x^m G_j^{(\theta)} D_{\bar{y}'}^{\eta} \bar{\zeta} \Delta_{t_i}^{t_{i-1}} D_{t_i}^{\mu_0} D_{\bar{y}'}^{\mu''-\eta} \bar{h} \, d\bar{y}'$$

(42)

$$- \sum_{i=0}^{\infty} \int_{\Gamma_{2i}'} D_x^m G_j^{(\theta)} D_{\bar{y}'}^{\eta} \bar{\zeta} D_{t_i}^{\mu_0} D_{\bar{y}'}^{\mu''-\eta} \bar{h} \cos(\vec{n}, \beta) \, d\Gamma_{2i}' \Big].$$

In (41) and (42) $\Gamma_{2i}'$ is the "curved" part of the boundary of $\mathcal{K}_{2i}'$, and $K_{t_i}$ is that part which lies in the hyperplane $t = t_i$. If $D_{t,\bar{x}'}^{\nu'-\mu'} = D_{\bar{x}'}^{2b}$, then we first must shift $2b\{p\}$ differentiations from $G_j^{(\theta)}$ to the other factors, and then we represent the result in a form analogous to (41).

Using the representations (40)–(42), it is easy to obtain the estimate (31) for $N_1$. The expressions $N_1'$ and $N_2'$ in (42) can be estimated in the same way as the corresponding terms of $J_\nu'$ in (27). Therefore we shall describe the derivation of the estimate for $N_1''$ in (41). In view of (37), (13.1), (25.1) and (26.1), the estimates for $\bar{h}$, and the fact that $|\nu'| = 2b\theta$, $|\mu'| = |m| - r_j$ and $t_i - \tau \leqq \chi\lambda^{2b}$ we have

$$|N_1''| \leqslant C \|h\| (t-\tau)^{\gamma(l+r_j-|m|)} \sum_{i=0}^{\infty} \Big\{ \iint_{\mathcal{K}_{2i}'} d^{-n-2b+1}(P,Q) \Psi_{c_2}(P,Q) |y-z|^{\alpha}$$

$$\times [d^{-n-r_j-l-\alpha}(t_i, y; M) \Phi_{c_1}(t_i, y; M) + d^{-n-r_j-l-\alpha}(t_i, z, M) \Phi_{c_1}(t_i, z; M)] \, d\bar{Q}'$$

$$+ (t_i - \tau)^{\gamma\alpha} d^{-n-r_j-l-\alpha}(t_i, z; M) \Phi_{c_1}(t_i, z; M) \Big[ \frac{1}{\lambda} \iint_{\mathcal{K}_{2i}'} d^{-n-2b+2}(P,Q)$$

$$\times \Psi_{c_2}(P,Q) \, d\bar{Q}' + \int_{\Gamma_{2i}'} d^{-n-2b+2}(P,Q) \Psi_{c_2}(P,Q) |\cos(\vec{n}, \bar{y}')| \, d\Gamma_{2i}' \Big] \Big\}.$$

In view of the definition of $z$ we have for $Q \in K_{2i}'$

$$|y - z| \leqslant 2|x-y| \leqslant 2d(P,Q),$$

$$\Psi_{c_2}(P,Q) \leqslant \Psi_{c_2}(P; t_i, z) \quad (\leqslant \Psi_{c_2}(P; t_i, y)),$$

$$d(t_i, y; M) \geqslant d_0 - d(P, t_i, y) \geqslant d_0 - 2^{\gamma} d(P,Q) \geqslant (1 - 2^{\gamma-1}) d_0,$$

$$d(t_i, z; M) \geqslant d_0 - d(P; t_i, z) \geqslant d_0 - d(P; t_i, y) \geqslant (1 - 2^{\gamma-1}) d_0,$$

and from (24.1) we have

$$\Psi_{c_2}(P; t_i, z) \Phi_{c_1}(t_i, z; M) \leqq \Phi_c(P, M).$$

Using all these inequalities, we obtain

$$|N_1''| \leqslant C\|h\|(t-\tau)^{\nu(l+r_j-|m|)} d_0^{-n-l-r_j-\alpha}\,\Phi_c(P,M)\cdot\Big[\iint\limits_{\mathcal{K}_2'} d^{-n-2b+1+\alpha}(P,Q)\,d\bar{Q}'$$

$$+(t-\tau)^{\nu\alpha}\frac{1}{\lambda}\iint\limits_{\mathcal{K}_2'} d^{-n-2b+2}(P,Q)\,d\bar{Q}' + (t-\tau)^{\nu(-n-2b+2+\alpha)}$$

$$\times\int\limits_{\Gamma_2'} |\cos(\vec{n},\bar{y}')|\,d\Gamma_2'\Big].$$

Estimating the first two integrals in the square brackets in the same way as the corresponding integral in the estimate of $L_1$, and using (39), we obtain from the last inequality the estimate (31) for $N_1''$.

Thus we have obtained the estimates for the derivatives of $\bar{v}_j$.

The estimates for the increments in these derivatives can be obtained in the same way as in Lemma 1: on the basis of (30), (38) and (40)–(42) we write down a special representation for the increments and derive the estimates for these expressions using this representation, analogous to the estimates given above.

**On the homogeneous Green matrix of the normal parabolic boundary value problem**. If we assume that the system of boundary conditions $B_j$ ($j = 1,\cdots,bN$) is also normal in the sense of [**29**], then we can obtain important information on the properties of the homogeneous Green matrix $G(t,x;\tau,\xi)$ as a function of the parametric variables, if we have the representation of the solution of the nonhomogeneous parabolic boundary value problem using the matrix $G(t,x;\tau,\xi)$.

The assumption that the boundary conditions are normal and that the system (1.1) has a Lagrange-adjoint system allows us to formulate the adjoint problem

$$L^*(\tau,\xi,D_\tau D_\xi)v = g(\tau,\xi),$$

$$B_j'(\tau,\xi,D_\xi)v|_\Gamma = 0 \quad (j=1,\ldots,bN), \tag{1}$$

$$v|_{\tau=T} = \psi(\xi)$$

and to write down the Green formula

$$\int_0^T dt\int_\Omega [Lu(t,x)]'\,\overline{v(t,x)}\,dx + \sum_{j=1}^{bN}\int_0^T dt\int B_jv(t,x)\,G_j'v(t,x)\,ds$$

$$+\int_\Omega u'(0,x)\,\overline{v(0,x)}\,dx = \int_0^T dt\int_\Omega u'(t,x)\,L^*v(t,x)\,dx + \tag{2}$$

$$+\sum_{j=1}^{bN}\int_0^T dt\int_S C_j u(t,x) B_j' v(t,x)\,dS+\int_\Omega u'(T,x)\overline{v(T,x)}\,dx,$$

where $u$ and $v$ are arbitrary matrix functions of appropriate dimensions from $C_{x,t}^{2b,1}(Q)$; $u'$ and $v'$ are matrices transposed to $u$ and $v$; $C_j$, $C_j'$ and $B_j'$ are differential expressions of order $m_j$, $m_j'$ and $r_j'$, respectively, defined on $\Gamma$; $r_j+m_j=2b-1$; $r_j'+m_j'=2b-1$; and $L^*$ is an expression formally adjoint to $L$.

Repeating the arguments of [**29**], we can show that the problem (1) is normal too. Its inverse parabolicity for one particular case has been established in [**31**], and in the general case it has been established by one of the authors in [**32**] as a result of a detailed study of the operator adjoint in the sense of $L_2(Q)$ to the Green operator of the nonhomogeneous boundary value problem.

From the inverse parabolicity of problem (1) and from Theorem 1 of this work it follows that problem (1) possesses a Green matrix $G^*(t,x;\tau,\xi)$ ($\xi$ and $\tau$ are the principal variables, and $t$ and $x$ are the parametric variables) with all properties described in Theorem 1.

Using the line of reasoning shown in [**25**], Russian pp. 91–94, we make the following assertions:

1. *The Green matrix of the homogeneous normal parabolic boundary value problem is normal:*

$$G^*(t,x;\tau,\xi)=G'(t,x;\tau,\xi) \tag{3}$$

2. *Any regular solution $u(t,x)$ of the problem*

$$Lu=f(t,x)\ B_j u|_\Gamma=g_j(t,x)=\varphi(x)$$

*can be represented by means of the homogeneous Green matrix in the form*

$$\begin{aligned}u(t,x)&=\int_\Omega G(t,x;0,\xi)\varphi(x)\,d\xi+\int_0^t d\tau\int_\Omega G(t,x,\tau,\xi)f(\tau,\xi)\,d\xi\\&+\sum_{j=1}^{bN}\int_0^t d\tau\int_S[\overline{C}_j'(\tau,\xi D_\xi)G'(t,x;\tau,\xi)]'g_j(\tau,\xi)\,dS_\xi\end{aligned} \tag{4}$$

3. *The following convolution formula is valid:*

$$G(t,x;\tau,\xi)=\int_\Omega G(t,x;\beta,y)G(\beta,y,\tau,\xi)\,dy \qquad (\tau<\beta<t) \tag{5}$$

From (3) there follow the differential properties of the matrix $G(t,x;\tau,\xi)$ as a function of $\tau$ and $\xi$, and from (5) there follow the differential properties of this matrix with respect to the set of all its variables.

Received 15 May 1969

## BIBLIOGRAPHY


[1] S. D. Èĭdel'man and S. D. Īvasišen, *Estimates of the derivatives of the Green matrix of a homogeneous parabolic boundary problem*, Proc. Internat. Congress Math. (Moscow, 1966), "Mir" Moscow, 1968, Abstracts, Section 7, pp. 62–63. (Russian)

[2] ———, *Estimates of derivatives of the Green's matrix for a parabolic boundary value problem in a half-space*, Dopovīdī Akad. Nauk Ukraïn. RSR **1966**, no. 7, 846–850. (Ukrainian) MR 34 # 4669.

[3] S. D. Īvasišen and S. D. Èĭdel'man, *Estimates of the Green's matrix for a homogeneous parabolic boundary value problem*, Dokl. Akad. Nauk SSSR **172** (1967), 1262–1265 = Soviet Math. Dokl. **8** (1967), 273–266. MR **34** # 7978.

[4] V. A. Solonnikov, *On boundary value problems for linear parabolic systems of differential equations of general form*, Trudy Mat. Inst. Steklov. **83** (1965) = Proc. Steklov Inst. Math. **83** (1965). MR **35** # 1965; # 1966.

[5] S. D. Īvasišen, *The Green's matrix of a nonhomogeneous parabolic boundary value problem*, Dokl. Akad. Nauk SSSR **187** (1969), 730–733 = Soviet Math. Dokl. **10** (1969), 928–931. MR **41** # 2225.

[6] Ju. P. Krasovskiĭ, *Investigation of potentials associated with boundary problems for elliptic equations*, Izv. Akad. Nauk SSSR Ser. Mat. **31** (1967), 587–640 = Math. USSR Izv. **1** (1967), 569–622. Mr **35** # 4585.

[7] ———, *Isolation of singularities of the Green's function*, Izv. Akad. Nauk SSSR Ser. Mat. **31** (1967), 977–1010 = Math. USSR Izv. **1** (1967), 935–966. MR **36** # 6788.

[8] ———, *Green function properties and generalized solutions of elliptic boundary value problems*, Izv. Akad. Nauk SSSR Ser. Mat. **33** (1969), 109-137 = Math. USSR Izv. **3** (1969), 105-130. MR **39** # 5925.

[9] P. E. Sobolevskiĭ, *Estimates of the Green function for second-order parabolic partial differential equations*, Dokl. Akad. Nauk SSSR **138** (1961), 313–316 = Soviet Math. Dokl. **2** (1961), 617–620. MR **23** # A3379.

[10] M. A. Krasnosel'skiĭ, P. P. Zabreĭko, E. I. Pustyl'nik and P. E. Sobolevskiĭ, *Integral operators in spaces of summable functions*, "Nauka", Moscow, 1966. (Russian) MR **34** # 6568.

[11] R. Armia, *On general boundary value problem for parabolic equations*, J. Math. Kyoto Univ. **4** (1964), 207–243. MR **33** # 6156.

[12] S. Mizohata and R. Arima, *Propriétés asymptotiques des valeurs propres des opérateurs elliptiques auto-adjoints*, J. Math. Kyoto Univ. **4** (1964), 245–254. MR **31** # 3712

[13] S. Mizohata, *Sur les propriétés asymptotiques des valeurs propres pour les opérateurs elliptiques*, J. Math. Kyoto Univ. **4** (1964/65), 399–428. MR **32** # 2738b.

[14] V. E. Šatalov and I. A. Šišmarev, *Dirichlet series for elliptic operators*, Dokl. Akad. Nauk SSSR **182** (1968), 1280–1282 = Soviet Math. Dokl. **9** (1968), 1286–1288. MR **40** # 1717.

[15] V. P. Lavrenčuk, *General boundary value problems for parabolic systems with increasing coefficients*, Dopovīdī Akad. Nauk Ukraïn. RSR Ser. A **1968**, 238–243. (Ukrainian) MR **38** # 1407.

[16] A. G. Kostjučenko, *Distribution of eigenvalues for singular differential operators*, Dokl. Akad. Nauk SSSR **168** (1966), 21–24 = Soviet Math. Dokl. **7** (1966), 584–587. MR **33** # 2947.

[17] S. D. Šmulevič, *On the distribution of the eigenvalues of an operator of a selfadjoint elliptic boundary-value problem in an unbounded region*, Dokl. Akad. Nauk SSSR **187** (1969), 959–962 = Soviet Math. Dokl. **10** (1969), 1551–1554.

[18] V. A. Il'in and I. A. Šišmarev, *On the equivalence of the systems of generalized and classical eigenfunctions*, Izv. Akad. Nauk SSSR Ser. Mat. **24** (1960), 757–774. (Russian) MR **23** # A1126.

[19] Ju. M. Berezanskiĭ, *Expansion in eigenfunctions of selfadjoint operators*, "Naukova Dumka", Kiev, 1965; English transl., Transl. Math. Monographs, vol. 17, Amer. Math. Soc., Providence, R. I., 1968. MR **36** # 5768; # 5769.

[20] Ju. M. Berezanskiĭ and Ja. A. Roĭtberg, *A theorem on homeomorphisms and Green's function for general elliptic boundary problems*, Ukrain. Mat. Ž. **19** (1967), no. 5, 3–32. (Russian) MR **36** # 1823.

[21] S. D. Èĭdel'man and S. D. Ĭvasišen, *The Green's matrix of a homogeneous parabolic boundary problem for a system with discontinuous coefficients*, Dokl. Akad. Nauk SSSR **183** (1968), 797–800 = Soviet Math. Dokl. **9** (1968), 1483–1487. MR **38** # 6241.

[22] S. D. Ĭvasišen, *Estimates of the Green's function of the homogeneous first boundary value problem for a second order parabolic equation in a noncylindrical region*, Ukrain. Mat. Ž. **21** (1969), 15–27. (Russian) MR **39** # 5947.

[23] I. I. Verenič and S. D. Ĭvasišen, *Green's matrix for general homogeneous parabolic value problems in noncylindrical domains*, Dopovidī Akad. Nauk Ukraïn. RSR **1970**, 1063–1066. (Ukrainian)

[24] T. Ja. Zagorskiĭ, *Mixed problems for systems of parabolic partial differential equations*, L'vov, 1961. (Russian)

[25] S. D. Èĭdel'man, *Parabolic systems*, "Nauka", Moscow, 1964; English transl., Walters-Noordhoff, Groningen; North-Holland, Amsterdam, 1969. MR **29** # 4998; **40** # 6023.

[26] S. D. Ĭvasišen and S. D. Èĭdel'man, *$\overrightarrow{2b}$-parabolic systems*, Proc. Sem. Functional Analysis (Kiev, 1968), no. 1, Akad. Nauk Ukrain. SSR Inst. Mat., Kiev, 1968, pp. 3–175, 271–273. (Russian) MR **41** # 5783.

[27] M. S. Agranovič and M. I. Višik, *Elliptic problems with a parameter and parabolic problems of general type*, Uspehi Mat. Nauk **19** (1964), no. 3 (117), 53–161 = Russian Math. Surveys **19** (1964), no. 3, 53–157. MR **33** # 415.

[28] O. A. Ladyženskaja, V. A. Solonnikov and N. N. Ural'ceva, *Linear and quasilinear equations of parabolic type*, "Nauka", Moscow, 1967; English transl., Transl. Math. Monographs, vol. 23, Amer. Math. Soc., Providence, R.I., 1968. MR **39** # 3159a, b.

[29] Ja. A. Roĭtberg and Z. G. Seftel', *A theorem on homeomorphisms for elliptic systems and its applications*, Mat. Sb. **78** (**120**) (1969), 446–472 = Math. USSR Sb. **7** (1969), 439–466. MR **40** # 539.

[30] V. A. Solonnikov, *Estimates in $L_p$ of solutions of elliptic and parabolic systems*, Trudy Mat. Inst. Steklov. **102** (1967), 137–160 = Proc. Steklov Inst. Math. **102** (1967), 157–181. MR **37** # 4388.

[31] B. Ja. Lipko and S. D. Èĭdel'man, *A contribution to the theory of parabolic potentials*, Dokl. Akad. Nauk SSSR **166** (1966), 1050–1053 = Soviet Math. Dokl. **7** (1966), 237–240. MR **33** # 7710.

[32] S. D. Ĭvasišen, *Adjoint Green operators. Generalized solutions of parabolic problems with normal boundary conditions*, Dokl. Akad. Nauk SSSR **197** (1971), 261–264 = Soviet Math. Dokl. **12** (1971), 423–427.

[33] V. A. Solonnikov, *On Green matrices of parabolic boundary value problems*, Zap. Naučn. Sem. Leningrad. Otdel. Mat. Inst. Steklov. (LOMI) **14**(1969), 256–287. (Russian)

Translated by:
A. Solan

Trudy Moskov. Mat. Obšč.
Tom 23 (1970)

Trans. Moscow Math. Soc.
Vol. 23 (1970)

# EXTREMALLY DISCONNECTED COMPACT SPACES AND ABSOLUTES

(On the centenary of the birth of Felix Hausdorff)

*B. A. EFIMOV*

## Contents



## Introduction

A topological space $X$ is called *extremally disconnected* if for any open set $U \subset X$ the closure of $U$ is again open. For example, the space associated with an ultrafilter, Stone spaces of complete Boolean algebras, and absolutes are extremally disconnected. We list those properties of extremally disconnected spaces, the class of which we denote by $\mathfrak{R}$, that will be used most frequently in this paper. Let $X \in \mathfrak{R}$. Then

($\alpha$) the Stone-Čech compactification $\beta X \in \mathfrak{R}$;

($\beta$) if $Y$ is dense in $X$, then $Y \in \mathfrak{R}$;

($\gamma$) $X$ contains no nontrivial convergent sequence.[1]

These and other properties of extremally disconnected spaces are set out in [**30**] and [**31**] and, in Russian, in [**25**] and [**23**], from which we quote without special reference.

---

*AMS (MOS) Subject classifications* (1970). Primary 54C55, 54G05.

[1] Properties ($\alpha$), ($\beta$) and ($\gamma$) were proved by Gleason [**35**].

The first part (§§3–6 and 10) is concerned with the study of extremally disconnected compact spaces. In [3] it is proved that the lattice number ($cX$) of an extremally disconnected compact space is always strictly less than the weight ($wX$). Further, if for any nonempty open set $U \subset X$ we have $cU \geqq \tau$, where $\tau$ is accessible or equal to $\aleph_0$, then $(wX)^\tau = wX$. Hence we obtain Pierce's theorem that the power of an infinite complete Boolean algebra is an $\aleph_0$-admissible cardinal and also that the following two conditions are equivalent:

(i) Every extremally disconnected compact space with the weight of the continuum satisfies Suslin's condition.

(ii) $\exp \aleph_0 < \exp \aleph_1$.

In §4 we establish a relation between the weight of an extremally disconnected compact space $X$ and the supremum of the characters of its points $\chi^*(X)$. Namely, we prove[2] that $wX \leqq \exp[\chi^*(X)]$. A deeper result is $wX = \chi^*(X)$, which is proved for absolutes of infinite dyadic compact spaces. Whether $wX = \chi^*(X)$ for arbitrary infinite extremally disconnected compact spaces remains an open question.

The basic result of §5 is a theorem which asserts that every compact space $X$ for which $cX \leqq \tau$ and $wX > \exp\exp\exp\tau$ contains all the extremally disconnected spaces of weights $\leqq (\exp\tau)^+$ (in particular $\beta T_\tau$). This theorem is based on the following two results of independent interest.[3]

(A) Let $X$ be a compact space which has a continuous map onto the Tihonov cube $\mathbf{I}^\tau$. Then the strength of $X \geqq \tau$. Conversely, if the strength of $X > \tau$, then $X$ maps onto $\mathbf{I}^{\log(\tau^+)}$.

(B) Let $X$ be a compact space where $cX \leqq \mathfrak{m}$, the massiveness of $X \geqq \tau$ and $\tau$ is an $\mathfrak{m}$-inaccessible cardinal. Then $X$ maps onto $\mathbf{I}^\tau$.

In §6 we analyze the case of a zero-dimensional extremally disconnected compact space in greater detail and give the corresponding conclusions for the dual category of Boolean algebras. In particular, we prove there that if $X$ is an extremally disconnected space and $\mathfrak{k}$ is an infinite cardinal such that

$$\mathfrak{k} < \log\log(wX),$$

then $X$ contains all extremally disconnected spaces of weight $\leqq \log(\mathfrak{k}^+)$.

Problems of homogeneity of extremally disconnected spaces are considered in §10. Here we prove that if the $\pi$-weight of an extremally disconnected compact space is less that $\Xi_1$ (the first uncountable, weakly inaccessible cardinal) and if $cX \geqq \log(\pi wX)$, then $X$ is not topologically homogeneous. This

---

[2] It is to be noted that from results recently obtained by Arhangel′skiĭ [46] this inequality is true for any compact space. However, I retain this theorem here since it was proved before Arhangel′skiĭ's theorem and the methods of proof are somewhat different, although basically (Lemma 4) they both reduce to a certain branching process.

[3] Definitions of strength, massiveness and $\mathfrak{m}$-accessibility are given in §5.

leads, in particular, to a solution of a problem posed by Arhangel′skiĭ [**5**]. Namely, if $\exp \aleph_0 < \Xi_1$, then the absolute of an arbitrary compact space with weight of the continuum is not homogeneous.

The second part (§§7–9) is devoted to absolutes. The idea of the absolute, or projective space, in general topology appeared comparatively recently in papers of Ponomarev and Gleason ([**7**], [**8**], [**35**]), although in the category of Boolean algebras the injectiveness of a complete Boolean algebra was proved by Sikorski in 1948 [**33**]. Recall that a map $f\colon X \overset{\text{onto}}{\Rightarrow} Y$ is called irreducible if there is no proper closed subset $F \subset X$ such that $f(F) = Y$. An irreducible, complete, extremally disconnected preimage of $X$ is called an absolute $pX$ of the space $X$. According to Iliadis and Fomin [**23**] the points of $pX$ are ultrafilters $\mathfrak{F}$ of open subsets of $X$ touching some given point $x \in X$. As a basis for open sets in $pX$ we can take sets of the form $\Gamma_U = \{\mathfrak{F} : U \in \mathfrak{F}\}$, where $U$ is open in $X$. If $X$ is compact, then the topology described above gives the Stone space associated with the Boolean algebra of canonically closed sets. It can be proved that the space $pX$ is extremally disconnected and is compact if and only if $X$ is compact. Further, by making each point $x \in pX$ (= an ultrafilter $\mathfrak{F}$) correspond to the unique adherent point in $X$ (which exists by definition) we obtain an irreducible perfect map of $pX$ onto $X$. In §7 we investigate the absolute $D^\tau$. The characteristic of $pD^\tau$ is given here in terms of dense dyadic systems, like the characteristic of Aleksandrov and Ponomarev [**3**] for dyadic compact spaces, and we also prove the universality of $pD^\tau$ (and consequently $\beta T_{\log \tau}$) for all extremally disconnected spaces of weight $\leqq \tau$, and we calculate many cardinal invariants of $pD^\tau$. In particular, we prove that $pD^\tau$ for every admissible cardinal ($\tau^{\aleph_0} = \tau$) is an extremally disconnected compact space with homogeneous character and weight $\tau$. Further, we show (§8) that the absolute of a compact topological group $G$ is homeomorphic to $T \times pD^\tau$. Finally, in §9 we solve a problem posed by Arhangel′skiĭ about the homogeneity of noncompact extremally disconnected spaces [**6**]. Here we prove that for each cardinal $\tau$ there exists an extremally disconnected, topologically homogeneous space with local power $\tau$. Further, it turns out that the absolute of any topological group decomposes into everywhere dense extremally disconnected orbits, which are semitopological groups.

The basic results of this paper were announced by the author in [**17**]–[**21**] and [**43**], and were presented in P. S. Aleksandrov's seminar at Moscow State University.

## §1. Cardinal invariants of topological spaces and the connections between them

In what follows we need certain global cardinal invariants of a topological space $X$. The weight ($wX$) of $X$ is the minimum of the powers of bases for open sets in $X$. The lattice number ($cX$) of $X$ is the supremum of the powers of systems of open disjoint sets in $X$. The density ($sX$) of $X$ is the minimum of powers of dense subsets in $X$. Finally, the $\pi$-weight ($\pi wX$ [**9**]) of $X$ is the minimum of the powers of $\pi$-bases for open subsets in $X$. Recall that a system $\mathfrak{B}$ is called a *$\pi$-basis* for $X$ if for every open set $U \subset X$ there exists $V \subset \mathfrak{B}$ such that $V \subset U$. Further we repeatedly use the theorem of Ponomarev which states that for a dyadic compact space $X$ we always have $\pi wX = wX$ [**9**]. The power of a set $X$ is denoted by $|X|$, and $\mathfrak{n}^+$ denotes the successor cardinal to $\mathfrak{n}$. Everywhere we write $\exp \mathfrak{n}$ instead of $2^{\mathfrak{n}}$, and propositions proved with the help of the continuum hypothesis (generalized continuum hypothesis) are denoted by (CH) or (GCH), respectively.

For $T_i$-spaces we have the following inequalities:

(1) $$(T_0) \quad cX \leqq sX \leqq \pi wX \leqq wX,$$

(2) $$(T_0) \quad |X| \leqq \exp(wX),$$

(3) $$(T_2) \quad sX \leqq |X| \leqq \exp\exp(sX),$$

(4) $$(T_3) \quad sX \leqq wX \leqq \exp(sX),$$

(5) $$(\text{compact}) \; wX \leqq |X|.$$

LEMMA 1. *For every regular space $X$ we have*

$$wX \leqq (\pi wX)^{cX}.$$

PROOF. Put $wX = \tau$ and $cX = \mathfrak{n}$. Since $X$ is regular, it is sufficient to prove that no canonically closed subset of $X$ has power greater than $\tau^{\mathfrak{n}}$. Let $\mathfrak{B}$ be a $\pi$-basis of $X$ with power $\tau$ and let $V$ be an arbitrary canonically closed subset of $X$ $(V = \operatorname{int} \overline{V})$. By induction we construct a system $\mathfrak{B} = \{V_\alpha\}$ consisting of disjoint open sets of $X$ and such that $V_\alpha \in \mathfrak{B}$, $\bigcup_\alpha V_\alpha \subset V$ and $\overline{\bigcup_\alpha V_\alpha} = \overline{V}$. In fact, since $\mathfrak{B}$ is a $\pi$-basis, there exists $V_1 \in \mathfrak{B}$ such that $V_1 \subset V$. Let us suppose that we have constructed in $X$ a sequence of disjoint open sets $V_1, V_2, \cdots, V_\alpha, \cdots$, such that $V_\alpha \subset V$ and $V_\alpha \in \mathfrak{B}$. Put $\Phi = \overline{\bigcup_{\alpha<\beta} V_\alpha}$. If $\Phi = \overline{V}$, then everything is proved. If $\Phi \neq \overline{V}$, then there exists $V_\beta \in \mathfrak{B}$ such that $V_\beta \in V - \Phi$. It is easy to see that $V_\alpha \cap V_\beta = \emptyset$ for all $\alpha < \beta$. In this way a sequence is constructed $V_0, V_1, \cdots, V_\beta, \cdots$; $\beta < w(\mathfrak{m})$, $\mathfrak{m} < cX$, (since the elements of this sequence are disjoint open sets) such that $V_\beta \in \mathfrak{B}$ and $\overline{\bigcup_\beta V_\beta} = \overline{V}$. Consequently to each canonically closed $\overline{V}$ there corresponds a certain subset of $\mathfrak{B}$ with power $\leqq \mathfrak{n}$. Conversely, to each such subset $\{V_\alpha\}$ there corresponds uniquely a canonically closed subset $\overline{\bigcup_\alpha V_\alpha}$. In this way there is a one-to-one transformation from the set of all subsets of $\mathfrak{B}$ with power $\leqq \mathfrak{n}$ onto the set of all canonically closed subsets of $X$, whence their power $\leqq \tau^{\mathfrak{n}}$.

COROLLARY 1. *If the regular space $X$ satisfies Suslin's condition and $\tau^{\aleph_0} = \tau$, then $\pi wX = wX$.*

Recall that the *character of a set $A$* in $X$, called $\chi(A, X)$, is the minimum of the powers of fundamental neighborhood systems for $A$ in $X$. The *pseudocharacter* $\psi(A, X)$ is the minimum of the powers of systems $\mathfrak{B}$ of open sets such that $A = \bigcap O_\alpha$, $O_\alpha \in \mathfrak{B}$. We introduce yet another character, which we call the $\pi$-character for $A$ in $X$, $[\pi\chi(A, X)]$. Namely, we call a system $\mathfrak{B}$ of open sets in $X$ a *local $\pi$-basis* for $A$ in $X$ if for any open set $U \supset A$ there exists a $V \in \mathfrak{B}$ such that $V \subset U$. The minimum of the powers of local $\pi$-bases for $A$ in $X$ is called the *$\pi$-character* of $A$ in $X$. Note that the local characteristic, the "$\pi$-character", is analogous to the global characteristic, the "$\pi$-weight". An essential role will be played later by the $\Delta$- and $\delta$-characters of closed sets, which were introduced by the author in [**18**].

A *$\Delta$-system for a closed set $F \subset X$* is a system $\Delta$ of canonically closed subsets of $X$ such that $F = \bigcap \overline{U}$, $\overline{U} \in \Delta$. We say that *the $\Delta$-system is fine* if for every open $V \supset F$ there exists a $\overline{U} \in \Delta$ such that $\overline{U} \subset V$. The minimum of the powers of (fine) $\Delta$-systems of $F$ is called the *$\Delta$-pseudocharacter* ($\Delta$-character) of the set $F$ in $X$. They are denoted by $\Delta\psi(F, X)$ and $\Delta(F, X)$.

A more subtle characteristic of $F$ is the $\delta$-character. A *$\delta$-system for a closed set* $F$ in $X$ is a system $\delta$ of canonically closed subsets of $X$ such that $F = \bigcap \overline{U}$ for all $\overline{U} \in \delta$ and for any finite collection $\overline{U}_1, \cdots, \overline{U}_n$ of elements of $\delta$ we have $\bigcap_{k=1}^n \operatorname{int} \overline{U}_k \neq \emptyset$. As always, $\delta$ is called *fine* if for any open $U \supset F$ there exists $\overline{V} \in \delta$ such that $\overline{V} \subset U$. The minimum of the powers of (fine) $\delta$-systems is called the *$\delta$-pseudocharacter* (*$\delta$-character*) *of the set* $F$ in $X$. These are denoted by $\delta\psi(F, X)$ and $\delta(F, X)$.

Note that the definitions of the above characters imply that for any regular topologic space $X$ and any point $x \in X$ we have

$$\pi\chi(x, X) \leqq \Delta(x, X) \leqq \delta(x, X) \leqq \chi(x, X) \leqq wX. \tag{6}$$

LEMMA 2. *Let $X$ be a regular space. Then*

$$\pi\chi(x, X) \leqq \delta(x, X) \leqq \exp[\pi\chi(x, X)]. \tag{7}$$

PROOF. The inequality $\pi\chi(x, X) \leqq \delta(x, X)$ is clear because the interiors of the elements of any fine $\delta$-system for a point $x$ are simultaneously a local $\pi$-basis for $x$. Let us prove the second inequality. Let $\mathfrak{B}_1$ be a fundamental system of neighborhoods for $x$ and let $\mathfrak{B}_2$ be a local $\pi$-basis for $x$. Let $|\mathfrak{B}_2| = \tau$. For any neighborhood $U \in \mathfrak{B}_1$ consider the set of all $V_\alpha(U) \in \mathfrak{B}_2$ such that $V_\alpha(U) \subset U$. Put $B(U) = \overline{\bigcup V_\alpha(U)}$. Then $B(U)$ is a canonically closed set, where $B(U) \ni x$. For if $B(U) \not\ni x$, then $U - B(U) = V$ is open and $x \in V$. But then by definition of a $\pi$-basis there would exist an open set $V_{\alpha_0} \in \mathfrak{B}_2$ such that $V_{\alpha_0} \subset V$. Consequently $V_{\alpha_0} \subset U$; that is, $V_{\alpha_0} \subset B(U)$, which is a contradiction. Thus $x \in B(U)$ for each $U \in \mathfrak{B}_1$. Consequently $x \in \bigcap_{U \in \mathfrak{B}_1} B(U) \subset \bigcap_{U \in \mathfrak{B}_1} U = x$. Further, let $B(U_1), \cdots, B(U_s)$ be an arbitrary finite collection of elements of $B(U)$. Since $U_i \in \mathfrak{B}_1$, we see that $\bigcap_{i=1}^s U_i = W$ is open and $x \in W$. Therefore there exists $V_{\alpha_0} \in \mathfrak{B}_2$ such that $V_{\alpha_0} \subset W$; that is, $V_\alpha \subset U_i$ for each $i = 1, \cdots, s$. Therefore $V_{\alpha_0} \subset \operatorname{int} B(U_i)$, whence $\bigcap_{i=1}^s \operatorname{int} B(U_i) \neq \emptyset$. Thus $\mathfrak{L} = \{B(U)\}$, $U \in \mathfrak{B}_1$, is a $\delta$-system for $x$. Next, this system is fine, since $\mathfrak{B}_1$ is a fundamental system of neighborhoods. Finally we estimate the power of $\mathfrak{L}$. We make each $B(U)$ correspond to the set of all $V_\alpha$ for which $V_\alpha \subset U$. This set $\{V_\alpha\}$ defines $B(U)$ uniquely. Consequently $|\mathfrak{L}| \leqq \tau^\tau = \exp \tau$.

Note that without using some supplementary cardinal invariants of $X$ it is impossible to improve (7). For let $T$ be a discrete space and let $\beta T$ be the Stone-Čech compactification of $T$. Then for each $x \in \beta T$ we have $\pi\chi(x, \beta T) \leqq |T|$. On the other hand, there exists a point $x$ in $\beta T - T$ (§4) for which $\chi(x, \beta T) = \delta(x, \beta T) = \exp|T|$.

**Lemma** 3. *Let $X$ be a compact and let $F$ be a closed subset of $X$. Then $\chi(F, X) = \psi(F, X)$ and $\delta\psi(F, X) = \delta(F, X)$.*

**Proof.** The first equality is well known [**2**]. For the proof of the second it is sufficient to show that $\delta(F, X) \leqq \delta\psi(F, X)$. Let $\delta$ be a certain $\delta$-system of power $\delta\psi(F, X)$. For each finite collection of elements $\overline{V}_1, \cdots, \overline{V}_s$ of $\delta$ let $U = \bigcap_{i=1}^s \operatorname{int} \overline{V}_i \neq \emptyset$ and $\overline{W} = \overline{U}$. Extend $\delta$ to a system $\delta'$ by introducing all possible such $W$. We claim that $\delta'$ is a fine $\delta$-system. In fact, since $\bigcap \delta = F$, for any open $U \supset F$ there exist finitely many elements of $\delta$ such that their intersection lies in $U$ (because $X$ is compact). Thus there exists $\overline{W} \in \delta'$ such that $F \subset \overline{W} \subset \overline{U}$.

**Lemma** 4. *Let $X$ be a Hausdorff space with $\delta(x, X) \leqq \mathfrak{n}$, $\mathfrak{n} \geqq \aleph_0$, for each point $x \in X$. If $|X| > \exp \mathfrak{n}$, then $cX \geqq \mathfrak{n}^+$.*

**Proof.** For each point $x \in X$ we fix some fine $\delta$-system of canonically closed sets

$$\mathfrak{L}_x = \{\Gamma_1(x), \Gamma_2(x), \cdots, \Gamma_\xi(x), \cdots, \xi < \omega(\mathfrak{n})\}, \tag{8}$$

where $\omega(n)$ is the least ordinal of power $\mathfrak{n}$. Then, even if $\delta(x, X) < \mathfrak{n}$, by repeating a certain canonical set $\Gamma_{\xi_0}(x)$ the necessary number of times, we obtain a sequence (8) of length $\omega(\mathfrak{n})$. Put

$$W_{\xi\eta}(x) = \mathrm{Cl}_X[\operatorname{int} \Gamma_\xi(x) \cap \operatorname{int} \Gamma_\eta(x)].$$

The latter intersection is not empty, since $\mathfrak{L}_x$ is a $\delta$-system. Denote by $f(x, y)$ a function defined on $X \times X$ with values in

$$Z = [\{\xi, \xi < \omega(\mathfrak{n})\} \times \{\eta, \eta < \omega(\mathfrak{n})\}] \cup \{(0,0)\}.$$

If $x \neq y$, then $f$ sends the pair $(x, y)$ into a certain pair $(\xi, \eta)$ for which $W_{\xi\eta}(x) \cap W_{\xi\eta}(y) = \emptyset$, and $(x, x)$ to $(0,0)$. Note that if $x \neq y$, then the required pair $(\xi, \eta)$ always exists. For since $X$ is Hausdorff, there exist neighborhoods $U(x)$ and $V(y)$ such that $U(x) \cap V(y) = \emptyset$. Since $\mathfrak{L}_x$ and $\mathfrak{L}_y$ are fine $\delta$-systems, there exist $\Gamma_\xi \in \mathfrak{L}_x$ and $\Gamma_\eta \in \mathfrak{L}_y$ such that $\Gamma_\xi \subset U(x)$ and $\Gamma_\eta \in V(y)$, and consequently $\Gamma_\xi \cap \Gamma_\eta = \emptyset$. Now $W_{\xi\eta}(x) \subseteq \Gamma_\xi$ and $W_{\xi\eta}(y) \subseteq \Gamma_\eta$; therefore $W_{\xi\eta}(x) \cap W_{\xi\eta}(y) = \emptyset$. The remainder of the proof of this lemma is completely analogous to that of Theorem 1 in the author's paper [**22**].

COROLLARY 2. *If a Hausdorff space $X$ satisfies Suslin's condition and if $\delta(x, X) \leqq \aleph_0$ for all $x \in X$, then $|X| \leqq \exp \aleph_0$.*

LEMMA 5. *Let $X = \prod_{\alpha \in A} X_\alpha$ be a Tihonov product of separable metric spaces and let $|A| \geqq \aleph_1$. Then*

$$\pi\chi(x, X) = \Delta(x, X) = \delta(x, X) = \chi(x, X) = wX = |A| \tag{9}$$

*for any point $x \in X$.*

PROOF. Note first of all that it follows from the definition of the Tihonov topology on $X$ and the condition $wX_\alpha \leqq \aleph_0$ that $wX = |A|$. Further, by (6) it is sufficient to prove that

$$|A| \leqq \pi\chi(x, X).$$

Suppose that $|A| > \pi\chi(x, X) = \tau$. Let $\mathfrak{B}$ be a local $\pi$-basis for $x$ of power $\tau$. Each element $U \in \mathfrak{B}$ is of the form $\varphi^{-1}(V)$, where $V$ is a basis open subset of $X_{\alpha_1} \times \cdots \times X_{\alpha_s}$ and $\varphi$ is the projection of $X$ onto this boundary. We make each $U \in \mathfrak{B}$ correspond to the collection $(V, \alpha_1, \cdots, \alpha_s)$ that defines it. Since $|\mathfrak{B}| = \tau$ and $wX_\alpha \leqq \aleph_0$, the power of the set of all such collections is not greater than $\aleph_0 \cdot \tau = \tau$. Let us call the indices in this collection the fixing indices. Obviously the power of the set of fixing indices is $\leqq \tau$. Since $|A| > \tau$, there exists an index $\alpha_0 \in A$ that is not a fixing index. Let $V$ be a proper open subset of $X_{\alpha_0}$ and let $W = \varphi_{\alpha_0}^{-1} V$, where $\varphi_{\alpha_0}$ is the projection of $X$ onto $X_{\alpha_0}$. Without loss of generality we may suppose that $x \in W$. Note that $W$ contains no elements of $\mathfrak{B}$ in spite of the fact that $\mathfrak{B}$ is a local $\pi$-basis. For if

$$\varphi^{-1} V \subset W, \; V \subset X_{\alpha_1} \times \cdots \times X_{\alpha_s}, \; \alpha_0 \notin (\alpha_1, \cdots, \alpha_s),$$

then the point $y = (y_\alpha)$ for which $(y_{\alpha_i}) \in V$, $i = 1, \cdots, s$, and $y_{\alpha_0} \in X_{\alpha_0} - V$ lies in $\varphi^{-1} V$ but not in $W$. The contradiction so obtained proves the lemma.

Let $\epsilon(x, X)$ be any local cardinal invariant of $X$ at $x$, for example one of the characters discussed above. Then $\epsilon^* X = \sup_{x \in X} \epsilon(x, X)$ is called the *upper $\epsilon$-character* and $\epsilon_* X = \inf_{x \in X} \epsilon(x, X)$ the *lower $\epsilon$-character*. For example, if $\epsilon(x, X) = \chi(x, X)$, we obtain the upper and lower $\chi$-characters of $X$, respectively.

Following Šanin [13], let us consider other cardinal invariants of topological spaces, the so-called "calibres". A cardinal number $\mathfrak{n} > 1$ is called the *calibre* of a topological space $X$ if every system $\mathfrak{U}$ of power $\mathfrak{n}$ consisting of nonempty open subsets of $X$ contains a subsystem $\mathfrak{U}_0 \subset \mathfrak{U}$ of power $\mathfrak{n}$ such that the intersection of all its elements is nonempty. For example, every regular number $\mathfrak{n} > sX$ is a calibre of $X$.

ŠANIN'S CONDITION [4]. *We say that the space $X$ satisfies Šanin's condition if every decreasing sequence of nonempty open subsets of $X$, totally ordered by* *inclusion and having empty intersection, contains a countable cofinal subsequence.*

LEMMA 6 (ARHANGEL'SKIĬ AND PONOMAREV). *A space $X$ satisfies Šanin's condition if and only if every regular cardinal $\mathfrak{n} \geqq \aleph_1$ is a calibre of $X$.*

PROOF. Denote by (Š) the statement "$X$ satisfies Šanin's condition" and by (C) the statement "every regular cardinal $\mathfrak{n} \geqq \aleph_1$ is a calibre of $X$".

a) (C) $\to$ (Š). Let

$$U_1 \supset U_2 \supset \cdots \supset U_\alpha \supset \cdots, \alpha < \omega(\tau)$$

be a totally ordered sequence of nonempty open subsets of $X$, of order type $\omega(\tau)$, and let $\bigcap_\alpha U_\alpha = \emptyset$. We pass to the smallest strictly decreasing subsequence cofinal with the given one, which, as before, we denote by

$$U_1 \supsetneqq U_2 \supsetneqq \cdots \supsetneqq U_\alpha \supsetneqq \cdots, \alpha < \omega(\tau_0).$$

Since the sequence is minimal, the number $\tau_0$ is regular, and since it is cofinal, $\bigcap_\alpha U_\alpha = \emptyset$. If $\tau_0 \geqq \aleph_1$, then by (C) we have $\bigcap_\alpha U_\alpha \neq \emptyset$ in spite of the result just obtained. Thus $\tau_0 \leqq \aleph_0$, as required.

b) (Š) $\to$ (C). Let $\mathfrak{C} = \{U_\alpha\}$, $\alpha \in A$, $|A| = \tau \geqq \aleph_1$, be an arbitrary family of nonempty open subsets with regular power $\tau$. We suppose that $\tau$ is not a calibre of $X$. This means that $\mathfrak{k}(x) = |\{U_\alpha \in \mathfrak{C}, x \in U_\alpha\}| < \tau$ for any point $x \in X$. Without loss of generality we may suppose that the elements of the family $\mathfrak{C}$ are indexed by all the transfinite numbers $\alpha < \omega(\tau)$. We put $V_\beta = \bigcup U_\alpha$, $\beta \leqq \alpha < \omega(\tau)$. Then the sequence $V_1 \supset V_2 \supset \cdots \supset V_\beta \supset \cdots$, $\beta < \omega(\tau)$, is decreasing, totally ordered with respect to inclusion, and moreover $\bigcap_{\beta < \omega(\tau)} V_\beta = \emptyset$. For since $\mathfrak{k}(x) < \tau$ for all $x \in X$, for any $x \in X$ we can find an ordinal $\beta$ such that $U_\alpha \not\ni x$ for all $\alpha > \beta$. Consequently $x \notin V_\beta$. Thus $\bigcap_{\beta < \omega(\tau)} V_\beta = \emptyset$. By (Š) there exists a countable cofinal subsequence $V_{\beta_1} \supset V_{\beta_2} \supset \cdots \supset V_{\beta_n} \supset \cdots$, where $\bigcap_{n=1}^\infty V_{\beta_n} = \emptyset$. Consequently $\mathrm{cf}(\tau) = \aleph_0$, which contradicts the regularity of $\tau$ and the assumption that $\tau \geqq \aleph_1$.

LEMMA 7. *Let $f\colon X \to Y$ be an irreducible closed map of $X$ onto $Y$. Then* 1) $cX = cY$; 2) $sX = sY$; 3) $\pi w X = \pi w Y$; 4) *if the regular cardinal $\mathfrak{k}$ is a calibre of $Y$, then $\mathfrak{k}$ is a calibre of $X$, and conversely*; 5) $\pi\chi(x, X) \geqq \pi\chi(fx, Y)$; 6) $\Delta(x, X) \geqq \Delta(fx, Y)$; 7) $\delta(x, X) \leqq \delta(fx, Y)$ *for all $x \in X$.*

PROOF. 1) The inequality $cX \geqq cY$ is obvious. Conversely, if $\{U_\alpha\}$ is a system of disjoint open sets in $X$, then $\{Y - f(X - U_\alpha)\}$ is an analogous system in $Y$ with the same power; that is, $cX \leqq cY$. 2) The inequality $sX \geqq sY$ is clear. Conversely, if $M$ is a dense subset of $Y$, then $M' = \{x \in f^{-1}y, y \in M\}$ is a dense subset of $X$ of the same power; that is, $sX \leqq sY$. 3) Since for an irreducible closed map the image of a canonical closed set is canonically closed, $\pi w X \leqq \pi w Y$. On the other hand, if $\mathfrak{L}$ is a $\pi$-basis, then $\{f^{-1}U, U \in \mathfrak{L}\}$ is a $\pi$-basis of $X$; that is $\pi w X \leqq \pi w Y$. 4) The implication "$\mathfrak{k}$ is a calibre of $X$" $\Rightarrow$ "$\mathfrak{k}$ is a calibre of $Y$" follows from the properties of continuous maps. Conversely, let $\{U_\alpha\}$ be a system of nonempty open sets of $X$ with power $\mathfrak{k}$. Then $\{Y - f(X - U_\alpha)\} = \{W_\alpha\}$ is a system of nonempty open subsets of $Y$, where $f^{-1}W_\alpha$ is dense in $U_\alpha$, since the map is irreducible. Two cases are possible: a) $|\{W_\alpha\}| < \mathfrak{k}$. Since $\mathfrak{k}$ is regular, there exists an $\alpha_0$ such that $|\{U_\alpha, U_\alpha \supset f^{-1}W_{\alpha_0}\}| = \mathfrak{k}$, as required. b) $|\{W_\alpha\}| = \mathfrak{k}$. Then by hypothesis there exists a subfamily $\mathfrak{C} \subset \{W_\alpha\}$ of power $\mathfrak{k}$ such that $\bigcap \mathfrak{C} \neq \emptyset$. Hence the family $\{U_\alpha, U_\alpha \supset f^{-1}W_\alpha, W_\alpha \in \mathfrak{C}\}$ is the required subfamily of $\{U_\alpha\}$ having nonempty intersection. 5) Let $\mathfrak{C} = \{U_\alpha\}$ be a local $\pi$-basis for the point $x \in X$. Then $\{\operatorname{int} f(U_\alpha)\}$ is a local $\pi$-basis for $f(x)$, since the map is irreducible and closed. 6) and 7) follow from the fact that for irreducible closed maps the image of a $\Delta$- or $\delta$-system is a $\Delta$- or $\delta$-system, because for these maps the images of canonically closed sets are canonically closed.

COROLLARY 3. *Suslin's condition and Šanin's condition are invariant under irreducible closed maps.*

LEMMA 8. *Let $f$: $X \to Y$ be a perfect map from $X$ to $Y$. Then there exists a closed subset $F \subset X$ such that $f|F$ is irreducible and $f(F) = Y$.*

This lemma is well known, and the proof is based on Brouwer's theorem [**28**] (see, for example, [**34**]).

## §2. Logarithms of cardinal numbers

*The logarithm of a cardinal number* $\mathfrak{m}$ *with respect to a base* $\mathfrak{n}$ is the smallest number $\mathfrak{k} = \log_{\mathfrak{n}} \mathfrak{m}$ such that $\mathfrak{n}^{\mathfrak{k}} \geqq \mathfrak{m}$. In particular, the logarithm of $\mathfrak{m}$ with respect to base 2 is the smallest cardinal $\mathfrak{k} = \log \mathfrak{m}$ such that $\exp \mathfrak{k} \geqq \mathfrak{m}$.

For example,

$$\log(\exp \aleph_0) = \aleph_0; \ \log \aleph_0 = \aleph_0.$$

The definition of a logarithm implies immediately the following basic inequality:

$$(\mathfrak{l}_1) \quad \exp \log \mathfrak{m} \geqq \mathfrak{m}.$$

Note that, as the second above example shows, this inequality may be strict. Recall that the cardinal number $\mathfrak{m}$ is said to be *dominant* if for any $\mathfrak{n} < \mathfrak{m}$ we have $\exp \mathfrak{n} < \mathfrak{m}$.

LEMMA 9. *A cardinal number* $\mathfrak{n}$ *is dominant if and only if* $\log \mathfrak{n} = \mathfrak{n}$.

PROOF. Let $\mathfrak{n}$ be dominant. If $\log \mathfrak{n} < \mathfrak{n}$, then $\exp \log \mathfrak{n} < \mathfrak{n}$, which contradicts $(\mathfrak{l}_1)$. Conversely, if $\log \mathfrak{n} = \mathfrak{n}$, then for any $\mathfrak{k} < \log \mathfrak{n}$ we have $\exp \mathfrak{k} < \mathfrak{n}$, by the definition of logarithm; that is, $\mathfrak{n}$ is dominant.

LEMMA 10. *For any two infinite cardinals* $\mathfrak{m}$ *and* $\mathfrak{n}$ *we have*

$$(\mathfrak{l}_2) \quad \mathfrak{m} \leqslant \mathfrak{n} \Rightarrow \log \mathfrak{m} \leqslant \log \mathfrak{n},$$

$$(\mathfrak{l}_3) \quad \log(\mathfrak{m}\,\mathfrak{n}) = \log \mathfrak{m} + \log \mathfrak{n}.$$

PROOF. ($\mathfrak{l}_2$) By the definition of the logarithm we have $\exp\log \mathfrak{m} \geqq \mathfrak{m}$ and $\exp\log \mathfrak{n} \geqq \mathfrak{n}$. Assume that $\log \mathfrak{m} > \log \mathfrak{n}$; then

$$\exp(\log \mathfrak{m}) \geqq \exp(\log \mathfrak{n}) \geqq \mathfrak{n} \geqq \mathfrak{m},$$

that is, $\exp(\log \mathfrak{n}) \geqq \mathfrak{m}$; hence $\log \mathfrak{n} \geqq \log \mathfrak{m}$ in spite of the assumption.

($\mathfrak{l}_3$) Let $\mathfrak{m} \geqq \mathfrak{n}$. Then by the above proof $\log \mathfrak{m} \geqq \log \mathfrak{n}$. Therefore $\mathfrak{m} \cdot \mathfrak{n} = \mathfrak{m}$ and $\log \mathfrak{m} + \log \mathfrak{n} = \log \mathfrak{m}$. Thus, $\log(\mathfrak{m} \cdot \mathfrak{n}) = \log \mathfrak{m}$, whence we obtain ($\mathfrak{l}_3$).

COROLLARY 4. *Let* $\mathfrak{n}$ *be a dominant number. Then for any* $\mathfrak{k}$ *such that* $\mathfrak{n} \leqq \mathfrak{k} \exp \mathfrak{n}$ *we have* $\log \mathfrak{k} = \mathfrak{n}$.

In fact, using ($\mathfrak{l}_2$) let us take logarithms in the given inequality. We obtain

$$\log \mathfrak{n} \leqq \log \mathfrak{k} \leqq \log(\exp \mathfrak{n}).$$

By Lemma 9 $\log \mathfrak{n} = \mathfrak{n}$, and by the definition of the logarithm we obtain $\log(\exp \mathfrak{n}) \leqq \mathfrak{n}$. Hence $\log \mathfrak{k} = \mathfrak{n}$, as required.

REMARK 1. It follows from results of Cohen [27] that the hypothesis $\exp \aleph_0 = \aleph_{\omega_1}$ is independent of the system $ZF$. Since $\aleph_0$ is dominant, using Corollary 4 we obtain

$$(\exp \aleph_0 = \aleph_{\omega_1}) \Rightarrow (\log \aleph_\alpha = \aleph_0, \text{ for any } \alpha < \omega_1(\mathfrak{c})).$$

LEMMA 11. *Let* $X = \prod_{\alpha \in A} X_\alpha$ *be a Tihonov product of separable metric spaces, where* $|A| \geqq \aleph_0$. *Then* $sX = \log|A|$; *in particular,* $sD^\tau = \log\tau$.

PROOF. Suppose $\mathfrak{k} = \log|A|$. Since $\exp \mathfrak{k} \geqq |A|$ and $sX \leqq \aleph_0 \leqq \mathfrak{k}$, by the Hewitt-Pondicherry theorem ([30], English p. 96) we obtain $sX \leqq \mathfrak{k}$. Let us assume that $sX < \mathfrak{k}$; then according to (4) and the definition of the logarithm we obtain

$$|A| = wX \leqq \exp(sX) < |A|.$$

Thus $sX = \mathfrak{k}$.

### §3. The relation between the weight and the lattice number in extremally disconnected compact spaces

Recall that the cofinal character $\mathrm{cf}(\mathfrak{n})$ of a cardinal number $\mathfrak{n}$ is the minimum of the powers of sets $A$ such that $\sum_{\alpha \in A} \mathfrak{m}_\alpha = \mathfrak{n}$, where $\mathfrak{m}_\alpha < \mathfrak{n}$. If $\mathrm{cf}(\mathfrak{n}) = \mathfrak{n}$, then $\mathfrak{n}$ is regular; if $\mathrm{cf}(\mathfrak{n}) < \mathfrak{n}$, then $\mathfrak{n}$ is singular.

LEMMA 12. *Let* $X$ *be a topological space.*[4] *If* $cX = \aleph_{\alpha+1}$ *or* $cX$ *is singular, there exists in* $X$ *a family of power* $cX$ *consisting of disjoint open sets of* $X$.

PROOF. Let $cX = \tau$. If $\tau = \aleph_{\alpha+1}$, then there exists in $X$ a system of disjoint open sets of power $\tau$. For if the power of every such system $< \aleph_{\alpha+1}$, then it

[4] No separation axioms are assumed.

is $\leqq \aleph_\alpha$, whereas $cX = \aleph_{\alpha+1}$. Let $cX = \tau$ be singular; then $\tau = \sum_{\alpha \in A} \mathfrak{n}_\alpha$ and $|A| = \mathrm{cf}(\tau)$ and $\mathfrak{n}_\alpha < \tau$. Two mutually exclusive cases are possible:

1) *There exists an open set* $V \subset X$ *such that for every open* $U \subset V$ *we have* $cU = cV = \tau$.

2) *For every open* $V \subset X$ *there exists an open* $U \subset V$ *such that* $cU < cV = \tau$.

*First case.* In this case we construct a system of power $\tau$ consisting of disjoint open sets in $V$ (and consequently open in $X$). Since $cV = \tau$ and $\mathrm{cf}(\tau) = \mathfrak{m} < \tau$, there exists a system $\mathfrak{B}$ of disjoint open sets in $V$ such that $\mathfrak{m} \leqq |\mathfrak{B}| \leqq \tau$. Let $\mathfrak{B}_0 \subset \mathfrak{B}$ and $|\mathfrak{B}_0| = \mathfrak{m}$. Let us establish any one-to-one correspondence between the elements of $\mathfrak{B}_0$ and the set of cardinals $\mathfrak{N} = \{\mathfrak{n}_\alpha\}$, $\alpha \in A$. Let $U_\alpha \in \mathfrak{B}$ correspond to $\mathfrak{n}_\alpha \in \mathfrak{N}$. Since, by hypothesis, for any open $U \subset V$ we have $cU = \tau$, in particular $cU_\alpha = \tau > \mathfrak{n}_\alpha$. Therefore there exists a system $\mathfrak{B}_\alpha$ in $U_\alpha$ consisting of disjoint open sets of $U_\alpha$, where $\mathfrak{n}_\alpha \leqq \mathfrak{B}_\alpha \leqq \tau$. Put $\mathfrak{B} = \bigcup_{\alpha \in A} \mathfrak{B}_\alpha$. Since $U_\alpha \cap U_\beta = \emptyset$ if $\alpha \neq \beta$,

$$\tau = \sum_{\alpha \in A} \mathfrak{n}_\alpha \leqslant |\mathfrak{B}| \leqslant \tau\mathfrak{m} = \tau.$$

Thus $\mathfrak{B}$ is the required system.

*Second case.* First of all, by induction we construct a system $\mathfrak{B} = \{U_\alpha\}$, $\alpha \in B$, of disjoint open sets in $X$, where $cU_\alpha < \tau$ for all $\alpha \in B$ and $\overline{\bigcup_{\alpha \in B} U_\alpha} = X$. Put $X = V$; then by hypothesis there exists an open $U_1 \subset X$ such that $cU_1 < \tau$. Suppose that for each $\alpha < \beta$ we have constructed the sequence

$$U_1, U_2, \cdots, U_\alpha, \cdots, \alpha < \beta$$

of disjoint open sets in $X$, where $cU_\alpha < \tau$ for all $\alpha < \beta$. Put $\Phi = \overline{\bigcup_{\alpha < \beta} U_\alpha}$. If $\Phi = X$, then our aim is achieved. If $X - \Phi = V \neq \emptyset$, then by hypothesis there exists an open $U_\beta \subset V$ such that $cU_\beta < \tau$. It is clear that $U_\alpha \cap U_\beta = \emptyset$ if $\alpha < \beta$. This transfinite process stops after not more than $\tau$ steps, and so the required system $\mathfrak{B} = \{U_\alpha\}$, $\alpha \in B$, has been constructed.

We assert that either $|\mathfrak{B}| = \tau$ or $\sup_{\alpha \in B}(cU_\alpha) = \tau$.

Suppose the contrary. Let $|\mathfrak{B}| = \mathfrak{n} < \tau$ and $\sup_{\alpha \in B}(cU_\alpha) = \mathfrak{l} < \tau$. Put $\mathfrak{k} = \max(\mathfrak{n}, \mathfrak{l})$. It is clear that $\mathfrak{k} < \tau$. Let $\mathfrak{U}$ be a system of disjoint open sets of $X$ such that $\mathfrak{k} < |\mathfrak{U}| \leqq \tau$. Such a system exists in $X$ since $cX = \tau > \mathfrak{k}$. With each $U_\alpha \in \mathfrak{B}$ we associate the set

$$\mathfrak{M}_\alpha = \{U \in \mathfrak{U},\ U \cap U_\alpha \neq \emptyset\}.$$

Since $\overline{\bigcup_{\alpha \in B} U_\alpha} = X$, for each $U \in \mathfrak{U}$ there exists a $U_\alpha \in \mathfrak{B}$ for which $U \cap U_\alpha \neq \emptyset$. Consequently $\mathfrak{U} = \bigcup_{\alpha \in B} \mathfrak{M}_\alpha$. On the other hand, $|\mathfrak{M}_\alpha| \leqq cU_\alpha \leqq \sup_{\alpha \in B}(cU_\alpha) \leqq \mathfrak{k}$; consequently $|\mathfrak{U}| \leqq |\bigcup_{\alpha \in B} \mathfrak{M}_\alpha| \leqq \mathfrak{k} \cdot \mathfrak{k} = \mathfrak{k} < |\mathfrak{U}|$; that is, $|\mathfrak{U}| < |\mathfrak{U}|$, which is a contradiction. Thus either $|\mathfrak{B}| = \tau$ or $\sup_{\alpha \in B}(cU_\alpha) = \tau$.

If the first holds, everything is proved. Let $|\mathfrak{B}| < \tau$; then $\sup_{\alpha\in B}(cU_\alpha) = \tau$. Note that by construction $cU_\alpha = \mathfrak{n}_\alpha < \tau$. We arrange the set of cardinals $\mathfrak{R} = \{\mathfrak{n}_\alpha\}$ in ascending order:

$$(*) \qquad \mathfrak{n}_{\alpha(1)} < \mathfrak{n}_{\alpha(2)} < \cdots < \mathfrak{n}_{\alpha(\beta)} < \cdots < \tau.$$

Since $\sup_{\alpha\in B}\{\mathfrak{n}_\alpha\} = \tau$, we have $\lim_\beta \mathfrak{n}_{\alpha(\beta)} = \tau$. Without loss of generality we suppose that the sequence (*) is strictly increasing. For each cardinal $\mathfrak{n}_{\alpha(\beta+1)}$ of (*) we consider the corresponding set $U_{\alpha(\beta+1)}$. Since $cU_{\alpha(\beta+1)} = \mathfrak{n}_{\alpha(\beta+1)} > \mathfrak{n}_{\alpha(\beta)}$, there exists a system $\mathfrak{B}_{\beta+1}$ of disjoint open sets in $U_{\alpha(\beta+1)}$, where $|\mathfrak{B}_{\beta+1}| = \mathfrak{n}_{\alpha(\beta)}$. Put $\mathfrak{B} = \bigcup_\beta \mathfrak{B}_{\beta+1}$. Then $|\mathfrak{B}| = \sum_\beta \mathfrak{n}_{\alpha(\beta)} = \lim_\beta \mathfrak{n}_{\alpha(\beta)} = \sup_\beta \mathfrak{n}_{\alpha(\beta)} = \tau$. $\mathfrak{B}$ is the required system.

REMARK 2. It has come to the author's notice that Lemma 12 follows from results in a paper by Erdös and Tarski [**36**] in which they obtain analogous results for partially ordered sets. Note further that the question of the existence of a family of disjoint open sets in $X$ with maximal power is closely related to the nature of the number $cX$ (also noted in [**36**]). Recall that a cardinal number $\tau$ is called a limit cardinal if for any $\mathfrak{n} < \tau$ there exists an $\mathfrak{m}$ such that $\mathfrak{n} < \mathfrak{m} < \tau$. A regular limit cardinal is called weakly inaccessible. The first uncountable weakly inacessible cardinal is denoted by $\Xi_1$. Every infinite cardinal less than $\Xi_1$ is of the form $\aleph_0$ or $\aleph_{\alpha+1}$, or is singular. If $cX = \aleph_0$ then it is easy to see that there always exists a countable family of disjoint open sets in $X$. If either $cX = \aleph_{\alpha+1}$ or $cX$ is singular, then such a family of power $cX$ exists by Lemma 12. In this way we obtain

COROLLARY 5. *Let $X$ be a topological space and let $cX < \Xi_1$. Then there exists a system $\mathfrak{B}$ of power $cX$ consisting of disjoint open sets in $X$.*

REMARK 3. It is well known [**26**] that if, for example, the Gödel system of axioms is consistent, it remains consistent after adjoining an axiom which asserts the accessibility of all uncountable cardinals. Thus we may postulate either the existence or the nonexistence of $\Xi_1$. Let $\tau = \Xi_1$. In this case we construct an extremally disconnected compact set $X$ for which $cX = \tau$ and there does not exist a system of disjoint open sets of $X$ of power $\tau$. For this we require the following theorem due to Šanin concerning the calibre of a topological space (§1).

THEOREM (ŠANIN). *A regular cardinal $\mathfrak{n} \geqq \aleph_1$ is a calibre of the Tihonov product $\prod_{\alpha\in A} X_\alpha$ of the topological spaces $X_\alpha$ if and only if $\mathfrak{n}$ is a calibre of each $X_\alpha$, $\alpha \in A$.*

We use this theorem to construct the required extremally disconnected compact space $\mathfrak{X}$. Let $\tau = \sum_{\alpha\in A}\mathfrak{n}_\alpha$, $\mathfrak{n}_\alpha < \tau$ and $|A| = \tau$. For each $\alpha \in A$ consider $\mathscr{A}(\mathfrak{n}_\alpha)$, the Aleksandrov compact space of power $\mathfrak{n}_\alpha$, and we put $Y = \prod_{\alpha\in A}\mathscr{A}(\mathfrak{n}_\alpha)$. Since $\tau$ is regular and greater than $|\mathscr{A}(\mathfrak{n}_\alpha)|$ for all $\alpha \in A$, we see that $\tau$ is a calibre of each $\mathscr{A}(\mathfrak{n}_\alpha)$, and consequently by Šanin's theorem $\tau$ is a calibre of $Y$. This means that $cY \leqq \tau$ and there does not exist a system of power $\tau$ consisting of disjoint open sets of $Y$. On the other hand, for every $\mathfrak{n}_\alpha < \tau$ the family $\mathfrak{B} = \{\pi_\alpha^{-1}x\}$, where $x$ runs over the set of isolated points of $\mathscr{A}(\mathfrak{n}_\alpha)$ and $\pi_\alpha$ is the projection of $Y$ on $\mathscr{A}(\mathfrak{n}_\alpha)$, is a family of power $\mathfrak{n}_\alpha$ consisting of disjoint open sets of $Y$. Let $\mathfrak{X} = pY$ be the absolute of $Y$. Then $\mathfrak{X}$ is extremally disconnected, and since $f$: $\mathfrak{X} \to Y$ is irreducible, by Lemma 7 we have $c\mathfrak{X} = cY = \tau$. Further, in view of Lemma 7, 1), there does not exist a system of disjoint open sets in $\mathfrak{X}$ of power $\tau$, which is what we wanted to prove.

**Lemma** 13. *Let* $\{\mathfrak{m}_\alpha, \alpha \in A\}$, $\mathfrak{m}_\alpha \geqq 2$, *be a certain infinite set of cardinals,*

$$\mathfrak{m}_1 \leqq \mathfrak{m}_2 \leqq \cdots \leqq \mathfrak{m}_\alpha \leqq \cdots, \alpha < \omega(|A|),$$

*indexed by the set* $A$, *where* $A$ *is totally ordered with the least ordinal type of power* $|A|$. *Further, let* $\tau = \prod_{\alpha \in A} \mathfrak{m}_\alpha$. *Then* $\tau^{|A|} = \tau$.

**Proof.** It is sufficient to prove that $\tau^{|A|} \leqq \tau$. Let $B \subset A$ and $|B| = |A|$. We prove that $\prod_{\beta \in B} \mathfrak{m}_\beta \geqq \tau$. For since $A$ is totally ordered with the least order type and $|A| = |B|$, type $A =$ type $B$. Let $f$: $A \to B$ be an order preserving map from $A$ to $B$. Since $B \subset A$, this map is an order preserving map from $A$ into $A$; therefore $f(\alpha) \geqq \alpha$. Hence $\mathfrak{m}_{f(\alpha)} \geqq \mathfrak{m}_\alpha$. Putting $f(\alpha) = \beta(\alpha)$, we obtain

$$\prod_{\beta \in B} \mathfrak{m}_\beta = \prod_{\alpha \in A} \mathfrak{m}_{\beta(\alpha)} \geqslant \prod_{\alpha \in A} \mathfrak{m}_\alpha = \tau.$$

Further, let $A = \bigcup_{\gamma \in C} B_\gamma$ be a partition of $A$ into $|A|$ disjoint subsets, each of which has power $|A|$. Then

$$\tau = \prod_{\alpha \in A} \mathfrak{m}_\alpha = \prod_{\gamma \in C} \prod_{\alpha \in B_\gamma} \mathfrak{m}_{\alpha,\gamma} \geqslant \Big(\prod_{\alpha \in A} \mathfrak{m}_\alpha\Big)^{|C|} = \tau^{|A|},$$

as required.

**Remark** 3.1. The hypothesis that the set $A$ has order type $\omega(|A|)$ is essential. For example, let

$$\mathfrak{m}_1 < \mathfrak{m}_2 < \cdots < \mathfrak{m}_k < \cdots < \mathfrak{n}$$

be a countable set of cardinals of order type $\omega_0 + 1$, where $\mathfrak{n} > \prod_{k=1}^\infty \mathfrak{m}_k$ and $\mathrm{cf}(\mathfrak{n}) = \aleph_0$. Then $(\prod_{k=1}^\infty \mathfrak{m}_k)\mathfrak{n} = \mathfrak{n}$, but $\mathfrak{n}^{\aleph_0} > \mathfrak{n}$.[5)]

**Definition** 1. Let $\tau$ and $\mathfrak{k}$ be infinite cardinals. The cardinal $\tau$ is called $\mathfrak{k}$*-admissible* if $\tau^{\mathfrak{k}} = \tau$. An $\aleph_0$-admissible cardinal is simply called *admissible*.

For example, $\tau = \exp \mathfrak{n}$ is $\mathfrak{k}$-admissible for any $\mathfrak{k} \leqq \mathfrak{n}$. On the other hand, by one of the corollaries of König's theorem ([**24**], Russian p. 234) $\tau = \sum_{n=1}^\infty \aleph_n$ is not admissible. In general, $\tau^{\mathfrak{k}} > \tau$ if $\mathfrak{k} \geqq \mathrm{cf}(\tau)$. On the other hand, we have the following

**Proposition** (GCH). *Every infinite cardinal is* $\mathfrak{k}$*-admissible if* $\mathfrak{k} < \mathrm{cf}(\tau)$.

The proof of this proposition follows easily from Tarski's recursive formula

[5)] In the statement of Lemma 2 in [**18**] the condition $\mathfrak{n}_1 \leqq \mathfrak{n}_2 \leqq \cdots \leqq \mathfrak{n}_k \leqq \cdots$ was omitted.

([**44**], English p. 288, Theorem 3). Note that it follows from this, for example, that $\aleph_{\omega_1} = \sum_{\alpha<\omega_1} \aleph_\alpha$ is an admissible cardinal.

Let the topological sum of spaces be denoted by the sign $\oplus$. ([**30**]) and let $\beta X$ denote the Stone-Čech compactification of $X$. Let $X = \oplus_{\alpha\in A} X_\alpha$ and $f_\alpha : X_\alpha \to Y_\alpha$. Then the map $f: X \to \oplus_{\alpha\in A} Y_\alpha$, having the restriction $f|X_\alpha = f_\alpha$, will be denoted by $f = \oplus_{\alpha\in A} f_\alpha$.

LEMMA 14. *Let* $X = \oplus_{\alpha\in A} X_\alpha$ *be the topological sum of zero-dimensional compact spaces, where* $|X_\alpha| \geqq 2$ *and* $|A| \geqq \aleph_0$. *Then*

$$w(\beta X) = \prod_{\alpha\in A} wX_\alpha.$$

PROOF. Put $wX_\alpha = \mathfrak{m}_\alpha$ and $\prod_{\alpha\in A} \mathfrak{m}_\alpha = \tau$, and let $C^*(X) = C^*$ be the ring of all bounded real valued functions on $X$. We prove first that $w(\beta X) \leqq \tau$. Since $wC^*(X) = w(\beta X)$ [**10**], it is sufficient to prove that $wC^* \leqq \tau$. Since $C^*$ is a metric space,

$$wC^* = sC^* = cC^*. \tag{10}$$

Therefore it is sufficient to show that $sC^* \leqq \tau$. Let $M_\alpha$ be dense in $C^*(X_\alpha)$. We prove that $\prod M_\alpha = \{f \in C, f|X_\alpha \in M_\alpha\}$ is dense in $C^*(X)$.[6] Let $f \in C^*(X)$ and $f_\alpha = f|X_\alpha$. Since $M_\alpha$ is dense in $C^*(X_\alpha)$ and $f_\alpha \in C^*(X_\alpha)$, there exists a sequence $\varphi_\alpha^n \in M_\alpha$ such that $\|f - \varphi_\alpha^n\| \to 0$ as $n \to \infty$. For each $\alpha \in A$ we choose a subsequence $\varphi_\alpha^{n_k}$ such that $\|f_\alpha - \varphi_\alpha^{n_k}\| < 1/2^k$. Put $\varphi^k = \oplus_{\alpha\in A} \varphi_\alpha^{n_k}$. Then $\|f - \varphi^k\| < 1/2^k$; that is, $\varphi^k \to f$ as $k \to \infty$, as required. Since $\varphi^k \in \prod_{\alpha\in A} M_\alpha$, we have $sC^* \leqq \tau$. Hence by (10) $wC^* = w(\beta X) \leqq \tau$. Let us prove now that $w(\beta X) = wC^*(X) \geqq \tau$. By (10) it is sufficient to prove that $cC^* \geqq \tau$. Since $X_\alpha$ is zero-dimensional and compact, there exists in $C^*(X_\alpha)$ a set $K_\alpha$, $|K_\alpha| = \mathfrak{m}_\alpha$, such that $\|f\| = 1$ for all $f \in K_\alpha$ and $\|f - g\| = 1$ for all $f, g \in K_\alpha$, $f \neq g$. For example, we could take for $K_\alpha$ the family of characteristic functions of open-and-closed subsets of $X$. Let $K = \prod_{\alpha\in A} K_\alpha = \{f \in C, f|X_\alpha \in K_\alpha\}$. Since $\|f_\alpha\| \leqq 1$ for all $\alpha \in A$, we have $K \subset C^*$. Further, for any $f, g \in K$ we have $\|f - g\| = \sup_{\alpha\in A} \|f_\alpha - g_\alpha\| \geqq 1$ if $f \neq g$. Thus $cC^*(X) \geqq \prod_{\alpha\in A} |K_\alpha| = \tau$, as required.

REMARK 3.2. Lemma 14 can also be proved by using Dwinger's theorem [**45**] which states that the Stone space of a direct sum of Boolean algebras $A_\alpha$ is the Stone-Čech compactification of the topological sum of the Stone spaces $X(A_\alpha)$. Note also that the proof of the analogous Lemma 3 in [**18**] has gaps; the proof restored here is more suitable for a sum of zero-dimensional compact spaces.

THEOREM 1. *Let* $X$ *be an extremally disconnected compact space, where*

[6] Only bounded functions are taken as belonging to $\prod_{\alpha\in A} M_\alpha$.

*$cU \geqq \tau \geqq \aleph_0$ for any nonempty open $U \subset X$. If $\tau$ is accessible or equal to $\aleph_0$, then the weight of $X$ is a $\tau$-admissible cardinal.*

PROOF. Since the weight of a space is a monotonic function on open subsets, we have $X = \overline{\bigcup_{\alpha \in A} U_\alpha}$, where each $U_\alpha$ is homogeneous with respect to the weight $\mathfrak{m}_\alpha$; that is, $wV = wU_\alpha$ for any open $V \subset U_\alpha$. Since $\tau$ is accessible or equal to $\aleph_0$, by Lemma 12 there exists in each $U_\alpha$ a system $\{U_{\alpha\beta}\}$ of disjoint open sets of power $\tau$. Without loss of generality we may suppose that $U_\alpha = \overline{\bigcup U_{\alpha\beta}}$, $\beta \in B$ and $|B| = \tau$. Since $U_\alpha$ has homogeneous weight, $wU_{\alpha\beta} = \mathfrak{m}$ for all $\beta \in B$. Hence by Lemma 14 $wU_{\alpha\beta} = \mathfrak{m}_\alpha = \mathfrak{m}_\alpha^\tau$, since $U_\alpha$ is extremally disconnected, $\bigcup_\beta U_{\alpha\beta}$ is dense in $U_\alpha$, and hence by Gleason's theorem $U_\alpha = \beta(\bigcup U_{\alpha\beta})$. Further, $X = \beta(\bigcup_{\alpha \in A} U_\alpha)$; consequently

$$wX = \prod_{\alpha \in A} \mathfrak{m}_\alpha = \prod_{\alpha \in A} \mathfrak{m}_\alpha^\tau = \Big(\prod_{\alpha \in A} \mathfrak{m}_\alpha\Big)^\tau = (wX)^\tau,$$

which is what we wanted to prove.

COROLLARY 6. *The weight of an infinite extremally disconnected compact space $X$ is an admissible cardinal.*

For every extremally disconnected compact space can be represented as a sum

$$X = X_1 \oplus X_2,$$

where $X_1$ has no isolated points and consequently $cU \geqq \aleph_0$ for all $U \subset X_1$ and $X_2 = \beta T$ or is finite. If $X_2 = \emptyset$, then by Theorem 1, $wX$ is admissible; if $X_1 = \emptyset$, then $X = \beta T$ and $|T| \geqq \aleph_0$; consequently $wX = \exp|T|$ and hence is admissible. Finally, if $X_1 \neq \emptyset$ and $X_2 \neq \emptyset$, then $wX = wX_1 wX_2 = \tau \exp|T|$ is admissible.

COROLLARY 7 (PIERCE [**32**]). *The power of an infinite complete Boolean algebra is an admissible cardinal.*

THEOREM 1-bis. *Let $X$ be an extremally disconnected compact space presented as the closure of a union of a disjoint collection of open-and-closed subsets*

$$X = \overline{\bigcup_{\alpha \in A} X_\alpha}, \quad wX_\alpha = \mathfrak{m}_\alpha \geqq 2,$$

*where the set $\{\mathfrak{m}_\alpha, \alpha \in A\}$, ordered by magnitude, is indexed by the set of all ordinals $< \omega(|A|)$. Then the weight of $X$ is an $|A|$-admissible cardinal.*

The proof of this theorem follows from Lemmas 13 and 14 and Gleason's theorem which states that an extremally disconnected compact space is the Stone-Čech compactification of any of its dense subsets. The following lemma is dual to results of Kantorovič, Fihtengol′c and Hausdorff (see [**32**]) about independence in complete Boolean algebras.

LEMMA 15. *Let $X$ be an extremally disconnected compact space containing* *a family $\sigma$ of power $\mathfrak{m} \geqq \aleph_0$ consisting of disjoint open subsets of $X$. Then $X$ can be mapped continuously onto $D^{\exp \mathfrak{m}}$.*

PROOF. Without loss of generality we may suppose that the elements of $\sigma$ are open-and-closed and $\overline{\bigcup U} = X$ for all $U \in \sigma$. Since by Lemma 11 $sD^{\exp \mathfrak{m}} = \log \exp \mathfrak{m} \leqq \mathfrak{m}$, there exists a dense subset $M \subset D^{\exp \mathfrak{m}}$, where $|M| = \mathfrak{m}$. Let $f: \sigma \to M$ be an arbitrary one-to-one transformation of $\sigma$ onto $M$. Then $f$ may be thought of as a continuous map of $V = \bigcup U$, $U \in \sigma$ onto $M$. Since $X$ is extremally disconnected and $V$ is dense in $X$, we see that $\beta V = X$ and consequently that there exists an extension $f: X \overset{\text{onto}}{\to} D^{\exp \mathfrak{m}}$.

COROLLARY 8. *Let $X$ be an extremally disconnected compact space, where $cX$ is accessible or equal to $\aleph_0$. Then $X$ can be mapped continuously onto $D^{\exp(cX)}$. But if $cX$ is not accessible, then $X$ can be mapped continuously onto $D^{\exp \tau}$ for any $\tau < cX$.*

This corollary follows from Lemmas 12 and 15.

COROLLARY 9. *If $cX$ is accessible or equal to $\aleph_0$, then the lattice number of an infinite extremally disconnected compact space is strictly less than its weight; that is, $cX < wX$.*

Suppose that $cX \geqq wX$. Then by Corollary 8 $X$ can be mapped onto $D^{\exp(cX)}$, and consequently

$$wX \geqq \exp(cX) > cX,$$

as required.

COROLLARY 10. *The following two conditions are equivalent:*

($\alpha$) $\exp \aleph_0 < \exp \aleph_1$.

($\beta$) *Each extremally disconnected compact space with the weight of the continuum satisfies Suslin's condition.*

Let us prove this. ($\alpha$) $\Rightarrow$ ($\beta$). If $cX \geqq \aleph_1$, then $wX \geqq \exp \aleph_1 > \exp \aleph_0$ in spite of the fact that $wX = \exp \aleph_0$.

($\beta$) $\Rightarrow$ ($\alpha$). Suppose that $\exp \aleph_0 = \exp \aleph_1$. Then $T^0 = \beta T_{\aleph_1}$ is an extremally disconnected compact space with the weight of the continuum and $cX = \aleph_1$, which is a contradiction.

REMARK 3.3. One can prove similarly the equivalence of

($\alpha'$) $\exp \tau < \exp \tau^+$,

($\beta'$) $cX \leqq \tau$ for each extremally disconnected compact space with weight $\exp \tau$.

## §4. The relation between the weight and the character in extremally disconnected compact spaces

Recall that *the upper character $\chi^*(X)$ of a topological space $X$ is* $\chi^*(X) = \sup_{x \in X} \chi(x, X)$.

LEMMA 16. *Let $X = \beta(\bigoplus_{\alpha \in A} Y_\alpha)$ and suppose that for each $\alpha \in A$ there* *exists a map $f_\alpha$ from $Y_\alpha$ to $D^{\mathfrak{m}_\alpha}$ with $\mathfrak{m}_1 < \mathfrak{m}_2 < \cdots < \mathfrak{m}_\alpha < \cdots$. Then there exists a map $f$ from $X$ onto $D^\tau$ with $\tau = \sum_{\alpha \in A} \mathfrak{m}_\alpha$.*

PROOF. Let $D^\tau = \prod_{\alpha < \iota} D_\alpha$. Further, let $H(\mathfrak{m}_\alpha)$ be the set of those points $x \in D^\tau$ at which the $\mathfrak{m}_\alpha$th coordinate is 1, all the coordinates with indices greater than $\mathfrak{m}_\alpha$ are 0, and the remainder are unrestricted. It is easy to see that each $H_\alpha$ is homeomorphic to $D^{\mathfrak{m}_\alpha}$ and all the $H_\alpha$ are disjoint [16]. We prove that $\bigcup_{\alpha < \tau} H_\alpha$ is dense in $D^\tau$. Let $H^{i_1, \cdots, i_s}_{\alpha_1, \cdots, \alpha_s}$ be an arbitrary elementary neighborhood of $D^\tau$. There exists an $\mathfrak{m}_\alpha > \alpha_i$, $i = 1, \cdots, s$. Then $H(\mathfrak{m}_\alpha) \cap H^{i_1, \cdots, i_s}_{\alpha_1, \cdots, \alpha_s} \neq \emptyset$, since in $H(\mathfrak{m}_\alpha)$ every coordinate with indices $\alpha_1, \cdots, \alpha_s$ is unrestricted. Hence $D^\tau = \overline{\bigcup_{\alpha \in \tau} H_\alpha}$. For each $\alpha$ let $f_\alpha : Y_\alpha \to D^{\mathfrak{m}_\alpha} \to H_\alpha$ be a map from $Y_\alpha$ onto $H_\alpha$. Then the map $f = \oplus_{\alpha \in A} f_\alpha$ from $\oplus_{\alpha \in A} Y_\alpha$ onto $\bigcup_{\alpha \in \tau} H_\alpha$ is continuous. Taking the Stone-Čech compactification we obtain $f : \beta(\oplus_{\alpha \in A} Y_\alpha) \xrightarrow{\text{onto}} D^\tau$, as required.

LEMMA 17. *Let $\tau \geqq \aleph_0$, and let $f : X \to D^\tau$ be a continuous map from $X$ onto $D^\tau$. Then there exists a closed subset $F \subset X$ such that $\delta(x, F) \geqq \tau$ for all $x \in F$.*

PROOF. By Lemma 8 there exists a closed $F \subset X$ such that $f|F$ is irreducible and $f(F) = D^\tau$. Then by Lemma 7, 7), we have $\delta(x, F) \geqq \delta(f(x), D^\tau)$ for any point $x \in F$. On the other hand, by Lemma 5 we have $\delta(y, D^\tau) = \tau$ for any $y \in D^\tau$. Consequently $\delta(x, F) \geqq \tau$ for all $x \in F$, as was required.

COROLLARY 11 (JUHASZ [34]). *Let $\tau \geqq \aleph_0$, and let $f : F \to D^\tau$ be a continuous map of the compact space $X$ onto $D^\tau$. Then*

$$\chi^*(X) \geqq \tau.$$

In fact, for any point $x \in F$

$$\tau \leqq \delta(x, F) \leqq \chi(x, F) \leqq \chi(x, X).$$

THEOREM 2.1. *Let $X$ be an infinite extremally disconnected compact space. Then $wX \leqq \exp \chi^*(X)$.*

PROOF. Suppose that $wX > \exp \chi^*(X)$. Then by the inequality $|X| \geqq wX$, which is valid for compact spaces, we have $|X| > \exp \chi^*(X)$. Since $X$ is extremally disconnected, $\chi(x, X) = \delta(x, X)$; hence from Lemma 4

$$cX \geqq [\chi^*(X)]^+. \tag{11}$$

If $cX$ is accessible or $\aleph_0$, then by Corollary 10 there exists an $f : X \to D^{\exp(cX)}$, so by Corollary 11 $\chi^*(X) \geqq \exp(cX)$. Combining this inequality with (11), we obtain

$$cX \geqq [\chi^*(X)]^+ > \chi^*(X) \geqq \exp(cX)$$

which is a contradiction. But if $cX$ is inaccessible, then by (11) $cX > [\chi^*(X)]^+$, since a successor cardinal is always accessible. Put $[\chi^*(X)]^+ = \aleph_{\alpha+1}$. Then there exists in $X$ a family of disjoint open sets of power $\tau = \aleph_{\alpha+1}$. Hence by Lemma 16 $X$ can be mapped onto $D^{\exp\tau}$ and consequently by Corollary 11

$$\chi^*(X) \geqq \exp\tau > \tau = [\chi^*(X)]^+ > \chi^*(X)$$

which is a contradiction.

COROLLARY 12. *If the weight of an extremally disconnected compact space $X$ is a dominant cardinal, then $wX = \chi^*(X)$.*

For if $\chi^*(X) < wX$, then by Theorem 2 $\exp[\chi^*(X)] \geqq wX$, which contradicts the fact that $wX$ is dominant.

COROLLARY 13.1 (**GCH**). *If the power of an extremally disconnected compact space is greater than its weight, then $wX = \chi^*(X)$.*

For if $wX$ is a limit cardinal, then (**GCH**) $wX$ is dominant; consequently by Corollary 12 $wX = \chi^*(X)$. Suppose that $wX$ is not a limit cardinal. Then $wX = \mathfrak{n}^+$; hence $|X| \geqq \mathfrak{n}^{++}$. Let us assume that $\chi^*(X) \leqq \mathfrak{n}$. Then we have

$$|X| \geqq \mathfrak{n}^{++} > \mathfrak{n}^+ = \exp\mathfrak{n}.$$

Hence by Lemma 4 $cX \geqq \mathfrak{n}^{++}$. Applying Lemma 16 and Corollary, we obtain

$$\chi^*(X) \geqq \exp\mathfrak{n}^+,$$

against the assumption.

COROLLARY 13.2. *Let $X$ be an extremally disconnected compact space. If $cX > \log(wX)$ then $wX = \chi^*(X)$.*

For in this case $cX \geqq \mathfrak{l}^+$ if $\mathfrak{l} = \log(wX)$. But $\mathfrak{l}^+$ is accessible; consequently there exists a system of disjoint open subsets of $X$ of power $\mathfrak{l}^+$, so $X$ can be mapped onto $D^{\exp(\mathfrak{l}^+)}$. Consequently there exists a point $x \in X$ such that

$$\chi(x, X) \geqq \exp(\mathfrak{l}^+) \geqq wX.$$

THEOREM 2.2 (**GCH**). *Let $X$ be an absolute infinite dyadic compact space. Then $wX = \chi^*(X)$. Further, there exists a point $x \in X$ for which $\chi(x, X) = wX$.*

Before proving this theorem we make several remarks which follow immediately from [**47**].

1) *Let $X$ be an absolute dyadic compact space, where $\pi wU = \pi wX = \tau \geqq \aleph_0$ for any open $U \subset X$. Then, given* (**GCH**), *$X$ can be mapped onto $D^\tau$.*

2) *An extremally disconnected compact space $X$ with no isolated points that satisfies Suslin's condition can be represented uniquely as $X = \overline{\bigcup_{k\in N} U_{\mathfrak{n}_k}}$, where the $U_{\mathfrak{n}_k}$ are disjoint open-and-closed subsets of homogeneous $\pi$-weight; that is, $\pi wV = \pi wU_{\mathfrak{n}_k} = \mathfrak{n}_k$ for any $V \subset U_{\mathfrak{n}_k}$, and if $k_1 \neq k_2$, then $\mathfrak{n}_{k_1} \neq \mathfrak{n}_{k_2}$.*

3) *If an extremally disconnected compact space $X$ can be mapped continuously onto $D^\tau$, where $\mathrm{cf}(\tau) = \aleph_0$, then there exists a point $x \in X$ for which $\chi(x, X) \geqq \tau^+$.*

4) *If the $\pi$-weight of an absolute dyadic compact space is $\tau$, where $\mathrm{cf}(\tau) = \aleph_0$, then given* (**GCH**) *its weight is $\tau^+$.*

PROOF OF THE THEOREM. Without loss of generality we may suppose that $X$ does not contain isolated points. Let $X = \overline{\bigcup_{k\in N} U_k}$ and let $U_k$ be as in 2). We consider two cases: a) there is a maximum, and b) there is no maximum, among the numbers $\{\mathfrak{n}_k\}$. In case a) let $\mathfrak{n}_0 = \max_{k\in N}\mathfrak{n}_k$. Consider $U_{\mathfrak{n}_0}$. By 1) there exists a map from $U_{\mathfrak{n}_0}$ onto $D^{\mathfrak{n}_0}$. Hence by Corollary 11 there exists an $x\in U_{\mathfrak{n}_0}$ such that $\chi(x,X) \geqq \mathfrak{n}_0$. If $\mathrm{cf}(\mathfrak{n}_0) \geqq \aleph_1$, then

$$\mathfrak{n}_0 \leqslant \chi(x, X) \leqslant wX \leqslant (\pi wX)^{\aleph_0} = \mathfrak{n}_0^{\aleph_0} = \mathfrak{n}_0.$$

The latter inequality follows from (GCH), as was required. If $\mathrm{cf}(\mathfrak{n}_0) = \aleph_0$, then by Lemma 22 $\chi(x,X) \geqq \mathfrak{n}_0^+$. Hence, applying 4), we obtain

$$\mathfrak{n}_0^+ \leqq \chi(x,X) \leqq wX = \mathfrak{n}_0^+,$$

as was required. In case b) put $\mathfrak{n}_0 = \sum_{k=1}^{\infty}\mathfrak{n}_k$; then $\pi wX = \mathfrak{n}_0$ and $\mathrm{cf}(\mathfrak{n}_0) = \aleph_0$. Hence, applying 4), we obtain $wX = \mathfrak{n}_0^+$. Further, for each $U_k$ there exists, by 1), a map from $X$ onto $D^{\mathfrak{n}_0}$. Consequently there exists a point $x\in X$ such that $\chi(x,X) \geqq \mathfrak{n}_0$. Therefore by Lemma 22 $\chi(x,X) \geqq \mathfrak{n}_0^+$. Hence

$$\mathfrak{n}_0^+ \leqq \chi(x,X) \leqq wX = \mathfrak{n}_0^+.$$

REMARK 4. The author has not found answers to the following questions:

1) Is $wX = \chi^*(X)$ true for arbitrary infinite extremally disconnected compact spaces $X$?

2) Could the cardinal of an infinite extremally disconnected compact space be a power of two (see [**49**])?

3) Let $x$ be an arbitrary point of an extremally disconnected compact space $X$. Is there a neighborhood $U$ of $x$ such that $wU \leqq \exp[\chi(x,X)]$?

## §5. A map from compact spaces onto the Tihonov cube $\mathbf{I}^\tau$

Note that one of the crucial points in the proof of Theorem 2 is the fact that there exists a continuous map from an extremally disconnected compact space $X$ with a "large" lattice number onto some $D^\tau$ with sufficiently large weight $\tau$. Of course, it is not exclusively $D^\tau$ that could play this role. For example, $D^\tau$ could be replaced by $\mathbf{I}^\tau$, since the latter is dyadic. The basic result of this section is the following theorem. A compact space $X$ with $cX \leqq \tau$ and $wX > \exp\exp\exp\tau$ can be mapped continuously onto $\mathbf{I}^{(\exp\tau)^+}$. This result enables us to embed an extremally disconnected space into any compact space with "sufficiently large" weight and a lattice number that is not "too large".

The space $D^\tau = \prod_{\alpha\in B}(0_\alpha, 1_\alpha)$ can be regarded as a naturally embedded subspace of $\mathbf{I}^\tau$, where $0_\alpha$ and $1_\alpha$ are the left- and right-hand end points of the unit interval $\mathbf{I}_\alpha$. We call this $D^\tau$ the set of "pseudovertices" of $\mathbf{I}^\tau$.[7)]

[7)] Note that by Keller's theorem [**37**] $\mathbf{I}^\tau$ is topologically homogeneous and consequently has no "vertices".

DEFINITION 2. A map $f: X \to \mathbf{I}^\tau$ is called *correct* if $f(X) \supset D^\tau$, where $D^\tau$ is the set of all pseudovertices of $\mathbf{I}^\tau$.

DEFINITION 3. A system of pairs of sets

$$\mathfrak{H}_\tau = \{ S_\alpha = (A_\alpha^0, A_\alpha^1) \}, \quad \alpha \in B, \quad |B| = \tau,$$

is called a *dyadic system of power* $\tau$ *in the space* $X$ if:

a) $A_\alpha^0$ and $A_\alpha^1$ are null sets of $X$ for all $\alpha \in B$,

b) $A_\alpha^0 \cap A_\alpha^1 = \emptyset$ for all $\alpha \in B$, and

c) for any finite collection $\alpha_1, \cdots, \alpha_s \in B$ and any collection $i_1, \cdots, i_s$ of zeros and ones

$$A_{\alpha_1, \ldots, \alpha_s}^{i_1, \ldots, i_s} = A_{\alpha_1}^{i_1} \cap \ldots \cap A_{\alpha_s}^{i_s} \neq \emptyset.$$

THEOREM 3. *A necessary and sufficient condition for the existence of a continuous map from a compact space* $X$ *onto* $\mathbf{I}^\tau$ *is that* $X$ *has a dyadic system* $\mathfrak{H}_\tau$ *of power* $\tau$.

PROOF. Let $f: X \to \mathbf{I}^\tau$ be a continuous map from $X$ onto $\mathbf{I}^\tau$ and let $\pi_\alpha : \mathbf{I}^\tau \to \mathbf{I}_\alpha$ be the projection of $\mathbf{I}^\tau$ onto the boundary $\mathbf{I}_\alpha$. Put

$$A_\alpha^0 = f^{-1} \pi_\alpha^{-1}(0_\alpha) \quad \text{and} \quad A_\alpha^1 = f^{-1} \pi_\alpha^{-1}(1_\alpha),$$

where $0_\alpha$ and $1_\alpha$ are the end points of the interval $\mathbf{I}_\alpha$. It is not difficult to show that the system $\mathfrak{H}_\tau = \{(A_\alpha^0, A_\alpha^1)\}$ is dyadic of power $\tau$. Conversely, let $\mathfrak{H}_\tau = \{S_\alpha\}$ be such a system in $X$. For each pair $S_\alpha$ we define a continuous real function $\varphi_\alpha : X \to \mathbf{I}_\alpha$ such that

$$\varphi_\alpha^{-1}(0) = A_\alpha^0 \quad \text{and} \quad \varphi_\alpha^{-1}(1) = A_\alpha^1.$$

The system of functions $\{\varphi_\alpha\}$ defines a map $f: X \to \mathbf{I}^\tau$ by the rule

$$x \in X \Rightarrow f(x) = \{\varphi_\alpha(x)\} \in \mathbf{I}^\tau.$$

Let $y = \{i_\alpha\} \in D^\tau$, $\alpha \in B$, where $D^\tau$ is the set of pseudovertices of $\mathbf{I}^\tau$. Then

$$f^{-1}(y) = \bigcap A_\alpha^{i_\alpha}, \quad \alpha \in B. \tag{12}$$

But the latter intersection is not empty in $X$, because the family $\{A_\alpha^{i_\alpha}\}$, $\alpha \in B$, has the finite intersection property by condition c), and $X$ is compact. Thus $f(X) \supset D^\tau$; that is, the map $f$ is correct. We put $Z = f^{-1}(D^\tau)$. Let $\psi: D^\tau \to \mathbf{I}^\tau$ be any map from $D^\tau$ onto $\mathbf{I}^\tau$; then $g = \psi f$ maps $Z \subset X$ onto $\mathbf{I}^\tau$. We prove that there exists an extension $\tilde{g}: X \to \mathbf{I}^\tau$ such that $\tilde{g} | Z = g$. For if $\pi_\alpha$ is the projection of $\mathbf{I}^\tau$ onto $\mathbf{I}_\alpha$, then $h_\alpha = \pi_\alpha f$ maps $Z$ onto $\mathbf{I}_\alpha$. By the Tietze-Uryson theorem $h_\alpha$ extends to the whole of $X$. Then the map defined by the rule

$$x \in X \Rightarrow \{\hat{h}_\alpha(x)\} \in \mathbf{I}^\tau,$$

is the required extension $\tilde{g}$. If $x \in Z$, then $h_\alpha(x) = \tilde{h}_\alpha(x)$, and therefore $\{h_\alpha(x)\} = \{\tilde{h}_\alpha(x)\}$; that is, $\tilde{g}(x) = g(x)$. Thus $\tilde{g}: X \to \mathbf{I}^\tau$ is a map from $X$ onto $\mathbf{I}^\tau$.

LEMMA 17. *Let $X$ be a regular space and $F$ a closed subset of $X$, where $\delta\psi(x,X) \geqq \tau \geqq \aleph_0$ for all $x \in F$ and $\delta\psi(F,X) \leqq \mathfrak{m} < \tau$. Then $wF \geqq \tau$.*

PROOF. Suppose that $wF \leqq \mathfrak{n} < \tau$. Let $\mathfrak{B}$ be a basis for $F$ of power $\leqq \mathfrak{n}$. The elements of $\mathfrak{B}$ are open in $F$. Consider an arbitrary point $x \in F$. Note that

(13) $$F-(x) = \bigcup_{\alpha \in A} \Phi_\alpha, \quad |A| \leqslant \mathfrak{n},$$

where the $\Phi_\alpha$ are closed in $X$. In fact, for each point $y \in F-(x)$ there exists a $V_\alpha \in \mathfrak{B}$ such that $y \in V_\alpha \subset \overline{V}_\alpha \subset F-(x)$, since $X$ and $F$ are regular. Then the $\overline{V}_\alpha = \Phi_\alpha$ are the required sets, which proves (13).

For each $\Phi_\alpha$, $\alpha \in A$, consider an open set $U_\alpha$ in $X$ such that $x \in U_\alpha \subset \overline{U}_\alpha \subset X - \Phi_\alpha$. Then $\bigcap_{\alpha \in A} \overline{U}_\alpha \cap F = (x)$. We write $\mathfrak{L} = \{U_\alpha\}$, $\alpha \in A$, and $\mathfrak{L}_0 = \{\operatorname{int} \overline{U}_\alpha\}$, $\alpha \in A$. Note that the system $\mathfrak{L}_0$ has the finite intersection property and consequently $\mathfrak{L}$ is a $\delta$-system. Since $\delta\psi(F,X) \leqq \mathfrak{m}$, there exists a $\delta$-system $\mathfrak{F} = \{\overline{H}_\alpha\}$, $\alpha \in B$, with nucleus $F$,[8] where $|\mathfrak{F}| \leqq \mathfrak{m}$. Put $\mathfrak{R} = \mathfrak{L} \cup \mathfrak{F}$. We prove that $\mathfrak{R}$ is a $\delta$-system with nucleus $(x)$. In fact,

$$\bigcap_{\alpha \in A} \overline{U}_\alpha \cap \bigcap_{\alpha \in B} \overline{H}_\alpha = \bigcap_{\alpha \in A} \overline{U}_\alpha \cap F = (x).$$

On the other hand, for each finite collection of indices $(\alpha_1, \cdots, \alpha_s, \alpha_{s+1}, \cdots, \alpha_n) \subset A \cup B$ we have

$$\bigcup_{i=1}^{s} \operatorname{int} \overline{H}_{\alpha_i} \cap \bigcap_{i=s+1}^{n} \operatorname{int} U_{\alpha_i} = \operatorname{int} \overline{W}_1 \cap \operatorname{int} \overline{W}_2,$$

where $W_1$ is canonically closed, $F \subset \overline{W}_1$ and $\overline{W}_2$ contains $x$ in its interior. Consequently

$$\operatorname{int} \overline{W}_1 \cap \operatorname{int} \overline{W}_2 \neq \emptyset.$$

Thus $\mathfrak{R}$ is a $\delta$-system of power $\leqq |A| + |B| = \mathfrak{n} + \mathfrak{m} < \tau$. Hence $\delta\psi(x,X) < \tau$, which contradicts the hypothesis. Consequently $wF \geqq \tau$.

DEFINITION 4. The *massiveness* $mX$ of a topological space $X$ is its lower $\delta$-character, namely

$$mX = \operatorname{int} \delta(x,X), \quad x \in X.$$

The *strength* $eX$ of the space $X$ is the supremum of the massivenesses of its subspaces; that is,

$$eX = \sup(mY), \quad Y \subset X.$$

For example, if $X = \beta N$, then $mX = 1$, but $eX = \exp \aleph_0$. Note the following inequality:

(14) $$mX \leqq eX \leqq \chi^* X \leqq wX.$$

LEMMA 18. *Let $f: X \to \mathbf{I}^{\mathfrak{m}}$ be a correct map from a compact space $X$ with massiveness $\geqq \tau$ onto $\mathbf{I}^{\mathfrak{m}}$, where $\aleph_0 \leqq \exp \mathfrak{m} < \tau$. Then there exists a disjoint*

---

[8] We call $F = \bigcap \overline{U}$, $\overline{U} \in \delta$, the *nucleus* of the $\delta$-system.

*pair of null sets* $(W_0, W_1)$ *in* $X$ *such that* $W_i \cap f^{-1}y \neq \emptyset$ *for all* $y \in D^{\mathfrak{m}}$, $i = 1, 2$, *where* $D^{\mathfrak{m}}$ *is the set of pseudovertices of* $\mathbf{I}^{\mathfrak{m}}$.

PROOF. Put $f(X) = Y$. Since $f$ is a correct map, $Y \supset D^{\mathfrak{m}}$. Let $Z = f^{-1}(D^{\mathfrak{m}})$, $Z \subset X$. Using Lemma 8 we find a closed subset $\Phi \subset Z \subset X$ such that $f|\Phi$ is irreducible and $f(\Phi) = D^{\mathfrak{m}}$. Using Lemma 7, we obtain

$$\pi w \Phi = \pi w D^{\mathfrak{m}} = w D^{\mathfrak{m}} = \mathfrak{m} \quad \text{and} \quad c\Phi = cD^{\mathfrak{m}} = \aleph_0.$$

Hence by Lemma 1

$$w\Phi \leqq (\pi w \Phi)^{c\Phi} = \mathfrak{m}^{\aleph_0} \leqq \mathfrak{m}^{\mathfrak{m}} = \exp \mathfrak{m} < \tau. \tag{15}$$

If we can find an open set $U_0 \supset \Phi$ such that $(X - U_0) \cap f^{-1}y \neq \emptyset$ for all $y \in D^{\mathfrak{m}}$, then by Uryson's theorem, since $X$ is normal, there exists a function $g$ on $X$ such that $g(\overline{V}) = 0$ and $g(X - U_0) = 1$, where $\Phi \subset V \subset \overline{V} \subset U_0$. In this case

$$(W_0 = g^{-1}(0), W_1 = g^{-1}(1))$$

is the required pair.

Indeed, we prove the stronger fact that there exists an open $U_0 \supset \Phi$ such that

$$(X - U_0) \cap f^{-1}y \neq \emptyset \quad \text{for all } y \in Y.$$

Assume the contrary. This means that for any open $U \supset \Phi$ there exists a certain $y \in Y$ for which $f^{-1}y \subset U$. Denote by $V(U)$ the set of all $f^{-1}y$, $y \in Y$, lying in $U$. Since $f: X \to Y$ is a continuous and closed map "onto", we have $V(U) = f^{-1}G$, where $G$ is open in $Y$. Let $\delta = \{\overline{V(U)}\}$, where $U$ runs over a certain fundamental system $\mathfrak{F}$ of neighborhoods of $\Phi$. We prove that $\delta$ is a $\delta$-system. For let $\overline{V(U_1)}, \cdots, \overline{V(U_s)}$ be an arbitrary finite set of elements in $\delta$. Then $M = \bigcap_{i=1}^{s} U_i$ is open and contains $\Phi$; consequently by hypothesis there exists an $f^{-1}y$ such that $f^{-1}y \subset M$. Again using the fact that $f$ is closed, we can find a nonempty open $G \subset Y$ such that $f^{-1}G \subset M$. Thus

$$\bigcap_{i=1}^{s} \operatorname{int} V(U_i) \supset f^{-1}G \neq \emptyset,$$

as required. Thus $\delta$ is a $\delta$-system. Since $X$ is compact, we have $L = \bigcap \overline{V(U)} \neq \emptyset$, $U \in \mathfrak{F}$. The set $L$ is the nucleus of $\delta$; hence $\delta\psi(L, X) \leqq |\delta|$. On the other hand, each element $\overline{V(U)} = \overline{f^{-1}G}$ of the family $\delta$ corresponds to an open $G \subset Y$. If $f^{-1}G \neq f^{-1}G_2$, then $G_1 \neq G_2$. Consequently this is a one-to-one correspondence. Therefore $|\delta|$ does not exceed the power of all open subsets of $Y$; but $Y \subset \mathbf{I}^{\mathfrak{m}}$, and hence $wY \leqq w\mathbf{I}^{\mathfrak{m}} = \mathfrak{m}$. Thus

$$\delta\psi(L, X) \leqq |\delta| \leqq \exp(wY) \leqq \exp \mathfrak{m} < \tau.$$

Further $L \subset \Phi$;[9)] consequently by (15)

---

[9)] $L = \bigcap \overline{V(U)} \subset \overline{U} = F$, if $U \in \mathfrak{F}$.

(16) $$wL \leqq w\Phi < \tau.$$

Finally, by assumption ($X$ is compact; see Lemma 3),

$$\delta\psi(x,X) = \delta(x,X) \quad \text{for all } x \in L.$$

Applying Lemma 17, we obtain $wL \geqq \tau$, which contradicts (16).

**Theorem** 4. *Any compact set $X$ of massiveness $\geqq \tau \geqq \aleph_0$ can be mapped onto $\mathbf{I}^{\log \tau}$.*

**Proof.** By Theorem 3 it is sufficient to construct a dyadic system in $X$ of power $\log \tau$. We construct the system $\mathfrak{H}_{\log \tau}$ by induction. Let $S_1$ be an arbitrary disjoint pair $(A_1^0, A_1^1)$ of null sets in $X$. Suppose that for all ordinals $\alpha < \xi < \omega(\log \tau)$ a pair $S_\alpha$, $\alpha < \xi$, has been constructed consisting of disjoint null sets of $X$ such that for any finite collection $\alpha_1, \cdots, \alpha_s < \xi$ and any collection $i_1, \cdots, i_s$ of zeros or ones we have $\bigcap_{k=1}^s A_{\alpha_k}^{i_k} \neq \emptyset$. Let $f_\xi : X \to \mathbf{I}^{|\xi|}$ be the map from $X$ into $\mathbf{I}^{|\xi|} = \prod_{\alpha<\xi} \mathbf{I}_\alpha$ described in Theorem 3. We have shown in the proof of Theorem 3 that $f_\xi$ is a correct map. Further, $f^{-1}(y) = \bigcap_{\alpha<\xi} A_\alpha^{i_\alpha}$ if $y \in D^{|\xi|}$ (compare (12)). Since $\xi < \omega(\log \tau)$, we have $|\xi| < \log \tau$.[10] Consequently $\exp|\xi| < \tau$. Further, by assumption $mX \geqq \tau$. Applying Lemma 18 with $\mathfrak{m} = |\xi|$, we conclude that there exists a disjoint pair of null sets $S_\xi = (A_\xi^0, A_\xi^1)$ such that $A_\xi^i \cap f^{-1}y \neq \emptyset$ for all $y \in D^{|\xi|} = \prod_{\alpha \leqq \xi} (0_\alpha, 1_\alpha)$. $S_\xi$ is the required pair. The inductive step is complete. Let us show that the system $\{S_\alpha\}$, $\alpha \leqq \xi$, is dyadic. The properties a) and b) hold by construction. We prove property c). Let $(\alpha_1, \cdots, \alpha_s, \xi)$ be an arbitrary finite collection of indices, where $\alpha_i < \xi$, and let $(i_1, \cdots, i_s, i_{s+1})$ be an arbitrary collection of zeros and ones. Note that $A_{\alpha_1,\cdots,\alpha_s}^{i_1,\cdots,i_s} = \bigcap_{k=1}^s A_{\alpha_k}^{i_k} \neq \emptyset$ by the inductive hypothesis; moreover, by (12) $A_{\alpha_1,\cdots,\alpha_s}^{i_1,\cdots,i_s} \supset f_\xi^{-1}(y)$, where $y$ is a point of $D^{|\xi|}$ having the values $i_1, \cdots, i_s$ at the places $\alpha_1, \cdots, \alpha_s$. But by the choice of $S_\xi$ we have $A_\xi^{i_{s+1}} \cap f_\xi^{-1}(y) \neq 0$, whence

$$\bigcap_{k=1}^{s+1} A_{\alpha_k}^{i_k} = A_{\alpha_1,\dots,\alpha_s}^{i_1,\dots,i_s} \cap A_\xi^{i_{s+1}} \supset f_\xi^{-1} y \cap A_\xi^{i_{s+1}} \neq \emptyset$$

and property c) is proved. Thus by the principle of transfinite induction we construct a dyadic system $\mathfrak{H}_{\log \tau}$ of power $\log \tau$.

Next we show that, given information about the lattice number of the compact space $X$, we can replace the $\log \tau$ in Theorem 4 by $\tau$ for certain cardinals $\tau$.

**Lemma** 19. *Let $X$ be compact and let $Y$ be a closed subset of $X$, where $\pi\chi(Y,X) \leqq \tau$ and $cX \leqq \mathfrak{n}$. Then there exists a nonempty closed set $F \subset Y$ such that*

$$\delta(F,X) \leqq \tau^{\mathfrak{n}}.$$

[10] $\omega(\log \tau)$ is the first ordinal of power $\log \tau$; $|\xi|$ is the power of all ordinals $< \xi$.

PROOF. Let $\mathfrak{L}$ be an arbitrary local $\pi$-base of $Y$ in $X$ of power $\leqq \tau$, and let $U$ be an arbitrary open set containing $Y$. Denote by $W(U)$ the set of all elements of $\mathfrak{L}$ lying in $U$. We prove that there exists a set

$$\{U_\alpha\},\ U_\alpha \in \mathfrak{L},\ \alpha \in A(U),\ |A(U)| \leqq \mathfrak{n},$$

such that $\overline{\bigcup U_\alpha} = \overline{W(U)}$, where $\alpha \in A(U)$.[11] Let $U_{\alpha_1} \in \mathfrak{L}$ and $U_{\alpha_1} \subset U$. If $\overline{U}_{\alpha_1} = \overline{W(U)}$, our aim is achieved. But if $\overline{W(U)} - \overline{U}_{\alpha_1} \neq \emptyset$, we proceed by induction. Suppose that for a certain ordinal $\xi$ we have constructed open sets $U_{\alpha_1}, \cdots, U_{\alpha_\beta}, \cdots, \beta < \xi$, such that $U_{\alpha_\beta} \in \mathfrak{L}$, $U_{\alpha_\beta} \subset U$ and $U_{\alpha_\beta} - \overline{\bigcup_{\gamma<\beta} U_{\alpha_\gamma}} \neq \emptyset$ for all $\beta < \xi$. If $\overline{\bigcup_{\beta<\xi} U_{\alpha_\beta}} = \overline{W(U)}$, our aim is achieved. But if $\overline{W(U)} - \overline{\bigcup_{\beta<\xi} U_{\alpha_\beta}} \neq \emptyset$, then there exists a $U_{\alpha_\xi} \in \mathfrak{L} \cap W(U)$ such that $U_{\alpha_\xi} - \overline{\bigcup_{\beta<\xi} U_{\alpha_\beta}} \neq \emptyset$, as required. The inductive step is complete. Put $V_\xi = U_{\alpha_\xi} - \overline{\bigcup_{\beta<\xi} U_{\alpha_\beta}}$. It is easy to see that $\{V_\xi\}$ is a family of disjoint nonempty open subsets of $X$. Since $xC \leqq \mathfrak{n}$, our transfinite process stops after not more than $\omega(\mathfrak{n})$ steps. Our immediate aim is achieved. Thus for any open $U \supset Y$ we have $\overline{W(U)} = \overline{\bigcup U_\alpha}$, $\alpha \in A(U)$, $U_\alpha \in \mathfrak{L}$ and $|A(U)| \leqq \mathfrak{n}$. To each $\mathfrak{M} \subset \mathfrak{L}$ for which $|\mathfrak{M}| \leqq \mathfrak{n}$ we assign the canonical set $\overline{V} = \overline{\bigcup U}$, where $U \in \mathfrak{M}$. Thus we obtain a single-valued (but not one-to-one) correspondence between the family

$$\{\mathfrak{M}\},\quad \mathfrak{M} \subset \mathfrak{L},\quad |\mathfrak{M}| \leqq \mathfrak{n}$$

and a family of canonical subsets of $X$ whose image covers the family

$$\{\overline{W(U)}\},\quad U \supset Y.$$

Hence $|\{\overline{W(U)}\}| \leqq \tau^{\mathfrak{n}}$. We now prove that the family $\{\overline{W(U)}\}$ is a $\delta$-system. Let $\overline{W(U_1)}, \cdots, \overline{W(U_n)}$ be a finite collection of such sets. Then $\emptyset \neq U_1 \cap \cdots \cap U_n = U \supset Y$. Consequently there exists $U_0 \in \mathfrak{L}$ such that $U_0 \subset U$. Hence $U_0 \subset U_i$, $i = 1, 2, \cdots, n$. Since $W(U_i)$ is the set of all $U_\alpha \in \mathfrak{L}$ such that $U_\alpha \subset U_i$ we have $U_0 \in W(U_i)$, $i = 1, \cdots, n$. This means that

$$\emptyset \neq U_0 \subset \bigcap_{i=1}^{n} \operatorname{int} \overline{W(U_i)}.$$

Thus $\{\overline{W(U)}\}$ is a $\delta$-system. Therefore, since $X$ is compact,

$$\emptyset \neq F = \bigcap_{U \supset Y} \overline{W(U)} \subset \bigcap_{U \supset Y} U = Y.$$

Since $|\{\overline{W(U)}\}| \leqq \tau^{\mathfrak{n}}$, we have $\delta\psi(F, X) = \delta(F, X) \leqq \tau^{\mathfrak{n}}$.

LEMMA 20. *Let $cX \leqq \mathfrak{n}$, $mX \geqq \tau$, and let $f: X \to \mathbf{I}^{\mathfrak{m}}$ be a correct map from the compact space $X$ into $\mathbf{I}^{\mathfrak{m}}$, where $\aleph_0 \leqq \mathfrak{m}^{\mathfrak{n}} < \tau$. Then there exists a disjoint pair of nullsets $(W_0, W_1)$ in $X$ such that $W_i \cap f^{-1}y \neq \emptyset$ for all $y \in D^{\mathfrak{m}}$, $i = 1, 2$, where $D^{\mathfrak{m}}$ is the set of pseudovertices of $\mathbf{I}^{\mathfrak{m}}$.*

[11] For simplicity we identify the set $W(U)$ with the union of all its elements.

The proof of this lemma is analogous to that of Lemma 18 and differs from it only in the choice of the family $\delta = \{\overline{V(U)}\}$, $U \in \mathfrak{F}$. In this case we have to suppose that $V(U) = f^{-1}G$, where $G$ is an element of a basis for open sets in $Y$ of power $wY$. Then for any open $U \supset \Phi$ there exists an element $G$ of the basis such that $U \supset f^{-1}G$. Consequently the family $\delta' = \{V(U)\}$ is a local $\pi$-basis for $\Phi$ of power $\leqq wY \leqq w\mathbf{I}^{\mathfrak{m}} = \mathfrak{m}$; that is, $\pi\chi(\Phi, X) \leqq \mathfrak{m}$. Hence by Lemma 19 there exists a nonempty closed $F \subset \Phi$ such that $\delta(F, X) \leqq \mathfrak{m}^{\mathfrak{n}} < \tau$.

Further by Lemma 17 $wF \geqq \tau$, a contradiction to the inequality (15) which can be rewritten in the following way:

$$wF \leqq w\Phi \leqq \mathfrak{m}^{\aleph_0} \leqq \mathfrak{m}^{\mathfrak{n}} < \tau.$$

Thus Lemma 20 is proved.

DEFINITION 5. A cardinal $\tau$ is called $\mathfrak{m}$-*inaccessible* if $\mathfrak{n}^{\mathfrak{m}} < \tau$ for any $\mathfrak{n} < \tau$.

For example, $(\exp \mathfrak{m})^+$ is $\mathfrak{m}$-inaccessible for any $\mathfrak{m} \geqq \aleph_0$. Every dominant cardinal $\tau$ is $\mathfrak{m}$-inaccessible for any $\mathfrak{m} \leqq \tau$. On the other hand, for example, $\aleph_1$ is not $\aleph_0$-inaccessible, because $\aleph_0^{\aleph_0} \geqq \aleph_1$. In this case we say that $\aleph_1$ is an $\aleph_0$-accessible cardinal. In general, if $\tau$ is an infinite cardinal, then $\tau^+$ is $\mathrm{cf}(\tau)$-accessible.

THEOREM 5. *Let $X$ be compact, where $cX \leqq \mathfrak{m}$, $mX \geqq \tau$ and $\tau$ is an $\mathfrak{m}$-inaccessible cardinal. Then there is a continuous map of $X$ onto $\mathbf{I}^{\tau}$.*

This theorem is proved in the same way as Theorem 4, with Lemmas 17 and 18 replaced by Lemmas 19 and 20, respectively.

REMARK 5. Theorems 4 and 5 are true for locally compact spaces. For let $X$ be a locally compact space and let $mX \geqq \tau$. Consider the Aleksandrov compactification $\alpha X$ of $X$. Let $U$ be a nonempty canonical subset of $\alpha X$ such that

$$\overline{U} \subset \alpha X - (\xi),$$

where $\xi = \alpha X - X$. Then for all $x \in \overline{U}$

$$\delta(x, \overline{U}) \geqq \delta(x, X) \geqq \tau \quad \text{and} \quad c\overline{U} \leqq cX \leqq \mathfrak{m}.$$

On the other hand, $\overline{U}$ is compact, and consequently there exists a map of $\overline{U}$ onto $\mathbf{I}^{\tau}$ (for example, by Theorem 5 if $\tau$ is $\mathfrak{m}$-inaccessible). Since $\alpha X$ is normal, there exists an extension of this map to the whole of $\alpha X$ and consequently to $X$. Further, since $\alpha X$ is the minimal compactification of $X$ we obtain

COROLLARY 14. *Let $X$ be compact, let $cX \leqq \mathfrak{n}$ and let $U \subset X$ be an open subset of $X$ of massiveness $\geqq \tau$, where $\tau$ is an $\mathfrak{n}$-inaccessible cardinal. Then $X$ can be mapped onto $\mathbf{I}^{\tau}$.*

THEOREM 6. *If the compact space $X$ can be mapped onto $\mathbf{I}^{\tau}$, then $eX \geqq \tau$. Conversely, if $eX > \tau$ then $X$ can be mapped onto $\mathbf{I}^{\log(\tau^+)}$.*

PROOF. Let $f: X \to \mathbf{I}^\tau$. By Lemma 8 there exists a closed $F \subset X$ such that $f|F$ is irreducible and $f(F) = \mathbf{I}^\tau$. Hence by Lemma 7 $\delta(X, F) \geqq \delta(fx, \mathbf{I}^\tau)$ for all $x \in F$. Further, by Lemma 5 $\delta(y, \mathbf{I}^\tau) = \tau$ for all $y \in \mathbf{I}^\tau$. Thus $\delta(x, F) \geqq \tau$ for all $x \in F$ or $mF \geqq \tau$. So $eX \geqq \tau$. Conversely, let $eX > \tau$; then $eX \geqq \tau^+$. Then there is a closed $F \subset X$ for which $mF \geqq \tau^+$. Consequently by Theorem 4 $F$ can be mapped onto $\mathbf{I}^{\log(\tau^+)}$. Extending this map to the whole of $X$ we obtain the required map from $X$ onto $\mathbf{I}^{\log(\tau^+)}$.

THEOREM 7. *Let $X$ be compact, and let $cX \leqq \tau$ and $wX > \exp\exp\exp\tau$. Then $X$ can be mapped onto $\mathbf{I}^{(\exp\tau)^+}$.*

PROOF. We write

$$M = \{x \in X, \delta(x, X) \leqq \exp\tau\}.$$

We prove that $U = X - \overline{M} \neq \emptyset$. Suppose that $X = \overline{M}$. Then $|M| > \exp\exp\tau$. For if $|M| \leqq \exp\exp\tau$, then by (4) we obtain

$$wX \leqq \exp(sX) \leqq \exp|M| \leqq \exp\exp\exp\tau,$$

against the assumption. Thus $|M| > \exp\exp\tau$. Put $\exp\tau = \mathfrak{n}$. Then, $\delta(x, M) \leqq \delta(x, X) \leqq \mathfrak{n}$ for all $x \in M$ and $|M| > \exp\mathfrak{n}$, because $M$ is dense in $X$. Applying Lemma 4, we obtain $cM \geqq \mathfrak{n}^+$, against the fact that $cM = cX \leqq \mathfrak{n}$ (and $M$ is dense in $X$!). Thus $U = X - \overline{M} \neq \emptyset$. Note that the massiveness of $U$ is greater than or equal to $(\exp\tau)^+$. On the other hand, $(\exp\tau)^+$ is a $\tau$-inaccessible cardinal and $cX \leqq \tau$. Hence by Corollary 14 $X$ can be mapped onto $\mathbf{I}^{(\exp\tau)^+}$.

THEOREM 8. *Let $X$ be compact where $cX \leqq \tau$ and $wX > \exp\exp\exp\tau$. Then $X$ contains every extremally disconnected space of weight $\leqq (\exp\tau)^+$; in particular, $X$ contains the Stone-Čech compactification $\beta T_\tau$ of the discrete space of power $\tau$.*

PROOF. Put $\mathfrak{n} = (\exp\tau)^+$. Note that $w(\beta T_\tau) = \exp\tau < \mathfrak{n}$. Consequently by Tihonov's theorem $\beta T_\tau$ (and, in general, every extremally disconnected space of weight $\leqq \mathfrak{n}$) lies in $\mathbf{I}^\mathfrak{n}$. Further, the compact space $X$ can be mapped onto $\mathbf{I}^\mathfrak{n}$ by Theorem 7, and so, applying Lemma 8 and the Ponomarev-Gleason theorem, we can lift any extremally disconnected space lying in $\mathbf{I}^\mathfrak{n}$ into $X$, which is what was required.

COROLLARY 15. *Let $X$ be a compact space satisfying Suslin's condition and one of the following conditions:*

a) *$X$ is a Fréchet-Uryson space;*

b) *$X$ is completely normal;*

c) *$X$ is a $\chi$-space* [**14**].

*Then* $wX \leqq \exp\exp\exp\aleph_0$.

PROOF. Otherwise, by Theorem 8, $X$ would contain $\beta N$, which is neither a Fréchet-Uryson space [**31**], nor completely normal [**31**], nor a $\chi$-space [**4**].

REMARK 6. Aleksandrov's compact space $\mathscr{A}_\tau$ is a Fréchet-Uryson space for any cardinal $\tau$, and $w\mathscr{A}_\tau = \tau$. This example shows that the restriction on $cX$ in Theorems 7 and 8 is necessary. Note that for Fréchet-Uryson spaces it can be proved by using $|X| \leqq (sX)^{\aleph_0}$ that $|X| \leqq \exp\exp\aleph_0$ if $cX \leqq \aleph_0$.[12] The author does not know of any examples to show that this bound is exact.

## §6. The embedding of free Boolean algebras in arbitrary complete Boolean algebras

In this section we study in more detail the maps of zero-dimensional and extremally disconnected compact spaces onto $D^\tau$, because precisely these compact spaces are the Stone spaces of arbitrary complete Boolean algebras.

THEOREM 4-bis. *Any zero-dimensional compact space $X$ of massiveness $\geqq \tau$ can be mapped onto $D^{\log\tau}$.*

THEOREM 5-bis. *Let $X$ be a zero-dimensional compact space, where $cX \leqq \mathfrak{m}$, $mX \geqq \tau$ and $\tau$ is an $\mathfrak{m}$-inaccessible cardinal. Then $X$ can be mapped onto $D^\tau$.*

THEOREM 7-bis. *Let $X$ be a zero-dimensional compact space, where $cX \leqq \tau$, and $wX > \exp\exp\exp\tau$. Then $X$ can be mapped onto $D^{(\exp\tau)^+}$.*

These theorems are consequences of Theorems 4, 5 and 7, if we recall that $D^\tau \subset \mathbf{I}^\tau$, put $F = f^{-1}(D^\tau)$, and use the following fact.

*Let $X$ be a zero-dimensional compact space, $F$ a closed subset of $X$, and $f: F \to D^\tau$ a map from $F$ onto $D^\tau$. Then there exists an extension $\tilde{f}$ of $f$ to the whole of $X$.*

DEFINITION 6. Let $A$ be an arbitrary Boolean algebra. The *lattice number* $cA$ *of* $A$ is the supremum of the powers of families consisting of disjunctive elements of $A$. We recall that elements $a, b \in A$ are called disjunctive if $a \cap b = \Lambda$.

THEOREM 9. *Let $A$ be a Boolean algebra, where $cA \leqq \tau$ and $|A| > \exp\exp\exp\tau$. Then $A$ contains a free subalgebra of power $(\exp\tau)^+$.*

This theorem is the dual to Theorem 7-bis.

THEOREM 10. *Let $X$ be an extremally disconnected compact space and $\mathfrak{k}$ an infinite cardinal such that*

$$\mathfrak{k} < \log\log(wX).$$

*Then there is a continuous map from $X$ onto $D^{\log(\mathfrak{k}^+)}$.*[13]

---

[12] The following result of Arhangel'skiĭ [48] has come to the attention of the author. If $X$ is sequentially compact and $cX \leqq \aleph_0$, then $|X| \leqq \exp\aleph_0$. This result is stronger than our inequality.

[13] By using Corollary 8 we can strengthen this theorem in the following way. Let $\mathfrak{n} = cX$, where $cX$ is accessible or equal to $\aleph_0$ (otherwise $\mathfrak{n}$ is an arbitrary number strictly less than $cX$). Further, let $\mathfrak{m} = \exp\mathfrak{n} + \log(\mathfrak{k}^+)$. Then there is a continuous map from $X$ onto $D^{\mathfrak{m}}$.

**Proof.** Put $M=\{x\in X, \chi(x,X)\leqq \mathfrak{k}\}$. Two cases are possible: a) $\overline{M}=X$ and b) $X-\overline{M}\neq\emptyset$. In the first case we prove that $|M|>\exp\mathfrak{k}$. For if $|M|\leqq \exp\mathfrak{k}$, then by (4)

$$wX\leqq \exp(sX)\leqq \exp|M|.$$

But by hypothesis $\mathfrak{k}<\log\log(wX)$, and consequently $\exp\exp\mathfrak{k}<wX$. Hence $wX<wX$. This is a contradiction. Thus $|M|>\exp\mathfrak{k}$. Further, since $M$ is dense in $X$, we see that $M$ is extremally disconnected; consequently its $\delta$-character coincides with the character at each point, and

$$\delta(x,M)=\chi(x,M)=\chi(x,X)\leqq\mathfrak{k}\quad\text{for any } x\in M.$$

Hence by Lemma 4 we obtain $cM\geqq\mathfrak{k}^+$. Since $M=X$, we have $cM=cX\geqq\mathfrak{k}^+$. So by Lemma 16 there is a continuous map from $X$ onto $D^{\exp(\mathfrak{k}^+)}$ and consequently onto $D^{\log(\mathfrak{k}^+)}$.

In the second case $U=X-\overline{M}$ is a nonempty open set. Let $V$ be an open-and-closed set lying in $U$. Then $V$ is extremally disconnected and consequently a zero-dimensional compact set, where

$$\delta(x,V)=\chi(x,V)=\chi(x,X)\geqq\mathfrak{k}^+\quad\text{for any } x\in V.$$

Hence by Theorem 4-bis $V$ can be mapped onto $D^{\log(\mathfrak{k}^+)}$. Thus $X$ also can be mapped onto $D^{\log(\mathfrak{k}^+)}$.

Corollary 16 (**GCH**). *Let $X$ be an extremally disconnected space.*

a) *If $wX=\aleph_{\alpha+3}$, then $X$ can be mapped onto $D^{\aleph_\alpha}$.*

b) *If $wX=\aleph_\alpha$ or $\aleph_{\alpha+1}$ or $\aleph_{\alpha+2}$, where $\alpha$ is a limit cardinal, then $X$ can be mapped onto $D^{\mathfrak{m}}$ for any $\mathfrak{m}<\aleph_\alpha$.*

For in case a) we have $\aleph_\alpha<\log\log\aleph_{\alpha+3}=\aleph_{\alpha+1}$. In b) we note that under the hypothesis (**GCH**) every limit cardinal is dominant, and consequently by Lemma 9 $\log\aleph_\alpha=\aleph_\alpha$. Thus we obtain

$$\log\log\aleph_{\alpha+2}=\log\log\aleph_{\alpha+1}=\log\log\aleph_\alpha=\aleph_\alpha.$$

Corollary 17. *Let $X$ be an extremally disconnected compact space of homogeneous character*[14] *and weight $\tau$, where $\tau$ is dominant. Then $X$ can be mapped onto $D^\tau$.*

For since $\tau$ is dominant, by Corollary 12 $\tau=wX=\chi^*(X)$. Since $X$ is of homogeneous character, $\chi^*(X)=\chi_*(X)$, and consequently $\chi_*(X)=\tau$. Since $X$ is extremally disconnected, $mX=\chi_*(X)=\tau$. Hence by Theorem 4-bis $X$ can be mapped onto $D^{\log\tau}$. But $\log\tau=\tau$ (Lemma 9). Thus $X$ can be mapped onto $D^\tau$.

---

[14] A space $X$ is said to be of homogeneous character if $\chi(x,X)=\chi(y,X)$ for any pair of points $x,y\in X$. Here we do not assume a priori that $\chi(x,X)=wX$.

THEOREM 10-bis. *Let $A$ be an infinite complete Boolean algebra and $\mathfrak{k}$ an infinite cardinal such that*

$$\mathfrak{k} < \log\log|A|.$$

*Then $A$ contains a free subalgebra of power* $\log\mathfrak{k}^+$.

This theorem is dual to Theorem 10.

REMARK 7. Theorems 4 and 4-bis cannot be strengthened by replacing $mX = \inf_{x\in X}\delta(x,X)$ by $\chi_*(X) = \inf_{x\in X}\chi(x,X)$. For let $\tau$ be an infinite regular cardinal. Denote by $\Lambda_\tau = 2^{\omega(\tau)}$ the set of all $\{0,1\}$-sequences of length $\omega(\tau)$, ordered lexicographically. Now, because $\Lambda_\tau$ is compact in the order topology there are no gaps in $\Lambda_\tau$. It is not difficult to show that $\Lambda_\tau$ is zero-dimensional and that $\chi(x,\Lambda_\tau) = \tau$ for every $x\in\Lambda_\tau$. This follows from Hausdorff's theorem [**29**] which states that $\Lambda_\tau$ cannot have two sequences of indices meeting from opposite directions

$$x_1 < x_2 < \cdots < x_\xi < \cdots | \cdots < y_\eta < \cdots < y_2 < y_1,$$

each of which is of order type strictly less than $\omega(\tau)$.

Thus $\chi_*(\Lambda_\tau) = \tau$. Nevertheless $\Lambda_\tau$ cannot be mapped onto $D^{\aleph_1}$, whatever $\tau \geqq \aleph_1$. This follows from a theorem of Mardešić and Papić [**39**]: every dyadic compact space that can be ordered continuously is metrizable. Note finally that (as in any ordered compact space without isolated points) there is an everywhere dense set of $\chi$-points in $\Lambda_\tau$. Consequently there is a dense subset $M_1$ in $\Lambda_\tau$ such that $\delta(x,\Lambda_\tau) = \aleph_0$ for every $x\in M_1$. On the other hand, since $\tau$ is regular, there is a dense subset $M_2$ in $\Lambda_\tau$ at each point of which the left-sided and right-sided characters coincide and are equal to $\tau$. Consequently $\delta(x,\Lambda_\tau) = \tau$ for any $x\in M_2$. This shows that in Theorems 4 and 4-bis it is impossible to replace $mX \geqq \tau$ by $\inf_{x\in M}\delta(x,X) \geqq \tau$, where $M$ is dense in $X$.

## §7. The absolute of $D^\tau$ and its properties

DEFINITION 7. Let $\mathfrak{H}_\tau$ be a dyadic system in a space $X$ (Definition 3). The system $\mathfrak{H}_\tau$ is called *dense* if for every open $U\subset X$ there is a collection of indices $\alpha_1,\cdots,\alpha_s\in B$ and a corresponding collection of zeros and ones $i_1,\cdots,i_s$ such that $A^{i_1,\cdots,i_s}_{\alpha_1,\cdots,\alpha_s}\subset U$. This system $\mathfrak{H}_\tau$ is called a system of coverings of $X$ if each $S_\alpha\in\mathfrak{H}_\tau$ is a covering of $X$.[15)]

THEOREM 11. *For an irreducible map from a zero-dimensional compact space $X$ onto $D^\tau$ to exist it is necessary and sufficient that there exist a dense dyadic system of coverings of power $\tau$.*

PROOF. Let $f: X\to D^\tau$ be an irreducible map of $X$ onto $D^\tau$, and let $\mathrm{pr}_\alpha$:

15) Since $A^0_\alpha\cap A^1_\alpha = \emptyset$, the condition that $S_\alpha$ is a covering of $X$ shows immediately that each $A_\alpha$ is open-and-closed.

$D^\tau \to D_\alpha$ be the projection of $D^\tau$ onto the two-point set $D_\alpha = (0,1)$. Then $\{S_\alpha = f^{-1}(\mathrm{pr}_\alpha)^{-1}(i), i = 0,1\}$ is the required system $\mathfrak{H}_\tau$. Conversely, let $\mathfrak{H}_\tau$ be a dense dyadic system of coverings of $X$. For each $S_\alpha \in \mathfrak{H}_\tau$ we consider the continuous function $f_\alpha$ taking the value 0 on $A_\alpha^0$ and 1 on $A_\alpha^1$. Then the map $f$ which sends the point $x \in X$ to the point $\{f_\alpha(x)\} \in D^\tau$ is continuous and "onto". Since $\mathfrak{H}_\tau$ is dense, $f$ is irreducible.

THEOREM 12. *The absolute $pD^\tau$ is (to within homeomorphism) the unique extremally disconnected compact space having a dense dyadic system of coverings of power $\tau$.*

PROOF. For in this case, by Theorem 11, $X$ is an irreducible extremally disconnected preimage of $D^\tau$, which, as is well known, completely characterizes the absolute $pD^\tau$.

LEMMA 21. *The weight of the absolute $pX$ of a compact space $X$ is equal to the power of all canonical closed subsets of $X$.*

PROOF. Let $f: pX \to X$ be an irreducible map from $pX$ onto $X$. With each canonical closed $\overline{U} \subset X$ we associate an open-and-closed subset $V = \overline{f^{-1}U}$ in $pX$. Since $f$ is perfect and irreducible, this rule defines a one-to-one correspondence between the open-and-closed subsets of $pX$ and the canonically closed subsets of $X$. Since $pX$ is infinite and zero-dimensional, the power of the open-and-closed subsets of $pX$ is the same as the power of any open-and-closed basis of $pX$, and consequently is equal to the weight of $pX$.

LEMMA 22. *Let $X$ be an extremally disconnected compact space and let $x \in X$, where $\chi(x, X) = \tau$. Then $\tau$ is not the union of countably many lesser cardinals; that is,* $\mathrm{cf}(\tau) \geqq \aleph_1$.

PROOF. Suppose that $\tau = \sum_{k=1}^{\infty} \mathfrak{n}_k$, $\mathfrak{n}_k < \tau$. Let $\mathfrak{C} = \{U_\alpha\}$, $\alpha < \omega(\tau)$, be a fundamental system of power $\tau$ consisting of open-and-closed sets containing the point $x$. Since $\mathrm{cf}(\tau) = \aleph_0$, there exists a countable sequence of ordinals $A = \{\alpha_n\}$ cofinal with $\omega(\tau)$. Put $F_n = \bigcap_{\alpha < \alpha_n} U_\alpha$. Without loss of generality we may suppose that the sequence $F_1 \supset F_2 \supset \cdots \supset F_n$ is strictly decreasing and $\bigcap_n F_n = (x)$. Let $x_n \in F_n - F_{n+1}$. Then the sequence $\{x_n\}$ tends to $x$; that is, $x$ is a $\chi$-point, which contradicts the fact that $X$ is extremally disconnected.

THEOREM 13. *For any cardinal $\tau \geqq \aleph_0$ the following assertions are valid.*
1) *$pD^\tau$ satisfies Šanin's condition, and consequently Suslin's condition.*
2) *Any open-and-closed subset of $pD^\tau$ is homeomorphic to $pD^\tau$.*
3) *The density of $pD^\tau$ is $\log \tau$.*
4) *The weight of $pD^\tau$ is $\tau^{\aleph_0}$.*
5) *The $\pi$-weight of $pD^\tau$ and the $\pi$-character of any point $x \in pD^\tau$ are both $\tau$.*
*If $\tau$ is an admissible cardinal ($\tau^{\aleph_0} = \tau$), then the following assertions are valid.*
6) *$\chi(x, pD^\tau) = \tau$ for all $x \in pD^\tau$.*
7) $|pD^\tau| = \exp \tau$.

*If $\tau$ is the sum of countably many lesser cardinals, the following assertions are valid.*

8) $\tau^+ \leqq \chi(x, pD^\tau) \leqq \tau^{\aleph_0}$ *for all* $x \in pD^\tau$;

9) $\exp(\tau^+) \leqq |pD^\tau| \leqq \exp(\tau^{\aleph_0})$.

PROOF. 1), 3) and 5) follow from the fact that Šanin's condition, the density and the $\pi$-weight are invariant under an irreducible closed map (Lemma 7) and from the equalities

$$sD^\tau = \log \tau,\ \pi w D^\tau = \tau,\ \pi w(pD^\tau) \geqq \pi\chi(x, pD^\tau) \geqq \pi\chi(fx, D^\tau) = \tau$$

of Lemmas 5, 7 and 11. We prove 2). Let

$$f: pD^\tau \to D^\tau$$

be an irreducible map from $pD^\tau$ onto $D^\tau$ and let $U$ be an arbitrary open-and-closed subset of $pD^\tau$. Then $V = f(U)$ is a canonically closed subset of $D^\tau$, and consequently $V$ is homeomorphic to $D^\tau$ ([**16**], Theorem 6). On the other hand, $U$ is extremally disconnected and $f|U$ is irreducible. Consequently $U$ is the absolute of $V = D^\tau$, and 2) is proved. To prove 4), put $w(pD^\tau) = \mathfrak{n}$. Obviously $\mathfrak{n} \geqq \tau$. On the other hand, by Lemma 21, $\mathfrak{n}$ is the power of the set of all canonical subsets of $D^\tau$. Each canonical subset of $D^\tau$ is of type $G^\tau$ ([**16**], Theorem 5). Since $D^\tau$ is compact and zero-dimensional, every closed $G_\delta$ set in $D^\tau$ is the intersection of countably many open-and-closed subsets of $D^\tau$. Consequently the power of the set of all canonical subsets of $D$ does not exceed $\tau^{\aleph_0}$, since the power of the set of all open-and-closed subsets of $D^\tau$ is equal to $\tau$. Hence

$$\tau \leqq \mathfrak{n} \leqq \tau^{\aleph_0}.$$

Raising this inequality to the power $\aleph_0$, we obtain

$$\tau^{\aleph_0} \leqq \mathfrak{n}^{\aleph_0} \leqq \tau^{\aleph_0}.$$

By Corollary 6, since $pD^\tau$ is an extremally disconnected compact space $\mathfrak{n}^{\aleph_0} = \mathfrak{n}$. Hence $\mathfrak{n} = \tau^{\aleph_0}$, and 4) is proved.

Now we prove 6). Since $pD^\tau$ is extremally disconnected, $\chi(x, pD^\tau) = \delta(x, pD^\tau)$ for all $f \in pD^\tau$. Since $f$ is irreducible and closed, by Lemma 7 we have $\delta(x, pD^\tau) \geqq \delta(fx, D^\tau)$ for all $x \in pD^\tau$. Applying Lemma 5, we obtain $\delta(fx, D^\tau) = \tau$. Hence

$$\tau = \tau^{\aleph_0} = w(pD^\tau) \geqq \chi(x, pD^\tau) = \delta(x, pD^\tau) \geqq \delta(fx, D^\tau) = \tau$$

and 6) is proved. Next we prove 7). Since $w(pD^\tau) = \tau^{\aleph_0} = \tau$, by (2) we obtain $|pD^\tau| \leqq \exp \tau$. On the other hand, since $f$ is "onto", $|pD^\tau| \geqq \exp \tau$. Hence we obtain the required result.

Now we prove 8). In proving 6) we have shown that $\chi(x, pD^\tau) \geqq \tau$ for all $x \in pD^\tau$. But $\chi(x, pD^\tau) = \tau$ is impossible, by Lemma 22 and the properties of $\tau$. Hence $\chi(x, pD^\tau) \geqq \tau^+$. Furthermore, $\chi(x, pD^\tau) \leqq w(pD^\tau) = \tau^{\aleph_0}$, which proves 8). Finally we prove 9). Since $\chi(x, pD^\tau) \geqq \tau^+$ for all $x \in pD^\tau$, by the Čech-Pospíšil theorem [38] we have $|pD^\tau| \geqq \exp \tau^+$. The second inequality follows from (2).

COROLLARY 18. *If $\tau$ is an admissible cardinal, then $pD^\tau$ is an extremally disconnected compact space of weight $\tau$ and of homogeneous character.*

LEMMA 23. *All extremally disconnected spaces of weight $\leqq \tau$ can be topologically embedded in the absolute of $pD^\tau$.*

PROOF. Let $X$ be an extremally disconnected space of weight $\leqq \tau$. In this case $X$ is zero-dimensional in the sense of ind, and therefore by Vedenisov's theorem [30] $X$ can be topologically embedded in $D^\tau$. Let $\overline{X}$ be the closure of $X$ in $D^\tau$ and let $f: pD^\tau \to D^\tau$ be an irreducible map from $pD^\tau$ onto $D^\tau$. Put $Y = \overline{f^{-1}X}$. Then $Y \subset pD^\tau$. By Lemma 8 there exists a closed set $Y^* \subset Y$ such that $f|Y^*$ is irreducible and $f(Y^*) = \overline{X}$. Put $Z = f^{-1}(X)$. Then $f|Z$ is irreducible and perfect, and consequently $Z$ is homeomorphic to $X$, which is what we wanted to prove.

We say that $X$ is a *universal extremally disconnected compact space of weight* $\tau$ if $wX = \tau$ and if every extremally disconnected space of weight $\leqq \tau$ can be topologically embedded in $X$. The question naturally arises (P. S. Aleksandrov): does there exist a universal extremally disconnected compact space of weight $\tau$ for each cardinal $\tau$?

Combining Lemma 23 and Corollary 18, we obtain

THEOREM 14. *For each admissible cardinal the absolute $pD^\tau$ is a universal extremally disconnected compact space of weight $\tau$.*

Since (Corollary 6) the weight of any infinite extremally disconnected compact space is an admissible cardinal, Theorem 14 completely answers the question for the class of compact spaces.

REMARK 8. As Katětov first observed, all extremally disconnected compact spaces of density $\leqq \tau$ can be topologically embedded in $\beta T_\tau$. It is not difficult to prove the following assertion.

*All extremally disconnected compact spaces of weight $\leqq \exp \tau$ can be topologically embedded in $\beta T_\tau$.*

For by Lemma 23 each such space can be embedded in $pD^{\exp \tau}$. On the other hand, $s(pD^{\exp \tau}) \leqq \tau$, and consequently there exists a continuous map from $\beta T_\tau$ onto $pD^{\exp \tau}$. By taking an irreducible perfect preimage we obtain the required homeomorphism by the Gleason-Ponomarev theorem.

However, $w(\beta T_\tau) = \exp \tau$ is a power of two. As was already noted, it follows from Lemma 14 (GCH) that, for example, $\aleph_{\omega_1}$ is an admissible cardinal and is not a power of two. Let us show that such cardinals exist even when the generalized continuum hypothesis is not assumed. Put

$$\mathfrak{n}_1 = \exp \aleph_0, \ldots, \aleph_{k+1} = \exp \aleph_k, \ldots, \mathfrak{n}_\omega = \sum_{k=1}^{\infty} \mathfrak{n}_k, \ \mathfrak{n}_{\omega+1} = \exp \mathfrak{n}_\omega, \ldots$$

$$\mathfrak{n}_{2\omega}=\sum_{k=1}\mathfrak{n}_{\omega+k}\,,\ \ldots\,,\ \mathfrak{n}_{\omega^2},\ \ldots\,,\mathfrak{n}_{\omega^3},\ \ldots\,,\mathfrak{n}_{\omega^\omega},\ \ldots\,,\mathfrak{n}_{\varepsilon},\ \ldots$$

Let $\mathfrak{k}=\sum\mathfrak{n}_\alpha$, $\alpha<\omega_1$, where the sum is taken over all countable ordinals. Then $\mathfrak{k}^{\aleph_0}=\mathfrak{k}$, $\mathfrak{k}\neq\exp\mathfrak{n}$. Thus among the compact spaces $\beta T_\tau$ there is none that is universal and of weight $\mathfrak{k}$ like the extremally disconnected compact space $pD^{\mathfrak{k}}$.

## §8. Absolutes of locally compact groups and their factor spaces

We prove here that the absolute of a compact topological group $G$ of weight $\tau\geqq\aleph_0$ is homeomorphic to $pD^\tau$, and so answer a question of Arhangel'skiĭ. This result implies that the absolute of every locally compact (but not compact) group $G$ is homeomorphic to the product $T\times pD^\tau$, where $T$ is a discrete space of power $cG$, $\tau=\chi(e,G)$, and $e$ is the identity in the group. An analogous result is true for factor spaces of locally compact groups.

LEMMA 24. *Let $X=\varprojlim\{X_\alpha,\varphi_\beta^\alpha\}$ and $Y=\varprojlim\{Y_\alpha,\psi_\beta^\alpha\}$ be inverse limits of compact spaces and "onto" maps. Further, let $\Phi=\{f_\alpha\}:\{X_\alpha,\varphi_\beta^\alpha\}\to\{Y_\alpha,\psi_\beta^\alpha\}$ be a spectral map such that $f_\alpha:X_\alpha\to Y_\alpha$ for any $\alpha$ is an irreducible map from $X_\alpha$ onto $Y_\alpha$. Then the limit $f=\lim_\alpha f_\alpha$ is an irreducible map from $X$ onto $Y$.*

PROOF. Recall that $f=\lim_\alpha f_\alpha$ is defined in the following way: if $x=(x_\alpha)\in X$, then $y=f(x)=\{f_\alpha(x_\alpha)\}\in Y$. Let us prove that $f$ is irreducible. It is sufficient for this to show that for any basic open set $\varphi_\alpha^{-1}(U_\alpha)\subset X$, where $U_\alpha\in X_\alpha$ and $\varphi_\alpha:X\to X_\alpha$ is the spectral map of $X$ onto $X_\alpha$, there exists a $y\in Y$ such that $f^{-1}(y)\subset\varphi_\alpha^{-1}(U_\alpha)$. Since $f_\alpha:X_\alpha\to Y_\alpha$ is irreducible, there exists a $y_\alpha\in Y_\alpha$ such that $U_\alpha\supset f_\alpha^{-1}(y_\alpha)=F_\alpha$. Since $\psi_\alpha f(x)=f_\alpha\varphi_\alpha(x)$ for any $x\in X$, we have $\varphi_\alpha^{-1}f_\alpha^{-1}(y_\alpha)=f^{-1}\varphi_\alpha^{-1}(y_\alpha)$. On the other hand, $\varphi_\alpha^{-1}(F_\alpha)\subset\varphi_\alpha^{-1}(U_\alpha)$. Therefore, taking an arbitrary point $y\in\psi_\alpha^{-1}(y_\alpha)$, we obtain $f^{-1}(y)\subset\varphi_\alpha^{-1}(U_\alpha)$, as required.

LEMMA 25. *Let $X=\varprojlim\{X_\alpha,\varphi_\beta^\alpha\}$ be an inverse limit of compact spaces indexed by all the transfinite ordinals $<\omega(\tau)$, and suppose that the following conditions hold:*

1) *$X_1$ is a metric compact space without isolated points.*

2) *For any consecutive ordinals $\alpha$ and $\alpha+1$ the map $\varphi_\alpha^{\alpha+1}$ is a locally trivial fibration with a fiber $K_\alpha$ that is metric and compact and for which $|K_\alpha|\geqq 2$.*

3) *If $\gamma$ is a limit ordinal, then $X_\gamma=\varprojlim\{X_\alpha,\varphi_\alpha^\beta\}$.*

*Then there is an irreducible map from $D^\tau$ onto $X$.*

PROOF. Consider $X_1$. Since $X_1$ is metric and compact and has no isolated points, by a theorem of P. S. Aleksandrov [1] there exists an irreducible map $f_1$ from a Cantor perfect set $C=D^{\aleph_0}=D^{\tau(1)}$ onto $X_1$. Suppose that for any ordinal $\alpha<\gamma$ we have constructed an irreducible "onto" map $f_\alpha:D^{\tau(\alpha)}\to X_\alpha$ and for any $\beta<\alpha<\gamma$ we have $f_\beta\pi_\beta^\alpha=\varphi_\beta^\alpha f_\alpha$, where $\pi_\beta^\alpha$ is the projection of

$D^{\tau(\alpha)} = \prod_{\zeta<\alpha} Z_\zeta$ onto the boundary $D^{\tau(\beta)} = \prod_{\zeta<\beta} Z_\zeta$, $Z_\zeta$ being a zero-dimensional compact metric space. If $\gamma$ is a limit ordinal, on taking inverse limits we obtain from property 3) that $X_\gamma = \varprojlim\{X_\alpha, \varphi^\alpha_\beta\}$. Further, by applying Lemma 3 we see that $f_\gamma = \lim f_\alpha$ is an irreducible map of $D^{\tau(\gamma)} = \prod_{\zeta<\gamma} Z_\zeta = \varprojlim\{D^{\tau(\alpha)}, \pi^\alpha_\beta\}$ onto $X_\gamma$. If $\gamma = \alpha+1$, consider $X_\alpha$. By the inductive hypothesis there exists an irreducible map $f_\alpha : D^{\tau(\alpha)} \to X_\alpha$ such that for any $\beta < \alpha$ we have $f_\beta \pi^\alpha_\beta = \varphi^\alpha_\beta f_\alpha$. The map $\varphi^{\alpha+1}_\alpha : X_{\alpha+1} \to X_\alpha$ is a locally trivial fibration. This means that for each $x \in X_\alpha$ there exists a neighborhood $V'(x)$ such that $X_{\alpha+1}$ is a direct product $V'(x) \times K_\alpha$ over $V'(x)$ and the map $\varphi^{\alpha+1}_\alpha$ is the projection of $V'(x) \times K_\alpha$ onto $V'(x)$. For each point $x \in X_\alpha$ consider a neighborhood $V(x) \subset \overline{V(x)} \subseteq V'(x)$. Since $X_\alpha$ is compact, there exists a finite covering $\xi_0 = \{V(x_1), \cdots, V(x_n)\}$. Consider the covering $\xi_1 = \{\overline{U}_1, \cdots, \overline{U}_n\}$ obtained from $\xi_0$ inductively: $\overline{U}_1 = \overline{V(x_1)}$; there exists $n_1 < n$ such that $\overline{U}_2 = \overline{V(x_{n_1}) - \overline{U}_1} \neq \emptyset$; and there exists $n_1 < n_2 < n$ such that $\overline{U}_3 = \overline{V(x_{n_2}) - (\overline{U}_1 \cup \overline{U}_2)} \neq \emptyset$, etc. The covering $\xi_1$ is a decomposition into canonical sets with nonintersecting interiors. We choose a zero-dimensional is a direct product over each $\overline{U}_i$. Consider $\xi_2 = \{\overline{W}_i = \overline{f^{-1}(\operatorname{int} \overline{U}_i)}\}$. Since $f_\alpha : D^{\tau(\alpha)} \to X_\alpha$ is irreducible, the covering $\xi_2$ is a decomposition of $D^{\tau(\alpha)}$ into canonical sets with nonintersecting interiors. We choose zero-dimensional metric compact space $Z_\alpha$ that is an irreducible preimage of the fiber $K_\alpha$ and assume that $g_\alpha : Z_\alpha \to K_\alpha$ is irreducible. For $D^{\tau(\alpha+1)} = D^{\tau(\alpha)} \times Z_\alpha$ we define $f_{\alpha+1} = f_\alpha \times g_\alpha$ according to the following rule: if $y = (y_\alpha, z_\alpha) \in D^{\tau(\alpha+1)}$, where $y_\alpha \in \overline{W}_i \subset D^{\tau(\alpha)}$ and $z_\alpha \in Z_\alpha$, we put $f_{\alpha+1}(y) = \{\bar{f}_\alpha(y_\alpha), g_\alpha(z_\alpha)\} \in X_{\alpha+1}$, where $\bar{f}_\alpha = f_\alpha | \overline{W}_i$. Further, put $\pi^{\alpha+1}_\alpha : D^{\tau(\alpha+1)} \to D^{\tau(\alpha)}$. Since $\bar{f}_\alpha$ is irreducible on each $\overline{W}_i$, we see that $\bar{f}_\alpha \times g_\alpha$ is an irreducible map from $\overline{W}_i \times Z_\alpha$ onto $\overline{U}_i$. Hence the map we have obtained is continuous and irreducible, and $f_\alpha \pi^{\alpha+1}_\alpha = \varphi^{\alpha+1}_\alpha f_{\alpha+1}$. Thus by induction on all $\alpha < \omega(\tau)$ we construct a spectral sequence $\{D^{\tau(\alpha)}, \pi^\alpha_\beta\}$ and a map $\Phi = \{f_\alpha\}$ from it into the spectral sequence $\{X_\alpha, \varphi^\alpha_\beta\}$ satisfying the conditions of Lemma 1. So we obtain that $X = \varprojlim\{X_\alpha, \varphi^\alpha_\beta\}$ has the image $D^\tau = \prod_\alpha K_\alpha = \varprojlim\{D^{\tau(\alpha)}, \pi^\alpha_\beta\}$, as required.

THEOREM A (SKLJARENKO [11]). *Let $G$ be a compact group, $H$ a closed subgroup, and $\theta$ the least ordinal of power $\tau$, where $\tau = \chi(e, G)$. Then*

$$G/H = \varprojlim\{B_\alpha, \varphi^\alpha_\beta, \beta < \alpha < \theta\}.$$

*Here the sequence of spaces $B_\alpha$ together with the maps $\varphi^\alpha_\beta$ satisfies the following conditions:*

1) *$B_1$ is a manifold.*

2) *For any $\alpha < \theta$ the map $\varphi^{\alpha+1}_\alpha$ is a locally trivial fibration whose fiber is a compact manifold $M_{\alpha+1}$.*

3) *For every limit ordinal $\alpha \leqq \theta$*

$$B_\alpha = \varprojlim\{B_\beta, \varphi^\beta_\gamma, \gamma < \beta < \alpha\}.$$

THEOREM B (SKLJARENKO [11]). *The space of any locally compact group $G$ is homeomorphic to a topological product $T \times D^{\tau} \times E^n \times G'$, where $E^n$ is Euclidean space and $G'$ is an arbitrary connected compact subgroup of $G$.*

THEOREM 15. *The absolute of an infinite factor space $G/H$ of a compact group $G$ with respect to a closed subgroup $H$ is homeomorphic to $pD^{\tau}$, where $\tau = w(G/H)$.*

PROOF. Put $X = G/H$. By Theorem A, $X$ may be represented as an inverse limit satisfying conditions 1), 2) and 3) of Lemma 25, and if $B_1$ is discrete, then since $wX \geqq \aleph_0$, we can put $B_1 = \varprojlim\{B_n, \varphi_n^{n+1}, n < \omega_0\}$. Thus $X$ is an irreducible image of $D^{\tau}$, and it follows that $pX = pD^{\tau}$, as required.

THEOREM 16. *The absolute of a nondiscrete locally compact noncompact group $G$ is homeomorphic to a topological product $T \times pD^{\tau}$, where $T$ is a discrete space of power $cG$ and $\tau = \chi(e, G)$ is the character of the identity element in the group.*

PROOF. By Theorem B we have

$$G = T \times D^{\tau_0} \times E^n \times G',$$

where $T$, $D^{\tau_0}$, $E^n$ and $G$ are the spaces described above. Consider the sphere $S^n$, the Aleksandrov compactification of $E^n$. Let $f_1 : D^{\aleph_0} \to S^n$ be an irreducible map from the Cantor perfect set onto $S^n$. Remove the set $f_0^{-1}(x_0)$ from $D^{\aleph_0}$, where $x_0$ is the point at infinity in the Aleksandrov compactification of $E^n$. Then

$$D^{\aleph_0} - f_1^{-1}(x_0) = N \times D^{\aleph_0} = Y,$$

where $N$ is the set of natural numbers and $f_1 | Y$ is an irreducible perfect map of $Y$ onto $E^n$. Further, as was proved earlier (Theorem 15), the compact group $G$ is an irreducible image of $D^{\tau_1}$. Let $f_2 : D^{\tau_1} \to G'$, where $\tau_1 = \chi(e, G')$. Thus there is an irreducible and perfect map from

$$T \times D^{\tau_0} \times (N \times D^{\aleph_0}) \times D^{\tau_1} = T \times D^{\tau}, \quad \tau = \tau_0 + \tau_1,$$

onto $G$. Hence $pG = T \times pD^{\tau}$.

The following theorem is proved analogously.

THEOREM 17. *The absolute of a locally compact nondiscrete factor space of a locally compact group $G$ is homeomorphic to $I \times pD^{\tau}$ for some $\tau$.*

## §9. Decomposition of the absolute of an arbitrary topological group into dense orbits

Let $\pi : pG \to G$ be an irreducible perfect map of the absolute $pG$ onto an arbitrary topological group $G$. Further, let $a \in G$ and let $\varphi_a$, $\varphi_a(x) = ax$, be the group homeomorphism taking the identity $e$ of $G$ to $a$. Since a homeomorphism is an irreducible perfect map, by the properties of the absolute it follows that there exists a homeomorphism $f_a : pG \to pG$ such that Diagram 1 commutes.

$$\begin{array}{ccc} pG & \xrightarrow{f_a} & pG \\ \pi\downarrow & & \downarrow\pi \\ G & \xrightarrow{\varphi_a} & G \end{array}$$

DIAGRAM 1

Put $F_e = \pi^{-1}(e)$. Choose an arbitrary point $z \in F_e$. Let $a$ run over $G$; then $\varphi_a$ runs over a certain set of group homeomorphisms, and consequently $f_a$ runs over a certain set of homeomorphisms of $pG$ onto itself.

DEFINITION 8. The *orbit* $\mathrm{Orb}(z, pG)$ of a point $z \in F_e$ in the absolute $pG$ is the set

$$\mathrm{Orb}(z, pG) = \{z_a \in pG, z_a = f_a(z), a \in G\}$$

in the topology induced from $pG$.

LEMMA 26. *An orbit is an everywhere dense extremally disconnected subspace of $pG$ for any $z \in F_e$; moreover,*

$$\mathrm{Orb}(z_1, pG) \cap \mathrm{Orb}(z_2, pG) = \emptyset \quad \textit{for } z_1 \neq z_2.$$

PROOF. Since $\pi$ is irreducible and perfect, for any open $U \subset pG$ there exists an $a \in G$ such that $U \supset \pi^{-1}(a)$. It follows from the commutativity of Diagram 1 that $\pi^{-1}(a) = f_a(\pi^{-1}(e))$. Since $z \in F_e$, we have $z_a \in \pi^{-1}(a)$; that is, $z_a \in U$, and consequently $\mathrm{Orb}(z, pG)$ is dense in $pG$. Since $pG$ is extremally disconnected, so is the orbit $\mathrm{Orb}(z, pG)$. Finally, if $M = \mathrm{Orb}(z_1, pG) \cap \mathrm{Orb}(z_2, pG) \neq \emptyset$, then for $z_a \in M$ we have $z = f_a^{-1}(z_a)$; that is, $z = z_1 = z_2$.

LEMMA 27. *For any $z \in F_e$ the orbit is topologically homogeneous.*

PROOF. Let $z_a, z_b \in \mathrm{Orb}(z, pG)$ and $a \neq b$. Let us construct a homeomorphism $f_a^b : pG \to pG$ such that

$$f_a^b(z_a) = z_b, \quad f_a^b(\mathrm{Orb}(z, pG)) = \mathrm{Orb}(z, pG).$$

Put $\varphi_a^b(x) = ba^{-1}x$. Then Diagram 2 commutes, where $\varphi_b(x) = bx$ and $\varphi_a(x) = ax$.

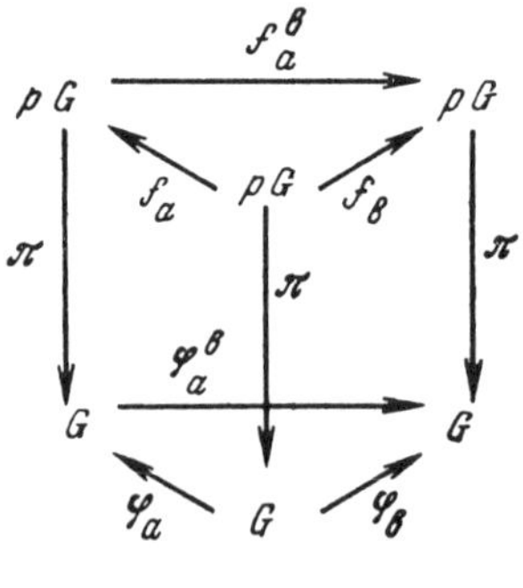

DIAGRAM 2

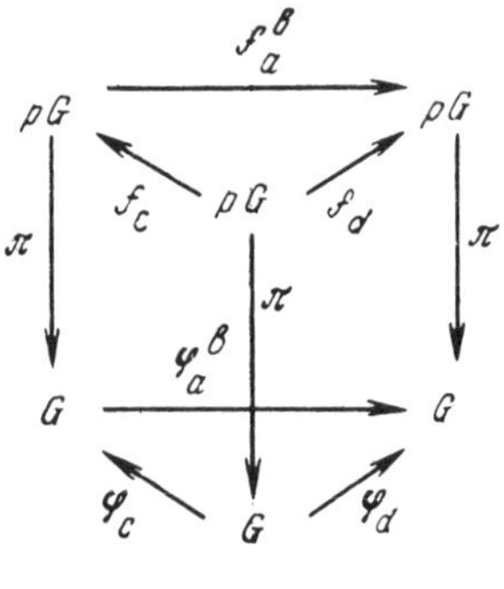

DIAGRAM 3

In fact, the bottom of the prism in Diagram 2 commutes because $\varphi_a^b\varphi_a(x) = ba^{-1}af = bx = \varphi_b(x)$. From the commutativity of the base and the sides it follows that $f_a^b f_a = f_b$. Hence $f_a^b f_a(z) = f_b(z)$ or $f_a^b(z_a) = z_b$, as required. Further, let $z_1 \in \mathrm{Orb}(z,pG)$ and $z_2 = f_a^b(z_1)$. We prove that $z_2 \in \mathrm{Orb}(z,pG)$. Since $z_1 \in \mathrm{Orb}(z,pG)$, we see that $z_1 = z_c$ for some $c \in G$. Put $d = ba^{-1}c$ and consider Diagram 3, which commutes since its sides and its base commute, $\varphi_a^b\varphi_c(x) = ba^{-1}cx = dx = \varphi_d(x)$. Hence $f_a^b f_c = f_d$; that is, $f_a^b f_c(z) = f_d(z)$ or $f_a^b(z_c) = f_a^b(z_1) = f_d(z) = z_d$, but $z_d \in \mathrm{Orb}(g,pG)$, which is what we wanted to prove.

Recall that a map $f\colon X \to Y$ is said to be *semi-open* if $\operatorname{int} f(U) \neq \emptyset$ for every open $U \subset X$. A transformation $f\colon X \to Y$ is said to be an *injection* if $|f^{-1}(y)| \leq 1$ for any $y \in Y$.

LEMMA 28. *For each point $z \in F_e$ there is a semi-open injection of* $\mathrm{Orb}(z,pG)$ *onto $G$.*

PROOF. Let $U$ be an open set in $\mathrm{Orb}(z,pG)$. Then $U = \mathrm{Orb}(z,pG) \cap V$, where $V$ is open in $pG$. Since $\pi\colon pG \to G$ is irreducible and closed, there exists a nonempty open set $W \subset G$ such that $\pi^{-1}(W) \subset V$. Further, $\mathrm{Orb}(z,pG)$ is dense in $pG$; consequently $U$ is dense in $V$, whence

$$\emptyset \neq U' = \mathrm{Orb}(z,pG) \cap \pi^{-1}(W) \subset U.$$

Therefore $\pi(U) \supset \pi(U')$. Since $\pi(\pi^{-1}(W)) = W$ and $\pi(\mathrm{Orb}(z,pG)) = G$, we have $\pi(U') = W$. Thus $\pi(U) \supset W$; that is, $\operatorname{int}\pi(U) \neq \emptyset$, as required. That this is an injection of $\mathrm{Orb}(z,pG)$ onto $G$ follows from the fact that there is precisely one element of the orbit in each $\pi^{-1}(a)$, $a \in G$.

Recall that the dispersion characteristic $dX$ of a topological space $X$ is $dX = \min|U|$, $U \subset X$, $U$ open.

THEOREM 18. *For any cardinal $\tau \leqq \aleph_0$ there exists an extremally disconnected topologically homogeneous space with dispersion character $\tau$.*

PROOF. Let $G$ be a subgroup of the generalized Cantor discontinuum in $D^\tau$ consisting of those points of $D^\tau$ for which only finitely many coordinates are nonzero. It is easy to see that $d(G) = |G| = \tau$. Consider the orbit $\mathrm{Orb}(z,pG) = X$ of some point $z \in \pi^{-1}(e)$. By Lemma 26 $X$ is extremally disconnected, and by Lemma 27 $X$ is topologically homogeneous. We prove that $dX = dG$. For by Lemma 28 there is a semi-open injection of $X$ onto $G$. Hence for any open $U \subset X$

$$|U| \geqq |\pi(U)| \geqq |\operatorname{int}\pi(U)| \geqq dG.$$

On the other hand, since $|X| = |G| = dG$, we have $dG = |X| \geqq |U|$. Hence $|U| = dG$, and since $U$ is arbitrary, $dX = dG = \tau$, which is what we wanted to prove.

THEOREM 19. *For any cardinal $\tau \geqq \aleph_0$ there exists a space $X$ having the following properties:*

1) *$X$ is extremally disconnected.*
2) *$X$ is topologically homogeneous.*
3) *$X$ is completely normal.*
4) *$X$ is completely paracompact.*
5) *$X$ is nondiscrete.*
6) *The power of $X$ is $\tau$.*

PROOF. Let $X$ be the topological product $\mathrm{Orb}(z,pG)\times T_\tau$, where $G$ is the group of rational numbers and $T_\tau$ is a discrete space of power $\tau$. Then $\mathrm{Orb}(z,pG)$ is a countable space and consequently has the properties 1) –5).

THEOREM 20. *The absolute $pG$ of any topological group $G$ decomposes into everywhere-dense disjoint topologically homogeneous absolutes*:

$$pG=\bigcup_{z\in F_e}\mathrm{Orb}(z,pG).$$

REMARK 9. Theorem 18 answers a question of Arhangel′skiĭ [**5**]: does there exist a topologically homogeneous nondiscrete extremally disconnected space? Note that Sirota has constructed a nondiscrete extremally disconnected group [**15**] by relying on the continuum hypothesis. The question naturally arises: when is the orbit $\mathrm{Orb}(z,pG)$ a group? For example, if $G$ is extremally disconnected, then $pG=G$ and $\mathrm{Orb}(z,pG)=\mathrm{Orb}(z,G)=G$. Sirota has proved [**14**] that if $G$ is not extremally disconnected, then $\mathrm{Orb}(z,pG)$ is never a topological group. On the other hand, it is not difficult to show that $\mathrm{Orb}(z,pG)$ is a semitopological group in the following sense. $\mathrm{Orb}(z,pG)$ is algebraically isomorphic to $G$ and the group operation in the orbit is continuous in each variable separately, but not continuous in both together.

## §10. Problems of homogeneity of extremally disconnected compact spaces

Let $T$ be a discrete space $|T|\geqq\aleph_0$, and let $\beta T$ be the Stone-Čech compactification of $T$, $x\in\beta T-T$. Then the subspace $P=x\cup T$ lying in $\beta T$ is called an *atom*, and the point $x\in P$ the *nucleus of the atom*. It is easy to see that each atom is the topological subspace associated with a certain ultrafilter $\mathfrak{F}$ on $T$ [**25**]. Two atoms $P_1$ and $P_2$ are called *equivalent* ($P_1\approx P_2$) if $P_1$ is homeomorphic to $P_2$. Note that if $P_1=T\cup x$ and $P_2=T\cup y$, then $P_1$ is equivalent to $P_2$ if and only if there exists a homeomorphism of $\beta T$ onto itself taking $x$ into $y$. Since each such homeomorphism is defined by a certain permutation of $T$, the power of each class of equivalent atoms $\leqq|T|^{|T|}=\exp|T|$. On the other hand, $|\beta T|=\exp\exp|T|$. Hence the power of all pairwise nonequivalent atoms is $\geqq\exp\exp|T|$. This was first noted by Parovičenko [**12**]. Frolík called the class of equivalent atoms the "ultrafilter type" [**40**], [**41**]. Two atoms $P_1$ and $P_2$ are called *weakly equivalent* if there exists two equivalent atoms $P_1'$ and $P_2'$ such that $P_1'\subset P_1$ and $P_2'\subset P_2$.

PROPOSITION. *Any two countable weakly equivalent atoms $P_1$ and $P_2$ are equivalent. On the other hand, there exist uncountable weakly equivalent atoms that are not equivalent.*

PROOF. Let $P_1\sim P_2$, where $P_1=N\cup x$ and $P_2=N\cup y$, $N$ denoting the natural numbers. Further, let

$$P_1'=N_1\cup x,\quad P_2'=N_2\cup y,\quad N_1\subset N,\quad N_2\subset N$$

and let $f: P_1' \to P_2'$ be a homeomorphism sending $P_1'$ onto $P_2'$. Split $N_1$ into two disjoint infinite sets $N_1'$ and $N_2'$. Since $P_1$ is extremally disconnected, one of the sets $N_1'$ or $N_2'$ is closed in $P_1'$. Let $N_1'$ be closed in $P_1'$; then $N_1'' = f(N_1')$ is closed in $N_2$ and $N_2' \cup x$ is homeomorphic to $N_2'' \cup y$, where $N_2'' = N_2 - N_1''$. Thus, $(N - N_1) \cup N_1' = M_1$ is closed in $N \cup x$ and $(N - N_2) \cup N_2'' = M_2$ is closed in $N \cup y$. Further, $M_1$ and $M_2$ are countable discrete subspaces. Let $\psi: M_1 \to M_2$ be an arbitrary one-to-one map of $M_1$ onto $M_2$. Then $g = \psi \oplus f^*$, where $f^* = f | N_2' \cup x$ is a homeomorphism from $P_1 = M_1 \cup N_2' \cup x$ onto $P_2 = M_2 \cup N_2'' \cup y$. This proves the first half of the proposition. To prove the second, consider $\beta T$, $|T| \geqq \aleph_1$. Let $N \subset T$ and $y \in \overline{N} - T$. Then $(T \cup y) \sim (N \cup y)$, whereas $(T \cup y) \not\approx (N \cup y)$ because $|T| > |N|$.

DEFINITION 9. A subspace $T = \{t_\alpha\}$ of a space $X$ is called *strongly discrete* if there exists a system $\mathfrak{B} = \{U_\alpha\}$ of disjoint open sets in $X$ for which $U_\alpha \ni t_\alpha$. An atom $P = T \cup x$ is said to be *properly embedded* in $X$ if $T$ is strongly discrete.

LEMMA 29. *Let $X$ be an extremally disconnected space, where $wX \leqq \tau$, $cX \leqq \mathfrak{n}$ and $x \in X$. Then the power of the set of all properly embedded pairwise weakly nonequivalent atoms with nucleus $x$ does not exceed $\tau^{\mathfrak{n}}$.*

PROOF. We make each atom $P = x \cup T$ correspond to a disjoint covering $\gamma(P)$ of the set $T$ consisting of basis elements. Such a covering exists, since $T$ is strongly discrete in $X$. Since $cX \leqq \mathfrak{n}$, we have $|\gamma(P)| \leqq \mathfrak{n}$. Further, $|\{\gamma(P)\}| \leqq \tau^{\mathfrak{n}}$, because each such covering consists of basis elements and $wX \leqq \tau$. Suppose that the assertion of the lemma does not hold. Then there exist two atoms $P = x \cup T$ and $P' = x \cup T'$ for which $\gamma(P) = \gamma(P')$. Since $P$ is not weakly equivalent to $P'$, we see that $x \notin \overline{T \cap T'}$. Each element of the system $\gamma = \gamma(P) = \gamma(P')$ intersects $T$ in precisely one point and $T'$ in precisely one point. Index the elements of the system $\gamma$ for which these points do not coincide by the ordinals $U_1, U_2, \cdots, U_\alpha, \cdots$, and put $t_\alpha = U_\alpha \cap T$ and $t_\alpha' = U_\alpha \cap T'$. From what has been said above, $t_\alpha \neq t_\alpha'$. From $\overline{T \cap T'} \not\ni x$ it follows that $x \in \overline{\bigcup t_\alpha}$ and $x \in \overline{\bigcup t_\alpha'}$. Choose for each $\alpha$ disjoint open neighborhoods $Ot_\alpha$ and $Ot_\alpha'$ of the points $t_\alpha$ and $t_\alpha'$ such that $Ot_\alpha \subset U_\alpha$ and $Ot_\alpha' \subset U_\alpha$. Then $\overline{\bigcup_\alpha Ot_\alpha} \ni x$ and $\overline{\bigcup_\alpha Ot_\alpha'} \ni x$ and $(\bigcup_\alpha Ot_\alpha) \cap (\bigcup_\alpha Ot_\alpha') = \emptyset$, which contradicts the fact that $X$ is extremally disconnected.

LEMMA 30. *Let $X$ be an infinite extremally disconnected compact space containing a family $\mathfrak{B} = \{U_\alpha\}$, of power $\mathfrak{n}$ consisting of disjoint open sets in $X$. Then $X$ contains* $\exp\exp\mathfrak{n}$ *properly embedded pairwise nonequivalent atoms.*

PROOF. Choose a point $t_\alpha \in U_\alpha$ for each $U_\alpha \in \mathfrak{B}$. Put $T = \bigcup t_\alpha$. Then $T$ is strongly discrete. We assert that $\overline{T} = \beta T$. For this it is sufficient to prove that $\overline{T}$ is extremally disconnected. Let $U$ and $V$ be open sets in $T$, where $T \cap V = \emptyset$. Put $T_1 = U \cap T$ and $T_2 = V \cap T$. Note that $\overline{T}_1 = \overline{U}$, $\overline{T}_2 = \overline{V}$ and $T_1 \cap T_2 = \emptyset$. Let $W(T_i) = \{U_\alpha \in \mathfrak{B}, U_\alpha \cap T_i \neq \emptyset\}$, $i = 1, 2$. Then $L(T_i) = \bigcup U_\alpha$, $U_\alpha \in W(T_i)$, $i = 1, 2$, are disjoint open sets in $X$ with $L(T_i) \supset T_i$.

Since $X$ is extremally disconnected, $\overline{L(T_1)} \cap \overline{L(T_2)} = \emptyset$; but $\overline{L(T_i)} \supset \overline{T}_i$; that is, $\overline{U} \cap \overline{V} = \emptyset$, as required. Thus, $\overline{T} = \beta T$. Let $x \in \beta T - T$; then $P = x \cup T$ is a properly embedded atom. We prove that the power of all properly embedded pairwise weakly nonequivalent atoms $P$ in $X$ having a nucleus in $\beta T - T$ is $\exp\exp \mathfrak{n}$. Consider two arbitrary weakly equivalent atoms $P_1 \doteq x_1 \cup T_1$ and $P_2 = x_2 \cup T_2$, $x_1 \neq x_2$, $T_1 \subset T$ and $T_2 \subset T$. We prove that to each such pair $(P_1, P_2)$ there corresponds a homeomorphism $\varphi: \beta T \to \beta T$. Let $P_1' \subset P_1$ and $P_2' \subset P_2$, and let $P_1'$ be homeomorphic to $P_2'$. If $P_1' = x_1 \cup T_1'$ and $P_2' = x_2 \cup T_2'$, then $T_1 - T_1'$ and $T_2 - T_2'$ are closed, since $P_1'$ and $P_2'$ are extremally disconnected. Let $f: P_1' \to P_2'$ be the homeomorphism from $P_1'$ onto $P_2'$; then by extending this to $T_1'$ we obtain a homeomorphism $\tilde{f}: \beta T_1' \to \beta T_2'$. Further, we obtain a homeomorphism $\varphi: \beta T \overset{\text{onto}}{\to} \beta T$ such that $\varphi(x_1) = x_2$ and $\overline{T}_0 = \overline{T - T_1} = \overline{T - T_2}$ remain fixed. But, as was stated at the beginning of this section, there are not more than $\exp|T| = \exp \mathfrak{n}$ such homeomorphisms. On the other hand, it is well known that $|\beta T| = \exp\exp \mathfrak{n}$. Consequently the power of the set of all pairwise weakly nonequivalent atoms in $X$ is not less than $\exp\exp \mathfrak{n}$.

Theorem 21 (Arhangel'skiĭ [5]). *An extremally disconnected compact space of weight* $\exp \aleph_0$ *is not topologically homogeneous.*

Proof. Let $X$ be an extremally disconnected compact space with $wX = \mathfrak{c}$. By Theorem 14 $X$ is a subspace of $pD^{\mathfrak{c}}$, and by Theorem 13 $w(pD^{\mathfrak{c}}) = \mathfrak{c}$. Since $c(pD^{\tau}) = \aleph_0$, we know that $pD^{\tau}$ contains a countable properly embedded discrete subspace. Therefore by Lemma 2 $X$ contains $\exp\exp \aleph_0$ properly embedded pairwise weakly nonequivalent atoms. But by Lemma 29 the power of the set of all properly embedded pairwise weakly nonequivalent atoms with nucleus $x$ does not exceed $\mathfrak{c}^{\aleph_0} = \mathfrak{c}$. This implies that there exist $\exp \mathfrak{c}$ points in $X$ that cannot be mapped into each other by a homeomorphism of $X$ onto itself.

Theorem 22. *Let* $X$ *be an extremally disconnected compact space with* $\pi w X = \tau$, *where* $\tau < \Xi_1$, *the first uncountable weakly inaccessible cardinal. If* $cX \geqq \log \tau$, *then* $X$ *is not topologically homogeneous.*

Proof. It is clear that $cX \leqq \pi w X$ and therefore $cX = \mathfrak{n} \leqq \tau < \Xi_1$. Applying Lemma 12, we see that there exists a system of disjoint open sets in $X$ of power $\mathfrak{n}$. In this case by Lemma 30 $X$ contains $\exp\exp \mathfrak{n}$ properly embedded pairwise weakly nonequivalent atoms. On the other hand, by Lemma 1, $wX \leqq (\pi w X)^{cX} = \tau^{\mathfrak{n}}$. Applying Lemma 29, we find that the power of the set of all properly embedded pairwise weakly nonequivalent atoms with nucleus at an arbitrary fixed point $x \in X$ does not exceed

$$(wX)^{cX} \leqq (\tau^{\mathfrak{n}})^{\mathfrak{n}} = \tau^{\mathfrak{n}^2} = \tau^{\mathfrak{n}}.$$

Since by hypothesis $\mathfrak{n} \geqq \log \tau$, we have $\exp \mathfrak{n} \geqq \tau$ and therefore

$$\tau^{\mathfrak{n}} \leqq (\exp \mathfrak{n})^{\mathfrak{n}} = \exp(\mathfrak{n} \cdot \mathfrak{n}) = \exp \mathfrak{n}.$$

Thus $\tau^{\mathfrak{n}} \leqq \exp\exp\mathfrak{n}$ and the number of properly embedded pairwise weakly nonequivalent atoms whose nucleus is an arbitrary fixed point $x \in X$ is strictly less than the number of all properly embedded pairwise weakly nonequivalent atoms in $X$. Hence it follows immediately that there exist two points in $X$ that cannot be mapped onto each other by a homeomorphism of $X$ onto itself.

COROLLARY 19. *If* $\mathfrak{c} < \Xi_1$, *then the absolute of an arbitrary compact space of weight* $\mathfrak{c}$ *is not homogeneous.*

For if $X$ is compact, $wX = \mathfrak{c}$ and $f: pX \to X$ is an irreducible map of $pX$ onto $X$, then, since the $\pi$-weight is invariant under such maps (Lemma 7),

$$\pi w(pX) = \pi wX \leqq wX = \mathfrak{c}.$$

On the other hand, $c(pX) \geqq \aleph_0 = \log \mathfrak{c}$ since $w(pX) = \mathfrak{c}$.

Hence by Theorem 22 $pX$ is not homogeneous.

REMARK 10. Corollary 19 answers a question of Arhangel'skiĭ [**5**]: Is the absolute of any compact space of weight $\mathfrak{c}$ homogeneous? Denote by $\mathbf{CH}(\mathfrak{m})$ the assertion: "There does not exist a cardinal between $\mathfrak{m}$ and $\exp\mathfrak{m}$". Frolík [**42**] has proved that if $\mathbf{CH}(\aleph_0)$ is true or $\mathbf{CH}(\exp\aleph_0)$ is not true, then any topologically homogeneous compact space is discrete. Note that in the theory of sets all three of the hypotheses

"$\mathbf{CH}(\aleph_0)$ is true"

"$\mathbf{CH}(\exp\aleph_0)$ is not true"

"$\exp\aleph_0 < \Xi_1$"

are mutually independent.

Received 30 May 1969

## BIBLIOGRAPHY

[1] P. S. Aleksandrov, *Introduction to the general theory of sets and functions*, GITTL, Moscow, 1948; German transl., VEB Deutscher Verlag, Berlin, 1956.

[2] P. S. Aleksandrov and P. S. Uryson, *Mémoire sur les espaces topologiques compactes*, Verh. Kon. Akad. Wetensch. Amsterdam. Afd. Natuurk. **14** (1929), no. 1; Russian revised transl., Trudy Mat. Inst. Steklov. **31** (1950). MR **13**, 264.

[3] P. S. Aleksandrov and V. I. Ponomarev, *On dyadic bicompacta*, Fund. Math. **50** (1961/62), 419-429. (Russian) MR **25** #1538.

[4] A. V. Arhangel'skiĭ and V. I. Ponomarev, *On dyadic bicompacts*, Dokl. Akad. Nauk SSSR **182** (1968), 993-996 = Soviet Math. Dokl. **9** (1968), 1220-1224. MR **38** #3827.

[5] A. V. Arhangel'skiĭ, *Every extremally disconnected bicompactum of weight c is inhomogeneous*, Dokl. Akad. Nauk SSSR **175** (1967), 751-754 = Soviet Math. Dokl. **8** (1967), 897-900. MR **36** #2122.

[6] Alexandre Arhangelski, *Groupes topologiques extrémalement discontinus*, C. R. Acad. Sci. Paris Sér. A-B **265** (1967), A822-A825. MR **36** #5259.

[7] V. I. Ponomarev, *Projective spectra and continuous mappings of paracompacta*, Mat. Sb.

**60** (**102**) (1963), 89-119; English transl., Amer. Math. Soc. Transl. (2) **39** (1964), 133-164. MR **27** # 2951.

[8] ———, *The absolute of a topological space,* Dokl. Akad. Nauk SSSR **149** (1963), 26-29 = Soviet Math. Dokl. **4** (1963), 299-302. MR **28** # 589b.

[9] ———, *Spaces co-absolute with metric spaces,* Uspehi Mat. Nauk **21** (1966), no. 4 (130), 101-132 = Russian Math. Surveys **21** (1966), no. 4, 87-114. MR **34** # 788.

[10] Ju. M. Smirnov, *On the weight of the ring of bounded continuous functions over a normal space,* Mat. Sb. **30** (**72**) (1952), 213-218. (Russian) MR **14,** 70.

[11] E. G. Skljarenko, *On the topological structure of locally bicompact groups and their quotient spaces,* Mat. Sb. **60** (**102**) (1963), 63-88; English transl., Amer. Math. Soc. Transl. (2) **39** (1964), 57-82. MR **26** # 5090.

[12] I. I. Parovičenko, *Some special classes of topological spaces and $\delta s$-operations,* Dokl. Akad. Nauk SSSR **115** (1957), 866-868. (Russian) MR **19,** 971.

[13] N. A. Šanin, *On the product of topological spaces,* Trudy Mat. Inst. Steklov. **24** (1948). (Russian) MR **8,** 334.

[14] S. M. Sirota, *Homogeneous subsets of absolutes,* Dokl. Akad. Nauk SSSR **189** (1969), 728-731 = Soviet Math. Dokl. **10** (1969), 1494-1498. MR **41** # 8573.

[15] ———, *The product of topological groups and extremal disconnectedness,* Mat. Sb. **79** (**121**) (1969), 179-192 = Math. USSR Sb. **8** (1969), 169-180. MR **39** # 4315.

[16] B. A. Efimov, *Dyadic bicompacta,* Trudy Moskov. Mat. Obšč. **14** (1965), 211-247 = Trans. Moscow Math. Soc. **1965**, 229-267. MR **34** # 1979.

[17] ———, *Extremal disconnectedness and dyadicity,* Proc. Second Prague Topological Sympos. (Prague, 1966), Publ. House Czech. Acad. Sci., Prague, 1967, pp. 129-130. MR **38** # 654.

[18] ———, *Extremally disconnected bicompacta,* Dokl. Akad. Nauk SSSR **172** (1967), 771-774 = Soviet Math. Dokl. **8** (1967), 168-171. MR **35** # 3630.

[19] ———, *Mappings of zero-dimensional bicompacta,* Dokl. Akad. Nauk SSSR **178** (1968), 525-528 = Soviet Math. Dokl. **9** (1968), 126-129. MR **36** # 7113.

[20] ———, *Absolutes of homogeneous spaces,* Dokl. Akad. Nauk SSSR **179** (1968), 271-274 = Soviet Math. Dokl. **9** (1968), 341-344. MR **37** # 3521.

[21] ———, *Extremally disconnected bicompacta of continuum $\pi$-weight,* Dokl. Akad. Nauk SSSR **183** (1968), 15-18 = Soviet Math. Dokl. **9** (1968), 1404-1407. MR **38** # 5172.

[22] ———, *A problem of de Groot and a topological theorem of Ramsey type,* Sibirsk. Mat. Ž. **11** (1970), 1280-1290 = Siberian Math. J. **11** (1970), 943-950.

[23] S. D. Iliadis and S. V. Fomin, *The method of centered systems in the theory of topological spaces,* Uspehi Mat. Nauk **21** (1966), no. 4 (130), 47-76 = Russian Math. Surveys **21** (1966), no. 4, 37-62. MR **34** # 3526.

[24] N. Bourbaki, *Théorie des ensembles,* 3rd rev. ed., Hermann, Paris, 1966; Russian transl., "Mir", Moscow, 1965. MR **34** # 7356.

[25] ———, *Topologie générale.* Chaps. I, II: *Structures fondamentales,* Actualités Sci. Indust., no. 1142, Hermann, Paris, 1961; Russian transl., "Nauka", Moscow, 1968. MR **24** # 4480; **39** # 6238.

[26] A. A. Fraenkel and Y. Bar-Hillel, *Foundations of set theory,* Studies in Logic and Foundations of Mathematics, North-Holland, Amsterdam, 1958; Russian transl., "Mir", Moscow, 1966. MR **21** # 648.

[27] P. J. Cohen, *Set theory and the continuum hypothesis,* Benjamin, New York, 1966; Russian transl., "Mir", Moscow, 1968. MR **38** # 999.

[28] F. Hausdorff, *Grundzüge der Mengenlehre,* 2nd rev. ed., de Gruyter, Leipzig, 1927; English transl., Chelsea, New York, 1962; Russian transl., ONTI, Moscow, 1937. MR **19,** 111.

[29] ———, *Summen von $\aleph_1$ Mengen,* Fund. Math. **24** (1936), 241-255.

[30] R. Engelking, *Outline of general topology,* PWN, Warsaw, 1965; English transl., North-Holland, Amsterdam, and Wiley, New York, 1968. MR **36** # 4508; **37** # 5836.

[31] L. Gillman and M. Jerison, *Rings of continuous functions,* Van Nostrand, Princeton, N. J., 1960. MR **22** # 6994.

[32] R. Sikorski, *Boolean algebras,* Academic Press, New York; Springer-Verlag, Berlin, 1964. MR **31** # 2178.

[33] ———, *A theorem on extension of homomorphisms*, Ann. Soc. Polon. Math. **21** (1948), 332-335. MR **11**, 76.

[34] I. Juhasz, *Remarks on a theorem of B. Pospíšil*, Comment. Math. Univ. Carolinae **8** (1967), 231-247. MR **35** # 7300.

[35] A. Gleason, *Projective topological spaces*, Illinois J. Math. **2** (1958), 482-489. MR **22** # 12509.

[36] P. Erdös and A. Tarski, *On families of mutually exclusive sets*, Ann. of Math. (2) **44** (1943), 315-329. MR **4**, 269.

[37] O. H. Keller, *Die Homomorphie det kompakten konvexen Mengen im Hilbertschen Raum*, Math. Ann. **105** (1931), 748-758.

[38] E. Čech and B. Pospíšil, I. *Sur les espaces compactes*. II. *Sur les caractères des points dans les espaces* $\mathfrak{P}$, Publ. Fac. Sci. Univ. Masaryk (Brno) **258** (1938).

[39] S. Mardešić and P. Papić, *Continuous images of ordered compacta, the Suslin property and diadic compacta*, Glasnik Mat.-Fiz. Astronom. Društvo Mat. Fiz. Hrvatske Ser. II **17** (1962), 3-25. MR **28** # 1591.

[40] Z. Frolík, *Types of ultrafilters on countable sets*, Proc. Second Prague Topological Sympos. (Prague, 1966), Publ. House Czech. Acad. Sci., Prague, 1967, pp. 142-143. MR **38** # 654.

[41] ———, *Sums of ultrafilters*, Bull. Amer. Math. Soc. **73** (1967), 87-91. MR **34** # 3525.

[42] ———, *Homogeneity problems for extremally disconnected spaces*, Comment. Math. Univ. Carolinae **8** (1967), 757-763. MR **41** # 9176.

[43] B. A. Efimov, *On embedding of Stone-Čech compactifications of discrete spaces in bicompacta*, Dokl. Akad. Nauk SSSR **189** (1969), 244-246 = Soviet Math. Dokl. **10** (1969), 1391-1394. MR **40** # 6505.

[44] K. Kuratowski and A. Mostowski, *Set theory*, PWN, Warsaw, 1952; English transl., PWN, Warsaw, North-Holland, Amsterdam, 1968. MR **14**, 960; **37** # 5100.

[45] P. Dwinger, *Remarks on the field representation of Boolean algebras*, Nederl. Akad. Wetensch. Proc. Ser. A **63** = Indag. Math. **22** (1960), 213-217. MR **27** # 1398.

[46] A. V. Arhangel'skiĭ, *On the cardinality of bicompacta satisfying the first axiom of countability*, Dokl. Akad. Nauk SSSR **187** (1969), 967-970 = Soviet Math. Dokl. **10** (1969), 951-955. MR **40** # 4922.

[47] B. A. Efimov, *Solution of some problems on dyadic bicompacta*, Dokl. Akad. Nauk SSSR **187** (1969), 21-24 = Soviet Math. Dokl. **10** (1969), 453-456. MR **40** # 6504.

[48] A. V. Arhangel'skiĭ, *Suslin number and cardinality. Characters of points in sequential bicompacta*, Dokl. Akad. Nauk SSSR **192** (1970), 255-258 = Soviet Math. Dokl. **11** (1970), 597-601. MR **41** # 7607.

[49] D. A. Vladimirov and B. A. Efimov, *On the power of extremally disconnected spaces and complete Boolean algebras*, Dokl. Akad. Nauk SSSR **194** (1970), 1247-1250 = Soviet Math. Dokl. **11** (1970), 1352-1356.

Translated by:
A. West

Trudy Moskov. Mat. Obšč.
Tom 23 (1970)

Trans. Moscow Math. Soc.
Vol. 23 (1970)

# MANY-VALUED MAPPINGS AND BOREL SETS. II[1)]

*M. M. ČOBAN*

## Contents



## §6. Weakly paracompact spaces and lower semicontinuous mappings

In this section we prove Theorem 6.1, which for paracompact spaces was established by Michael.

In principle, our method does not differ from that of Michael.

THEOREM 6.1. *For a regular space $X$ the following conditions are equivalent:*

I) *$X$ is weakly paracompact.*

II) *For every lower semicontinuous mapping $\phi$: $X \to \mathscr{F}(Y)$, where $Y$ is a complete metrizable space, there exists a lower semicontinuous mapping $f$: $X \to C(Y)$ such that $fx \subset \phi x$ for all $x \in X$.*

PROOF. I $\to$ II. First we construct:

a) pointwise finite coverings $\omega_n = \{U_\alpha | \alpha \in A_n\}$ of $X$;

b) sets of points

$$L_n = \{y_\alpha(x) \in \varphi x \,]\, x \in U_\alpha,\ \alpha \in A_n\};$$

c) mappings $\pi_n$: $A_{n+1} \to A_n$

satisfying the following conditions:

1) If $\alpha \in A_n$, there exist points $x_\alpha$ and $y_\alpha$ in $\phi x_\alpha$ such that

$$[U_\alpha] \subseteq \left\{x \in X \mid \varphi x \cap O\left(y_\alpha, \frac{1}{2^{n+1}}\right) \neq \varnothing\right\}.$$

2) If $\alpha \in A_n$,

$$U_\alpha = \bigcup \{U_\beta \mid \beta \in \pi_n^{-1}\alpha\}.$$

---

*AMS (MOS) subject classifications* (1970). Primary 54C60; Secondary 54C05, 54C10, 54H05.

[1)] The first part of this article was published in Trudy Moskov Mat. Obšč. **22** (1970), 229–250 = Trans. Moscow Math. Soc. 22 (1970), 258–280.

3) If $\alpha \in A_n$, $\beta \in \pi_n^{-1}\alpha$ and $x, x' \in U_\beta$,

$$\rho(y_\alpha(x), y_\beta(x)) < \frac{1}{2^n} \quad \text{and} \quad \rho(y_\alpha(x), y_\alpha(x')) < \frac{1}{2^{n-1}}.$$

4) If $\alpha \in A_n$, then $y_\alpha(x) \in O(y_\alpha, 1/2^{n+1})$ for all $x \in U_\alpha$.

We can assume that the diameter of $Y$ does not exceed $\frac{1}{2}$, in which case we put

$$\omega_1 = \{U_\alpha = X\},\ A_1 = \{\alpha = 1\},\ L_1 = \{y_\alpha(x) \in \varphi(x) \mid x \in X\}.$$

We suppose that the coverings $\omega_i$ and the sets $A_i$ and $L_i$ have already been constructed for $i = 1, 2, \cdots, n$.

We take an arbitrary element $\alpha \in A_n$. By 1) there exist points $x_\alpha \in U_\alpha$ and $y_\alpha \in \phi x_\alpha$ for which

$$[U_\alpha] \subset \varphi^{-1} O\left(y_\alpha, \frac{1}{2^{n+1}}\right).$$

We take the system

$$\left\{\varphi^{-1} O\left(z, \frac{1}{2^{n+2}}\right) \middle| z \in O\left(y_\alpha, \frac{1}{2^{n+1}}\right)\right\}.$$

This is an open covering of $[U_\alpha]$.

Since $[U_\alpha]$ is regular and weakly paracompact, there exists an open pointwise finite covering $\tilde{\omega}^\alpha_{n+1} = \{V_\beta \mid \beta \in A^\alpha_{n+1}\}$ of $[U_\alpha]$ such that for each $\beta \in A^\alpha_{n+1}$ there are points $x_\beta \in V_\beta \cap U_\alpha$ and $y_\beta \in \phi x_\beta \cap O(y_\alpha, 1/2^{n+1})$ for which

$$[V_\beta] \subset \varphi^{-1} O\left(y_\beta, \frac{1}{2^{n+1}}\right).$$

We put

$$U_\beta = V_\beta \cap U_\alpha,$$
$$A_{n+1} = \bigcup \{A^\alpha_{n+1} \mid \alpha \in A_n\},$$
$$\omega_n = \{U_\beta \mid \beta \in A_{n+1}\},$$
$$\pi_n : A_{n+1} \to A_n, \quad \text{where } \pi_n^{-1}\alpha = A^\alpha_{n+1},$$
$$L_{n+1} = \left\{y_\beta(x) \in \varphi x \cap O\left(y_\beta, \frac{1}{2^{n+2}}\right) \middle| x \in U_\beta,\ \beta \in A_{n+1}\right\}.$$

It is easy to see that the conditions 1)–4) are satisfied. Now we put

$$Ax = \{\alpha = \{\alpha_1, \ldots, \alpha_n, \ldots\} \mid \alpha \in A_n,\ x \in \bigcap_{n=1}^{\infty} U_{\alpha_n} \text{ and } \pi_n\alpha_{n+1} = \alpha_n\}.$$

To every $\alpha \in Ax$ there corresponds a Cauchy sequence $\{y_{\alpha_n}(x) \in L_n \mid n = 1, 2, \cdots\}$. Since $Y$ is complete, this sequence has a limit $y_\alpha(x) \in \phi x$. We put $fx = \{y_\alpha(x) \mid \alpha \in Ax\}$.

**Lemma** 6.1. *The set $fx$ is compact for every $x \in Ax$.*

We put $A_n(x) = \{\alpha \in A_n \mid x \in U_\alpha\}$. Then $A_n(x)$ is finite; we give it the discrete topology. Now

$$Ax = \lim \{A_n(x) \mid \pi_k^m = \pi_k \circ \pi_{k+1} \circ \ldots \circ \pi_{m-1}\}.$$

It is well known (see [1]) that $Ax$ is compact.

We define the mapping $\psi$: $Ax \to fx$ by $\psi\alpha = y_\alpha(x)$ for all $\alpha \in Ax$. Then $\psi$ is continuous.

Indeed, let $\alpha \in Ax$ and $\epsilon > 0$ by arbitrary. We take an $n$ such that $1/2^{n-2} < \epsilon$. We put

$$O_\alpha^n = \{\alpha' = \{\alpha_1', \ldots, \alpha_n', \ldots\} \mid \alpha_i' = \alpha_i \text{ when } i \leqslant n\}.$$

For $\alpha' \in O_\alpha^n$, by (3) we have

$$\rho(\psi\alpha, \psi\alpha') = \rho(y_\alpha(x), y_{\alpha'}(x)) \leqslant \rho(y_\alpha(x), y_{\alpha_n}(x)) + \rho(y_{\alpha_n'}(x), y_{\alpha_n}(x))$$

$$+ \rho(y_{\alpha_n'}(x), y_{\alpha'}(x)) \leqslant \frac{1}{2^{n-1}} + 0 + \frac{1}{2^{n-1}} < \varepsilon.$$

This proves Lemma 6.1.

By construction we have

LEMMA 6.2. *$fx \subset \phi x$ for every $x \in X$.*

LEMMA 6.3. *The mapping $f$: $X \to C(Y)$ is lower semicontinuous.*

We choose $x \in X$, $y \in fx$ and $\epsilon > 0$ arbitrarily. We must find a neighborhood $Ox$ of $x$ such that $\rho(y, f(x')) < \epsilon$ for all $x' \in Ox$. It is clear that $y = y_\alpha(x)$ for some $\alpha \in Ax$. We choose an $n$ such that $1/2^{n-3} < \epsilon$ and put $Ox = U_{\alpha_n}$. Let $x' \in Ox = U_{\alpha_n}$. Then $y_{\alpha'}(x') \in fx'$ for $\alpha_n' = \alpha_n$ and by (3) we have

$$\rho(y, f(x')) \leqslant \rho(y_\alpha(x), y_{\alpha_n}(x)) + \rho(y_{\alpha_n}(x), y_{\alpha_n'}(x))$$

$$+ \rho(y_{\alpha_n'}(x), y_{\alpha'}(x')) \leqslant \frac{1}{2^n} + \frac{1}{2^{n-1}} + \frac{1}{2^{n-1}} < \varepsilon.$$

This completes the proof of the first part of Theorem 6.1.

II $\to$ I. Let $\omega = \{U_\alpha \mid \alpha \in A\}$ be an open covering of $X$. We put $Y = A$ with the discrete topology and define $\phi$: $X \to \mathscr{F}(Y)$ by $\phi x = \{\alpha \in A \mid x \in U_\alpha\}$ for all $x \in X$.

Since $\phi^{-1}\alpha = U_\alpha$ for each $\alpha \in Y$, $\phi$ is lower semicontinuous. By assumption there exists a lower semicontinuous mapping $f$: $X \to C(Y)$ such that $fx \subset \phi x$ for all $x \in X$.

As $\phi^{-1}\alpha = U_\alpha$ and $fx \subset \phi x$, we have $f^{-1}\alpha \subset U_\alpha$ for every $\alpha \in A$. Consequently the open covering $\gamma = \{f^{-1}\alpha \mid \alpha \in A\}$ is a refinement of $\omega$. As $fx$ is compact for every $x$ (in fact, it consists of a finite number of points of $Y$), $\gamma$ is pointwise finite. This completes the proof of Theorem 6.1.

From Theorems 6.1 and 5.2 we obtain

THEOREM 6.2. *Let $f$: $X \to \mathscr{F}(Y)$ be a lower semicontinuous mapping of a regular weakly paracompact space $X$ into a complete metrizable space $Y$. Then there exists a single-valued mapping $\phi$: $X \to Y$ for which*

1) *$\phi x \in fx$ for all $x \in X$;*
2) *$f^{-1}U \in L(x)$ for every set $U$ that is open in $Y$.*

The following class of mappings was considered in [2]: a mapping $f$: $X \to Y$ is said to be inductively open if there exists a subspace $X_1 \subseteq X$ such that $fX_1 = Y$ and $f_1 = f|X_1$ is open. Theorem 6.1 enables us to prove the following theorem, which is the definitive generalization of one of Michael's theorems.

THEOREM 6.3. *Let $f$: $X \to Y$ be an inductively open mapping of a metric space $X$ onto a regular weakly paracompact space $Y$ for which the inverse image of every point is complete (in the metric on $X$). Then there exists a subspace $X_1 \subseteq X$ such that the mapping $f_1 = f|X_1$ satisfies all the following conditions:* 1) *it is open;* 2) *inverse images of points are compact;* 3) *$fX_1 = Y$.*

PROOF. Let $\widetilde{X}$ be the completion of the metric space $X$ and let $X' \subset X$ be such that $fX' = Y$ and $\tilde{f} = f|X'$ is open. The mapping $\theta$: $Y \to \mathcal{F}(X)$ given by $\theta y = [\tilde{f}^{-1}y]|_{\widetilde{X}}$ is lower semicontinuous. Since $f^{-1}y$ is closed in $\widetilde{X}$ for every $y \in Y$, we have $\theta y \cap \theta y' = \emptyset$ for $y \neq y'$. By Theorem 6.1 there exists a lower semicontinuous mapping $g$: $X \to C(\widetilde{X})$ such that $gy \subset \theta y$ for every $y \in Y$. $X_1 = \bigcup \{gy | y \in Y\}$ is the required set.

## §7. Some remarks on $\rho$-continuous many-valued mappings

Let $X$ be a metric space with a bounded metric $\rho$. $\mathcal{F}(X\colon \rho)$ denotes the space of nonempty closed subsets of $X$ with the Hausdorff metric:

$$d(L,B) = \min\{r | O(L,r) \supset B,\ O(B,r) \supset L\}.$$

It is well known that for compact spaces the Hausdorff metric generates the Vietoris topology.

PROPOSITION 7.1. *The mapping $h$: $\mathcal{F}(X\colon \rho) \to \mathcal{F}(X)$ given by $hL = L$ for every $L \in \mathcal{F}(X\colon \rho)$ is lower semicontinuous.*

PROOF. It is obvious that if $d(L,B) \leqq d$, then for every $\epsilon > 0$ and every point $a \in L$ there exists a point $b(a,\epsilon) \in B$ such that $\rho(a, b(a,\epsilon)) < d + \epsilon$. Suppose that $U$ is an open subset of $X$ and that $L \cap U \neq \emptyset$. Then there exists a point $a \in L \cap U$ and an $l > 0$ such that $O(a,l) \subset U$. We prove that

$$O^*(L,l) = \{B \in \mathcal{F}(X:\rho) | d(L,B) < l\} \subseteq h^{-1}U = \{B \in \mathcal{F}(X:\rho) | B \cap U \neq \emptyset\}.$$

For let

$$d(L,B) = l - \epsilon, \text{ where } 0 \leqq \epsilon < l.$$

Then there exists a point $b \in B$ for which $\rho(a,b) < l$; consequently $b \in U$, and hence $B \in H^{-1}U$. The proof of the proposition is complete.

From Proposition 7.1 and Theorem 5.3 we obtain

THEOREM 7.1. *Let $X$ be a complete metrizable space. Then for any bounded* *metric $\rho$ on $X$ there exists a single-valued mapping $\phi$: $\mathscr{F}(X:\rho)\to X$ such that 1) $\phi L\in L$ for every $L\in\mathscr{F}(X:\rho)$; 2) $\phi^{-1}U$ is an $F_\sigma$-set for every set $U$ that is open in $X$.*

A continuous single-valued mapping $f$: $Y\to\mathscr{F}(X:\rho)$, where $X$ is a metric space with the bounded metric $\rho$, is called a *$\rho$-continuous many-valued mapping of $Y$ into $X$.*

From Theorem 7.1 we obtain

THEOREM 7.2. *Let $f$: $Y\to\mathscr{F}(X:\rho)$ be a $\rho$-continuous mapping, where $X$ is a complete metrizable space. Then there exists a single-valued mapping $\phi$: $Y\to X$ for which*

*1) $\phi y\in fy$ for all $y\in Y$;*

*2) for every complete open set $U$ in $X$ the set $f^{-1}U$ is an $F_\sigma$-set.*

THEOREM 7.3. *Let $f$: $Y\to\mathscr{F}(X:\rho)$ be a $\rho$-continuous mapping of a normal space $Y$ into a complete metrizable space $X$. If* $\dim Y=0$, *then there exists a single-valued continuous mapping $\phi$: $Y\to X$ such that $\phi y\in fy$ for every $y\in Y$.*

PROOF. By Pasynkov's factorization theorem (see [**4**]), there exist a zero-dimensional metric space $Z$ and single-valued continuous mappings $g$: $Y\to Z$ and $\psi$: $Z\to\mathscr{F}(X:\rho)$ such that $f=\psi\circ g$. By Proposition 7.1 $\psi$: $Z\to\mathscr{F}(X:\rho)$ is lower semicontinuous. By Michael's well-known theorem, there exists a single-valued continuous mapping $\xi$: $Z\to X$ such that $\xi z\in\psi z$ for all $z\in Z$. We set $\phi$: $Y\to X$, where $\phi=\xi\circ g$. Theorem 7.3 is proved.

## §8. The case of spaces with a countable base

Let $X$ be a topological space and let $B_\lambda''(X)$ be the class of sets of the form $\langle U_1,\cdots,U_n,X\rangle=\{L\in\mathscr{F}(X)\mid L\cap U_i\neq\emptyset\}$, where the $U_i$ are open subsets of $X$.

Let $B_\lambda^0(X)$ be the class of sets of the form $L\setminus B$, where $L$, $B\in B_\lambda''(X)$.

Now we denote by $B_\lambda(X)$ the class of sets of the form $\bigcup_{i=1}^\infty L_i$, where $L_i\in B_\lambda^0(X)$ and $i=1,2,\cdots$. Similarly we define the family $B_\kappa(X)$, only instead of $B_\lambda''(X)$ we take the system $B_\kappa''(X)$ of sets of the form $\langle F_1,\cdots,F_n,X\rangle=\{L\in\mathscr{F}(X)\mid L\cap F_i\neq\emptyset\}$, where the $F_i$ are closed subsets of $X$.

It is easy to prove

LEMMA 8.1. $B_\lambda(X)\subseteq B_\kappa(X)$ *for any completely normal space $X$.*

LEMMA 8.2. *Suppose that a mapping $\phi$: $Y\to\mathscr{F}(X)$ is such that $\{y\in Y\mid\phi y\cap U\neq\emptyset\}$ is a Borel set of class $\alpha$ for every set $U$ that is open in $X$. Then $\phi^{-1}L$ is a Borel set of class $\alpha$ for every $L\in B_\lambda(X)$.*

THEOREM 8.1. *Let $X$ be a complete metric space with a countable base. There exists a single-valued mapping $\phi$: $\mathscr{F}(X)\to X$ such that*

*1) $\phi L\in L$ for every $L\in\mathscr{F}(X)$;*

2) $\phi^{-1}U \in B_\lambda(X)$ *for every set* $U$ *that is open in* $X$.

PROOF. Let $\rho$ be a complete metric on $X$ for which $\rho(x,y) \leqq \frac{1}{2}$ for any pair of points.

We construct a countable set of mappings $\phi_i$: $\mathscr{F}(X) \to X$ and sets $H_{ik}$, $k$, $i = 1, 2, \cdots$, satisfying the following conditions:

1) $\rho(\phi_i, \phi_{i+1}) < 1/2^i$, $i = 1, 2, \cdots$.

2) $\rho(\phi_i L, L) < 1/2^i$, $i = 1, 2, \cdots$.

3) $H_{ik} \in B^0_\lambda(X)$, $i$, $k = 1, 2, \cdots$.

4) $\mathscr{F}(X) \bigcup_{k=1}^{\infty} H_{ik}$, $i = 1, 2, \cdots$.

5) The set $\phi_i H_{ik} = \{a_{ik}\}$ consists of a single point.

6) $H_{ik} \cap H_{ij} = \emptyset$ for $k \neq j$.

The mapping $\phi_1$: $\mathscr{F}(X) \to X$ is defined by $\phi_1 L = a_{11}$, where $a_{11}$ is a fixed point in $X$. Further, we put $H_{11} = \mathscr{F}(X)$ and $H_{1k} = \emptyset$ for $k = 2, 3, \cdots$. Suppose that $\phi_i$ and $H_{ik}$ have been constructed for $i = 1, \cdots, n$ and $k = 1, 2, \cdots$. Let $k$ be fixed. There exists a countable system $\{W_j^{nk} \mid j = 1, 2, \cdots\}$ of open subsets of $X$ such that

a) diam $W_j^{nk} < 1/2^{n+1}$,

b) $O(a_{nk}, 1/2^n) = \bigcup_{j=1}^{\infty} W_j^{nk}$.

By assumption we have $H_{nk} = A \setminus B$, where $A$, $B \in B''_\lambda(X)$; that is, $A = \{L \in \mathscr{F}(X) \mid L \cap U_i \neq \emptyset,\ i = 1, 2, \cdots, m\}$. We set

$$V_j^{nk} = \{L \in \mathscr{F}(X) \mid L \cap W_j^{nk} \neq \varnothing,\ L \cap U_i \neq \varnothing,\ i = 1, \ldots, m\}.$$

By (2) and (5) we have $H_{nk} = \bigcup V_j^{nk}$.

Let $a_{nkj}$ be a fixed point in $W_j^{nk}$. We put

$$M_{nkj} = V_j^{nk} \setminus B = \{L \in \mathscr{F}(X) \mid L \cap W_j^{nk} \neq \varnothing\} \cap H_{nk}.$$

It is obvious that $M_{nkj} \in B^0_\lambda(X)$ for $j = 1, 2, \cdots$. Now we define $\phi_{n+1}$: $\mathscr{F}(X) \to X$ by $\phi_{n+1} L = a_{nkj}$ for $L \in M_{nkj}$. Changing the notation and the indices of the $M_{nkj}$, we obtain the required mapping $\phi_{n+1}$ and sets $H_{(n+1)k}$ satisfying the conditions (1)–(6). We obtain $\phi$: $\mathscr{F}(X) \to X$ by putting $\phi L = \lim_{n \to \infty} \phi_n L$. By (1), (2), (3) and Hausdorff's lemma, $\phi$ is the required mapping.

Theorem 8.1 is equivalent in content to a theorem of Kuratowski and Ryll-Nardzewski [7], but it is a good supplement to the present work and so is appropriate here. Also the proof is clearer than that given in [7].

Lemma 8.2 and Theorem 8.1 give us

THEOREM 8.2 [7]. *Let* $\phi$: $Y \to \mathscr{F}(X)$ *be such that* $\{y \in Y \mid \phi y \cap U \neq \emptyset\}$ *is a Borel set of class* $\alpha$ *for every set* $U$ *that is open in* $X$. *If* $X$ *is a complete separable metrizable space, then there exists a single-valued mapping* $f$: $Y \to X$ *such that*

1) $fy \in \phi y$ *for every* $y \in Y$;

2) $f^{-1}U$ *is a Borel set of class* $1 + \alpha$ *for every set* $U$ *that is open in* $X$.

## §9. Upper semicontinuous mappings

THEOREM 9.1. *Let $\theta: X \to C(Y)$ be an upper semicontinuous mapping of a space $X$ into a metric space $Y$. Then there exists a single-valued mapping $f: X \to Y$ such that $fx \in \theta x$ for all $x \in X$ and $f^{-1}U \in L(X)$ for every set $U$ that is open in $Y$.*

First we prove

PROPOSITION 9.2. *Let $\omega = \{F_\alpha | \alpha \in A\}$ be a closed locally finite covering of a space $X$. Then there exist refinements $\xi_m = \{C_\beta | \beta \in B_m\}$ of $\omega$ satisfying the following conditions:*

1) $\bigcup \{C_\beta | \beta \in B'_m\} \in L_1(X)$ *for any* $B'_m \subset B_m$.

2) $\xi_m$ *is discrete in the subspace* $M_m = \bigcup \{C_\beta | \beta \in B_m\}$ *for each natural number* $m$.

3) $X = \bigcup_{m=1}^{\infty} M_m$.

PROOF. Let $M_m$ denote the set of all points of $X$ that are contained in exactly $m$ different elements of $\omega$. We put

$$\omega_m = \{\Phi_\beta = F_{\alpha_1} \cap \cdots \cap F_{\alpha_m} | \alpha_1, \ldots, \alpha_m \in A, \ \alpha_i \neq \alpha_j \text{ for } i \neq j\},$$

$$\Phi_m = \bigcup \{\Phi_\beta \in \omega_m\}, \quad m = 1, 2, \ldots.$$

$$\xi_m = \{C_\beta = \Phi_\beta \cap M_m | \Phi_\beta \in \omega_m\}.$$

Since $\Phi_{m+1}$ is closed in $X$ and $X \setminus \Phi_{m+1} = \bigcup_{i=1}^{m} M_i$, we have

LEMMA 9.1. $\bigcup_{i=1}^{m} M_i$ *is open in* $X$ *and* $M_m \in L_1(X)$ *for each* $m$.

The fact that the $\omega_m$ are locally finite and the $\xi_m$ are disjoint gives us

LEMMA 9.2. *The system* $\xi_m$ *is discrete in* $M_m$ *for each* $m$.

Suppose now that $\xi'_m$ is a subsystem of $\xi_m$. Since $M_m = \Phi_m \setminus \Phi_{m+1}$ and $\xi_m$ is discrete in $M_m$, we have

$$\bigcup \{C_\beta \in \xi'_m\} = \bigcup \{\Phi_\beta \in \omega'_m\} \setminus \Phi_{m+1}.$$

This completes the proof of the proposition.

PROOF OF THE THEOREM. As in the proof of Theorem 4.4, we can reduce the general case to the case when $\dim Y = 0$. Then there exist discrete coverings $\gamma_n = \{U_\alpha | \alpha \in A_n\}$ of $Y$ such that

a) for each $m$ the covering $\gamma_{m+1}$ is a refinement of $\gamma_m$ and $\operatorname{diam} U_\alpha < 1/2^m$ for every $\alpha \in A_m$.

Then there exist mappings $\pi_n: A_{n+1} \to A_n$, $n = 1, 2, \cdots$, such that

b) $U_\alpha = \bigcup \{U_\beta | \beta \in \pi_n^{-1}\alpha\}$ for every $\alpha \in A_n$.

It is easy to see that $A_n$ can be well ordered in such a way that

c) $\pi_n\alpha > \pi_n\beta$ implies $\alpha > \beta$ for $\alpha, \beta \in A_{n+1}$.

We shall construct mappings $f_n: X \to Y$ and sets $X_{nk}$, $n, k = 1, 2, \cdots$, satisfying the conditions

1) $\rho(f_n, f_{n+1}) < 1/2^n, n = 1, 2, \cdots$.
2) $\rho(f_n, \theta) < 1/2^n, n = 1, 2, \cdots$.
3) $X_{nk} \in L_1(X_1), n, k = 1, 2, \cdots$.
4) $X_{nk} \cap X_{nm} = \emptyset$, if $k \neq m$.
5) $f_{nk} = f_n | X_{nk}$ is continuous.
6) $X = \bigcup_{k=1}^{\infty} X_{nk}, n = 1, 2, \cdots$.

We can assume that $\rho(y, z) \leqq \frac{1}{2}$ for all $y, z \in Y$.

Now we define $f_1$: $X \to Y$ by $f_1 x = y_0$ for all $x \in X$ and put $X_{11} = X$ and $X_{1k} = \emptyset$ for $k = 2, 3, \cdots$; the conditions 1)–6) are satisfied.

Suppose that we have constructed the mappings $f_i$ and sets $X_{ik}$ for $i = 1, 2, \cdots, n$ and $k = 1, 2, \cdots$. Let $k$ be fixed. We set

$$\theta_{nk}: X_{nk} \to C(Y), \quad \text{where} \quad \theta_{nk} x = \theta x \cap \gamma_n(f_{nk} x)$$

for all $x \in X_{nk}$. By (2), $\theta_{nk} x \neq \emptyset$ for any $x \in X$. By Proposition 10.5, $\theta_{nk}$ is upper semicontinuous. The covering $\omega_{n+1}^{-1} = \{\theta_{nk}^{-1} U_\beta | \beta \in A_{n+1}\}$ is closed and locally finite on $X_{nk}$. The closure follows from the upper semicontinuity of $\theta_{nk}$ and the local finiteness from the fact that $\theta_{nk} x$ is compact for every $x \in X$.

By Proposition 9.2 there exist systems $\xi_m = \{C_\beta | \beta \in B_m\}$ satisfying conditions 1)–3) of that proposition. For clarity, we put

$$M_{nkm} = \bigcup \{C_\beta | \beta \in B_m\}.$$

We can assume that $M_{nkm} \cap M_{nkj} = \emptyset$ for $m \neq j$.

For each $\beta \in B_m$ there exists an $\alpha(\beta) \in A_{n+1}$ such that $C_\beta \subset \theta_{nk}^{-1} U_{\alpha(\beta)}$. Let $y_\alpha$ be a fixed point in $U_\alpha$. We set

$$f_{nkm}: M_{nkm} \to Y, \quad \text{where} \quad f_{nkm} x = y_{\alpha(\beta)}$$

for $x \in C_\beta$. Then $f_{nkm}$ is continuous. Now we define

$$f_{n+1}: X \to Y, \quad \text{where} \quad f_{n+1} x = f_{nkm} x, \quad \text{for} \quad x \in M_{nkm}.$$

Changing the notation and the indices of the $M_{nkm}$, we obtain the required mapping $f_{n+1}$ and sets $X_{(n+1)k}$, $k = 1, 2, \cdots$.

The conditions (3), (4) and (5) imply that $f_n^{-1} U \in L(X)$ for every $n$ and every open set $U \subset Y$. Then it follows from (1), (2) and Hausdorff's lemma that $f = \lim_{n \to \infty} f_n$ is the required mapping. The theorem is proved.

From a preprint which R. Engelking has kindly sent to me I have learnt of the following result of his: let $\theta$: $X \to \mathscr{F} Y$ be an upper semicontinuous mapping of a completely normal paracompact space $X$ into a complete metric space $Y$ such that $\theta x$ is separable for every $x$. Then there exists a single-valued mapping $f$: $X \to Y$ such that $fx \in \theta x$ for every $x \in X$ and $f^{-1} U$ is an $F_\sigma$-set whenever $U$ is open in $Y$. This led me to prove the following proposition.

THEOREM 9.3. *Let $\theta$: $X \to \mathscr{F}(Y)$ be an upper semicontinuous mapping of a $\sigma$-space $X$ into a complete metric space $Y$ such that $\theta x$ is separable for every*

$x \in X$. *Then there exists a single-valued mapping* $f$: $X \to Y$ *such that* $fx \in \theta x$ *for every* $x \in X$ *and* $f^{-1}U$ *is an* $F_\sigma$*-set for every set* $U$ *that is open in* $Y$.[2)]

PROOF. By Proposition 10.4 there exist a complete metric $\rho$ on $Y$ and locally finite coverings

$$\gamma_n = \{U_\alpha \mid \alpha \in A_n\} \qquad (n = 1, 2, \cdots)$$

of $Y$ satisfying the following conditions:

1) $\rho(y, z) \leqq \frac{1}{2}$ for all $y, z \in Y$.
2) diam $U_\alpha = 1/2^n$ for all $\alpha \in A_n$.

As in the proofs of the other theorems, we construct mappings $f_n$: $X \to Y$ and sets $X_{nk}$ $(n,\ k = 1, 2, \cdots)$ satisfying the following conditions:

1) $\rho(f_n, f_{n+1}) < 1/2^n$, $n = 1, 2, \cdots$.
2) $\rho(f_n, \theta) < 1/2^n$, $n = 1, 2, \cdots$.
3) $X_{nk}$ is an $F_\sigma$-set, $n, k = 1, 2, \cdots$.
4) $X_{nk} \cap X_{nj} = \emptyset$ for $k \neq j$.
5). $f_{nk} = f_n \mid X_{nk}$ is continuous.
6) $X = \bigcup_{k=1}^{\infty} X_{nk}$, $n = 1, 2, \cdots$.

We define $f_1$: $X \to Y$ by putting $f_1 x = y_0 \in Y$ for all $x \in X$, and we put $X_{11} = X$ and $X_{1k} = \emptyset$ for $k = 2, 3, \cdots$. It is obvious that 1)–6) are satisfied. Suppose that we have constructed the mappings $f_i$ and sets $X_{ik}$ for $i = 1, 2, \cdots, n$, and $k = 1, 2, \cdots$. Let $k$ be fixed. We define $\theta_{nk}$: $X_{nk} \to \mathscr{F}(X)$ by $\theta_{nk}(x) = \theta x \cap \bar{\gamma}_n(f_{nk}x)$ for all $x \in X_{nk}$, where $\bar{\gamma}_n = \{[U_\alpha] \mid \alpha \in A_n\}$. By (2), $\theta_{nk}x \neq \emptyset$ for any $x \in X_{nk}$. Now by Proposition 10.5, $\theta_{nk}$ is upper semicontinuous. For each point $x \in X_{nk}$ we put

$$U_x^n = \bigcup \{U_\alpha \in \gamma_n \mid \theta_{nk}x \cap U_\alpha \neq \emptyset\},$$
$$A_{x(n)} = \{\alpha \in A_n \mid \theta_{nk}x \cap U_\alpha \neq \emptyset\},$$
$$V_x^n = \{z \in X_{nk} \mid \theta_{nk}z \subset U_x^n\}.$$

Since $\theta_{nk}$ is upper semicontinuous, $\omega_{n+1} = \{V_x^{n+1} \mid x \in X_{nk}\}$ is an open covering of $X_{nk}$, Therefore there exist systems $\xi_m = \{F_\beta^m \mid \beta \in B_m\}$ satisfying the following conditions:

1) $\xi_m$ is closed and discrete in $X$ for every $m$;
2) $\xi_m$ is a refinement of $\omega_{n+1}$;
3) $X_{nk} = \bigcup_{m=1}^{\infty} \bigcup \{F_\beta^m \mid \beta \in B_m\}$.

For each $\beta \in B_m$ there exists a point $x(\beta)$ for which $F_\beta^m \subset V_{x(\beta)}^{n+1}$.

The separability of $\theta_{nk}x$ implies that $A_{x(n+1)}$ is countable for all $x \in X_{nk}$ and all $n$. Let $A_{x(n+1)} = \{\alpha_x^1, \cdots, \alpha_x^l, \cdots\}$ and let $a_\alpha$ be a fixed point in $U_\alpha$. We put

$$\xi_{ml} = \{F_\beta^{ml} = F_\beta^m \cap \{x \in X_{nk} \mid \theta_{nk}x \cap [U_{\alpha^l_{x(\beta)}}] \neq \emptyset\} \mid \beta \in B_m\},$$

---

[2)] For the definition of a $\sigma$-space see §5. Note that a subspace of a $\sigma$-space is a $\sigma$-space.

$m, l = 1, 2, \cdots$. It is easy to see that $\xi_{ml}$ is closed and discrete in $X$. Therefore the sets $M_{nkml} = \bigcup \{F_\beta^{ml} \mid \beta \in B_m\}$, $k, m, l = 1, 2, \cdots$, are closed in $X$. We define mappings

$$f_{nkml}\colon M_{nkml} \to Y$$

by $f_{nkml}z = a^l_{\alpha_{x(\beta)}}$ for every $z \in F_\beta^{ml}$. It is clear that $f_{nkml}$ is continuous,

$$\rho(f_{nkml}, \theta) < \frac{1}{2^{n+1}} \text{ and } \rho(f_n, f_{nkml}) < \frac{1}{2^{n-1}}.$$

Since $X = \bigcup_{k=1}^{\infty} \bigcup_{m=1}^{\infty} \bigcup_{l=1}^{\infty} M_{nkml}$ and since the $M_{nkml}$ are closed in $X$, there exist pairwise disjoint $F_\sigma$-sets $X_{nkml}$ such that

1) $X = \bigcup_{k=1}^{\infty} \bigcup_{m=1}^{\infty} \bigcup_{l=1}^{\infty} X_{nkml}$;

2) $X_{nkml} \subset M_{nkml}$, $k, m, l = 1, 2, \cdots$.

We define the mapping $f_{n+1}\colon X \to Y$ by $f_{n+1}x = f_{nkml}x$ for $x \in X_{nkml}$. Changing the indices of the sets $X_{nkml}$, we obtain the required mapping $f_{n+1}$ and sets $X_{(n+1)k}$, $k = 1, 2, \cdots$. Now we put $f = \lim_{n\to\infty} f_n$. Then it follows from (1), (2) and Hausdorff's lemma that $f$ is the required mapping. This completes the proof.

EXAMPLE 9.1. In Theorem 9.3, complete normality was essential for $f^{-1}U$ to be an $F_\sigma$-set for every set $U$ that is open in $Y$. We show the necessity by means of an example. We let $\Omega(\omega_1) = \{\alpha \leqq \omega_1\}$ with the ordinal topology and let $D = \{0, 1\}$ be the usual discrete two-point set.

We identify the points $(\omega_1, 0)$ and $(\omega_1, 1)$ in $\Omega(\omega_1) \times D$. The resulting factor space is denoted by $X$. The sets $A = \{(\alpha, 0) \mid \alpha < \omega_1\}$ and $B = \{(\alpha, 1) \mid \alpha < \omega_1\}$ are open in $X$ and pairwise disjoint, but are not $F_\sigma$-sets. We define $\theta\colon X \to \mathscr{F}(D)$ by

$$\theta x = \begin{cases} 0, & \text{when } x \in A, \\ 1, & \text{when } x \in B, \\ D, & \text{when } x \in X \setminus (A \cup B). \end{cases}$$

The mapping $\theta$ is upper semicontinuous, and $\theta x$ is compact for every $x \in X$. We observe that $X \setminus (A \cup B) = a$ consists of a single point. We take a section $f\colon X \to D$ for $\theta$. Then $fx = \theta x$ for $x \in A \cup B$ and $fa$ is equal to 0 or 1.

Suppose $fa = 1$. Then $f^{-1}\{0\} = A$, but $A$ is not an $F_\sigma$-set. If $fa = 0$, then $f^{-1}\{1\} = B$, which also fails to be an $F_\sigma$-set.

## §10. Some general theorems

The purpose of this section is to indicate why we have been unable to obtain a general result analogous to the theorem of Kuratowski and Ryll-Nardzewski.

In the proof of all our theorems Borel sections were constructed either explicitly or implicitly by means of an approximation method. Namely, we constructed mappings $\phi_n$: $X \to Y$ and sets $X_{nk}$ ($n$, $k = 1, 2, \cdots$ and $f$: $X \to \mathscr{F}(Y)$ is a many-valued mapping) satisfying the following conditions:

1) $\rho(\phi_n, \phi_{n+1}) < 1/2^n$, $n = 1, 2, \cdots$;
2) $\rho(\phi_n, f) < 1/2^n$, $n = 1, 2, \cdots$;
3) $\phi_{nk} = \phi_n | X_{nk}$ is continuous;
4) $X = \bigcup_{k=1}^{\infty} X_{nk}$, $n = 1, 2, \cdots$;
5) $X_{nk}$ is an $F_\sigma$-subset of $X$.

The conditions (1)–(5) can be replaced by

1*) $\lim_{n\to\infty} \rho(\phi_n x, fx) = 0$ for all $x \in X$;
2*) $\phi_{nk} = \phi_n | X_{nk}$ is continuous;
3*) $X = \bigcup_{k=1}^{\infty} X_{nk}$;
4*) each $X_{nk}$ is closed in $X$.

THEOREM 10.1. *Let $X$ be a completely normal space. The following conditions are equivalent:*

1) *Every open covering* $\omega = \{U_\alpha | \alpha \in \theta, \operatorname{card} \theta \leqq \tau\}$ *has a $\sigma$-discrete closed refinement.*

2) *For every lower semicontinuous mapping $f$: $X \to \mathscr{F}(Y)$, where $Y$ is a complete metrizable space of cardinality $\leqq \tau$, there exist mappings $\phi_n$ and sets $X_{nk}$ satisfying conditions* (1)–(5).

3) *For every lower semicontinuous mapping $f$: $X \to \mathscr{F}(Y)$, where $Y$ is a complete metrizable space of cardinality $\leqq \tau$, there exist mappings $\phi_n$ and sets $X_{nk}$ satisfying conditions* (1*)–(4*).

PROOF. The fact that (1) implies (2) is proved as Theorem 5.1, only here we operate with systems of cardinality $\leqq \tau$.

The fact that (2) implies (3) is trivial. Let us prove that (3) implies (1). Let $\omega = \{U_\alpha | \alpha \in \theta, \operatorname{card} \theta \leqq \tau\}$ be an open covering of $X$. We put $Y = \theta$ with the discrete metric $\rho(\alpha_1, \alpha_2) = 1$ for $\alpha_1 \neq \alpha_2$ and $f$: $X \to \mathscr{F}(Y)$, where $U_\alpha = f^{-1}\alpha$ for every $\alpha \in \theta$.

It is obvious that $f$ is lower semicontinuous. Consequently there exist sets $X_{nk}$ and mappings $\phi_n$ satisfying conditions (1)–(4). Since $Y$ is discrete, for each $x \in X$ there exists an $m$ such that

$$\rho(\phi_m x, fx) = 0; \text{ that is, } \phi_m x \in fx.$$

We put

$$F_m = \{x \in X | \phi_m x \in fx\}.$$

It is obvious that $X = \bigcup_{m=1}^{\infty} F_m$.

Now we put

$$\xi_{nk} = \{F_\alpha^{nk} = X_{nk} \cap F_n \cap \phi_n^{-1}\alpha | \alpha \in \theta\}.$$

By (2*) and (4*) the system $\xi_{nk}$ is discrete on $X$ and $F_\alpha^{nk} = [F_\alpha^{nk}] \subseteq U_\alpha$ for all $\alpha \in \theta$. By construction, the system $\xi = \bigcup_{k=1}^{\infty} \bigcup_{n=1}^{\infty} \xi_{nk}$ covers $X$, is $\sigma$-

discrete, and is a refinement of $\omega$. This completes the proof of Theorem 10.1. Thus, the conditions (1)–(5) are equivalent to (1*)–(4*).

Theorem 10.1 shows that the approximation method cannot lead to a result more general than Theorem 5.1. Since every countable covering of a completely normal space always contains a $\sigma$-discrete closed refinement, it is clear why the most general result can be obtained for spaces with a countable base.

DEFINITION. A many-valued mapping $f$: $X \to \mathcal{F}(Y)$ is said to be a *$\sigma$-mapping* if for every locally finite system $\omega = \{G_\alpha \mid \alpha \in A\}$ is $\sigma$-hereditarily conservative.

THEOREM 10.2. *Let $f$: $X \to \mathcal{F}(Y)$ be a lower semicontinuous $\sigma$-mapping of a completely normal space $X$ into a complete metrizable space $Y$. Then there exists a single-valued mapping $\phi$: $X \to Y$ such that*

a) *$\phi x \in fx$ for every $x \in X$, and*

b) *$\phi^{-1}U$ is an $F_\sigma$-set for every set $U$ that is open in $Y$.*

THEOREM 10.2′. *Let $f$: $X \to \mathcal{F}(Y)$ be an upper semicontinuous $\sigma$-mapping of a completely normal space $X$ into a complete metrizable space $Y$. Then there exists a single-valued mapping $\phi$: $X \to Y$ such that*

a) *$\phi x \in fx$ for all $x \in X$, and*

b) *$\phi^{-1}U$ is an $F_\sigma$-set for every set $U$ that is open in $Y$.*

The proof of these theorems does not differ in principle from the proofs of Theorems 5.1 and 4.4. Only in the construction of the $\phi_n$ and the $X_{nk}$ we must now construct $\sigma$-discrete systems starting out from the fact that $f$ is a $\sigma$-mapping. Obviously Theorem 10.2′ is a formal generalization of Theorem 4.4. The condition that the mappings are $X$-closed seems to us to be more natural than the condition that they are $\sigma$-mappings.

The following theorem is an indication why $X$-closure is essential (or why $f$ has to be a $\sigma$-mapping).

THEOREM 10.3. *Let $X$ be a completely normal space. Then the following conditions are equivalent:*

a) *Every closed conservative covering of $X$ has a closed $\sigma$-discrete closed refinement.*

b) *For every upper semicontinuous mapping $f$: $X \to \mathcal{F}(Y)$, where $Y$ is a complete metrizable space, there exist mappings $\phi_n$ and sets $X_{nk}$ satisfying conditions* (1)–(5).

c) *For every upper semicontinuous mapping $f$: $X \to \mathcal{F}(Y)$, where $Y$ is a complete metrizable space, there exist mappings $\phi_n$ and sets $X_{nk}$ satisfying the conditions* (1*)–(4*).

First we prove

PROPOSITION 10.4. *On a complete metrizable space $Y$ there exist a complete* *metric* $\rho$ *and open locally finite coverings* $\gamma_n = \{U_\alpha | \alpha \in A_n\}$ *such that* $\operatorname{diam} \gamma_n x = 1/2^n$ *for all* $x \in Y$ *and* $n = 1, 2, \cdots$.

PROOF. Let $d$ be an arbitrary complete metric on $Y$ and let $\gamma_n = \{U_\alpha | \alpha \in A_n\}$ be arbitrary locally finite coverings, $\gamma_{n+1}$ being a refinement of $\gamma_n$ for all $n$, such that $\operatorname{diam} U_\alpha < 1/2^n$ for every $\alpha \in A_n$. For any $\alpha \in A_n$ there exists a continuous function $f_\alpha$ such that a) $f_\alpha(y) = 0$ if and only if $y \in Y \setminus U_\alpha$, and b) $\sum_{\alpha \in A_n} f_\alpha(y) = 1/2^n$. Putting

$$\rho(x, y) = \sum_{n=1}^{\infty} \sum_{\alpha \in A_n} |f_\alpha(x) - f_\alpha(y)|$$

we obtain the required metric $\rho$ and the required coverings $\gamma_n$.

PROPOSITION 10.5. *Let* $\phi: X \to \mathscr{F}(Y)$ *be an upper semicontinuous mapping, where* $Y$ *is a normal space, and let* $f: X \to Y$ *be a continuous single-valued mapping. Further, let* $\omega = \{F_\alpha | \alpha \in A\}$ *be a locally finite system of closed subsets of* $Y$. *If* $\omega f x \cap \phi x = \bigcup \{F_\alpha \ni fx\} \cup \phi x \neq \emptyset$ *for any* $x \in X$, *then the mapping* $\psi: X \to \mathscr{F}(Y)$ *given by* $\psi x = \omega f x \cup \phi x$ *is upper semicontinuous.*

PROOF. As $\omega$ is locally finite, the mapping $h: X \to \mathscr{F}(Y)$ given by $hx = \bigcup \{F_\alpha \in fx\}$ is upper semicontinuous. By [8], Theorem 1, Russian p. 189, the mapping $\psi x = hx \cap \phi x$ is upper semicontinuous.

Now we prove Theorem 10.3. The mappings $\phi_1$ and the sets $X_{1k}$ $(k = 1, 2, \cdots)$ are constructed as before. Suppose that $\phi_n$ and the $X_{nk}$ $(k = 1, 2, \cdots)$ have been constructed. We can assume that $X_{nk} \cap X_{nj} = \emptyset$ for $k \neq j$. We put $\phi_{nk} = \phi_n | X_{nk}$ and $\phi^*_{nk} x = \bigcup \{[U_\alpha] \ni \phi_{nk} x | \alpha \in A_n\}$. By Proposition 10.5, the mapping $f_{nk}: X_{nk} \to \mathscr{F}(Y)$ given by $f_{nk} x = \phi^*_{nk} x \cap fx$ is upper semicontinuous. As $X_{nk}$ is an $F_\sigma$-subset of $X$, we see that $X_{nk}$ satisfies condition a). By the upper semicontinuity of the $f_{nk}$ the system of sets $\gamma^{-1}_{n+1} = \{f^{-1}_{nk}[U_\alpha] | \alpha \in A_{n+1}\}$ is conservative in $X_{nk}$. Therefore there exists a $\sigma$-discrete covering

$$\xi = \{\xi_j = \{F^j_\alpha | \alpha \in \alpha \in A_{n+1}\} \quad (j = 1, 2, \cdots)\},$$

such that $F^j_\alpha \subset f^{-1}_{nk}[U_\alpha]$ for all $\alpha \in A_{n+1}$ and $j = 1, 2, \cdots$ . We put

$$M_{nk1} = \bigcup \{F^1_\alpha | \alpha \in A_{n+1}\} \text{ and } M_{nkj} = \bigcup \{F^j_\alpha | \alpha \in A_{n+1}\} \setminus \bigcup_{i=1}^{j-1} M_{nki},$$

$j = 2, 3, \cdots$. It is clear that every $M_{nkj}$ is an $F_\sigma$-subset of $X$. Let $a_\alpha$ be a fixed point in $U_\alpha$. We define $\phi_{nkj}: M_{nkj} \to Y$ by $\phi_{nkj} x = a_\alpha$ for $x \in F^j_\alpha \cap M_{nkj}$. It is obvious that $\phi_{nkj}$ is continuous. Now we obtain $\phi_{n+1}: X \to Y$ by putting $\phi_{n+1} x = \phi_{nkj} x$ for $x \in M_{nkj}$. By construction, we have

$$\rho(\varphi_n, \varphi_{n+1}) < \frac{1}{2^n} \text{ and } \rho(\varphi_{n+1}, f) < \frac{1}{2^{n+1}}.$$

After changing the notation for the $M_{nkj}$ in an appropriate way we can claim that the proof of the first part of the theorem is complete.

The fact that (b) implies (c) is trivial.

Let us prove that (c) implies (a). Let $\omega = \{F_\alpha \mid \alpha \in A\}$ be a closed conservative covering of $X$. We take the discrete metric

$$\rho(x, y) = \begin{cases} 0, & \text{when } x = y, \\ 1, & \text{when } x \neq y. \end{cases}$$

We define $f\colon X \to \mathscr{F}(Y)$ by $f^{-1}\alpha = F_\alpha$ for all $\alpha \in A$.

As $\omega$ is closed and conservative, $f$ is upper semicontinuous. By hypothesis there exist mappings $\phi_n$ and sets $X_{nk}$ satisfying (1*)-(4*). For each point $x \in X$ there exists an $n(x)$ such that $\phi_{n(x)}x \in fx$. We put

$$X_n = \{x \in X \mid \varphi_n x \in fx\},$$

$$\xi_{nk} = \{\Phi_\alpha^{nk} = \varphi_n^{-1}\alpha \cap X_n \cap X_{nk} \mid \alpha \in A\}.$$

The system $\xi_{nk}$ is closed and discrete, and $\Phi_\alpha^{nk} \subset F_\alpha$ for every $\alpha \in A$. Consequently $\xi = \{\xi_{nk} \mid n, k = 1, 2, \cdots\}$ is the required $\sigma$-discrete refinement of $\omega$

Suppose that mappings $\phi_n\colon X \to Y$ and sets $X_{nk}$ satisfy conditions (1)–(5). We define $\phi\colon X \to Y$ by $\phi x = \lim_{n\to\infty}\phi_n x$. Then $\phi$ satisfies the following condition:

(A) *For any discrete system $\omega = \{U_\alpha \mid \alpha \in \theta\}$ of open subsets of $Y$ there exist discrete systems $\xi_m = \{F_\alpha^m \mid \alpha \in \theta\}$ in $X$ such that $\phi^{-1}U_\alpha = \bigcup_{m=1}^{\infty} F_\alpha^m$ for every $\alpha \in \theta$.*

We can assume that every $X_{nk}$ is closed in $X$.

We put

$$\xi_{nk} = \left\{F_\alpha^{nk} = \varphi_n^{-1}\left(Y \setminus O\left(Y \setminus U_\alpha, \frac{1}{2^n}\right)\right) \cap X_{nk} \mid \alpha \in \theta\right\}.$$

It is obvious that the systems $\xi_{nk}$ are closed and discrete in $X$. It is sufficient to prove that

$$\varphi^{-1}U_\alpha = \bigcup_{n=1}^{\infty}\bigcup_{k=1}^{\infty} F_\alpha^{nk}.$$

Suppose $x \in \phi^{-1}U_\alpha$. Then there exists an $n$ such that $\rho(Y \setminus U_\alpha, \phi x) \geq 1/2^{n-1}$. Suppose $x \in X_{nk}$. Then $\rho(\phi_n x, \phi x) < 1/2^n$, since $\rho(\phi_n x, \phi x) > 1/2^n$ and $\rho(\phi x, Y \setminus U_\alpha) > 1/2^{n-1}$. Suppose now that $x \in F_\alpha^{nk}$. Then $\rho(\phi x, Y \setminus U_\alpha) \geq 1/2$, and as $\rho(\phi_n x, \phi x) < 1/2^n$, we have $\phi x \in U_\alpha$; that is, $x \in \phi^{-1}U_\alpha$. This proves property (A).

**Proposition** 10.6 *Suppose that the single-valued mapping $\phi\colon X \to Y$, where $Y$ is a metrizable space, satisfies condition* (A). *Then there exist a sequence of mappings $\phi_n\colon X \to Y$ converging uniformly to $\phi$, and sets $X_{nk}$ satisfying conditions* (1)–(5). *$X$ is assumed to be completely normal.*

**Proof.** We take an arbitrary $n$. Let $\omega = \{\omega_j = \{U_\alpha^j \mid \alpha \in \theta_j\}\colon j = 1, 2, \cdots\}$ be a $\sigma$-discrete open covering of $Y$ such that $\operatorname{diam} U_\alpha^j < 1/2^{n+2}$ for all $\alpha \in \bigcup_{j=1}^{\infty}\theta_j$. By condition (A), for each $j$ there exist discrete systems

$\xi_{jm} = \{F_\alpha^{jm} \mid \alpha \in \theta_j\}$ $(m = 1, 2, \cdots)$ such that $\phi^{-1}U_\alpha^j = \bigcup_{m=1}^{\infty} F_\alpha^{jm}$ for every $\alpha \in \theta_j$. We put

$$M_{mj} = \bigcup \{F_\alpha^{jm} \mid \alpha \in \theta_j\} \setminus \bigcup \{M_{ik} \mid i + k < m + j\}.$$

The $M_{mj}$ are pairwise disjoint and are $F_\sigma$-subsets of $X$. Let $a_\alpha^j$ be a fixed point in $U_\alpha^j$. We define $\phi_n \colon X \to Y$ by $\phi_n x = a_\alpha^j$ if for some $m$ and $j$

$$x \in M_{mj} \cap F_\alpha^{jm}.$$

It is obvious that $\phi_n \mid M_{mj}$ is continuous. Since $\phi_n x \in U_\alpha^j$ and $\phi x \in U_\alpha^j$ for every $x \in F_\alpha^{jm}$, we see that $\rho(\phi, \phi_n) < 1/2^{n+2}$ and

$$\rho(\varphi_n, \varphi_{n+1}) < \frac{1}{2^{n+2}} + \frac{1}{2^{n+3}} < \frac{1}{2^{n+1}}.$$

After changing the notation and the indices of the $M_{mj}$ we can claim that conditions (3)–(5) are satisfied.

In the case of a separable $Y$ any single-valued mapping $\phi \colon X \to Y$ for which $\phi^{-1}U$ is an $F_\sigma$-set whenever $U$ is open in $Y$ satisfies condition (A). Hence for separable spaces all Borel sections can be obtained by the approximation method.

Thus the condition that $X$ should be a $\sigma$-space is necessary and sufficient for every lower semicontinuous mapping $f \colon X \to \mathscr{F}(Y)$, where $Y$ is a complete metrizable space, to have a section satisfying condition (A).

In conclusion we give a simple example to show that Corollary 2.3 cannot be extended even to the class of one-dimensional spaces.

Let $X = \{(1, \phi) \mid 0 \leqq \phi \leqq 2\pi\}$ be the unit circle in the two-dimensional plane. We define a single-valued mapping $f \colon X \xrightarrow{\text{onto}} X$ by

$$f(1, \varphi) = \begin{cases} (1, 2\varphi), & \text{when } \varphi \leqslant \pi, \\ (1, 2(\varphi - \pi)), & \text{when } \pi < \varphi \leqslant 2\pi. \end{cases}$$

Then $f$ is open-and-closed and continuous and $f^{-1}x$ consists of exactly two points for every $x \in X$.

Since $X$ is a circle, it is easy to see that there is no subset $X_1 \subset X$ for which $fX_1 = X$ and $f \mid X_1$ is a homeomorphism. The fact that $X$ is a circle in this example is no accident. It turns out that it is impossible to find a suitable space $X$ among ordered spaces.

THEOREM 10.7. *Let $X$ be an ordered space. Let $C(X)$ be the space of nonempty compact subspaces of $X$ with the Vietoris topology. Then there exists a single-valued continuous mapping $\phi \colon C(X) \to X$ such that $\phi L \in L$ for every $L \in C(X)$.*

PROOF. For every $L \in C(X)$ we put $\phi L = \inf\{x \in L\}$. As $X$ is ordered, this is well defined. We take an arbitrary $L \in C(X)$ and an arbitrary neighborhood $O\phi L$. The following cases are possible:

a) $O\phi L = (a_1, a_2)$, where $a_1 < \phi L < a_2$.

We put $OL = \langle (a_1, a_2), \{x \in X \mid x > a_1\} \rangle$.

b) $O\phi L = [\phi L, a_2)$. Here $OL = \langle [\phi L, a_2), \{x \geqq \phi L\}\rangle$.
c) $O\phi L = (a_1, \phi L]$. Here $OL = \langle (a_1, \phi L], \{x > a_1\}\rangle$.
d) $O\phi L = \phi L$. Here $OL = \langle \phi L, \{x \geqq \phi L\}\rangle$.
It is easy to see that in every case $\phi OL = O\phi L$.
Theorem 9.7 gives us

COROLLARY 10.8. *Let $f$: $X \to Y$ be an open complete single-valued mapping on an ordered space $X$. Then there exists a subset $X_1 \subset X$ such that $fX_1 = Y$ and $f|X_1$ is a homeomorphism.*

COROLLARY 10.9. *An open-and-closed image of the straight line $E^1$ must have one of the following forms:*
I) $(a, b) \subset E^1$;
II) $(a, b] \subset E^1$;
III) $[b, a) \subset E^1$;
IV) $a \in E^1$.

PROOF. It is easy to see that every set of the form I–IV is an open-and-closed image of the real line.

Let $f$: $E^1 \to X$ be an open-and-closed continuous mapping. By a theorem in [**5**], $f$ is peripherally compact. If $X$ does not consist of a single point, then $Frf^{-1}x = f^{-1}x$ for all $x \in X$. Consequently $f$ is complete. $X$ is connected and is homeomorphic to a subset $B \subset E^1$ by Corollary 9.8. Since $X$ is not a single point, it is not compact, and so is of the form I–IV. Now we extend the class of metric spaces for which the theorems of §§5 and 6 are true.

DEFINITION. The class $\mathscr{A}$ is the *minimal class* of spaces satisfying the following conditions:
1°. Every complete metric space belongs to the class.
2°. An open image of a space in the class belongs to the class.
This class of topological spaces was studied in [**2**], [**3**] and [**21**].

THEOREM 10.10. *Let $f$: $X \to \mathscr{F}(Y)$ be a lower semicontinuous mapping, where $Y$ is a topological space belonging to $\mathscr{A}$, and let $X$ satisfy one of the following conditions:*
a) *$X$ is paracompact.*
b) *$X$ is a $\sigma$-space.*
*Then there exists a single-valued mapping $\phi$: $X \to Y$ such that $\phi x \in fx$ for all $x \in X$ and $\phi^{-1}U$ is an $F_\sigma$-set for every set $U$ that is open in $Y$.*

PROOF. Let $h$: $Z \to Y$ be an open single-valued mapping, where $Z$ is a complete metric space. We define $\psi$: $X \to \mathscr{F}(Z)$ by $\psi x = h^{-1}(fx)$ for all $x \in X$.

As $\psi^{-1}M = f^{-1}(hM)$ for every set $M \subseteq Z$, the mapping $\psi$ is lower semicontinuous. By Theorems 5.1 and 5.3 there exists a single-valued mapping $\xi$: $X \to Z$ such that
c) $\xi x \in \psi x$ for all $x \in X$;
d) $\xi^{-1}U$ is an $F_\sigma$-set for every set $U$ that is open in $Z$.

We define $\phi : X \to Y$ by $\phi x = h(\xi x)$ for all $x \in X$. As $h$ is single-valued, $\phi$ is also single-valued. Since $\phi^{-1}L = \xi^{-1}(h^{-1}L)$ for every set $L \subseteq Y$, b) is satisfied. This completes the proof of Theorem 10.10. Similarly the other theorems on lower semicontinuous mappings carry over to the class $\mathscr{A}$.

NOTE. One can consider mappings $f$: $X \to \mathscr{A}(Y)$, where $\mathscr{A}(Y)$ is the space of all nonempty subsets of $Y$. The theorems of the present article are true for such many-valued mappings; however, the condition on the single-valued mapping $\phi$: $X \to Y$ is not "$\phi x \in fx$ for all $x \in X$" but "$\phi x \in [fx]$ for all $x \in X$".

## §11. Operations on many-valued mappings

In this section we strengthen Michael's results in the following ways.

1°) We prove the analog of Theorem 1.1 of [13] for $\tau$-paracompact spaces.

2°) We treat the analog of Theorem 1 of [12] for finite-dimensional spaces.

3°) We prove the converses of Michael's theorems.

A space $X$ is said to be *$\tau$-paracompact*[3)] if every open covering $\omega = \{U_\alpha\}$ of cardinality $\leqq \tau$ has an open locally finite refinement.

PROPOSITION 11.1. *Let $\omega_n = \{U_\alpha | \alpha \in A_n\}$ be locally finite coverings of a normal space $X$ and $\pi_n : A_{n+1} \to A_n$ $(n = 1, 2, \cdots)$ mappings for which $U_\alpha = \bigcup \{U_\beta | \beta \in \pi_n^{-1}\alpha\}$. If $\dim X \leqq m$, then there exist locally finite coverings $\xi_n = \{F_\alpha | \alpha \in A_n\}$ satisfying the following conditions*:

1) *$F_\alpha \subseteq U_\alpha$ for all $\alpha \in \bigcup_{n=1}^{\infty} A_n$.*

2) *The multiplicity of $\xi_n$ is $\leqq m+1$ for all $n$.*

3) *For every $n$ and $\alpha \in A_n$, we have*

$$F_\alpha = \bigcup \{F_\beta | \beta \in \pi_n^{-1}\alpha\}.$$

PROOF. First we prove the proposition for metric spaces without the assumption $\omega_1, \cdots, \omega_n, \cdots$ are locally finite. If $X$ is a metric space and $\dim X \leqq m$, then there exist a zero-dimensional metric space $Z$ and a complete mapping $f$: $Z \to X$ such that $f^{-1}x$ consists of not more than $m+1$ points for each $x \in X$ (see [25]). As $\dim Z = 0$, there exist discrete coverings $\tilde{\xi}_n = \{\Phi_\alpha | \alpha \in A_n\}$ satisfying the following condition:

a) for every $n$ and all $\alpha \in A_n$

$$\Phi_\alpha \subseteq f_\alpha^{-1} U_\alpha \text{ and } \Phi_\alpha = \bigcup \{\Phi_\beta | \beta \in \pi_n^{-1}\alpha\}.$$

Here some of the $\Phi_\alpha$ may be empty. Clearly the coverings $\xi_n = \{F_\alpha = f\Phi_\alpha | \alpha \in A_n\}$ are what we are looking for. Now let $X$ be an arbitrary normal space. Then it is easy to construct coverings $\tilde{\omega}_n = \{V_\alpha | \alpha \in A_n\}$ consisting of open $F_\sigma$-sets that satisfy the condition: for every $n$ and for $\alpha \in A_n$

3) $\tau$ is always assumed to be an infinite cardinal number.

$$V_\alpha \subseteq U_\alpha \quad \text{and} \quad V_\alpha = \bigcup \{ V_\beta \mid \beta \in \pi_n^{-1}\alpha \}.$$

For each $\alpha \in \bigcup_{n=1}^\infty A_n$ we construct a continuous nonnegative function $f_\alpha(x)$ for which 1°) $f_\alpha^{-1}\{0\} = X \setminus V_\alpha$; 2°) $\sum_{\alpha \in A_n} f_\alpha(x) = 1$ for every $x \in X$.

For each $n$ we define the continuous mapping[4] $f_n$: $X \to L(A_n)$ by $f_n(x) = \{ f_\alpha(x) \mid \alpha \in A_n \}$ for all $x \in X$.

We put $\Gamma_\alpha = \{ g \in L(A_n) \mid g(\alpha) > 0 \}$. $\Gamma_\alpha$ is open in $L(A_n)$ and $f_n^{-1}\Gamma_\alpha = V_\alpha$ for all $\alpha \in A_n$. We define $f$: $X \to \prod_{n=1}^\infty L(A_n)$ by $fx = \{ f_1(x), \cdots, f_n(x), \cdots \}$ for all $x \in X$. We put

$$\gamma_n = \{ W_\alpha = \Gamma_\alpha \times \prod_{i \neq n} L(A_i) \mid \alpha \in A_n \} \quad (n = 1, 2, \ldots).$$

For every $n$, the system $\gamma_n$ is open in $\prod_{n=1}^\infty L(A_n)$ and $\tilde{\omega}_n = \{ f^{-1} W_\alpha \mid \alpha \in A_n \}$. By the factorization theorem (see [**4**] and [**25**]) there exist a metric space of dimension $\leqq m$ and continuous mappings $g$: $X \overset{\text{onto}}{\Longrightarrow} Y$ and $h$: $Y \overset{\text{onto}}{\Longrightarrow} \prod_{n=1}^\infty L(A_n)$ for which $f = h \circ g$.

For each $n$ we put $\eta_n = \{ G_\alpha = h^{-1} W_\alpha \mid \alpha \in A_n \}$. Since $gX = Y$ and $\tilde{\omega}_n = \{ g^{-1} G_\alpha \mid \alpha \in A_n \}$, for every $n$ we have:

1°) $\eta_n$ covers $Y$;

2°) $G_\alpha = \bigcup \{ G_\beta \mid \beta \in \pi_n^{-1}\alpha \}$ for every $\alpha \in A_n$.

Since $Y$ is a metric space and $\dim Y \leqq m$, there exist locally finite closed coverings $\tilde{\xi}_n = \{ \Phi_\alpha \mid \alpha \in A_n \}$ of $Y$ for which

1) $\Phi_\alpha \subseteq G_\alpha$ for every $\alpha \in A_n$, $n = 1, 2, \cdots$;

2) the order of $\tilde{\xi}_n$ is $\leqq m + 1$;

3) $\Phi_\alpha = \bigcup \{ \Phi_\beta \mid \beta \in \pi_n^{-1}\alpha \}$ for every $\alpha \in A_n$, $n = 1, 2, \cdots$.

Now we put $\xi_n = \{ F_\alpha = g^{-1}\Phi_\alpha \mid \alpha \in A_n \}$, and this completes the proof.

The fundamental result is

**Theorem** 11.1. *For a $T_1$-space $X$ the following conditions are equivalent:*

a) *$X$ is normal and $\tau$-paracompact, and $\dim X \leqq m$.*

b) *For every lower semicontinuous mapping $\theta$: $X \to \mathscr{F}(Y)$, where $Y$ is a complete metric space of cardinality $\leqq \tau$, there exists an upper semicontinuous mapping $\phi$: $X \to C(Y)$ such that $\phi x \subseteq \theta x$ and $\phi x$ consists of not more than $m+1$ points for every $x \in X$.*

**Proof.** (a) $\to$ (b). First we construct locally finite coverings $\omega_n = \{ U_\alpha \mid \alpha \in A_n \}$, open systems of sets $\gamma_n = \{ V_\alpha \subset Y \mid \alpha \in A_n \}$ and mappings $\pi_n$: $A_{n+1} \to A_n$ such that for all $n$ and all $\alpha \in A_n$

a) $U_\alpha = \bigcup \{ U_\beta \mid \beta \in \pi_n^{-1}\alpha \}$;

b) $[U_\alpha] \subseteq \theta^{-1} V_\alpha$;

c) $\operatorname{diam} V_\alpha \leqq 1/2^n$;

d) $V_\alpha \supseteq \bigcup \{ V_\beta \mid \beta \in \pi_n^{-1}\alpha \}$.

---

[4] $L(A_n)$ is the generalized Hilbert space generated by $A_n$ (see [**11**]).

We can assume that the complete metric $\rho$ on $Y$ satisfies the condition $\rho(y,z) \leqq \frac{1}{2}$ for all $y, z \in Y$. We put $\omega_1 = \{U_1 = X\}$, $\gamma_1 = \{V_1 = Y\}$ and $A_1 = \{1\}$. Suppose that the coverings $\omega_k$, the systems $\gamma_k$ and the sets $A_k$ have been constructed for $k = 1, 2, \cdots, n$. Let $\alpha \in A_n$ be fixed. We put $\gamma_\alpha = \{O(y, 1/2^{n+2}) \mid y \in V_\alpha\}$. It is obvious that there exists a locally finite refinement $\eta_\alpha = \{V_\beta \mid \beta \in B_\alpha\}$ of $\gamma_\alpha$ such that $V_\alpha = \bigcup \{V_\beta \mid \beta \in B_\alpha\}$. By conditions (a) and (b), $[U_\alpha] \subseteq \bigcup \{\theta^{-1} V_\beta \mid \beta \in B_\alpha\}$. It is clear that the cardinality of $B_\alpha$ is $\leqq \tau$. Therefore $\eta_\alpha^{-1} = \{\theta^{-1} V_\beta \mid \beta \in B_\alpha\}$ has a locally finite refinement $\tilde{\omega}_{n+1}^\alpha = \{\widetilde{U}_\beta \mid \beta \in B'_\alpha \subseteq B_\alpha\}$ such that

$$[\widetilde{U}_\beta] \subseteq \theta^{-1} V_\beta \quad \text{and} \quad [U_\alpha] \subseteq \bigcup \{\widetilde{U}_\beta \mid \beta \in B'_\alpha\}.$$

We put

1) $\omega_{n+1}^\alpha = \{U_\beta = \widetilde{U}_\beta \cap U_\alpha \mid \beta \in B'_\alpha\}$;
2) $A_{n+1} = \bigcup \{B'_\alpha \mid \alpha \in A_n\}$;
3) $\omega_{n+1} = \{U_\beta \mid \beta \in A_{n+1}\}$;

and define

4) $\pi_n \colon A_{n+1} \to A_n$ by $\pi_n^{-1}\alpha = B'_\alpha$ for all $\alpha \in A_n$.

Clearly the covering $\omega_{n+1} = \{U_\beta \mid \beta \in A_{n+1}\}$ and the systems $\gamma_{n+1} = \{V_\alpha \mid \alpha \in A_{n+1}\}$ are required objects.

Now we use the fact that $\dim X \leqq m$. By Proposition 11.1 there exist coverings $\xi_n = \{F_\alpha \mid \alpha \in A_n\}$ satisfying the following conditions:

1) $F_\alpha \subseteq U_\alpha$ for all $\alpha \in A_n$, $n = 1, 2, \cdots$.
2) $F_\alpha = \bigcup \{F_\beta \mid \beta \in \pi_n^{-1}\alpha\}$ for all $\alpha \in A_n$, $n = 1, 2, \cdots$.
3) For every $n$ the multiplicity of $\xi_n$ is $\leqq m + 1$.

For each $x \in X$, we put

$$Bx = \{\alpha = \{\alpha_1, \ldots, \alpha_n, \ldots\} \mid \alpha_n \in A_n;\ \pi_n \alpha_{n+1} = \alpha_n;\ x \in \bigcup_{n=1}^{\infty} F_{\alpha_n}\}.$$

LEMMA 11.1. *For each $\alpha_n \in A_n$ and $x \in F_{\alpha_n}$ there exists an $\alpha(x, \alpha_n) \in Bx$, where $\alpha(x, \alpha_n) = \{\alpha'_1, \cdots, \alpha'_{n-1}, \alpha_n, \alpha'_{n+1}, \cdots\}$.*

$\alpha(x, \alpha_n)$ is constructed as follows: for $k < n$ we take

$$\alpha'_k = \pi_k(\pi_{k+1}(\cdots(\pi_{n-1}\alpha_n)\cdots))$$

and if $\alpha'_m$ $(m \geqq n)$ has been constructed, then $\alpha'_{m+1}$ is taken to be some element of $\pi_m^{-1}\alpha'_m$ for which $x \in F_{\alpha_{m+1}}$.

LEMMA 11.2. *For every $x \in X$ the set $Bx$ is not empty and consists of not more than $m + 1$ elements.*

The nonemptiness follows from Lemma 11.1.

Suppose that for some point $x \in X$, $Bx$ contains $m + 2$ different points $\alpha^i = \{\alpha_1^i, \cdots, \alpha_n^i, \cdots\}$, $i = 1, 2, \cdots, m + 2$. Then since $\pi_1, \cdots, \pi_n, \cdots$ are single-valued mappings, there exists an $n$ such that $\alpha_n^1, \cdots, \alpha_n^{m+2}$ are all distinct. Consequently the multiplicity of $\xi_n$ is greater than or equal to $m + 2$. This is a contradiction.

LEMMA 11.3. *For every* $\alpha \in Bx$ *the point* $y_\alpha(x) = \bigcap_{n=1}^{\infty}[V_{\alpha_n}] \in \theta x$.

This follows from the fact that $V_{\alpha_n} \cap \theta x = \emptyset$ for all $n$. We put $\phi x = \{ y_\alpha(x) \mid \alpha \in Bx \}$. By Lemmas 11.2 and 11.3, $\phi x \subseteq \theta x$ and $\phi x$ contains not more than $m+1$ points. Now we prove.

LEMMA 11.4. *The mapping* $\phi$: $X \to C(Y)$ *is upper semicontinuous.*

We choose $\phi x$ and $O\phi x$ arbitrarily. There exists an $n$ for which $O(\phi x, 1/2^n) \subseteq O\phi x$. We put $Ox = X \setminus \bigcup \{ F_\alpha \mid \alpha \in A_n,\ x \notin F_\alpha \}$. We take an arbitrary $x' \in Ox$ and an arbitrary $y' \in \phi x'$. Now $y' = y_\alpha(x')$, where $\alpha = \{\alpha_1, \cdots, \alpha_n, \cdots\}$. It is clear that $x \in F_{\alpha_n}$. Therefore $y_{\alpha(x,\alpha_n)}(x) \in \phi x$ and $y_\alpha(x')$, $y_{\alpha(x,\alpha_n)} \in V_{\alpha_n}$. Consequently $\phi x' \subset O\phi x$. This completes the proof of the first part of Theorem 11.1.

The second part follows from the following theorems.

THEOREM 11.2. *For a* $T_1$*-space* $X$ *the following conditions are equivalent*:

1) *$X$ is normal and $\tau$-paracompact.*

2) *For any lower semicontinuous mapping* $\theta$: $X \to \mathscr{F}(Y)$, *where $Y$ is a complete metric space of cardinality* $\leqq \tau$, *there exist mappings* $\phi$: $X \to C(Y)$ *and* $\psi$: $X \to C(Y)$ *for which*:

a) $\phi x \subseteq \psi x \subseteq \theta f$ *for all* $x \in X$;

b) $\phi$ *is lower semicontinuous*;

c) $\psi$ *is upper semicontinuous.*

3) *For any lower semicontinuous mapping* $\theta$: $X \to \mathscr{F}(Y)$, *where $Y$ is a complete metric space of cardinality* $\leqq \tau$, *there exists an upper semicontinuous mapping* $\psi$: $X \to C(Y)$ *such that* $\psi x \subseteq \theta x$ *for all* $x \in X$.

4) *Every open covering of cardinality* $\leqq \tau$ *has a closed conservative refinement.*

PROOF. (1) $\to$ (2). The locally finite coverings $\omega_n = \{ U_\alpha \mid \alpha \in A_n \}$, open systems $\gamma_n = \{ V_\alpha \mid \alpha \in A_n \}$ and mappings $\pi_n$: $A_{n+1} \to A_n$ constructed above are such that for all $n$ and $\alpha \in A_n$ we have

a) $U_\alpha = \bigcup \{ U_\beta \mid \beta \in \pi_n^{-1}\alpha \}$;

b) $[U_\alpha] \subseteq \theta^{-1} V_\alpha$;

c) diam $V_\alpha \leqq 1/2^n$;

d) $V_\alpha \supseteq \bigcup \{ V_\beta \mid \beta \in \pi_n^{-1}\alpha \}$.

We put

$$Bx = \{\alpha = \{\alpha_1, \ldots, \alpha_n, \ldots\} \mid \alpha_n \in A_n,\ \pi_n\alpha_{n+1} = \alpha_n,\ x \in \bigcap_{n=1}^{\infty} [U_{\alpha_n}]\},$$

$$Ax = \{\alpha \in Bx \mid x \in \bigcap_{n=1}^{\infty} U_{\alpha_n}\},$$

$$B_n x = \{\alpha \in A_n \mid x \in [U_{\alpha_n}]\},$$

$$A_n x = \{\alpha \in A_n \mid x \in U\alpha_n\}.$$

The sets $B_n x$ are finite for all $x \in X$.

The following is proved in the same way as Lemma 11.1.

LEMMA 11.5. *For every $\alpha_n \in A_n$, $x \in [U_{\alpha_n}]$ and $z \in U_{\alpha_n}$, there exist an $\alpha(x, \alpha_n) \in Bx$ and an $\alpha(z, \alpha_n) \in Az$, where*

$$\alpha(x,\alpha_n) = \{\alpha_1^1, \cdots, \alpha_n, \alpha_{n+1}^1, \cdots\} \text{ and } \alpha(z,\alpha_n) = \{\alpha_1^1, \cdots, \alpha_n, \alpha_{n+1}^2, \cdots\}.$$

We put $y_\alpha(x) \bigcap_{n=1}^{\infty} [V_{\alpha_n}]$ for all $x \in X$ and all $\alpha \in Bx$.

LEMMA 11.6. *The sets $\phi x = \{y_\alpha(x) \mid \alpha \in Ax\}$ and $\psi x = \{y_\alpha(x) \mid \alpha \in Bx\}$ are compact, and $\phi x \subseteq \psi x \subseteq \theta x$ for all $x \in X$.*

The fact that $\phi x \subseteq \psi x \subseteq \theta x$ is proved in the same way as Lemma 11.3. We give the sets $B_n x$ the discrete topology. Then $Bx = \lim\{B_n x \mid \pi_n^j = \pi_{j-1} \circ \cdots \circ \pi_n\}$ is compact (see [1]). We define $f$: $Bx \to \psi x$ by $f\alpha = y_\alpha(x)$ for all $\alpha \in Bx$. Now $f$ is continuous. This follows from the fact that for any two points $\alpha^1 = \{\alpha_1, \cdots, \alpha_n, \alpha_{n+1}^1, \cdots\}$ and $\alpha^2 = \{\alpha_1, \cdots, \alpha_n, \alpha_{n+1}^2, \cdots\}$ in $Bx$ we have $\rho(y_{\alpha_1}(x), y_{\alpha_2}(x)) < 1/2^n$. Hence $\psi x$ is compact. That $\phi x$ is compact can be proved in a similar way.

The following is proved in the same way as Lemma 11.4.

LEMMA 11.7. *$\psi$: $X \to C(Y)$ is upper semicontinuous.*

Now we prove

LEMMA 11.8. *$\phi$: $X \to C(Y)$ is lower semicontinuous.*

We choose an arbitrary $x \in X$, $y_\alpha(x) \in \phi x$ and $\epsilon > 0$. We must find a neighborhood $Ox$ such that $\rho(y_\alpha(x), \phi x') < \epsilon$ for all $x' \in OX$. We choose an $n$ such that $1/2^n < \epsilon$. We put $Ox = U_{\alpha_n}$, where $\alpha = \{\alpha_1, \cdots, \alpha_n, \cdots\}$. By Lemma 11.5 we have

$$\rho(y_\alpha(x), \varphi x') \leqslant \rho(y_\alpha(x), y_{\alpha(x'\alpha_n)}(x) \leqslant \frac{1}{2^n} < \varepsilon.$$

This proves the first part of Theorem 11.2.

(2) $\to$ (3). This is obvious.

(3) $\to$ (4). Let $\omega_n = \{U_\alpha \mid \alpha \in A, \operatorname{card} A \leqq \tau\}$ be an arbitrary covering of cardinality $\leqq \tau$. We give $A$ the discrete topology. We define $\theta$: $X \to \mathscr{F}(A)$ by $\theta x = \{\alpha \in A \mid x \in U_\alpha\}$ for all $x \in X$. Since $\theta^{-1}\alpha = U_\alpha$ for all $\alpha \in A$, $\theta$ is lower semicontinuous. By assumption there exists an upper semicontinuous mapping $\psi$: $X \to \mathscr{F}(A)$ such that $\psi x \subseteq \theta x$ for all $x \in X$. Now $\psi^{-1}\alpha \subseteq \theta^{-1}\alpha = U_\alpha$ for all $\alpha \in A$. Consequently $\gamma = \{F_\alpha = \psi^{-1}\alpha \mid \alpha \in A\}$ is a closed refinement of $\omega$. Since for any set $A' \subseteq A$ we have $\theta^{-1}A = \bigcup\{F_\alpha \mid \alpha \in A'\}$, it follows that $\gamma$ is conservative.

(4) $\to$ (1). Let $A$ and $B$ be arbitrary disjoint closed subsets of $X$. We put $\omega = \{U = X \setminus A,\ V = X \setminus B\}$. By assumption this covering has a closed conservative refinement $\gamma = \{F_\alpha \mid \alpha \in \Omega\}$. We put

$$OA = X \setminus \bigcup\{F_\alpha \in \gamma \mid F_\alpha \cap A = \emptyset\}, \quad OB = X \setminus \bigcup\{F_\alpha \in \gamma \mid F_{\alpha\alpha} \cap B = \emptyset\}.$$

Since

$$OB \subseteq X \setminus \bigcup \{F_\alpha \in \gamma \mid F_\alpha \cap A \neq \emptyset\} \text{ and } OA \subseteq \bigcup \{F_\alpha \in \gamma \mid F_\alpha \cap A \neq \emptyset\},$$

we have $OA \cap OB = \emptyset$. This proves that $X$ is normal and consequently countably paracompact (see [**28**]). Now using an argument similar to Michael's (see [**29**] or [**3**], Russian p. 155) we see that any open covering $\omega = \{U_\alpha \mid \alpha \in A\}$ of cardinality $\leqq \tau$ has a $\sigma$-discrete open refinement $\gamma = \{V_\alpha \mid \alpha \in A\}$. The fact that $X$ is normal and countably paracompact implies that $\gamma$ has an open locally finite refinement. This completes the proof of Theorem 11.2.

THEOREM 11.3. *For a $T_1$-space $X$ the following conditions are equivalent*:

a) *$X$ is normal.*

b) *For any lower semicontinuous mapping $\theta$: $X \to C(Y)$, where $Y$ is a separable metrizable space, there exists an upper semicontinuous mapping $\phi$: $X \to C(Y)$ such that $\phi x \subseteq \theta x$ for all $x \in X$.*

PROOF. (a) $\to$ (b). The proof is similar to the part (1) $\to$ (2) in the proof of Theorem 11.2, only here we use the fact that $\eta_\alpha^{-1}$ (constructed as in the proof of Theorem 11.1 (a) $\to$ (b)) is countable and pointwise finite; but it is well known (see [**28**]) that any countable pointwise finite covering of a normal space has a locally finite refinement.

(b) $\to$ (a). Let $A$ and $B$ be arbitrary closed disjoint subsets of $X$. We put $I = [0, 1]$ with the usual topology. We define $\theta$: $X \to C(I)$ by

$$\theta x = \begin{cases} 0, & \text{when } x \in A, \\ 1, & \text{when } x \in B, \\ I, & \text{when } x \in X \setminus (A \cup B). \end{cases}$$

It is easy to check that $\theta$ is lower semicontinuous. By hypothesis there exists an upper semicontinuous mapping $\phi$: $X \to C(I)$ such that $\phi x \subseteq \theta x$ for all $x \in X$.

We put

$$U = \{x \in X \mid \varphi x \subseteq [0, 1/2)\},$$
$$V = \{x \in X \mid \varphi x \subseteq (1/2, 1]\}.$$

By their construction, $U$ and $V$ are open and disjoint. Also $U \supseteq A$ and $V \supseteq B$. This completes the proof of Theorem 11.3.

THEOREM 11.4. *For any $T_1$-space $X$ the following conditions are equivalent*:

a) *$X$ is normal and* $\dim X \leqq m$.

b) *For any lower semicontinuous mapping $\theta$: $X \to C(Y)$, where $Y$ is a separable metrizable space, there exists an upper semicontinuous mapping $\phi$: $X \to C(Y)$ such that $\phi x \subseteq \theta x$ and $\phi x$ consists of not more than $m + 1$ points.*

PROOF. (a) $\to$ (b). The proof is similar to the proof of Theorem 11.1, only here the observation made in Theorem 11.3 is essential.

(b) $\to$ (a). Normality follows from Theorem 11.3. Let $\omega = \{U_{\alpha_1}, \cdots, U_{\alpha_n}\}$ be an arbitrary open finite covering of $X$. We put $Y = \{1, 2, \cdots, n\}$ with the discrete topology. We define $\theta$: $X \to C(Y)$ by $\theta x = \{i \mid x \in U_i\}$ for all $x \in X$. Since $\theta^{-1} i = U_i$, $\theta$ is lower semicontinuous. By hypothesis there exists an upper semicontinuous mapping $\phi$: $X \to C(Y)$ such that $\phi x \subseteq \theta x$ and $\phi x$ consists of not more than $m+1$ points. It is easy to see that the covering $\{F_i = \phi^{-1} i \mid i = 1, 2, \cdots, n\}$ is a closed refinement of $\omega$ and has multiplicity not greater than $m+1$. This completes the proof of Theorems 11.1 and 11.4.

Theorem 11.1 yields

COROLLARY 11.1. *Let $f$: $X \to Y$ be an inductively open mapping of a metric space $X$ onto a paracompact space $Y$ such that the inverse image of every point is complete* (*in the metric given on $X$*). *If* $\dim Y \leqq m$, *then there exists a subspace $X_1 \subseteq X$ such that $fX_1 = Y$ and $f_1 = f \mid X_1$ is perfect and* $(m+1)$*-fold.*

In particular, we have

COROLLARY 11.2. *Let $f$: $X \to Y$ be an inductively open mapping of a metric space $X$ onto a paracompact space $Y$ such that the inverse image of every point is complete* (*in the metric given on $X$*). *If* $\dim Y = 0$, *then there exists a subspace $X_1 \subseteq X$ such that $fX_1 = Y$ and $f_1 = f \mid X_1$ is a homeomorphism.*

Results more general than Corollaries 11.1. and 11.2 are also valid.

THEOREM 11.5. *Let $f$: $X \to Y$ be an inductively open mapping of a metric space $X$ onto a metric space $Y$ such that the inverse image of every point is complete* (*in the metric given on $X$*). *For any continuous mapping $g$: $Z \to Y$ of a zero-dimensional metric space $Z$ onto $Y$ there exists a mapping $h: Z \to X$ such that $g = f \circ h$.*

PROOF. Pasynkov suggested the following construction to me: we put $S = \{(z, x) \in Z \times X \mid gz = fx\}$. Then we have the following diagram:

$$\begin{array}{ccc} S & \xrightarrow{\psi} & X \\ {\scriptstyle\varphi}\downarrow & & \downarrow{\scriptstyle f} \\ Z & \xrightarrow{g} & Y \end{array}$$

in which $g \circ \phi = f \circ \psi$. If $f$ is open here, then $\phi$ is also open.

For let $U \subseteq Z$ and $V \subseteq X$ be arbitrary open subsets. That $\phi$ is open follows from

$$\phi((U \times V) \cap S) = U \cap g^{-1} f V.$$

Therefore, if $f$ is inductively open, $\phi$ is also.

Let $\rho_1$ be an arbitrary metric on $X$ for which all sets of the form $f^{-1} y$ are complete, and let $\rho_2$ be a metric on $Y$. On $S$ we introduce the metric

$$\rho((z_1, x_1), (z_2, x_2)) = \rho_2(x_1, x_2) + \rho_2(z_1, z_2).$$

Since $\phi^{-1}z=\{z\}\times f^{-1}gz$, the set $\phi^{-1}z$ is complete in the metric $\rho$ for all $z\in Z$. By Corollary 11.2 there exists a subspace $S_1\subseteq S$ homeomorphic to $Z$. We put $h=\psi|S_1$, where $Z=S_1$. This completes the proof of Theorem 11.5.

We remark that Theorem 11.1 (b) cannot be expressed in the same way as Theorem 11.2 (2) because of the following theorem.

**Theorem** 11.6. *Let $f\colon X\to Y$ be an open finitely multiple continuous mapping of a normal weakly paracompact space $X$ into a normal space $Y$. Then* $\dim X=\dim Y$.

**Proof.** We put $Y_n=\{y\in Y|f^{-1}y$ consists of exactly $n$ points$\}$, $X_n=f^{-1}Y_n$ and $f_n=f|X_n$. It is well known (see [**23**]) that $\bigcup_{i=1}^n Y_i$ is closed in $Y$ and $f_n$ is locally a homeomorphism for all $n$. It is easy to see that $Y$ is weakly paracompact.

Suppose that $\dim X\leqq m$. We show that $\dim\bigcup_{i=1}^n Y_i\leqq m$ for all $n$. Since $f_1$ is a homeomorphism and $X_1$ is closed in $X$, we have $\dim Y_1=\dim X_1\leqq m$. Suppose that we know that $\dim\bigcup_{i=1}^{n-1}Y_i\leqq m, n>1$. Let $F$ be a closed subset of $Y$ such that $F\subseteq Y_n$. Then $f^{-1}F\subseteq X_n$ and is closed in $X$. As $f_n$ is a local homeomorphism, $\operatorname{loc}\dim F\leqq m$. By [**24**], Theorem 1, we have $\dim F\leqq m$, and by [**26**], Lemma 2.1, we have $\dim\bigcup_{i=1}^n Y_i\leqq m$. Consequently $\dim Y\leqq m$. So now let $\dim Y\leqq m$. Since $\bigcup_{i=1}^n Y_i$ is closed in $Y$ and $Y_n$ is open in $\bigcup_{i=1}^n Y_i$, we see that $\operatorname{loc}\dim X_n\leqq m$ for all $n$. The fact that $f_n$ is a local homeomorphism implies that $\operatorname{loc}\dim X_n\leqq m$. Since $X$ is weakly paracompact, we have $\dim F\leqq m$ for every $n$ and every closed subset $F$ of $X$ satisfying $F\subseteq X_n$, as a result of [**24**], Theorem 1. By [**26**], Lemma 2.1, $\dim\bigcup_{i=1}^n X_i\leqq m$ for all $n$, and consequently $\dim X\leqq m$. This proves Theorem 11.6.

**Theorem** 11.7. *Let $f\colon X\to Y$ be an open finitely multiple mapping of a metric space $X$ onto a normal space $Y$. Then* $\dim Y=\operatorname{Ind}Y=\dim X$.

**Proof.** In this case, $X$ has a uniform basis (see [**2**], Theorem 2.2) and hence is completely normal. Also $Y_n$ is metrizable for every $n$. Thus $\dim Y_n=\operatorname{Ind}Y_n\leqq\dim X$. By Lemma 2.1 of [**27**] and Theorem 11.6 we have $\dim Y=\operatorname{Ind}Y=\dim X$.

## Bibliography


[1] P. S. Aleksandrov, *On the concept of space in topology*, Uspehi Mat. Nauk 2 (1947), no. 1 (17), 5–57. (Russian) MR **10**, 389.

[2] A. V. Arhangel'skiĭ, *Open and near open mappings. Connections between spaces*, Trudy Moskov. Mat. Obšč. **15** (1966), 181–223 = Trans. Moscow Math. Soc. **1966**, 204–250. MR **34** #6725.

[3] ———, *Mappings and spaces*, Uspehi Mat. Nauk 21 (1966), no. 4 (130), 133–184 = Russian Math. Surveys **21** (1966), no. 4, 115–162. MR **37** #3534.

[4] ———, *Factorization of mappings according to weight and dimension*, Dokl. Akad. Nauk SSSR **174** (1967), 1243–1246 = Soviet Math. Dokl. **8** (1967), 731–734. MR **35** #7305.

[5] I. A. Vaĭnšteĭn, *On closed mappings*, Moskov. Gos. Univ. Uč. Zap. **155** Mat. **5** (1952), 3–53. (Russian) MR **17**, 992.

[6] L. V. Keldyš, *Sur les transformations ouvertes des ensembles A*, C. R. (Dokl.) Acad. Sci. URSS **49** (1945), 622–624. MR **8**, 16.

[7] K. Kuratowski and C. Ryll-Nardzewski, *A general theorem on selectors*, Bull. Acad. Polon. Sci. Sér. Sci. Math. Astronom. Phys. **13** (1965), 397–403. MR **32** #6421.

[8] K. Kuratowski, *Topologie*. Vol. I, Monografie Mat., Tom 20, PWN, Warsaw, 1948; Russian transl., "Mir", Moscow, 1966; English transl., Academic Press, New York; PWN, Warsaw, 1966. MR **10**, 389.

[9] N. S. Lašnev, *Closed images of metric spaces*, Dokl. Akad. Nauk SSSR **170** (1966), 505–507 = Soviet Math. Dokl. **7** (1966), 1219–1221. MR **34** #3547.

[10] E. A. Michael, *Topologies on spaces of subsets*, Trans. Amer. Math. Soc. **71** (1951), 152–182. MR **13**, 54.

[11] ———, *Continuous selections*. I, Ann. of Math. (2) **63** (1956), 361–382. MR **17**, 990.

[12] ———, *Selected selection theorems*, Amer. Math. Monthly **63** (1956), 233–238.

[13] ———, *A theorem on semi-continuous set-valued functions*, Duke Math. J. **26** (1959), 647–651. MR **22** #229.

[14] V. I. Ponomarev, *A new space of closed sets and multivalued continuous mappings of compacts*, Mat. Sb. **48 (90)** (1959), 191–212; English transl., Amer. Math. Soc. Transl. (2) **38** (1964), 95–118. MR **22** #8473.

[15] ———, *Properties of topological spaces preserved under multivalued continuous mappings*, Mat. Sb. **51 (93)** (1960), 515–536; English transl., Amer. Math. Soc. Transl. (2) **38** (1964), 119–140. MR **22** #12512.

[16] A. H. Stone, *Non-separable Borel sets*, Rozprawy Mat. **28** (1962). MR **27** #2435.

[17] B. A. Pasynkov, *On open mappings*, Dokl. Akad. Nauk SSSR **175** (1967), 292–295 = Soviet Math. Dokl. **8** (1967), 853–856. MR **36** #862.

[18] A. D. Taĭmanov, *On open mappings of Borel sets*, Mat. Sb. **37 (79)** (1955), 293–300. (Russian) MR **17**, 772.

[19] ———, *On closed mappings*. II, Mat. Sb. **52 (94)** (1960), 579-588. (Russian) MR **22** #9957.

[20] M. M. Čoban, *On exponential topology*, Dokl. Akad. Nauk SSSR **186** (1969), 272–274 = Soviet Math. Dokl. **10** (1969), 597–600. MR **39** #7553.

[21] H. H. Wicke and J. M. Worrell, Jr., *Open continuous mappings of spaces having bases of countable order*, Duke Math. J. **34** (1967), 255-271. MR **35** #979.

[22] R. Engelking, *Selectors of the first Baire class for semicontinuous set-valued functions*, Bull. Acad. Polon. Sci. Sér. Sci. Math. Astronom. Phys. **16** (1968), 277–282. MR **38** #2748.

[23] P. S. Aleksandrov, *Über abzählbar-fache offene Abbildungen*, C. R. (Dokl.) Akad Sci. URSS **13** (1963), 295–299.

[24] A. V. Zarelua, *On a theorem of Hurewicz*, Mat. Sb. **60 (102)** (1963), 17–28; English transl., Amer. Math. Soc. Transl. (2) **55** (1966), 141–152. MR **26** #4317.

[25] B. A. Pasynkov, *Universal bicompacta and metric spaces of given dimension*, Fund. Math. **60** (1967), 285–308. (Russian) MR **36** #7108.

[26] C. H. Dowker, *Local dimension of normal spaces*, Quart. J. Math. Oxford Ser. (2) **6** (1955), 101-120. MR **19**, 157.

[27] ———, *Inductive dimension of completely normal spaces*, Quart. J. Math. Oxford Ser. (2) **4** (1953), 267–281. MR **16**, 157.

[28] ———, *On countably paracompact spaces*, Canad. J. Math. **3** (1951), 219–224. MR **13**, 264.

[29] E. A. Michael, *Another note on paracompact spaces*, Proc. Amer. Math. Soc. **8** (1957), 822–828. MR **19**, 299.

Translated by:
R. K. Thomas